Plato

The Timaeus of Plato

Plato

The Timaeus of Plato

ISBN/EAN: 9783743373174

Manufactured in Europe, USA, Canada, Australia, Japa

Cover: Foto ©Thomas Meinert / pixelio.de

Manufactured and distributed by brebook publishing software (www.brebook.com)

Plato

The Timaeus of Plato

ΠΛΑΤΩΝΟΣ ΤΙΜΑΙΟΣ

THE TIMAEUS OF PLATO

EDITED

WITH INTRODUCTION AND NOTES

BY

R. D. ARCHER-HIND, M.A.
FELLOW OF TRINITY COLLEGE, CAMBRIDGE

London
MACMILLAN AND CO.
AND NEW YORK
1888

[The Right of Translation is reserved]

Cambridge:
PRINTED BY C. J. CLAY, M.A. AND SONS,
AT THE UNIVERSITY PRESS.

PREFACE.

THE present appears to be the first English edition of the *Timaeus*. Indeed since the sixteenth century, during which this dialogue was published separately no less than four times, it had not, so far as I am aware, been issued apart from the rest of Plato's works until the appearance of Lindau's edition, accompanied by a Latin translation, in 1828. Lindau's commentary, though here and there suggestive, does not afford much real help in grappling with the main difficulties of the dialogue; and sometimes displays a fundamental misapprehension of its significance. Ten years later came Stallbaum's edition; concerning which it were unbecoming to speak with less than the respect due to the zeal and industry of a scholar who has essayed the gigantic enterprise of editing with elaborate prolegomena and commentary the entire works of Plato, and it would be unfair to disparage the learning which the notes display: none the less it cannot be denied that in dealing with this dialogue the editor seems hardly to have realised the nature of the task he has undertaken. Stallbaum was followed in 1841 by Th. H. Martin, whose work, published under the modest title of 'Études sur le Timée de Platon,' is far and away the ablest and completest edition of the *Timaeus* which exists. As an exposition of the philosophical import of the dialogue I should not be disposed to rate it so very highly; but so far as it deals with the physical and other scientific questions discussed and with the numerous grave difficulties of detail, it is invaluable: the acuteness and ingenuity, the luminous clearness, and (not least) the unfailing candour of the editor, deserve all admiration. The debt owed to Martin by any subsequent editor must needs be very great. The most recent edition known to me was published in 1853 in the useful series issued by Engelmann at Leipzig, including text, German translation, and rather copious notes. Böckh's 'Specimen editionis' unfortunately is but a small fragment.

The only English translations with which I am acquainted are Thomas Taylor's and Prof. Jowett's: in German there are several. Martin's edition includes a clear and close French rendering, considerably more accurate than Cousin's.

Among the most valuable and important contributions to the explanation of the *Timaeus* are some writings of August Böckh, especially his admirable treatise 'Ueber das kosmische System des Platon.' It is much to be regretted that so excellent a scholar did not give us a complete edition of the dialogue.

The chief ancient exponent is Proklos, of whose commentary, θείᾳ τινὶ μοίρᾳ, only perhaps one third, a fragment of some 850 octavo pages, is extant, breaking off at 44 D. This disquisition is intolerably verbose, often trivial, and not rarely obscure: nevertheless one who has patience to toil through it may gain from it information and sometimes instruction; and through all the mists of neoplatonic fantasy the native acuteness of the writer will often shine.

The principal object of this edition is to examine the philosophical significance of the dialogue and its bearing on the Platonic system. At the same time, seeing that so few sources of aid are open to the student of the *Timaeus*, I have done my best to throw light upon the subsidiary topics of Plato's discourse, even when they are of little or no philosophical importance; nor have I willingly neglected any detail which seemed to require explanation. But as in the original these details are subordinate to the ontological teaching, so I have regarded their discussion as subordinate to the philosophical interpretation of this magnificent and now too much neglected dialogue.

A translation opposite the text has been given with a view to relieving the notes. The *Timaeus* is one of the most difficult of Plato's writings in respect of mere language; and had all matters of linguistic exegesis been treated in the commentary, this would have been swelled to an unwieldy bulk. I have hoped by means of the translation to show in many cases how I thought the Greek should be taken, without writing a grammatical note; though of course it has been impossible to banish such subjects entirely.

My obligation to Dr Jackson's essays on the ideal theory will be manifest to any one who reads both those essays and my commentary. I am as fully as ever convinced of the high importance of his contribution to the interpretation of Plato. In his essay on the *Timaeus* indeed there are some statements to which I can by no means assent; but as that paper in its present form does not contain Dr Jackson's final expression of opinion, I have not thought it necessary to discuss divergencies of view, which may prove to be very slight, and which do not affect the main thesis for which he is contending.

Lastly I must thank my friend Dr J. W. L. Glaisher for his kindness in examining my notes on the arithmetical passage at the beginning of chapter VII, and for mathematical information in other respects.

TRINITY COLLEGE, CAMBRIDGE,
17 *January*, 1888.

ERRATUM.

P. 204, 1st col. of notes, line 11, cancel as erroneous the words 'And if...as the first.'

INTRODUCTION.

§ 1. OF all the more important Platonic writings probably none has less engaged the attention of modern scholars than the *Timaeus*. Nor is the reason of this comparative neglect far to seek. The exceeding abstruseness of its metaphysical content, rendered yet more recondite by the constantly allegorical mode of exposition; the abundance of *a priori* speculation in a domain which experimental science hás now claimed for its own; the vast and many-sided comprehensiveness of the design—all have conspired to the end that only a very few of the most zealous students of Plato's philosophy have left us any considerable work on this dialogue. It has been put on one side as a fantastic, if ingenious and poetical, cosmogonical scheme, mingled with oracular fragments of mystical metaphysic and the crude imaginings of scarcely yet infant science.

<small>Vindication of the importance attached to the *Timaeus* by ancient authorities.</small>

But this was not the position assigned to the *Timaeus* by the more ancient thinkers, who lived 'nearer to the king and the truth.' Contrariwise not one of Plato's writings exercised so powerful an influence on subsequent Greek thought; not one was the object of such earnest study, such constant reference. Aristotle criticises it more frequently and copiously than any other dialogue, and perhaps from no other has borrowed so much: Cicero, living amid a very stupor and paralysis of speculative philosophy, was moved to translate it into Latin: Appuleius gives for an account of the Platonic philosophy little else but a partial abstract of the *Timaeus*, with some ethical supplement from the *Republic:* Plutarch has sundry more or less elaborate disquisitions on several of the subjects handled in it. As for the neoplatonic school, how completely their thought was dominated by the metaphysic of the *Timaeus*, despite the incongruous and almost monstrous accretions which some of them superimposed, is manifest to any reader of Plotinos or Proklos. Such being the concordance of ancient authorities, is it not worth while to inquire whether they be not justified in attaching so profound a significance to this dialogue?

The object of this essay is to establish that they were justified. No one indeed can read the *Timaeus*, however casually, without perceiving that in it the great master has given us some of his profoundest thoughts and sublimest utterances: but my aim is to show that in this dialogue we find, as it were, the focus to which the rays of Plato's thought converge; that by a thorough comprehension of it (can we but arrive at this) we may perceive the relation of various parts of the system one to another and its unity as a whole: that in fact the *Timaeus*, and the *Timaeus* alone, enables us to recognise Platonism as a complete and coherent scheme of monistic idealism.

I would not be understood to maintain that Plato's whole system is unfolded in the *Timaeus*; there is no single dialogue of which that could be said. The *Timaeus* must be pieced together with the other great critical and constructive dialogues of the later period, if we are rightly to apprehend its significance. But what I would maintain is that the *Timaeus* furnishes us with a master-key, whereby alone we may enter into Plato's secret chambers. Without this it is almost or altogether impossible to find in Platonism a complete whole; with its aid I am convinced that this is to be done. I am far from undervaluing the difficulty of the task I have proposed: but it is worth the attempt, if never so small a fraction may be contributed to the whole result.

With this end in view, it is necessary to consider Plato's intellectual development in relation to certain points in the history of previous Greek philosophy. These points are all notorious enough, but it seems desirable for our present purpose to bring them under review.

Preplatonic basis of Platonism: Herakleitos, Parmenides, Anaxagoras.

§ 2. Now it seems that if we would rightly estimate the task which lay before Plato at the outset of his philosophical career and appreciate the service he has rendered to philosophy, we must throw ourselves back into his position, we must see with his eyes and compute as he would have computed the net result of preplatonic theorising. What is the material which his predecessors had handed down for him to work upon? what are the solid and enduring verities they have brought to light? and how far have they amalgamated these into a systematic theory of existence?

In the endeavour to answer these questions I think we can hardly fail to discern amid the goodly company of those early pioneers certain men rising by head and shoulders above their fellows: Herakleitos, Parmenides, Anaxagoras, these three. Each one of these bequeathed to his successors a great principle peculiarly his own; a principle of permanent importance, with which Plato was bound to deal and has dealt. And save in so far as the Pythagorean theory of numbers may have influenced the outward form of his exposition, there is hardly anything in the early philosophy before Sokrates, outside the teaching of these three men, which has seriously contributed to Plato's store of raw material. The synthesis of their one-sided truths required nothing less than the whole machinery of Plato's metaphysical system: it is from their success and their failure that he takes his start—the success of each in enunciating his own truth, the failure of each to recognise its relations.

Since these three men, as I conceive, furnished Plato with his base of operations—or, more correctly perhaps, raised the problems which he must address himself to solve, it is incumbent on us to determine as precisely as we can the nature of the contributions they severally supplied.

§ 3. The old Ionian physicists were all unknowingly working their way to the conception of Becoming. They did not know this, because they knew not that matter, with which alone they were concerned[1], belonged altogether to the realm of Becoming. Nor yet did they reach this conception, for they had not been able to conceive continuity in change—that is to say, they had not conceived Becoming. They imagined the indefinite diversity of material nature to be the complex manifestations of some uniform underlying element, which, whether by condensation and expansion or by some more fundamental modification of its substance, transmuted itself into this astonishing multiplicity of dissimilar qualities. But according to their notion this underlying element, be it water or air or some indefinable substrate, existed at any given place now in one form, now in another; that is, it abode for a while in one of its manifestations, then changed and abode for a while in another. Air *is* air for a time; then it is

The Ionians and Herakleitos.

[1] Of course the antithesis of matter and spirit had not yet presented itself to Greek thought.

condensed and turns to water. Thus the notion of continuity is absent, and consequently the notion of Becoming. Yet, for all that, Thales, Anaximandros, and Anaximenes were on the path to Becoming.

The penetrating intellect of Herakleitos detected the shortcoming of his predecessors. All nature is a single element transmuting itself into countless diversities of form: be it so. But the law or force which governs these transmutations must be omnipresent and perpetually active. For what power is there that shall hold it in abeyance at any time? or how could it intermit its own activity without perishing altogether? Therefore there can be no abiding in one form; transmutation must be everywhere ceaseless and continuous, since nature will not move by leaps. Motion is all-pervading, and rest is there nowhere in the order of things. And this privation of rest is not a matter of degree nor to be measured by intervals of time. Rest during an infinitesimal fraction of the minutest space which our senses can apprehend were as impossible and inconceivable as though it should endure for ages. We must see the ὁδὸς ἄνω κάτω as Herakleitos saw it: all nature is a dizzy whirl of change without rest or respite, wherein there is no one thing to which we can point and say 'See, it is this, it is that, it is so.' For in the moment when what we call 'it' has begun to be 'this' or 'that' or 'so,' at that very moment it has begun to pass from the state we thus seek to indicate: there is nowhere a fixed point. And thus Herakleitos attains to the conception of continuity and Becoming. He chose appropriately enough fire, the most mobile and impalpable of the four reputed elements, to be the vehicle of this never resting activity of nature: but it matters nothing what was his material substrate. His great achievement is to have firmly grasped and resolutely enunciated the principle of continuity and hence of Becoming: for continuity is a mode of Becoming, or Becoming a mode of continuity, according as we may choose to view it. Moreover, Herakleitos introduces us to the antithesis of ὄν and μὴ ὄν. We cannot say of any object 'it is so,' or use any other phrase which implies stability. Yet the thing in some sense or other *is*, else it would be nothing; it is at any rate a continuity of change. So then the thing is and is not; that is to say, it becomes. Or if, as we watch a falling drop of rain, we take any spot in its course which it would just fill, we can never say 'it is there,' for it never rests: yet, by the time the drop reaches the earth, that spot has been filled by it. The drop has a 'where,' though we can never define the 'where.' Thus throughout the teaching of Herakleitos the 'is' is confronted by 'is not.'

§ 4. In the preceding paragraph I have confined myself within the limits of the actual teaching of Herakleitos: the Platonic developments of it will occupy our attention later on. What then is the actual result—the contribution to the philosophical capital with which Plato had to start? We have conceived change as continuous, that is, we have conceived Becoming. And Becoming is negation of stable Being. Also since change is a transition, it involves motion: therefore in affirming Becoming we affirm Motion. And since change is a transition from one state to another, it involves plurality. So in affirming Becoming we affirm Multitude. Becoming, Motion, Multitude—these are three aspects of one and the same fact: and this is the side of things which Herakleitos presents to us as the truth and reality of nature. The importance of this aspect cannot be exaggerated, neither can its insufficiency. *Result of Herakleiteanism.*

§ 5. For where does this doctrine leave us in regard to the acquisition of knowledge? Surely of all men most hopeless. Let us set aside for the present the question of the relation between subject and object as elaborated in the *Theaetetus*, and confine ourselves simply to the following considerations. The object of knowledge must exist: of that which is not there can be no knowledge. But we have seen that according to Herakleitos it is as true to say of everything that it is not as to say that it is: therefore at best it is as true that there is no knowledge as that there is. Again the object of knowledge must be abiding: how can the soul have cognisance of that which unceasingly slips away and glides from her grasp? For it is not possible that we cognise our elemental substrate now in one form, now in another, since change is continuous: there is no footing anywhere; for each thing the beginning of birth is the beginning of dissolution; every new form in the act of supplanting the old has begun its own destruction. In this utter elusiveness of fluidity where is knowledge to rest? Plato sums up the matter in these words: εἰ μὲν γὰρ αὐτὸ τοῦτο, ἡ γνῶσις, τοῦ γνῶσις εἶναι μὴ μεταπίπτει, μένοι τε ἂν ἀεὶ ἡ γνῶσις καὶ εἴη γνῶσις· εἰ δὲ καὶ αὐτὸ τὸ εἶδος μεταπίπτει τῆς γνώσεως, ἅμα τ' ἂν μεταπίπτοι εἰς ἄλλο εἶδος γνώσεως καὶ οὐκ ἂν εἴη γνῶσις· εἰ δὲ ἀεὶ μεταπίπτει, ἀεὶ οὐκ ἂν εἴη *Impossibility of knowledge the necessary inference from Herakleitean teaching.*

γνῶσις· καὶ ἐκ τούτου τοῦ λόγου οὔτε τὸ γνωσόμενον οὔτε τὸ γνωσθησόμενον ἂν εἴη. *Cratylus* 440 A.

Thus the teaching of Herakleitos tends to one inevitable end—none can know, for nothing can be known.

Parmenides.
§ 6. Seeing then that Becoming and Multitude are unknowable, are we therefore forced to abandon in despair all striving after knowledge? Or is it perchance possible that there exists Being or Unity, which abides for ever sure and can be really and certainly known? Such at least was the conviction of Parmenides.

This great philosopher, who may be considered as the earliest herald of the idealism which should come but yet was not, set about his work by a method widely different from that of the Ionian physicists[1]. The Ionians indeed, and even Herakleitos himself, in a certain sense sought unity, inasmuch as they postulated one single element as the substrate of material phenomena. But such a unity could not content Parmenides. What, he may have asked, do we gain by such a unity? If there is one element underlying the appearances of material nature, why choose one of its manifestations as the fundamental form in preference to another? If the same substance appears now as fire, now as air, now as water, what is the use of saying that fire, air, or water is the ultimate element? And if with Anaximandros we affirm that the ultimate substance is an undefined unlimited substrate, this is only as much as to say, we do not know the substrate of things. In any case the supposition of a material substrate leaves us just where we were. The unity that pervades nature must be one of a totally different sort; not a material element which is transformed into multitudinous semblances, but a principle, a formative essence, distinct from the endless variety of visible nature. It must be no ever-changing substrate, but an essence simple, immutable, and eternal, far removed from the ken of sensation and to be reached by reason alone. And not only must it be verily existent, it must be the sum-total of existence; else would it fail of its own nature and fall short of itself. Since then the One is and is the whole, it must needs follow that the Many are not at all. Material nature then, with all her processes and appearances, is utterly non-existent, a vain delusion of the senses: she is Not-being, and Not-being exists in no wise—only Being is. And since

[1] I take Parmenides as the representative of Eleatic thought, regarding Xenophanes as not, properly speaking, a philosopher at all, and Zeno as merely developing one aspect of Parmenidean teaching.

Not-being is not, neither is there Becoming; for Becoming is the synthesis of Being and Not-being. Again if there is not Becoming, Motion exists not either, for Becoming is a motion, and all motion is becoming. Multitude, Motion, Becoming—all these are utterly obliterated and annihilated from out of the nature of things: only the One exists, abiding in its changeless eternity of stillness[1].

§ 7. Such is the answer returned by Parmenides and his school to the question asked at the beginning of our previous section. Material nature is in continual flux, you say, and cannot be known: good—then material nature does not exist. But Being or the One does exist and can be known, and it is all there is to know.

The Eleatic theory, taken by itself, is as inadequate as that of Herakleitos.

Now it is impossible to conceive a sharper antithesis than that which exists at all points between the two theories I have just sketched. The Herakleiteans flatly deny all unity and rest, the Eleatics as flatly deny all plurality and motion. If then either of these schools is entirely right, the law of contradiction is peremptory—the other must be entirely wrong. Is then either entirely right or wrong?

We have already admitted that Herakleiteanism presents us with a most significant truth, and also that it remorselessly sweeps away all basis of knowledge. Therefore we conclude that, though Herakleitos has given us a truth, it is an incomplete and onesided truth. Let us notice next how the Eleatics stand in this respect.

About the inestimable value of the Eleatic contribution there can be no doubt. Granted that the phenomena of the material world are ever fleeting and vanishing and can never be known—what of that? The material world does not really exist: it is not there that we must seek for the object of knowledge, but in the eternally existent Unity. Thus they oppose the object of reason

---

[1] This sheer opposition of the existent unity to the non-existent plurality led Parmenides to divide his treatise on Nature into two distinct portions, dealing with Truth and Opinion. I am not disposed to contest Dr Jackson's affirmation that 'Parmenides, while he denied the real existence of plurality, recognised its apparent existence, and consequently, however little value he might attach to opinion, was bound to take account of it'. That Parmenides was perfectly consistent in embracing the objects of Opinion in his account, I admit. But none the less does his language justify the statements in the text: he emphatically affirms the non-existence of phenomena, and has no care to explain why they appear to exist.

to the object of sensation. This is good, so far as it goes: it points to the line followed by Plato, who said, if material nature cannot be known, the inference is, not that knowledge is impossible, but that there is some immaterial existence, transcending the material, which is the true object of knowledge. But the further we examine the Eleatic solution, the more reason we shall see to be dissatisfied with it. First the problem of the material world is not answered but merely shelved by the negation of its existence. Here are we, a number of conscious intelligences, who perceive, or fancy we perceive, a nature which is not ourselves. What then are we, what is this nature, why do we seem to perceive it, and how can there be interaction between us and it? A bald negation of matter will not satisfy these difficulties. Again, the Eleatics are bound to deny not merely the plurality of objects, but the plurality of subjects as well. What then are these conscious personalities, which seem so real and so separate, and which yet on Eleatic principles must, so far as their plurality and their separation is concerned, be an idle dream? Secondly, if we ask Parmenides what is this eternally existent One, no satisfactory answer is forthcoming. On the one hand his description of the ἓν ὂν πᾶν is clogged with the forms of materiality: it is 'on all sides like unto the globe of a well-rounded sphere, everywhere in equipoise from the centre:' on the other, it is a mere aggregate of negations, and, as Plato has shown, an idle phantom of the imagination, an abstraction without content, whereof nothing can be predicated, which has no possible mode of existence, which cannot be spoken, conceived, or known. This is all Parmenides has to offer us for veritable existence. If it is true that on Herakleitean principles nothing can be known, it is equally true that on Eleatic principles there is nothing to know.

The Herakleitean and Eleatic theories are in fact alike incomplete, but potentially complementary one to the other: the fusion of the two is the work left to Plato.

§ 8. How is it then that either of these most opposite theories leads to an equally hopeless deadlock? It is because each of them presents us with one side of a truth as if it were the whole. For opposite as the doctrines of Herakleitos and Parmenides may appear, they are in fact mutually complementary, and neither is actually true except in conjunction with its rival. Herakleitos did well in affirming Motion; but he forgot that, if Motion is to be, there must likewise be Rest: for opposite requires opposite. So too Parmenides in denying plurality saw not that he thereby abolished unity: for One and Many can exist only in mutual correlation—each is meaningless without the other. Both must exist, or neither: the two are as inseparable as concave and convex.

Here then lies the radical difference between Parmenides and Plato. Parmenides said, Being is at rest, therefore Motion is not; Being is one, therefore Multitude is not; Being is, therefore Not-being is not at all. Plato said, since there is Rest, there must be Motion; since Being is one, it must also be many; that Being may really be, Not-being must also be real. The chasm between the two sides must be bridged, the antinomy conciliated: Rest must agree with Motion, Unity with Multitude, Being with Not-being.

But, it may be objected, is not this the very thing we just now said that the theory of Herakleitos achieved? is not his great merit to have shown that each thing becomes, that is to say, it is at once and is not? True, Herakleitos shows this in the case of particulars: he exhibits 'is' and 'is not' combined in the processes of material nature. But as his universal result he gives us the negation of Being, just as Parmenides gives us the negation of Not-Being: each in the universal is oné-sided. This Becoming, to which Herakleitos points in the material world, must be the symbol of a far profounder truth, of which Herakleitos never dreamed, which even Plato failed at first to realise.

So then these are our results up to the present point. On the one side we have Multitude, Motion, Becoming; on the other Unity, Rest, Being. The two rival principles confront each other in sheer opposition, stiff, unyielding, impracticable. And till they can be reconciled, human thought is at a standstill. The partisans of either side waste their strength in idle wrangling that ends in nothing. And indeed, as we have them so far, these two principles are hopelessly conflicting: some all-powerful solvent must be found which shall be able to subdue them and hold them in coalescence. Now this very thing is the contribution of the last of the three great thinkers who are at present under consideration: he brought into the light, though he could not use, the medium wherein the fundamental antithesis of things was to be reconciled.

§ 9. Anaxagoras belongs to the Ionian school of thought and mainly concerned himself with physics. But such was the originality of his genius and such the importance of his service to philosophy that he stands forth from the rest, as prominent and imposing a figure as Herakleitos himself. With his physical

Anaxagoras.

theories we are not now concerned, since it is the development of Greek metaphysic alone which we are engaged in tracing. Anaxagoras distinguished himself by the postulation of Mind as an efficient cause: therefore it is that Aristotle says he came speaking the words of soberness after men that idly babbled. All was chaos, says Anaxagoras, till Mind came and ordered it. Now what is the meaning of this saying, as he understood it?

First we must observe that the teaching of Anaxagoras is not antithetical to that of either Herakleitos or Parmenides, as these two are to each other: he takes up new ground altogether. His doctrine of νοῦς is antagonistic to the opinions of Empedokles and of the atomists. Empedokles assumes Love and Hate as the causes of union and disunion. But herein he really introduces nothing new; he merely gives a poetical half-personification to the forces which are at work in nature. The atomists, conceiving their elemental bodies darting endlessly through infinite space, assigned as the cause of their collision τύχη or ἀνάγκη, by which they meant an inevitable law operating without design, a blind force inherent in nature. This is what Anaxagoras gainsaid: to him effect required a cause, motion a movent. Now he observed that within his experience individual minds are the cause of action: what more likely then, he argued, than that the motions of nature as a whole are caused by a universal mind? It did not seem probable to him that a universe ordered as this is could be the chance product of blindly moving particles; he thought he saw in it evidence of intelligent design. He knew of but one form of intelligence—the mind of living creatures, and chiefly of man. Mind then, he thought, must be the originator of order in the universe—a mind transcending the human intelligence by so much as the operations of nature are mightier than the works of man. Thus then he postulated an efficient cause distinct from the visible nature which it governed.

Anaxagoras and causation.
This leads us briefly to compare his attitude towards causation with that of Herakleitos and Parmenides. Herakleitos sought for no efficient cause. The impulse of transmutation is inherent in his elemental fire, and he looks no further. Why things are in perpetual mutation is a question which he does not profess to answer; it is enough, he would say, to have affirmed a principle that will account for the phenomena of the universe: it is neither necessary nor possible to supply a reason why the universe exists on this principle. And in fact every philosophy

must at some point or other return the same reply. Herakleitos then conceives a motive force to exist in matter, but seeks not any ulterior cause thereof.

The Eleatics simply abolished causation altogether. Since the One alone exists and changes never, it is the cause of nothing either to itself or to anything else. Causation in fact implies Becoming, and is thus excluded from the Eleatic system. No attempt is made to establish any relation of causality between the One and the Many, since the latter are absolutely negated. Nor does Parmenides in his treatise on the objects of Opinion make any effort to account for the apparent existence of the multitude of material particulars.

Anaxagoras is thus the first with whom the conception of an efficient cause came to the front; and herein, however defective may have been his treatment of the subject, his claim of originality is indefeasible.

§ 10. The shortcomings of the Anaxagorean theory have been dwelt upon both by Plato and by Aristotle. Plato found indeed much in Anaxagoras with which he could sympathise. His conception that Intelligence, as opposed to the atomistic ἀνάγκη, is the motive cause in nature, is after Plato's own heart. But after advancing so far, Anaxagoras stops short. Plato complains that he employs his Intelligence simply as a mechanical cause, as a source of energy, whereby he may have his cosmical system set in motion. But if, says Plato, the ἀρχή of the universe is an intelligent mind, this must necessarily be ever aiming at the best in its ordering of the universe—no explanation can be adequate which is not thoroughly teleological. But Anaxagoras does not represent 'the best' as the cause why things are as they are: having assumed his νοῦς as a motive power, he then, like all the rest, assigns only physical and subsidiary causes. The final cause has in fact no place in the philosophy of Anaxagoras. Nor does he ever regard Mind as the indwelling and quickening essence of Nature, far less as her substance and reality. On the contrary Mind is but an external motive power supplying the necessary impetus whereby the universe may be constructed on mechanical principles. Material phenomena stand over against it as an independent existence; they are ordered and controlled by Mind, but are not evolved from it, nor in any way conciliated with it. Thus we see how far Anaxagoras was from realising the immeasurable importance of the principle which he

Deficiencies of Anaxagoras.

Results.

himself contributed to metaphysics, the conception of a causative mind. And so his philosophy ends in a dualism of the crudest type.

§ 11. And now we have lying before us the materials out of which, with the aid of a hint or two gained from Sokrates, Plato was to construct an idealistic philosophy. These materials consist of the three principles enunciated by the three great teachers whose views we have been considering[1]. These principles we may term by different names according to the mode of viewing them— Motion, Rest, Life; Multiplicity, Unity, Thought; Becoming, Being, Soul: all these triads amount to the same. But however pregnant with truth these conceptions may prove to be, they are thus far impotent and sterile to the utmost. Each is presented to us in helpless isolation, incapable by itself of affording an explanation of things or a basis of knowledge. To bring them to light was only for men of genius, rightly to conciliate and coordinate them required the supreme genius of all. Like the bow of Odysseus, they await the hand of the master who alone can wield them. The One of Parmenides and the Many of Herakleitos must be united in the Mind of Anaxagoras: that is to say, unity and plurality must be shown as two necessary and inseparable modes of soul's existence, before a philosophy can arise that is indeed worthy of the name. And it is very necessary to realise that to all appearance nothing could be more hopeless than the deadlock at which philosophical speculation had arrived: every way seemed to have been tried, and not one led to know-

---

[1] It may be thought strange that I here make no mention of the Pythagoreans. But the Pythagorean influence on Platonism has been grossly overrated. Far too much importance has often been attached to the statements of late and untrustworthy authorities, or to fragments attributed on most unsubstantial grounds to Pythagorean writers. All that we can safely believe about Pythagorean philosophising is to be found, apart from what Plato tells us, in Aristotle: and from his statements we may pretty fairly infer that they had no real metaphysical system at all. There is indeed some superficial resemblance between the Pythagorean theory of numbers and the Platonic theory of ideas—a resemblance sufficient to induce Aristotle to draw a comparison between them in the first book of the *metaphysics*. But that the similarity was merely external is plain from Aristotle's own account, and also that the significance to be attached to the Pythagorean numbers had been left in an obscurity which probably could not have been cleared up by the authors of the theory. We may doubtless accept the verdict of Aristotle in a somewhat wider sense than he meant by the words—λίαν ἁπλῶς ἐπραγματεύθησαν.

ledge. The natural result was that men despaired of attaining philosophic truth.

§ 12. Before we proceed further, perhaps a few words are due to Empedokles. For he seems to have been dimly conscious of the necessity to amalgamate somehow or other the principles which Herakleitos and Parmenides had enunciated, the principles of Rest and Motion. But of any scientific method whereby this should be done he had not the most distant conception. His scheme is crudely physical, a mere mechanical juxtaposition of the two opposites—$\mu\hat{\iota}\xi\acute{\iota}s\ \tau\epsilon\ \delta\iota\acute{\alpha}\lambda\lambda\alpha\xi\acute{\iota}s\ \tau\epsilon\ \mu\iota\gamma\acute{\epsilon}\nu\tau\omega\nu$: a real ontological fusion of them was utterly beyond his thought. Still, although he really contributes nothing to the solution of the problem concerning the One and Many, the fact that he did grope as it were in darkness after it is worthy of notice.

Empedokles.

§ 13. The hopelessness of discovering any certain verity concerning the nature of things found an expression in the sophistic movement. This phase of Greek thought need not detain us long, since it did nothing directly for the advancement of metaphysical inquiry. It is possible enough that the new turn which the sophists gave to men's thoughts may have done something to prepare the way for psychological introspection, and their studies in grammar and language can hardly have been other than beneficial to the nascent science of logic. From our present point of view however the only member of the profession that need be mentioned is Protagoras, who was probably the clearest and acutest thinker among them all, and who is interesting because Plato has associated his name with some of his own developments of the Herakleitean theory. The historical Protagoras probably did little or nothing more in this direction than to popularise some of the teaching of Herakleitos and to give it a practical turn. What seems true to me, he said, is true for me; what seems true to you is true for you : there is no absolute standard—$\pi\acute{a}\nu\tau\omega\nu\ \chi\rho\eta\mu\acute{a}\tau\omega\nu\ \mu\acute{\epsilon}\tau\rho\sigma\nu\ \acute{a}\nu\theta\rho\omega\pi\sigma s$. Therefore let us abandon all the endeavours to attain objective truth and turn our minds to those practical studies which really profit a man. The genuine interest of the doctrine of the relativity of knowledge, which Protagoras broached, is to be found in Plato's development of it; and this will be considered in its proper place. So far as concerns our present study, we see in Protagoras only a striking representative of the reaction against the earlier dogmatic philosophy.

The sophists, especially Protagoras.

Sokrates.

§ 14. Into the question whether Sokrates was a sophist or not we are not concerned to enter. And, deep as was the mark which he left on his time, we need not, since our inquiry deals with metaphysics, linger long with him: for whatever metaphysical importance Sokrates possesses is indirect and may be summed up in a very few words. With Sokrates the ultimate object of inquiry is, not the facts given in experience, but our judgments concerning them. Whereas the physicists had thought to attain knowledge by speculation upon the natural phenomena themselves, Sokrates, by proceeding inductively to a classification and definition of various groups of phenomena, substituted concepts for things as the object of cognition. By comparing a number of particulars which fall under the same class, we are enabled to strip off whatever accidental attributes any of them may possess and retain only what is common and essential to all. Thus we arrive at the concept or universal notion of the thing: and since this universal is the sole truth about the thing, so far as we are able to arrive at truth, it follows that only universals are the object of knowledge, so far as we are able to attain it. This Sokratic doctrine, that knowledge is of universals is the germ of the Platonic principle that knowledge is of the ideas: and though, as we shall see, a too close adherence to it led Plato astray at first, it remained, since there was a Plato to develop it, a substantial contribution to philosophical research.

Plato: two stages to be distinguished in his treatment of the metaphysical problem.

§ 15. We are now in a position to appreciate the nature of the work which lay before Plato and of the materials which he found ready to his hand. We have seen that philosophy, properly speaking, did not yet exist, though the incomposite elements of it were there ready for combination. Now it would be a very improbable supposition that Plato realised at first sight the full magnitude and the exact nature of the problem he had to encounter: and a careful study of his works leads, I believe, to the conclusion that such a supposition would be indefensible[1]. If then this is so—if Plato first dealt with the question incompletely and with only a partial knowledge of what he had to do, but afterwards revised and partly remodelled his theory, after he had fully realised the nature of the problem—

[1] For a full statement of the reasons for holding that in Plato's dialogues are to be found two well-defined phases of his thought, I must refer to Dr Jackson's essays on the later theory of ideas.

obviously our business is to investigate his mode of operation at both stages: we must see how he endeavoured in the first instance to escape from the philosophical scepticism which seemed to be the inevitable result of previous speculation, what were the deficiencies he found in the earlier form of his theory, and how he proposed to remedy its faults. We must see too how far his conception of the nature of the problem may have altered in the interval between the earlier and the later phase of the ideal theory.

To this end it will be necessary to examine Plato's metaphysical teaching as propounded in a group of dialogues, whereof the most important metaphysically are the *Republic* and *Phaedo*—with which are in accordance the *Phaedrus, Symposium, Meno*, and apparently the *Cratylus*—and next the amended form of their teaching, as it appears in four great dialogues of the later period, *Parmenides, Sophist, Philebus, Timaeus;* especially of course the last. The Sokratic dialogues may be dismissed as not bearing upon our question.

§ 16. Plato had thoroughly assimilated the physical teaching of Herakleitos. He held no less strongly than the Ionian philosopher the utter instability and fluidity of material nature. We are not perhaps at liberty to allege the very emphatic language of the *Theaetetus* as evidence that this was his view in the earlier phase of his philosophy, with which we are at present dealing: but there is abundant proof within the limits of the *Republic* and *Phaedo:* see *Republic* 479 B, *Phaedo* 78 B. He therefore, like Protagoras, was bound to draw his inference from the Herakleitean principle. The inference drawn by Protagoras was that speculation is idle, knowledge impossible. The inference drawn by Plato was that, since matter cannot be known, there must be some essence transcending matter, which alone is the object of knowledge. And furthermore this immaterial essence must be the cause and sole reality of material phenomena. Thus it was Plato's acceptance of the Herakleitean πάντα ῥεῖ, together with his refusal to infer from it the impossibility of knowledge, that led him to idealism.

Plato starts from a Herakleitean standpoint.

At this point the hint from Sokrates is worked in. What manner of immaterial essence is it which we are to seek as the object of knowledge? Plato cordially adopted the Sokratic principle that universals alone can be known. But the Sokratic universal, being no substantial existence but merely a conception in our own mind, will not meet Plato's demand for a

The contribution of Sokrates, and the ideal theory as presented in the Republic.

self-existent intelligible essence. Plato therefore hypostatises the Sokratic concept, declaring that every such concept is but our mental adumbration of an eternal and immutable idea. Thus in every class of material things we have an idea, whereof the particulars are the material images, and the concept which we form from observation of the particulars is our mental image of it. Immaterial essence then exists in the mode of eternal ideas or forms, one of which corresponds to every class, not only of concrete things, but of attributes and relations,—of all things in fact which we call by the same class-name (*Republic* 596 A). The particulars exist, so far as they may be said to exist, through inherence of the ideas in them—at least this is the way Plato usually puts it, though in *Phaedo* 100 D he declines to commit himself to a definition of the relation. These ideas are arranged in an ascending scale: lowest we have the ideas of concrete things, next those of abstract qualities, and finally the supreme Idea of the Good, which is the cause of existence to all the other ideas, and hence to material nature as well.

Now since, as we have seen, there is an idea corresponding to every group of particulars, we may note the following classes of ideas in the theory of the *Republic*: (1) the idea of the good; (2) ideas of qualities akin to the good, καλόν, δίκαιον and the like; (3) ideas of natural objects, as man, horse; (4) ideas of σκευαστά, such as beds or tables; (5) ideas of relations, as equal, like; (6) ideas of qualities antagonistic to good, ἄδικον, αἰσχρόν, and so forth (*Republic* 476 A).

Thus then we have the multitude of particulars falling under the above six classes deriving their existence from a number of causative immaterial essences, which in turn derive their own existence from one supreme essence, to wit, the idea of the good. The particulars themselves cannot be known, because they have no abiding existence: but by observation and classification of the particulars we may ascend from concept to concept until we attain to the apprehension of the αὐτὸ ἀγαθόν, whence we pass to the cognition of the other ideas. Thus Plato offers us a theory of knowledge which shall enable us to escape from metaphysical scepticism. But he also offers us in the theory of ideas his solution of a pressing logical difficulty—the difficulty raised by Antisthenes and others as to the possibility of predication. The application of the ideal theory to this question is to be found in *Phaedo* 102 B. Predication signifies that the idea of the quality predicated is

Predication.

inherent in the subject whereof it is predicated: if we say 'Sokrates is small,' we do not, as Antisthenes would have it, identify 'Sokrates' and 'small,' but simply indicate that Sokrates partakes of the inherent idea of smallness. Thus we find in the doctrine of ideas on the metaphysical side a theory of knowledge, on the logical side a theory of predication.

§ 17. Such is Plato's first essay to solve the riddle bequeathed him by his predecessors. Let us try to estimate the merits and deficiencies of his solution. The advance made in Plato's earlier theory.

The bold originality of Plato's theory is conspicuous at a glance. In the first place, by proclaiming the Absolute Good as the sole source of existence, he identifies the ontological with the ethical first principle, the formal with the final cause. Thus he makes good the defect whereof he complained in the philosophy of Anaxagoras. For in the Platonic system a theory of being is most intimately bound up with a theory of final causes: ontology and teleology go hand in hand. Everything exists exactly in proportion as it fulfils the end of being as perfect as possible; for just in that degree it participates in the idea of the good, which is the ultimate source of all existence. In just the same way he escapes from the utilitarian doctrine of Protagoras, by deducing his ethical teaching from the very fount of existence itself. Thus he finds one and the same cause for the existence of each thing and for its goodness. A good thing is not merely good relatively to us: as it exists by participating in the idea of the good, so it is good by resembling the idea; the participation is the cause of the resemblance. Hence good is identified with existence, evil with non-existence; and, as I have said, each thing exists just in so far as it is good, and no further. The source of Being and of Good the same.

Again in the ideal theory we for the first time reach a conception, and a very distinct conception, of immaterial existence. Perhaps we are a little liable to be backward in realising what a huge stride in advance this was. I will venture to affirm that there is not one shadow of evidence in all that we possess of preplatonic utterances to show that any one of Plato's predecessors had ever so remote a notion of immateriality. Parmenides, who would gladly have welcomed idealism, is as much to seek as any one in his conception of it. And when we see such a man as Parmenides 'the reverend and awful' with all his 'noble profundity' hopelessly left behind, we may realise what an invincible genius it was that shook from its wings the materialistic bonds that clogged Conception of immaterial existence.

both thought and speech and rose triumphant to the sphere of the 'colourless and formless and intangible essence which none but reason the soul's pilot is permitted to behold.'

Distinction between perceiving and thinking.

And as the material and immaterial are for the first time distinguished, so between perception and thought is the line for the first time clearly drawn. Perception is the soul's activity as conditioned by her material environment; thought her unfettered action according to her own nature: by the former she deals with the unsubstantial flux of phenomena, by the latter with the immutable ideas.

Plato works in whatever is valid in Herakleitos, Parmenides and Anaxagoras.

Plato then recognises and already seeks to conciliate the conflicting principles of Herakleitos and Parmenides. He satisfies the demand of the Eleatics for a stable and uniform object of cognition, while he concedes to Herakleitos that in the material world all is becoming, and to Protagoras that of this material world there can be no knowledge nor objective truth. He also affirms with Anaxagoras that mind or soul is the only motive power in nature—soul alone having her motion of herself is the cause of motion to all things else that are moved. Thus we see that Plato has taken up into his philosophy the great principles enounced by his forerunners and given them a significance and validity which they never had before.

Deficiencies of the earlier Platonism.

§ 18. Now had Plato stopped short with the elaboration of the philosophical scheme of which an outline has just been given, his service to philosophy would doubtless have been immense and would still probably have exceeded the performance of any one man besides. But he does not stop short there—nay, he is barely half way on his journey. We have now to consider what defects he discovered in the earlier form of his theory, and how he set about amending them.

Herakleitos and Parmenides not yet conciliated.

First we must observe that the conciliation of Herakleitos and Parmenides is only just begun. It is in fact clear that Plato, although recognising the truth inherent in each of the rival theories, had, when he wrote the *Republic*, no idea how completely interdependent were the two truths. For in the *Republic* his concern is, not how he may harmonise the Herakleitean and Eleatic principles as parts of one truth, but how, while satisfying the just claims of Becoming, he may establish a science of Being. He simply makes his escape from the Herakleitean world of Becoming into an Eleatic world of Being. And the world of Becoming is for him a mere superfluity, he does not recognise it as an inevitable concomitant of the world of Being. This amounts to saying that he does not yet recognise the Many as the inevitable counterpart to the One.

Plato is in fact still too Eleatic. He does not roundly reject the material world altogether: he sees that some explanation of it is necessary, and endeavours to explain it as deriving a kind of dubious existence from the ideas. But this part of his theory was, as he himself seems conscious, quite vague and shadowy: the existence or appearance of material nature is left almost as great a mystery as ever. And, as we shall see, the nature of the ideas themselves is not satisfactorily made out, still less their relation to the αὐτὸ ἀγαθόν. *Phenomena inadequately explained.*

Plato is also too Sokratic. He allows the Sokratic element in his system to carry weight which oversets the balance of the whole. We have seen that, owing to his admission of a hypostasis corresponding to every Sokratic concept, we have among the denizens of the ideal world ideas of σκευαστά, of relations, and of things that are evil. In the first place the proposition that there exist in nature eternal types of artificial things seems very dubious metaphysic. Again, we have only to read the *Phaedo* in order to perceive what perplexities beset the ideas of relations[1]. Finally, the derivation from the supreme good of ideal evil is a difficulty exceeding in gravity all the rest. Clearly then the list of ideas needs revision. *Necessity of revising the list of ideas.*

Moreover but scant justice is done to the Anaxagorean principle of νοῦς. Plato had indeed supplied the teleological deficiency of Anaxagoras; but we have no hint yet of soul as the substance and truth of all nature, spiritual and material, nor of the conciliation of unity and multitude as modes of soul's existence. Nor have we any adequate theory to explain the relation of particular souls to phenomena and to the ideas. Even the Herakleitean principle itself is not carried deep enough. It is not sufficient to recognise its universal validity in the world of matter. For if there be any truth in Becoming, this must lie deeper than the mere mutability of the material world: the changefulness of matter must be some expression of changeless truth. *Principle of νοῦς undeveloped.*

I conceive then we may expect to see in Plato's revised theory (1) a more drastic treatment of the problem concerning the One and the Many, (2) a searching inquiry into the relation between ideas and particulars, (3) a large expurgation of the list *Summary.*

[1] For instance *Phaedo* 102 B.

of ideas, (4) a theory of the relation of soul, universal and particular, to the universe. The answer to these problems may be latent in the earlier Platonism: but Plato has not yet realised the possibilities of his theory. By the time he has done this, we find most important modifications effected in it. Still they are but modifications: Plato's theory remains the theory of ideas, and none other, to the end.

The Parmenides.

§ 19. The severe and searching criticism to which Plato subjects his own theory is begun in the *Parmenides*. This remarkable dialogue falls into two divisions of very unequal length. In the first part Parmenides criticises the earlier form of the theory of ideas; in the second he applies himself to the investigation of the One, and of the consequences which ensue from the assumption either of its existence or of its non-existence. The discussion of the ideal theory in the first part turns upon the relation between idea and particulars. Sokrates offers several alternative suggestions as to the nature of this relation, all of which Parmenides shows to be subject to the same or similar objections. The purport of his criticisms may be summed up as follows: (1) if particulars participate in the idea, each particular must contain either the whole idea or a part of it; in the one case the idea exists as a number of separate wholes, in the other it is split up into fractions; and, whichever alternative we accept, the unity of the idea is equally sacrificed: (2) we have the difficulty known as the τρίτος ἄνθρωπος—if all things which are like one another are like by virtue of participation in the same idea, then, since idea and particulars resemble each other, they must do so by virtue of resembling some higher idea which comprehends both idea and particulars, and so forth εἰς ἄπειρον: (3) if the ideas are absolute substantial existences, there can be no relation between them and the world of particulars: ideas are related to ideas, particulars to particulars; intelligences which apprehend ideas cannot apprehend particulars, and *vice versa*. It may be observed that the second objection is not aimed at the proposition that particulars resemble one another because they resemble the same idea, but against the hypothesis that because particulars in a given group resemble each other it is necessary to assume an idea corresponding to that group.

Sokrates is unable to parry these attacks upon his theory, but in the second part of the dialogue Plato already prepares a way of escape. In the eight hypotheses comprised in this section of the dialogue Parmenides examines τὸ ἕν, conceived in several different

INTRODUCTION.

senses with the view of ascertaining what are the consequences both of the affirmation and of the negation of its existence to τὸ ἓν itself and to τἆλλα τοῦ ἑνός. The result is that in some cases both, in other cases neither, of two strings of contradictory epithets can be predicated of τὸ ἓν or of τἆλλα. If both series of epithets can be predicated, τὸ ἓν can be thought and known, if neither, it cannot be thought nor known[1]. Now in the latter category we find a conception of ἓν corresponding to the Eleatic One and to the idea of the earlier Platonism.

The positive result of the *Parmenides* then is that the ideal theory must be so revised as to be delivered from the objections formulated in the first part: the second part points the direction which reform is to take. We must give up looking upon One and Many, like and unlike, and so forth, as irreconcilable opposites: we must conceive them as coexisting and mutually complementary. Thus is clearly struck the keynote of the later Platonism, the conciliation of contraries. In this way Plato now evinces his perfect consciousness of the necessity to harmonise the principles of his Ionian and Eleate forerunners, giving to each its due and equal share of importance.

§ 20. It will be convenient to take the *Theaetetus* next[2]. This dialogue, starting from the question what is knowledge, presents us with Plato's theory of perception—a theory which entirely harmonises with the teaching of the *Timaeus* and in part supplements it. This theory Plato evolves by grafting the μέτρον ἄνθρωπος of Protagoras upon the πάντα ῥεῖ of Herakleitos and developing both in his own way. As finally stated, it is as complete a doctrine of relativity as can well be conceived. What is given in our experience is no objective existence external to us; between the percipient and the object are generated perception on the side of the percipient and a percept on the side of the object: e.g. on the part of the object the quality of whiteness, on the part of the subject the perception of white. And subject and object are inseparably correlated and exist only in mutual connexion—subject cannot be percipient without object, nor object

The *Theaetetus*.

---

[1] For a detailed investigation of the intricate reasoning contained in this part of the dialogue see Dr Jackson's excellent paper in the *Journal of Philology*, vol. XI p. 287.

[2] Dr Jackson's arguments for including the *Theaetetus* in its present form among the later dialogues appear to me irresistible, although parts of the dialogue have such decided literary affinity to some of the earlier series that I am disposed to entertain the supposition that what we possess is a second and revised edition.

generate a percept without subject. And subject as well as object is undergoing perpetual mutation: thus, since a change either of object or of subject singly involves a change in the perception, every perception is continually suffering a twofold alteration. Perception is therefore an ever-flowing stream, incessantly changing its character in correspondence with the changes in subject and in object. Nothing therefore can be more complete than the absolute instability of our sensuous perceptions. The importance of this theory will be better realised when we view it in the light of the *Timaeus*.

The Sophist.

§ 21. More important than even the *Parmenides* is the *Sophist*, one of the most profound and far-reaching of Plato's works. Plato starts with an endeavour to define the sophist, who, when accused of teaching what seems to be but is not knowledge, turns upon us, protesting the impossibility of predicating not-being: it is nonsense to say he teaches what is not, for τὸ μὴ ὂν can neither be thought nor uttered. Hereupon follows a truly masterly examination into the logic of being and not-being. The result is to show that either of the two, viewed in the abstract and apart from the other, is self-contradictory and unthinkable. And as being cannot exist without not-being, so unity also, if it is to have any intelligible existence, must contain in itself the element of plurality; one is at the same time one and not-one, else it has no meaning. The failure to grasp this truth is the fundamental flaw in Eleatic metaphysics and consequently in the earlier ideal theory. It seems to me hardly open to doubt that the εἰδῶν φίλοι of 248 A represent Plato's own earlier views. The strictures he passes upon these εἴδη are just those to which we have seen that the incomplete ideal theory is liable. He shows that the absolute immobility of the εἴδη, to which all action and passion are denied, renders them nugatory as ontological principles—they are empty and lifeless abstractions: yet, says Plato, a principle of Being must surely have life and thought—249 A. Next he takes five of the μέγιστα γένη, as he calls them, Rest, Motion, Same, Other, Being; and he demonstrates their intercommunicability, total or partial. The deduction from this is that such relations are not αὐτὰ καθ' αὑτὰ εἴδη, or self-existing essences, but forms of predication, or, as we might say, categories. Thus the ideas of relations which gave us so much trouble are swept away; for were these γένη substantial ideas, they could not thus be intercommunicable. Finally, the sophistic puzzle about μὴ ὂν

INTRODUCTION.

is disposed of by resolving the notion of negation into that of difference: μὴ ὄν is simply ἕτερον.

The foregoing statement, brief and general as it is, will suffice, I think, for enabling us to estimate the extent of the contribution made by this dialogue towards building up the revised system. We have (1) the overthrow of the Eleatic conception of being and unity, which warns us that the ideal theory, if it would stand, must abandon its Eleatic character, (2) the most important declaration that Being must have life and thought—this of course implies that the only Being is soul, and points to the universal soul of the *Timaeus*, (3) the deposition of relations from the rank of ideas, (4) the dissipation of all the fogs that had gathered about the notion of μὴ ὄν, and the affirmation that there is a sense in which not-being exists. The *Sophist*, it may be observed, does for the logical side of Being and Not-being very much what the *Timaeus* does for the metaphysical side. There is much besides which is important and instructive in this dialogue, but I believe I have summed up its main contributions to the later metaphysic.

§ 22. The *Sophist* then has expunged relations from the list of ideas. But there is another class of ideas included in the earlier system which is not expressly dealt with in any one of the later dialogues, and which it may be as well to mention here. We have seen reason to desire the abolition of ideas of σκευαστά. Now so far as Plato's own statements are concerned, the abandonment of these ideas is only inferential. There is continual reference to such ideas in the earlier dialogues, but absolutely none in the later. This would perhaps sufficiently justify us in deducing the absence of σκευαστά in the revised list of ideas. But we have in addition the distinct testimony of Aristotle on this point. See *metaphysics* Λ iii 1070ᵃ 18 διὸ δὴ οὐ κακῶς ὁ Πλάτων ἔφη ὅτι εἴδη ἐστὶν ὁπόσα φύσει, with which compare A ix 991ᵇ 6 οἷον οἰκία καὶ δακτύλιος, ὧν οὔ φαμεν εἴδη εἶναι. We know that in the earlier period Plato did recognise ideas of οἰκία and δακτύλιος: therefore Aristotle, in denying such ideas, must have the later period in his mind. In just the same way we read in *metaphysics* A ix 990ᵇ 16 οἱ μὲν τῶν πρός τι ποιοῦσιν ἰδέας, ὧν οὔ φαμεν εἶναι καθ᾽ αὑτὸ γένος. Relations were undoubtedly included among the ideas of the earlier period; yet, since, as we have seen, they are rejected in the later, Aristotle simply denies their existence without reference to the earlier view.

Ideas of σκευαστά abolished.

Thus then, sweeping away all ideas of σκευαστά, we are able to affirm that in Plato's later metaphysic there are ideas corresponding only to classes of particulars which are determined by nature, and none corresponding to artificial groups.

The Philebus. § 23. In the *Philebus* we come for the first time to constructive ontology. We have the entire universe classed under four heads—Limit, πέρας—the Unlimited, ἄπειρον—the Limited, μικτόν—the Cause of limitation, αἰτία τῆς μίξεως. In this classification πέρας is form, as such; ἄπειρον is matter, as such; μικτὸν is matter defined by form; αἰτία τῆς μίξεως is the efficient cause which brings this information to pass: and this efficient cause is declared to be the universal Intelligence or νοῦς. The objects of material nature are the result of a union between a principle of form and a formless substrate, the latter being indeterminate and ready to accept impartially any determination that is impressed upon it. It is not indeed correct to say that the ἄπειρον of the *Philebus* is altogether formless: it is indeterminately qualified, and the πέρας does but define the quantity. For example, ἄπειρον is 'hotter and colder,' that is, indeterminate in respect of temperature : the effect of the πέρας is to determine the temperature. The result of this determination is μικτόν, i.e. a substance possessing a definite degree of heat. The analysis of the material element given in the *Philebus* therefore falls far short, as we shall see, of the analysis in the *Timaeus*.

It is not however the πέρας itself which informs the ἄπειρον: Plato speaks of the informing element as πέρας ἔχον, or πέρατος γέννα. This it is which enters into combination with matter, not the πέρας itself. What then is the πέρας ἔχον? I think we cannot err in identifying it with the εἰσιόντα καὶ ἐξιόντα of the *Timaeus*; i.e. the forms which enter into the formless substrate, generating μιμήματα of the ideas, and which vanish from thence again. The πέρας ἔχον will then be the Aristotelian εἶδος—the form inherent in all qualified things and having no separate existence apart from things. Every sensible thing then consists of two elements, logically distinguishable but actually inseparable, form and matter. Nowhere in the material universe do we find form without matter or matter without form. Form then or limit, as manifested in material objects, must be carefully distinguished from the absolute πέρας itself, which does not enter into communion with matter : but every πέρας ἔχον possesses the principle of limitation, which it imposes upon the ἄπειρον wherewith it is combined.

But what is the πέρας itself? I think we are not in a position to answer this question until we have considered the *Timaeus*. But the nature of the reply has been indicated by a hint given us in the *Parmenides*, viz. that the ideas are παραδείγματα ἑστῶτα ἐν τῇ φύσει. For the πέρας ἔχον, by imposing limit, so far assimilates the ἄπειρον to the πέρας; consequently the μικτὸν is the μίμημα of the πέρας as παράδειγμα. We may therefore regard the πέρας as the ideal type to which the particulars approximate. Thus we derive from the *Philebus* a hint of the paradeigmatic character of the idea, which assumes its full prominence in the *Timaeus*. This part of the theory however cannot be adequately dealt with until we have examined the latter dialogue.

The most important metaphysical results of the *Philebus* may thus, I conceive, be enumerated: (1) the assertion of universal mind as the efficient cause, and as the source of particular minds, (2) the distinction of the formal and material element in things, (3) the theory of matter as such, rudimentary as it is, which is given us in the ἄπειρον.

§ 24. Besides this, the *Philebus* enables us to make another very important deduction from the number of ideas. We now regard the particular as resembling the idea in virtue of its information by the πέρας ἔχον. And in so far as this information is complete the particular is a satisfactory copy of the idea. Now let us represent any class of particulars or μικτὰ by the area of a circle. The centre of this circle would be marked by the particular, if such could be found, which is a perfect material copy of the idea—that particular in which the formal and material elements are blended in exactly the right way. Let us suppose the other particulars to be denoted by various points within the circle in every direction at different distances from the centre. Now in so far as the particulars approximate to the centre, they are like the idea, and by virtue of their common resemblance to the idea they resemble each other. Such particulars then as resemble each other because of their common resemblance to the idea are called by the class-name appropriate to the idea. But it is clear that particulars may also resemble each other because of a similar divergence from the idea: we may have a number of them clustering round a point within our circle far remote from the centre and therefore very imperfectly representing the idea. Such particulars have a class-name not derived from the idea, but denoting a similar divergence from the idea. A word denoting

Ideas of evil no longer admitted.

divergence from the idea denotes evil. Therefore there are class-names of evil things; but such class-names do not presuppose a corresponding idea: they simply indicate that the particulars comprehended by them fall short of the idea in a similar manner.

For example: a human being who should exactly represent the αὐτὸ δ ἐστιν ἄνθρωπος would be perfectly beautiful and perfectly healthy. But in fact humanity is sometimes afflicted with deformity and sickness: we have accordingly class-names for these evils. But one who is deformed or sick fails, to the extent of his sickness or deformity, in representing the ideal type: these class-names then do not represent an idea but a certain falling-off from the idea. Hence we have no idea of fever, because fever is only a mode of deviation from the type[1]; and the same is true of all other imperfections. Thus at one stroke we are rid of all ideas of evil.

Advance made in the four dialogues on the metaphysic of the earlier period.

§ 25. Let us now pause to consider how far these four dialogues have carried us in the work of reconstruction, and how much awaits accomplishment.

In the first place, the elimination of spurious ideas is fully achieved. The *Sophist* frees us from ideas of relations, the *Philebus* from those of evil; while σκευαστά are rejected on the strength of Aristotle's testimony, confirmed by the total absence of reference to them in the later dialogues: accordingly we have now ideas corresponding only to classes naturally determined. It seems to me manifest that ideas of qualities must also be banished from the later Platonism; and on this point too we have the negative evidence that they are never mentioned in the later dialogues; but there is no direct statement respecting them.

We have also a clear recognition, especially prominent in the *Parmenides*, of the indissoluble partnership between One and Many, Rest and Motion, Being and Not-being. The necessity for reconciling these apparent opposites is distinctly laid down, though the conciliation is not yet worked out. The acknowledgement of soul as the one existence, from which all finite souls are derived, and as the one efficient cause is a notable advance, as is also the theory of the *Theaetetus* concerning the relation between particular souls and material nature. And finally we have the analysis of ὄντα into their formal and material elements, and the still immature conception of matter as a potentiality.

[1] In the *Phaedo*, on the contrary, we definitely have an idea of fever: see 105 C.

Moreover, putting the *Theaetetus* and *Philebus* together, we obtain a result of peculiar importance. From the latter we learn that finite souls are derived from the universal soul, from the former that material objects are but the perceptions of finite souls. The conclusion is inevitable, since the objects which constitute material nature do not exist outside the percipient souls, and since these percipient souls are part of the universal soul, that material nature herself is a phase of the universal soul, which is thus the sum total of existence. Thus we have the plainest possible indication of the ontological theory which is set forth in the *Timaeus;* though, as usual, Plato has not stated this doctrine in so many words, but left us to draw the only possible inference from his language.

§ 26. Yet great as is the progress that has been made, even more remains to be achieved; and it is to the *Timaeus* that we must look for fulfilment. *Deficiencies still to be supplied.*

Although the fundamental problem of the One and the Many is now fairly faced, the solution is not yet worked out. Nor is the relation between the universal efficient Intelligence and the world of matter clearly established: the failure of Anaxagoras in this regard remains still unremedied. Also (what is the same thing viewed in another way) the relation between ideas and particulars is left undefined. Nay, in this respect we seem yet worse off than we were in the *Republic*. For the old unification, such as it was, has disappeared, and no new one has taken its place. Formerly we were content to say that the particulars participated in the ideas, and from the ideas derived their existence. But now this consolation is denied us. We have the ideas entirely separate from the particulars, as types fixed in nature; and no explanation is offered as to how material nature came to exist, or seem to exist, over against them. We have the 'subjective idealism' of the *Theaetetus*, and that is all. In fact, while we vindicate the idea as a unity, we seem to sacrifice it as a cause.

Furthermore we desiderate a clearer account of the relation between the supreme idea and the inferior ideas, and also between limited intelligences and the infinite intelligence: nor can we be satisfied without a much more thorough investigation into the nature of materiality. And the answers to all these questions must be capable of being duly subordinated to one comprehensive system.

Now if the *Timaeus* supplies in any reasonable degree a solution of the aforesaid problems, it seems to me that no more need be said about the importance of the dialogue.

The central metaphysical doctrine of the Timaeus.

§ 27. In the *Timaeus* Plato has given us his ontological scheme in the form of a highly mystical allegory. I propose in the first place to give a general statement of what I conceive to be the metaphysical interpretation of this allegory, reserving various special points for after consideration[1]. The ontological teaching of this dialogue, though abounding in special difficulties, can in my belief be very clearly apprehended, if we but view it in the light afforded us by the other writings of this period; on which in turn it sheds an equal illumination.

In the *Timaeus* then the universe is conceived as the self-evolution of absolute thought. There is no more a distinction between mind and matter, for all is mind. All that exists is the self-moved differentiation of the one absolute thought, which is the same as the Idea of the Good. For the Idea of the Good is Being, and the source of it; and from the *Sophist* we have learnt that Being is Mind. And from the *Parmenides* we have learnt that Being which is truly existent must be existent in two modes: it must be one and it must be many. For since One has meaning only when contrasted with Many, Being, forasmuch as it is One, demands that Many shall be also. But since Being alone exists, Being must itself be that Many. Again, Being is the same with itself; but Same has no meaning except as correlated with Other; so Being must also be Other. Once more, Being is at rest; Rest requires its opposite, Motion; therefore Being is also in motion. Seeing then that Being is All, it is both one and many, both same and other, both at rest and in motion: it is the synthesis of every antithesis. The material universe is Nature manifesting herself in the form of Other: it is the one changeless thought in the form of mutable multitude. Thus does dualism vanish in the final identification of thought and its object: subject and object are but different sides of the same thing. Thought must think: and since Thought alone exists, it can but think itself[2].

[1] Considering that the exposition here offered deals with matters of much controversy, my statement may be thought unduly categorical and dogmatic. In reply I would urge that difficulties of interpretation and the manner in which Plato's meaning comes out are pretty copiously discussed in the commentary. At present I am aiming at making my story as clear as possible, to which end I have given results rather than processes. What I conceive to be the justification for the views advanced will, I hope, appear in the course of my exposition.

[2] It is easy to see that Aristotle's

Yet, though matter is thus resolved into a mode of spirit, it is not therefore negated. It is no longer contemptuously ignored or dismissed with a metaphor. Matter has its proper place in the order of the universe and a certain reality of its own. Though it has no substantial being, it has a meaning. For Nature, seeing that she is a living soul, evolves herself after a fixed inevitable design, in which all existence, visible and invisible, finds its rightful sphere and has its appointed part to play in the harmony of the universe. But there is more to be said ere we can enter upon the nature of matter.

§ 28. The universal mind, we say, must exist in the form of plurality as well as in the form of unity. How does this come to pass? The hint for our guidance is to be found in the *Philebus*, where we learn that, as the elements which compose our bodies are fragments of the elements which compose the universe, so our souls are fragments, as it were, of the universal soul. Hence we see how the one universal intelligence exists in the mode of plurality: it differentiates itself into a number of finite intelligences, and so, without ceasing to be one, becomes many. These limited personalities are of diverse orders, ranging through all degrees of intellectual and conscious life; those that are nearest the absolute mind, if I may use the phrase, possessing the purest intelligence, which fades into deeper and deeper obscurity in the ranks that are more remote. First stands the intelligence of gods, which enjoys in the highest degree the power of pure unfettered thought; next comes the human race, possessing an inferior but still potent faculty of reason. Then as we go down the scale of animate beings, we see limitation fast closing in upon them—intelligence grows ever feebler and sensation ever in proportion stronger, until, passing beyond the forms in which sensation appears to reign alone, we come in the lowest organisms of animal and vegetable life to beings wherein sensation itself seems to have sunk to some dormant state below the level of consciousness. Yet all these forms of life, from the triumphant intellect of a god to the green scum that gathers on a stagnant pool, are modes of one universal all-pervading Life. Reason may degenerate to sensation, sensation to a mere faculty of growth;

Pluralisation of the universal mind in the form of finite existences.

νόησις νοήσεως is directly derived from the *Timaeus:* though his very fragmentary utterances on this subject leave us in doubt how far he had proceeded on Plato's lines in conceiving of material nature as one mode of the eternal thought.

but all living things are manifestations of the one intelligence expanding in ever remoter circles through the breadth and depth of the universe: each one is a finite mode of the infinite—a mould, so to speak, in which the omnipresent vital essence is for ever shaping itself.

<small>The nature of matter and its place in the Platonic ontology.</small>

§ 29. So far as the theory has yet taken us, we have on the one hand the universal soul, on the other finite existences into which the universal evolves itself. Matter has not yet made its appearance in our system. But Plato is not wanting in an account of matter; and here the theory of perception in the *Theaetetus* will come to our aid.

In the pluralisation of universal soul finite souls attain to a separate and independent consciousness. But for this independent consciousness every soul has to pay a fixed price. The price is limitation, and the condition of limitation is subjection to the laws of what we know as time and space. But the degree of subjection varies in different orders of existence; and in the higher forms is tempered with no mean heritage of freedom. The object of cognition for finite souls is truth as it is in the universal soul. Now intelligences of the higher orders have two modes of apprehending this universal truth—one direct, by means of the reason, one symbolical, by means of the senses. And when we speak of soul acting by the reason and through the senses, we mean by these phrases that in the one case the soul is exercising the proper activity of her own nature, *qua* soul; in the other that she is acting under the conditions of her limitation, *qua* finite soul: which conditions we saw were time and space. Now the direct apprehension, which we call reasoning, exists to any considerable degree only in gods and in the human race. In the inferior forms of animation the direct mode grows ever feebler, until, so far as we can tell, it disappears altogether, leaving the symbolical mode of sensuous perception alone remaining. Time and space then are the peculiar adjuncts of particular existence, and material objects, i.e. sensuous perceptions, are phenomena of time and space—in other words symbolical apprehensions of universal truth under the form of time and space. Thus the material universe is, as it were, a luminous symbol-embroidered veil which hangs for ever between finite existences and the Infinite, as a consequence of the evolution of one out of the other. And none but the highest of finite intelligences may lift a corner of this veil and behold aught that is behind it.

But we must beware of fancying that this material nature has any independent existence of its own, apart from the percipient —it has none[1]. All our perceptions exist in our own minds and nowhere else; the only existence outside particular souls is the universal soul. Material nature is but the refraction of the single existent unity through the medium of finite intelligences: each separate soul is, as it were, a prism by which the white light of pure being is broken up into a many-coloured spectacle of ever-changing hues. Matter is mind viewed indirectly. Yet this does not mean the negation of matter: matter has a true reality in our perceptions; for these perceptions are real, though indirect, apprehensions of the universal. And since universal Nature evolves herself according to some fixed law and order, there is a certain stability about our perceptions, and a general agreement between the perceptions of beings belonging to the same rank. But none the less are we bound to affirm that matter has no separate existence outside the percipient soul. Such objectivity as it possesses amounts to this: it is the same eternal essence which is thus symbolically apprehended by all finite intelligences. Mind is the universe, and beside Mind is there nothing.

§ 30. But all this time what has become of the ideas? So far they have not even been mentioned in our exposition. Yet their existence is most strenuously upheld in this dialogue, and therefore their place in the theory must be determined. Our duty then plainly is to search the ontology of the *Timaeus* for the ideas.

The ideal theory in the Timaeus.

It is notable that in the *Timaeus* we hear less than usual of the plurality of ideas; nor is that surprising, when so much stress is laid upon a comparatively neglected principle, the unity of the Idea. But the plurality of ideas is not only reaffirmed in the most explicit language, it is a metaphysical principle especially characteristic of the dialogue. The paradeigmatic aspect of the ideas now comes into marked prominence: they are the eternal

---

[1] The teaching of the *Theaetetus*, viewed in relation with the space-theory of the *Timaeus*, seems to me perfectly conclusive on this point. It may indeed be argued that only the αἴσθησις is purely subjective, according to the theory of the *Theaetetus*; the object generating the αἰσθητόν, although existing in correlation with the subject, has an existence external to it. But this is no real objection. For if Soul is the sum-total of existence, all that exists independently of finite soul is the universal soul. Therefore, so far as the object exists outside the subject, that object is the universal soul itself: that is, as said above, our sense-perceptions are perceptions of the universal under the condition of space.

and immaterial types on which all that is material is modelled. 'Alles Vergängliche ist nur ein Gleichniss' might be adopted as the motto of the *Timaeus*.

In order to make clear the position of the ideas in Plato's maturest ontology, I fear I must to some extent repeat what has been said in the preceding section. The supreme idea, αὐτὸ ἀγαθόν, we have identified with universal νοῦς, for which τὸ ὄν, τὸ ἕν, and τὸ πᾶν are synonyms. This universal thought then realises itself by pluralisation in the form of finite intelligences. These intelligences possess a certain mode of apprehending the universal, which we term sensuous perception. By means of such perception true Being cannot be apprehended as it is in itself; what is apprehended is a multitude of symbols which shadow forth the reality of existence, and which constitute the only mode in which such existence can present itself to the senses. These symbols or likenesses we call material objects, which come to be in space, and processes, which take place in time. They have no substantial existence, but are subjective affections of particular intelligences: what is true in them is not the representation in space and time, but the reality of existence which they symbolise. But these symbols do not arise at random nor assume arbitrary forms. Since the evolution of absolute thought is not arbitrary, but follows the necessary and immutable law of its own nature, it may be inferred that all finite intelligences of the same rank have, within a certain margin, similar perceptions. Now the unity of Being presents itself to diverse kinds of sense and to each sense in manifold wise. Each of these presentations is the εἰκών, or image, of which that unity is the παράδειγμα, or original; and the accuracy of the image varies according to the clearness of the presentation. A perfectly clear presentation is a perfect symbol of the truth, the εἰκών exactly reflects the παράδειγμα: a dimmer presentation is a more imperfect image. The παράδειγμα then is the perfect type, to which every particular more or less approximates. Now were this approximation quite successfully accomplished, in every class the particulars, since they all exactly reflected the type, would be all exactly alike. Deviations from the type and consequent dissimilarities among the particulars are due to the imperfect degree in which our senses are capable of apprehending, even in this indirect way, the eternal type.

Since then we see that different classes of material phenomena are so many different forms in which the eternal unity presents itself to the senses, it follows that the types or ideas corresponding to such classes are simply determinations of the universal essence or αὐτὸ ἀγαθόν itself: that is to say, each idea is the idea of the good specialised in some particular mode or form—blueness is the mode in which the good reveals itself to the faculty which perceives blue. So then everything in nature which we hear or see or perceive by any perception means the idea of the good. There is thus nothing partial or fractional in Nature: she reveals herself to us one and entire in each of her manifestations. Diversity is of us. We are all beholding the same truth with a variety of organs: it is as though we looked at a flame through a many-faceted crystal, which repeats it on every surface. And since the unity is eternal and inexhaustible, inexhaustible is the number of forms in which it may present itself to every sense.

§ 31. Furthermore, if it were in the nature of finite intelligences to receive through the senses accurate symbols of the good, all things must be perfectly fair; foulness is due to defect of presentation. Hence there can be no ideas of ugliness and dirt, of injustice and evil: all these things arise from failure in representing the idea and consequent failure in existence. For in all things that exist there must be a certain degree of good, else they could not exist at all: even in visible objects that are most hideous there is some fairness; the likeness to the type is there, however marred and scarce discernible. Evil is nothing positive, it is but defect of existence; and this defect is due to the limitations of finite intelligence and of finite modes of being. *Evil is defective presentation of the type.*

To sum up: the one universal Thought evolves itself into a multitude of finite intelligences, which are so constituted as to apprehend not only by pure reason, but also by what we call the senses, with all their attendant subjective phenomena of time and space. These sensible phenomena group themselves into a multitude of kinds, each kind representing or symbolising the universal Thought in some determinate aspect. It is the Universal itself which in each of these aspects constitutes an idea or type, immaterial and eternal, whereof phenomena are the material and temporal representations: the phenomena do in fact more or less faithfully express the timeless and spaceless in terms of space and time. Thus the αὐτὸ ἀγαθόν is the ideas, and the ideas are the phenomena, which are merely a mode of their manifestation to finite intelligence. The whole universe, then, ideal and material, *Summary.*

is seen to be a single Unity manifesting itself in diversity. Such I conceive to be the theory of ideas in its final form.

The plurality of ideas a necessary corollary of the pluralisation of universal thought.

§ 32. One thing more should be added. It is plain from what has been said, that the plurality of ideas is the inevitable consequence of the pluralisation of absolute thought into finite minds. For the various classes of phenomena, to which we need corresponding ideas, are part of our consciousness as limited beings, and arise from our limitation. It is because universal Being is presented to us in this sensuous manner, in groups of material phenomena, that universal Being must determine itself into types of such phenomena. If we were not constituted so as to see roses, there would be no idea of roses. We should then be contemplating the eternal unity directly, as it is in itself: differentiation would neither be necessary nor possible. But this may not be, for pluralisation without limitation is inconceivable: and limitation to us involves space and time. Therefore—paradoxical, nay profane as the statement would have appeared in the days of the *Republic*—ideas can no more exist without particulars than particulars can exist without the ideas.

Question raised: are there ideas of ζῷα only?

§ 33. Before we leave this subject, a question suggests itself to which it is perhaps impossible to return a decisive answer. We have seen that in the mature Platonism ideas are restricted to classes which are naturally determined. Ought we to go a step further and confine the ideas to classes of living things? It appears to me that there are good grounds for an affirmative answer; but Plato has left his intention uncertain.

All the ideas mentioned in the *Timaeus*, with the exception of one passage, are ideas of ζῷα—a term which includes plants as well as animals. The exceptional passage is 51 B, where we hear of πῦρ αὐτὸ ἐφ' ἑαυτοῦ and, by implication, of ideas of the other three elements also. Now that ideas should be confined to ζῷα seems reasonable on the following grounds. The supreme idea is expressed in the *Timaeus* as αὐτὸ ὅ ἐστι ζῷον, and this includes all other ideas that exist. If then the supreme universal idea is ζῷον, it would seem that the more special ideas, which are subordinate to it, ought to be ζῷα likewise. Or let us put it in another way. We have been led to identify the supreme intelligence with the αὐτὸ ἀγαθόν. We have said too that this supreme intelligence or idea pluralises itself into finite existences, and that it determines itself into special ideas. Now do not this pluralisation and this determination constitute one act? Is not the evolution of Mind in the form of human minds the same process as the determination of the idea of Man? If this be so, then, since Mind can only pluralise itself in the form of living beings, it can only determine itself into ideas of ζῷα. Aristotle[1] indeed seems to account πῦρ, σάρξ, κεφαλή, as natural classes whereof there are ideas: but I very much doubt whether Plato would have admitted ideas of these. The idea of Star involves in its material representation πῦρ, even as the idea of Man involves in its material representation σάρξ and κεφαλή: but it in no way requires the existence of any ideas of these things.

There is however the passage 51 B, in which an idea of fire is distinctly mentioned. I think it probable that this passage ought not to be pressed too hard. After he has been speaking of the four material elements, Plato raises the question whether these material substances alone are existent, or whether there is such a thing as immaterial essence: and the four elements being in possession of the stage, it naturally occurs to contrast them with ideal types of the elements. I do not think we are forced to conclude from this that Plato deliberately meant to postulate such ideas. If this explanation be not admitted, I should say that we have in this passage a relic of the older theory, which Plato ought to have eliminated, and would have eliminated, had his attention been drawn to the subject. Practically then I believe that we should regard the ideal world as confined to ideas of ζῷα.

§ 34. The foregoing account of the metaphysical teaching contained in the *Timaeus* suffices, I think, to show that in this dialogue, taken in conjunction with the other later writings, Plato does offer us a solution of the problems enumerated in § 26 as yet unsolved. We now have his theory (1) as to the relation of the efficient mind to material nature, the latter arising from the pluralisation of the former; (2) the relation of the supreme idea to the other ideas, which are determinations of it; (3) the relation of ideas and particulars—that the particular is the symbolical presentation of the idea to limited intelligence under the conditions of space and time; (4) the relation between the supreme intelligence and the finite intelligences, into which it differentiates itself; (5) the relation between the finite intelligence and material nature, involving an account of matter itself;

Summary of results.

[1] *Metaphysics* Λ iii 1070ᵃ 19.

and (6) we have the fundamental antithesis of One and Many treated with satisfying completeness. Plato is indeed far more profoundly Herakleitean than Herakleitos himself. Not content, like the elder philosopher, with recognising the antithesis of ὄν and μὴ ὄν as manifested in the world of matter, he shows that this is but the visible symbol of the same antithesis existing in the immaterial realm. True Being itself is One and Many, is Same and Other. Were there not a sense in which we could say that Being is not, there were no sense in which we could say that it verily is. Matter in its mobility, as in all besides, is a likeness of the eternal and changeless type.

It now remains to deal with some special features of the dialogue, and to discuss certain objections and difficulties which may seem to us to threaten our interpretation.

<small>Difficulty arising from the allegorical style of the *Timaeus*.</small>

§ 35. The form which Plato gives to his thoughts in this dialogue has greatly multiplied labour to his interpreters. For all his clearness of thought and lucidity of style Plato is always the most difficult of authors: and in the *Timaeus* we have the added difficulty of an allegorical strain pervading the whole exposition of an ontological theory in itself sufficiently abstruse. And if we would rightly comprehend the doctrine, we must of course interpret the allegory aright. Plato is the most imaginative writer produced by the most imaginative of nations; and he insists on a certain share of imagination in those who would understand him. A blind faithfulness to the letter in this dialogue would lead to a most woful perversion of the spirit. Here, more than in any of Plato's other writings, the conceptions of his reason are instantly decked in the most vivid colours by his poetic fancy. And of all poetical devices none is dearer to Plato than personification. Hence it is that he represents processes of pure thought, which are out of all relation to time and space, as histories or legends, as a series of events succeeding one another in time. In conceiving the laws and relations of mind and matter, the whole thing rises up before his imagination as a grand spectacle, a procession of mighty events passing one by one before him. First he sees the unity of absolute thought, personified as a wise and beneficent creator, compounding after some mysterious law the soul that shall inform this nascent universe: next he descries a doubtful and dreamlike shadow, formless and void, which under the creator's influence, gradually shapes itself into visible existence and is interfused with the world-soul which controls and orders it, wherewith it forms a harmonious whole, a perfect sphere, a rational divine and everlasting being. Next within this universe arise other divine beings, shining with fire and in their appointed orbits circling, which measure the flight of time and make light in the world. Finally, the creator commits to these gods, who are the work of his hands, the creation of all living things that are mortal: for whom they frame material bodies and quicken them with the immortal essence which they receive from the creator.

All this is pure poetry, on which Plato has lavished all the richness of imagery and splendour of language at his command. But beneath the veil of poetry lies a depth of philosophical meaning which we must do what in us lies to bring to light. And there is not a single detail in the allegory which it will be safe to neglect. For Plato has his imagination, even at its wildest flight, perfectly under control: the dithyrambs of the *Timaeus* are as severely logical as the plain prose of the *Parmenides*.

Most of the details of this myth are considered in the notes as they arise; but there are one or two of its chief features which must be examined here.

§ 36. The central figure in what may be called Plato's cosmological epic is the δημιουργός, or Artificer of the universe. It is evidently of the first importance to determine whether Plato intends this part of his story to be taken literally; and if not, how his language is to be interpreted. *How is the δημιουργός to be understood?*

The opinions which have been propounded on this subject may fairly be arranged under three heads.

According to the first view the δημιουργὸς is a personal God, external to the universe and actually prior to the ideas: to this appertains one form of the opinion that the ideas are 'the thoughts of God.' *(1) is he a personal God, external to the universe and prior to the ideas?*

There is but one passage in all Plato's works which can give the slightest apparent colour to the theory that the ideas are in any sense created or caused by God. This is in *Republic* 597 B—D, where God is described as the φυτουργὸς of the ideal bed. But a little examination will show that no stress can really be laid upon this. For to the three beds, the ideal, the particular, and the painted, Plato has to assign three makers. For the two latter we have the carpenter and the artist: then, if the series is to be completed, who could possibly be named as the creator of the ideal bed save God? And the series must needs be completed to attain Plato's immediate purpose, in order that the carpenter and the artist may be placed in their proper order of merit. The postulation of God as the creator of the ideal bed is merely an expedient designed to serve a temporary end, not a principle of the Platonic philosophy. If we take any other view we bring the passage into direct conflict with the statement beginning 508 E, where it is declared in the plainest language that the Idea of the Good is the cause of all existence whatsoever. Moreover to maintain that the ideas are the thoughts of a personal God is utterly to ignore Plato's emphatic and constantly iterated affirmation of the self-existent substantiality of the ideas. Even could these declarations be explained away, we should have to face Aristotle's criticism of the ideal theory—nay, Plato's own criticism in the *Parmenides;* neither of which would have any meaning were not the ideas independent essences: the argument of the τρίτος ἄνθρωπος, for instance, would be irrelevant. The hypothesis then that a personal God is in any sense the cause of the ideas must be dismissed as incompatible with Platonic principles.

(2) is he a personal creator, external to the universe and to the ideas?

§ 37. Secondly, it is held that the δημιουργὸς is a personal creator, external to the universe and to the ideas, on the model of which he fashioned material nature. This view demands the most careful consideration, since it is the literal statement of the *Timaeus*. But it will prove, I think, to be totally untenable. In the first place it makes Plato offer us, instead of an ontological theory, a theological dogma: it is an evasion, not a solution of the problem. For we are asked to suppose that after constructing an elaborate ontology which is to unfold the secret of nature, Plato suddenly cuts the knot with a hypothesis which has absolutely no connexion with his ontology. Again, however much opinions may differ as to the extent of Plato's success in eliminating dualism, it will hardly be disputed that to do this was his aim. But here we have not merely dualism, but a triad: the ideas, the creator, and matter. All these are distinct and independent, nor is there any evolution of one from another. Can we seriously believe that Plato's speculations ended in this? And there remain yet more cogent considerations. In this story we find the δημιουργὸς represented as creating ψυχή. But ψυχή, we know, is eternal. Her creation must then be purely mythical: and if the creation, surely the creator also. Or if not, since ψυχὴ and the δημιουργὸς are alike eternal, are we to suppose that there are two separate and distinct Intelligences—that is, inasmuch as νοῦς exists in ψυχή alone, two ψυχαί to all eternity existing? What could be gained by such a reduplication? Moreover, if two such ψυχαί exist, there ought to be an idea of them—a serious metaphysical complication[1]. If on the other hand it be maintained that the cosmic soul is an emanation or effluence of the δημιουργός, this is practically abandoning the present hypothesis in favour of that which is next to be considered. Finally, if the δημιουργός is a personal creator, he is certainly ζῷον, and νοητὸν ζῷον. What then is his relation to the αὐτὸ ὃ ἔστι ζῷον? Either he is identical with it or contained in it: in either case the hypothesis falls to the ground. The literal interpretation of Plato's words must therefore be abandoned for the reason that its acceptance would reduce Plato's philosophy to a chaos of wild disorder.

§ 38. Lastly, the δημιουργός is identical with the αὐτὸ ἀγαθόν. This view, properly understood, I conceive to be in a sense correct: but it needs the most careful defining, and, in the form in which it is sometimes propounded, is unsatisfactory. We can only accept it by realising that the αὐτὸ ἀγαθόν is the infinite intelligence, which is manifested in the visible universe: and we shall approach the question better if we identify the δημιουργός, not in the first place with the ἀγαθόν, but with ψυχή, which comes to the same in the end.

(3) is he identical with the αὐτὸ ἀγαθόν?

Now the position of the δημιουργός in the *Timaeus* is precisely that of νοῦς βασιλεὺς in the *Philebus*: see *Philebus* 26 E—28 E. Therefore the δημιουργός is the universal intelligence from which all finite intelligences are derived. But intelligence or νοῦς is nothing else than ψυχή pure and simple, apart from any conjunction with matter. What then is the relation of pure intelligence to the cosmic soul which informs the universe? Let us turn once more to the *Philebus*. In 29 E—30 A νοῦς is definitely identified with the cosmic soul; it is the universal ψυχή whereof all visible nature is σῶμα. So then the δημιουργός of the *Timaeus* must be identical with the world-soul. This is so: but the statement is not yet complete. For the δημιουργός is pure reason, while the world-soul, being in conjunction with matter, is ψυχή in all her aspects, con-

---

[1] Compare *Timaeus* 31 A τὸ γὰρ περιέχον πάντα, ὁπόσα νοητὰ ζῷα, μεθ' ἑτέρου δεύτερον οὐκ ἄν ποτ' εἴη· πάλιν γὰρ ἂν ἕτερον εἶναι τὸ περὶ ἐκείνω δέοι ζῷον, οὗ μέρος ἂν εἴτην ἐκείνω. The argument is the same as in *Republic* 597 C.

taining the element not only of the Same, but of the Other also. In other words the δημιουργὸς is to the world-soul as the reasoning faculty in the human soul is to the human soul as a whole, including her emotions and desires. But the reasoning faculty is nothing distinct from the human soul; it is only a mode thereof. The δημιουργὸς then is one aspect of the world-soul: he is the world-soul considered as not yet united to the material universe— or more correctly speaking, since time is out of the question, he is the world-soul regarded as logically distinguishable from the body of the universe. And since the later Platonism has taught us to regard matter as merely an effect of the pluralisation of mind or thought, the δημιουργὸς is thought considered as not pluralised— absolute thought as it is in its primal unity. As such it is a logical conception only; it has not any real existence as yet, but must exist by self-evolution and consequent self-realisation[1]. These two notions, thought in unity and thought in plurality, are mythically represented in the *Timaeus*, the first by the figure of the creator, the second by the figure of the creation: but the creator and the creation are one and the same, and their self-conscious unity in the living κόσμος is the reality of both.

Application to the αὐτὸ ἀγαθόν.

§ 39. Now we may apply what has been said to the αὐτὸ ἀγαθόν. In § 27 we identified the αὐτὸ ἀγαθὸν with absolute thought or universal spirit. The identity of νοῦς with the ἀγαθὸν is plainly affirmed in *Philebus* 22 C: compare too the language used of νοῦς in *Philebus* 26 E with that used of the ἀγαθόν in *Republic* 508 E. We are justified then in identifying the δημιουργὸς with the αὐτὸ ἀγαθόν, so long as the ἀγαθὸν is conceived as not yet realised by pluralisation. For the realisation of the Good or of Thought comes to pass by the evolution of the One into the Many and the unification of both as a conscious whole. Thus Plato's system is distinctly a form of pantheism: any attempt to separate therein the creator from the creation, except logically, must end in confusion and contradiction.

Creation not an arbitrary exercise of will, but the fulfilment of eternal law.

§ 40. Thus we see that the process which is symbolised in the creation of the universe by the Artificer is no mere arbitrary exercise of power: it is the fulfilment of an inflexible law. The creator does not exist but in creating: or, to drop the metaphor, absolute thought does not really exist unless it is an object to

[1] I must guard against being supposed to mean that the pluralised thought is more real than the primal unity: only that the existence of both is essential to the reality of either.

itself. So then the creator in creating the world creates himself, he is working out his own being. Considered as not creating he has neither existence nor concrete meaning. Thus we have not far to seek for the motive of creation: it is so, because it must be so. A creator who does not create is thought which does not think, being which does not exist: it is no more than the lifeless abstraction of Eleatic unity.

After what has been said, it is almost a truism to affirm that the process represented in the *Timaeus* is not to be conceived as occupying time or as having anything whatsoever to do with time. Yet so potent is the spell of Plato's ποτανὰ μαχανά, that it may not be amiss to insist upon this once more. The whole story is but a symbolisation of the eternal process of thought, which is and does not become. All succession belongs to the phenomena of thought pluralised; it is part of the apparatus pertaining to them: but with the process of thought itself time has no more to do than space. It seems therefore vain to discuss, as has often been done, the eternity of the material universe in Plato's system. Considered as one element in the evolution of thought, material nature is of course eternal; but its phenomena, considered in themselves, belong to the sphere of Becoming and have no part in eternity: although, viewed in relation to the whole, time itself is a phase of the timeless, or, as Plato calls it, 'an eternal image of eternity.'

The process symbolised in the Timaeus independent of time and space.

§ 41. Only if we adopt the interpretation of the δημιουργός which I have been defending can we understand Plato's statement in 92 C that the universe is 'the image of its maker'—for the reading ποιητοῦ is better authenticated than νοητοῦ. If the κόσμος is the image of its maker, the maker must be identical with the αὐτὸ ὃ ἔστι ζῷον. Now since the κόσμος is πᾶν, the ζῷον cannot be anything outside it: rather it must be the notion which is realised in the universe; a type not separate from the copy, but fulfilled in the copy and in that fulfilment existing. It must be the unity whereof the κόσμος is the expression in multiplicity. Unity is the type, multiplicity the image thereof: and it is necessary that unity, if it is really to exist, must appear also in the form of multiplicity. Thus then must it be with the ζῷον. But this is exactly the position we have seen reason for assigning to the δημιουργός, so that Plato is fully justified in identifying the two. So if we say that the universe is the likeness of its creator, we mean that it is unity manifested in plurality and so realised.

The universe as the likeness of its creator.

The type and the likeness are the same thing viewed on different sides.

It is perhaps worth noticing that our view harmonises with Plato's statement in *Parmenides* 134 c, that as absolute knowledge cannot belong to man, so the knowledge of finite things cannot appertain to God. But if God be distinct from the universe, and so far limited, there seems no reason why he should not have knowledge of finite things. A God who is not the All, however much his knowledge may transcend human knowledge, would surely have the same kind of knowledge. But a God whose knowledge is of the absolute alone is a God whose knowledge is of himself alone; and such a God must be the universe, not a deity external to the universe.

<small>The κόσμος and the ψυχὴ τοῦ κόσμου.</small>

§ 42. Having thus investigated the relation of the δημιουργὸς to the cosmic soul and to the material universe, it behoves us to make a similar inquiry concerning the relation of the κόσμος and the ψυχὴ τοῦ κόσμου. The ψυχογονία of the *Timaeus* has been treated with some fulness in the commentary, so that a comparatively brief statement may here suffice by way of supplement.

The cosmic soul, like finite souls, consists of three elements— of ταὐτόν, θάτερον, and οὐσία: that is to say, the principle of Same, i.e. of unity and rest, of Other, i.e. of variety and motion, of Essence, which signifies the identification of these two in one conscious intelligence. The terms ταὐτὸν θάτερον and οὐσία have distinct applications, according to the side from which we regard the subject: these applications I have endeavoured to distinguish in the note on the passage which deals with the question. Let us first look at it thus. The world-soul consists (1) of absolute undifferentiated thought, (2) of this thought differentiated into a multitude of finite existences, and (3) it unites these two elements in a single consciousness. Now of what consists the material part, the body of the κόσμος? Simply of the perceptions of finite consciousnesses. And as these perceptions exist only in the consciousness of the percipient souls, so these souls are comprehended in the universal soul, whereof we have seen that finite souls are, as it were, fractional parts. Therefore the cosmic soul comprehends within her own nature all that exists, whether spiritual or material. Thus the only reality of the universe is the soul thereof, which is the one totality of existence. Matter is nothing but the revelation to finite consciousness, in the innumerable modes of its apprehension, of the universal spirit. All that is material is the expression in terms of the visible of the invisible, in terms of space and time of the spaceless and timeless, in terms of Becoming of Being. All sensible Nature is a symbol of the intelligible, and she is what she symbolises. So are all things at last resolved into an ultimate unity, which yet contains within itself all possible multiplicity; and Plato's philosophy, shaking off the last remnants of duality, reaches its final culmination in an absolute idealism.

§ 43. But is the cosmic soul herself percipient of matter, or is such perception confined to limited intelligences? I think the true answer is that the cosmic soul is percipient of matter through the finite souls into which she evolves herself. We may regard her elements, ταὐτόν, θάτερον, οὐσία, either as modes of her existence or as modes of her activity. As a mode of her existence, θάτερον signifies the multitude of finite souls in which she is pluralised. As a mode of her activity, θάτερον is sensible perception. But both modes must belong to the same sphere, so that perception of matter must belong to that phase of the universal soul which appears as a number of finite souls. Thus then the aggregate of perceptions experienced by all finite souls constitute the perception of matter in the cosmic soul: there is no such perception by the cosmic soul apart from the perceptions of finite souls. We must observe that in the region which is θάτερον relatively to the ψυχὴ τοῦ κόσμου, ταὐτόν and οὐσία reappear relatively to the finite souls which constitute that region. Each separate soul must have ταὐτόν also, else it would not have οὐσία, it would not substantially exist: and hence the element of θάτερον in the cosmic soul, and by consequence the cosmic soul herself, would be nonexistent. So each finite soul is a complete miniature copy of the great soul. Accordingly in Plato's similitude we find that the Circle of the Other is constructed of soul which is composed both of Same and of Other.

The cosmic soul and material perception.

§ 44. There is yet another question, the answer to which is indeed to be inferred from what has been already said, but which ought perhaps to receive explicit treatment: how are the ideas related to the cosmic soul?

Relation between the cosmic soul and the ideas.

Since we have seen our way to identifying the δημιουργός both with the αὐτὸ ἀγαθόν and with one element of the ψυχὴ τοῦ κόσμου, the simple unity of thought, conceived as still undifferentiated, it follows that whatever relation we have established between the αὐτὸ ἀγαθόν and the other ideas will hold good as between the cosmic soul and the ideas. But perhaps it may serve to render the matter clearer, if we put it in some such way as this.

The ideas, we know, are self-existing, substantial realities. But they can in no wise be essences external to the world-soul, else would the world-soul cease to be All: they must therefore be included in it or identical with it. Now the body of the universe is the material image of the soul thereof: also all material things are images of the ideas. Thus then, being παραδείγματα of the same εἰκόνες, the ideas and the cosmic soul coincide. The ideas, I say—not an idea. For every single idea is the type of one class of material images; the ideal tree is the type of material trees, and of nothing else. The material trees then represent the cosmic soul in so far as that can be expressed in terms of trees—they represent, so to speak, the δενδρότης of it. Accordingly the idea of tree is one determinate aspect of the cosmic soul—that aspect which finds its material expression in a particular tree. And so the sum total of the ideas will be the sum total of the determinations of the cosmic soul—the soul in all her aspects and significations. Also the supreme idea, the αὐτὸ ἀγαθόν, will be the soul herself as such, considered as not in any way specially determined: the material copy of which is not anything in the universe, but the material universe as a whole, which is fairer, Plato says, than aught that is contained within it.

Thus by following up this line we arrive at a result which precisely tallies with that which we reached when considering the relation between the αὐτὸ ἀγαθὸν and the inferior ideas. And so is the substantial existence of the ideas preserved intact, since each idea is the universal soul in some special determination. So too is the unity of the eternal essence maintained; for all the ideas are the same verity viewed in different aspects. And here, as everywhere in the mature Platonism, do the principles of Unity and Multitude go hand in hand, mutually supporting one another and never to be parted.

Θάτερον as space.

§ 45. We have seen that the universal soul is constituted of ταὐτὸν θάτερον and οὐσία, and the general significance of these terms has been discussed. But there is one special application of θάτερον which has not yet occupied our attention. This is Plato's conception of χώρα, or Space.

Plato's identification of the material principle in nature with space—than which there is no more masterly piece of analysis in ancient philosophy—has also been very copiously dealt with

INTRODUCTION. 45

in the notes; but it is too important to be entirely passed over in this place.

It has been seen that in the *Philebus* the analysis of the material element in things was manifestly incomplete. The ἄπειρον was not altogether ἀπαθές, but possessed ἐναντιότητες, such as hotter and colder, quicker and slower, which were quantified and defined by the πέρας ἔχον. But only the quantity or limit is imposed upon the ἄπειρον from without; the quality, though in an undefined form, is still resident in it. Now however, in the *Timaeus*, all quality and attribute is withdrawn : we have an absolutely formless ὑποδοχή, or substrate, potentially receptive of all quality, but possessing none. So far, this may be identified with Aristotle's πρώτη ὕλη. But Plato takes a further step, which was not taken by Aristotle: the ὑποδοχή is expressly identified with Space. How is this done?

The ὑποδοχή is absolutely without form and void : no sense can apprehend it. The sensible objects of perception are the εἴδη εἰσιόντα καὶ ἐξιόντα—the images thrown off in some mysterious way by the ideas and localised in the ὑποδοχή. All attributes which belong to our perceptions are due to these εἴδη, save one alone, which is extension. The ὑποδοχή, submissive in all besides, is peremptory on this one point—of whatever kind a material object may be, it must be extended. So then, if we abstract from matter all the attributes conferred by the εἰσιόντα καὶ ἐξιόντα, we have remaining just a necessity that the objects composing material nature shall be extended. Thus we see θάτερον in another way playing its part as the principle of Difference. For, as Plato says, if the type and the image are to be different, if they are to be two and not one, they must be apart, not inherent one in the other: the copy must exist in something which is not the type, οὐσίας ἁμωσγέπως ἀντεχομένη. Hereupon θάτερον steps in and provides that something, to wit, the law of our finite nature which ordains that we shall perceive all objects as extended in space. Space then is the differentiation of the type and its image.

But extension is nothing independently and objectively existing. For all our perceptions of things are within our own souls, which are unextended; and the things exist not but in these perceptions. Extension then exists only subjectively in our minds. All the objectivity it has is as a universal law binding on finite intelligences, that they should all perceive in this way. It is a consequence and condition of our limitation as finite souls.

The significance of θάτερον as space is thus but a corollary of its significance as pluralisation of mind; since this pluralisation carries with it sensuous perception, which in its turn involves extension as an attribute of its objects. In like manner is time another consequence of this pluralisation: so that we may regard space and time as secondary forms of θάτερον. And so are all the aspects in which we view the element of θάτερον necessarily contingent upon its primary significance of Being in the form of Other, the principle of Multitude inevitably contained in the principle of Unity.

<small>Plato's motive for devoting so much space to physical speculation.</small>

§ 46. Up to this point I have dwelt exclusively upon the metaphysical significance of the dialogue: this being of course incomparably more important than all the other matters which are contained in it. Nevertheless the larger portion of the work is occupied with physical and physiological theories, with elaborate explanations of the processes of nature and the structure and functions of the human body. This being the case, it would seem advisable to say a few words on this subject also.

It might excite not unreasonable surprise that Plato, so strongly persuaded as he was that of matter there can be no knowledge, has yet devoted so much attention to the physical constitution of nature; more especially as he repeatedly declares that concerning physics he has no certainty to offer us, but at most 'the probable account.' It is perhaps worth while to see if we can discover any motives which may have influenced him.

In the first place it is to be observed that the restriction of ideas to classes of natural objects tended in some degree to raise the importance of physical study. If it is true that of natural phenomena themselves there can be no knowledge, it is yet possible that the investigation of these phenomena may serve to place us in a better position for attaining knowledge (or approximate knowledge) of the ideas, which are the cause and reality of the phenomena. For from the knowledge of effects we may hope to rise to the cognition of causes. If then ideas are of natural classes alone, we may at least gain thus much from the study of nature: we may by the observation of particulars ascertain what classes naturally exist in the material world, and thence infer what ideas exist in the intelligible world. As Plato says in 69 A, we ought to study the ἀναγκαῖον for the sake of the θεῖον: that is to say, we must investigate the laws of matter in the hope that we may more clearly ascertain the laws of spirit. Physical speculation is not an end in itself: at best it is a recreation for the philosopher when wearied by his more serious studies: but considered as a means of attaining metaphysical truth, it is worthy of his earnest attention. For this cause the study of material nature was encouraged in Plato's school; though Plato would have been scornful enough of the disproportionate importance attached to it by some of his successors. And since he thought it deserving of his scholars' attention, it was fitting that the master should declare the results of his own scientific speculation.

It must be remembered too how Plato had found fault with Anaxagoras for not introducing τὸ βέλτιστον in his physical theories as the final cause. In the physical part of the *Timaeus* he seeks to make good this defect. He strives to show in detail how the formative intelligence disposed all matter so as to achieve the best result of which its nature was capable; to show that the hypothesis of intelligent design was borne out by facts. He is careful to point out that the physical processes he expounds are but subsidiary causes, subordinate to the main design of Intelligence; for example, after explaining the manner in which vision is produced, he warns us that all this is merely a means to an end: the true cause of vision is the design that we may look upon the luminaries of heaven and thence derive the knowledge of number, which is the avenue to the greatest gift of the gods, philosophy. Now of course on Platonic principles such a teleological account of Nature can have no completeness, unless it be based upon ontology; since everything is good in so far as it represents the αὐτὸ ἀγαθόν. Plato describes phenomenal existence as materially expressing the truth of intelligible existence; and in so far as this expression is perfectly accomplished, the phenomena are fair and good. So then Plato, from the teleological side seeks to show that the material universe is ordered as to all its details in the best possible way, and demonstrates, from the ontological side, that this is so because all the phenomena of the universe are symbols of the eternal idea of good. Plato's contention is that there is an exact correspondence between the ideal and phenomenal worlds, that material Nature is not a mere random succession of appearances, but has a meaning and a truth. And if material Nature has this significance, she cannot be unworthy of the philosopher's attention; she must be studied that her meaning may be revealed. Viewed in this light, the physical portions of the *Timaeus* have a genuine bearing on philosophy; and the very minuteness with which Plato has treated the subject proves that he attached no slight importance to it.

The scientific value of these speculations is naturally but small: many of them are however very interesting, both intrinsically, for their ingenuity and scientific insight, and historically, as showing us how a colossal genius, working without any of the materials accumulated by modern science, and without the instruments which it employs, endeavoured to explain to himself the constitution of the material universe in which he lived.

<small>Plato's final opinions concerning knowledge.</small>

§ 47. From the question that has just been raised, concerning the bearing of physical inquiry upon metaphysical knowledge, naturally arises another question which should not be left altogether unnoticed. What did the Plato of the *Timaeus* conceive to be the province of human knowledge, and what sort of knowledge did he conceive to be attainable? We have already seen reason to believe that he had more or less altered his position with regard to this point since the *Republic* and *Phaedo* were written. This was to be expected: for, as the *Theaetetus* showed, ontology must precede epistemology; before we can say definitely what knowledge is, we must find out what there is to know. Therefore, since Plato's ontology has been modified, it may well be that this modification had its effect on his views of knowledge.

The object of knowledge is plainly the same as ever. Only the really existent can be known: and the only real existence is the ideas, and ultimately the αὐτὸ ἀγαθόν. Knowledge then, in the truest and fullest sense of the word, signifies only the actual cognition of the supreme idea as it is in itself. Now in the days of the *Phaedo* and *Republic* we know that Plato actually aimed at such cognition. However remote the consummation might be, however despondingly the Sokrates of the *Phaedo* may speak of it, that and that alone was the end of the philosopher's labours— an end regarded as one day attainable by man. But now, both in the *Parmenides* and in the *Timaeus*, Plato disclaims such absolute knowledge as lying beyond the sphere of finite intelligence. And he is right. For he who should know the Absolute would *ipso facto* be the Absolute. Only the All can comprehend the All. And if the supreme idea cannot be absolutely known, neither can

INTRODUCTION.

the other ideas. For since every idea is, as has been said, a determination of the supreme idea, a complete knowledge of any one idea would amount to a complete knowledge of every other idea and of the supreme idea itself. From such ambitious dreams we must refrain ourselves. But we are not therefore left beggared of our intellectual heritage. Absolute knowledge of universal truth may be beyond our reach, but an approximation to such knowledge is in our power, an approximation to which no bounds are set. We have said that the supreme idea determines itself into a series of subordinate ideas. The more of these subordinate ideas we contemplate, the more comprehensive will be our conception of the supreme idea: and in proportion as our vision of the subordinate ideas gains in clearness, even so will our conception of the highest advance in truth. For since Truth is one and simple, every mode of truth is an access to the whole. This then is what Plato now holds up as the philosopher's hope—an ever brightening vision of universal truth, attained by industrious study of particular forms of truth. Thus in place of the complete fruition of knowledge, once for all, of which we once dreamed, we have the prospect of a perpetual advance therein. And whatever increment of knowledge we may win, although it is necessarily incomplete, it is real: the ladder has no summit, but we have gained one step above our former place. And there seems certainly nothing discouraging in the reflection that, however much we may succeed in learning, behind all our knowledge there lies something in wait to be known—that though the truth which we know is true, there is always a truth beneath it that is truer still.

Knowledge then is now as ever for Plato to be found in the ideal world: and there alone. Material nature is still to him a realm of mists and shadows, where nothing stable is nor any truth, where we grope doubtfully by the dim light of opinion. But through these mists lies the road to the bright sphere of reason, where abide the ideal archetypes, which are the true objects of our thought, and which have lost none of that lustre that once was chanted in the *Phaedrus*. There is no recession here: still the immaterial and eternal only can be known. All that is changed is the extension of the word *knowledge*. We know the ideas but as finite minds may know them; that is, partially, with never perfect yet ever clearer vision: being ourselves incomplete, completeness of knowledge is beyond our scope. This restriction of the bounds of human knowledge must needs have presented itself to Plato's mind along with the clear conception of an infinite universal soul which is the sum and substance of all things. For only in the endeavour to grasp the boundlessness of the infinite would he become fully alive to the limitation of the finite.

Concluding remarks.

§ 48. The account I have thought it necessary to give of the philosophical doctrines contained in the *Timaeus* is now completed. There are indeed divers matters of high importance handled in the dialogue which I have either left unnoticed or dismissed with brief mention. The theory of space propounded in the eighteenth chapter, although its profound originality and importance can hardly be overestimated, has been only partially examined: further treatment being reserved for the commentary on the said chapter, since it involves too much detail to be conveniently included in a general view of the subject such as I have here sought to give. The same will apply to the very interesting ethical disquisition towards the end of the dialogue, and to the psychological theories advanced in the thirty-first and thirty-second chapters.

In the foregoing pages my aim has been to trace the chief currents of earlier Greek speculation to their union in the Platonic philosophy, and to follow the ever widening and deepening stream through the region of Platonism itself, until it is merged in the ocean of idealism into which Plato's thought finally expands. In particular I have sought to follow the history of the fundamental antithesis, the One and the Many, from the lisping utterance of it (as Aristotle would say) by the preplatonic thinkers to its clear enunciation as the central doctrine of the later Platonism. And however imperfectly this object may have been accomplished, I trust I have at least not failed in justifying the affirmation that the *Timaeus* is second in interest and importance to none of the Platonic writings.

Of course it is not for a moment maintained that all the teaching I have ascribed to this dialogue is to be found fully expanded and explicitly formulated within its limits. To expect this would argue a complete absence of familiarity with Plato's method. Plato never wrote a handbook of his own philosophy,

INTRODUCTION.

nor will he do our thinking for us: he loves best to make us construct the edifice for ourselves from the materials with which he supplies us. And this we can only do by careful combination of his statements on the subject in hand, spread, it may be, over several dialogues, and by sober interpretation of his figurative language, availing ourselves at the same time of whatever light we may be able to derive from ancient expositors of Plato, and chiefly from Aristotle. Consequently no theory we may thus form is a matter of mathematical demonstration: if we can find one which combines Plato's various statements into a systematic whole and reveals a distinct sequence of his thought, all reasonable expectation is satisfied. In evolving the opinions which have in this essay been offered concerning the interpretation of the *Timaeus*, I have made but two postulates—that Plato does not talk at random, and that he does not contradict himself. To any who reject one or both of these postulates the arguments adduced in the foregoing are of course not addressed, since there is no common ground for arguing. But of those who accept them, whoever has an interpretation to propound which more thoroughly harmonises all the elements of Plato's thought than I have been able to do, and which more readily and directly arises from his language, ἐκεῖνος οὐκ ἐχθρὸς ὢν ἀλλὰ φίλος κρατεῖ.

---

§ 49. It remains to say a few words about the text. In this edition I have rather closely adhered to the text of C. F. Hermann, which on the whole presents most faithfully the readings of the oldest and best manuscript, Codex Parisiensis A. The authority of this ninth century ms. is such that recent editors have frequently accepted its readings in defiance of a *consensus* among the remainder; an example which I have in general followed. In departing from Hermann I have usually had some manuscript support on which to rely, and sometimes that of A itself: but in a very few cases (about six or seven, I believe, in all) I have introduced emendations, or at least alterations, of my own; none of which are very important. In order that the reader may have no trouble in checking the text here presented to him, I have added brief critical notes in Latin, wherein are recorded the readings of the Paris manuscript (quoted on Bekker's testimony), of C. F. Hermann, of Stallbaum, and of the Zürich edition by Baiter Orelli and Winckelmann, wherever these differed from my own. These authorities are denoted respectively by A, H, S, and Z. The readings of other manuscripts have not been cited. Fortunately the text of the *Timaeus* is for the most part in a fairly satisfactory condition.

There are some small points of orthography in which this edition systematically differs from Hermann's spelling; but I have deemed it superfluous to record these.

ΤΙΜΑΙΟΣ

ΤΙΜΑΙΟΣ

[ἡ περὶ φύσεως· φυσικός.]

ΤΑ ΤΟΥ ΔΙΑΛΟΓΟΥ ΠΡΟΣΩΠΑ

ΣΩΚΡΑΤΗΣ, ΚΡΙΤΙΑΣ, ΤΙΜΑΙΟΣ, ΕΡΜΟΚΡΑΤΗΣ.

St.
III. p.

I. ΣΩ. Εἷς, δύο, τρεῖς· ὁ δὲ δὴ τέταρτος ἡμῖν, ὦ φίλε Τίμαιε, 17 A
ποῦ τῶν χθὲς μὲν δαιτυμόνων, τὰ νῦν δὲ ἑστιατόρων;
ΤΙ. Ἀσθένειά τις αὐτῷ συνέπεσεν, ὦ Σώκρατες· οὐ γὰρ ἂν
ἑκὼν τῆσδε ἀπελείπετο τῆς συνουσίας.
5 ΣΩ. Οὐκοῦν σὸν τῶνδέ τε ἔργον καὶ τὸ ὑπὲρ τοῦ ἀπόντος
ἀναπληροῦν μέρος;
ΤΙ. Πάνυ μὲν οὖν, καὶ κατὰ δύναμίν γε οὐδὲν ἐλλείψομεν· B
οὐδὲ γὰρ εἴη ἂν δίκαιον, χθὲς ὑπὸ σοῦ ξενισθέντας, οἷς ἦν πρέπον
ξενίοις μὴ οὐ προθύμως σε τοὺς λοιποὺς ἡμῶν ἀνταφεστιᾶν.

8 εἴη ἄν: εἶναι A. ἂν εἴη SZ.   9 ἀνταφεστιᾶν: ἀντεφεστιᾶν AZ.

17 A—19 B, c. i. Sokrates meets by appointment three of the friends to whom he has on the previous day narrated the conversation recorded in the *Republic*. After the absence of the fourth member of the party has been explained, he proceeds to summarise the social and political theories propounded in that dialogue.

It will be observed that the unusually long introductory passage, extending to 27 C, has its application not to the *Timaeus* only, but to the whole trilogy, *Republic, Timaeus, Critias*. The recapitulation of the *Republic* indicates the precise position of that work in the series; while the myth of Atlantis marks the intimate connexion which Plato intended to exist between the *Timaeus* and *Critias*: it is indeed artistically justifiable only in relation to Plato's projected, not to his accomplished work. It is obvious that when the *Republic* was written no such trilogy was in contemplation.

The supposed date of the present discussion is two days after the meeting in the house of Kephalos. The latter, as we learn from the beginning of the *Republic*, took place on the day of the newly established festival of the Thracian deity

PERSONS OF THE DIALOGUE:

SOKRATES, TIMAEUS, HERMOKRATES, KRITIAS.

I. *Sokrates.* One, two, three—what is become of the fourth, my dear Timaeus, of our yesterday's guests and our entertainers of to-day?

Timaeus. He has fallen sick, Sokrates: he would not willingly have been missing at this gathering.

Sokrates. Then it is for you and your companions, is it not, to fulfil the part of our absent friend?

Timaeus. Unquestionably; and we will omit nothing that lies in our power. For indeed it would not be fair, seeing how well we were entertained by you yesterday, that the rest of us should not heartily requite you with a fitting return of hospitality.

Bendis, a goddess whom the Athenians seem to have identified with their own Artemis. The festival took place on the 19th or 20th Thargelion (=about 22nd or 23rd May). On the following day Sokrates reports to the four friends what passed at the house of Kephalos; and on the next the present dialogue takes place.

1. εἰς δύο τρεῖς] This very simple opening has given rise to a strange amount of animadversion, as may be seen by any one who struggles through the weary waste of words which Proklos has devoted to its discussion. Quintilian (IX iv 78) attacks it for beginning with part of a hexameter. It is quoted in Athenaeus IX 382 A, where there is a story of a man who made his cooks learn the dialogue by heart and recite it as they brought in the dishes.

ὁ δὲ δὴ τέταρτος] Some curiosity has been displayed as to the name of the absentee; and Plato himself has been suggested. But seeing that the conversation is purely fictitious, the question would seem to be one of those ἀναπόδεικτα which are hardly matter of profitable discussion.

2. δαιτυμόνων] i.e. guests at the feast of reason provided by Sokrates.

56   ΠΛΑΤΩΝΟΣ   [17 B—

ΣΩ. Ἆρ' οὖν μέμνησθε, ὅσα ὑμῖν καὶ περὶ ὧν ἐπέταξα εἰπεῖν;
ΤΙ. Τὰ μὲν μεμνήμεθα, ὅσα δὲ μή, σὺ παρὼν ὑπομνήσεις· μᾶλλον δέ, εἰ μή τί σοι χαλεπόν, ἐξ ἀρχῆς διὰ βραχέων πάλιν ἐπάνελθε αὐτά, ἵνα βεβαιωθῇ μᾶλλον παρ' ἡμῖν.
5 ΣΩ. Ταῦτ' ἔσται. χθές που τῶν ὑπ' ἐμοῦ ῥηθέντων λόγων περὶ πολιτείας ἦν τὸ κεφάλαιον, οἷά τε καὶ ἐξ οἵων ἀνδρῶν ἀρίστη C κατεφαίνετ' ἄν μοι γενέσθαι.
ΤΙ. Καὶ μάλα γε ἡμῖν, ὦ Σώκρατες, ῥηθεῖσα πᾶσι κατὰ νοῦν.
ΣΩ. Ἆρ' οὖν οὐ τὸ τῶν γεωργῶν ὅσαι τε ἄλλαι τέχναι πρῶτον
10 ἐν αὐτῇ χωρὶς διειλόμεθα ἀπὸ τοῦ γένους τοῦ τῶν προπολεμησόντων;
ΤΙ. Ναί.
ΣΩ. Καὶ κατὰ φύσιν δὴ δόντες τὸ καθ' αὑτὸν ἑκάστῳ πρόσφορον ἓν μόνον ἐπιτήδευμα καὶ μίαν ἑκάστῳ τέχνην τούτους, οὓς D
15 πρὸ πάντων ἔδει πολεμεῖν, εἴπομεν ὡς ἄρα αὐτοὺς δέοι φύλακας εἶναι μόνον τῆς πόλεως, εἴ τέ τις ἔξωθεν ἢ καὶ τῶν ἔνδοθεν ἴοι κακουργήσων, δικάζοντας μὲν πράως τοῖς ἀρχομένοις ὑπ' αὐτῶν καὶ φύσει φίλοις οὖσι, χαλεποὺς δὲ ἐν ταῖς μάχαις τοῖς ἐντυγχά- 18 A νουσι τῶν ἐχθρῶν γιγνομένους.
20 ΤΙ. Παντάπασι μὲν οὖν.
ΣΩ. Φύσιν γὰρ οἶμαί τινα τῶν φυλάκων τῆς ψυχῆς ἐλέγομεν ἅμα μὲν θυμοειδῆ, ἅμα δὲ φιλόσοφον δεῖν εἶναι διαφε-

13 δόντες: διδόντες A.   14 μίαν ἑκάστῳ τέχνην: sic SZ e Bekkeri coniectura. ἀφ' ἑκάστου τῇ τέχνῃ A, quae uncis inclusa retinuit H.   16 ἔνδοθεν: ἔνδον SZ.

1. ὅσα ὑμῖν] This is doubtless the right reading. Sokrates had bargained with his friends, as we may learn from 20 B, that they should supply the sequel to his discourse: and this they had consented to do. Thus in recapitulating his own contribution Sokrates recalls to their minds what is expected of them.

6. περὶ πολιτείας] Sokrates in his summary of the *Republic* deals with it solely as a political treatise, totally ignoring its metaphysical bearings. This, while very significant of the change in Plato's views, is due to the fact that it is only on its political side that the *Republic* is connected with the rest of the trilogy. Its metaphysical teaching is superseded by the more advanced ontology of the *Timaeus*; and were the dialogue actually incorporated in a trilogy, it would stand in need of sundry important modifications. But the ideal commonwealth is maintained intact: the laws of the καλλίπολις are agreeable to the ontological and physical principles set forth in the *Timaeus* and find their counterpart in the institutions of ancient Athens as they are to be depicted in the *Critias*. Now it seems to me highly important to notice that the political theories of the *Republic* are thus stamped with Plato's deliberate approval in a work belonging to the ripest maturity of his thought—μάλα γε ἡμῖν ῥηθεῖσα πᾶσι κατὰ νοῦν. We ought

18 A]   ΤΙΜΑΙΟΣ.   57

Sokrates. Do you remember the extent and scope of the subjects I appointed for your discussion?

Timaeus. In part we remember; and whatever we have forgotten, you are here to aid our memory. But I should prefer, if it is not troublesome, that you should briefly recapitulate them from beginning to end, that they may be more firmly fixed in our minds.

Sokrates. I will. The main subject of my discourse yesterday was a political constitution, and the kind of principles and citizens which seemed to me likely to render it most perfect.

Timaeus. Yes, and what you said, Sokrates, was very much to the satisfaction of us all.

Sokrates. Was not our first step to separate the agricultural class and tradesmen in general from those who were to be the defenders of our state?

Timaeus. It was.

Sokrates. And in assigning on natural principles but one single pursuit or craft which was suited to each citizen severally, we declared that those whose duty it was to fight on behalf of the community must be guardians only of the city, in case any one whether without or within her walls should seek to injure her, and that they should give judgment mercifully to their subjects and natural friends, but show themselves stern to the enemies they met in battle.

Timaeus. Quite true.

Sokrates. For we described, I think, a certain temperament which the souls of our guardians must possess, combining in a peculiar degree high spirit and thoughtfulness, that they might

not then to regard the *Laws* as indicating any abandonment by Plato of his political ideal, but simply as offering a working substitute so long as the attainment of that ideal was impracticable. Plato remains all his life long a true citizen of that city 'whereof the pattern is preserved in heaven'.

7. **κατεφαίνετ' ἄν**] *ἄν* belongs to γενέσθαι.

9. **τὸ τῶν γεωργῶν**] *Republic* 370 E foll.

15. **φύλακας**] The distinction between φύλακες and ἐπίκουροι is here neglected, cf. *Republic* 414 A ἆρ' οὖν ὡς ἀληθῶς ὀρθότατον καλεῖν τούτους μὲν φύλακας παντελεῖς τῶν τε ἔξωθεν πολεμίων τῶν τε ἐντὸς φίλων, ὅπως οἱ μὲν μὴ βουλήσονται, οἱ δὲ μὴ δυνήσονται κακουργεῖν, τοὺς δὲ νέους, οὓς νῦν δὴ φύλακας ἐκαλοῦμεν, ἐπικούρους τε καὶ βοηθοὺς τοῖς τῶν ἀρχόντων δόγμασιν;

22. **ἅμα μὲν θυμοειδῆ**] *Republic* 375 D foll.

ρόντως, ἵνα πρὸς ἑκατέρους δύναιντο ὀρθῶς πρᾷοι καὶ χαλεποὶ γίγνεσθαι.

ΤΙ. Ναί.

ΣΩ. Τί δὲ τροφήν; ἆρ' οὐ γυμναστικῇ καὶ μουσικῇ μαθήμασί τε, ὅσα προσήκει τούτοις, ἐν ἅπασι τεθράφθαι;

ΤΙ. Πάνυ μὲν οὖν.

ΣΩ. Τοὺς δέ γε οὕτω τραφέντας ἐλέχθη που μήτε χρυσὸν μήτε ἄργυρον μήτε ἄλλο ποτὲ μηδὲν κτῆμα ἑαυτῶν ἴδιον νομίζειν δεῖν, ἀλλ' ὡς ἐπικούρους μισθὸν λαμβάνοντας τῆς φυλακῆς παρὰ τῶν σῳζομένων ὑπ' αὐτῶν, ὅσος σώφροσι μέτριος, ἀναλίσκειν τε δὴ κοινῇ καὶ ξυνδιαιτωμένους μετὰ ἀλλήλων ζῆν, ἐπιμέλειαν ἔχοντας ἀρετῆς διὰ παντός, τῶν ἄλλων ἐπιτηδευμάτων ἄγοντας σχολήν.

ΤΙ. Ἐλέχθη καὶ ταῦτα ταύτῃ.

ΣΩ. Καὶ μὲν δὴ καὶ περὶ γυναικῶν ἐπεμνήσθημεν, ὡς τὰς φύσεις τοῖς ἀνδράσι παραπλησίας εἴη ξυναρμοστέον, καὶ τὰ ἐπιτηδεύματα πάντα κοινὰ κατά τε πόλεμον καὶ κατὰ τὴν ἄλλην δίαιταν δοτέον πάσαις.

ΤΙ. Ταύτῃ καὶ ταῦτα ἐλέγετο.

ΣΩ. Τί δὲ δὴ τὸ περὶ τῆς παιδοποιίας; ἢ τοῦτο μὲν διὰ τὴν ἀήθειαν τῶν λεχθέντων εὐμνημόνευτον, ὅτι κοινὰ τὰ τῶν γάμων καὶ τὰ τῶν παίδων πᾶσιν ἁπάντων ἐτίθεμεν, μηχανώμενοι, ὅπως μηδείς ποτε τὸ γεγενημένον αὐτῷ ἰδίᾳ γνώσοιτο, νομιοῦσι δὲ πάντες πάντας αὐτοὺς ὁμογενεῖς, ἀδελφὰς μὲν καὶ ἀδελφοὺς ὅσοιπερ ἂν τῆς πρεπούσης ἐντὸς ἡλικίας γίγνωνται, τοὺς δ' ἔμπροσθεν καὶ ἄνωθεν γονέας τε καὶ γονέων προγόνους, τοὺς δ' εἰς τὸ κάτωθεν ἐκγόνους παῖδάς τε ἐκγόνων;

ΤΙ. Ναί, καὶ ταῦτα εὐμνημόνευτα, ᾗ λέγεις.

ΣΩ. Ὅπως δὲ δὴ κατὰ δύναμιν εὐθὺς γίγνοιντο ὡς ἄριστοι τὰς φύσεις, ἆρ' οὐ μεμνήμεθα, ὡς τοὺς ἄρχοντας ἔφαμεν καὶ τὰς ἀρχούσας δεῖν εἰς τὴν τῶν γάμων σύνερξιν λάθρᾳ μηχανᾶσθαι

20 τί δέ : τί δαί AH.   22 μηχανώμενοι : μηχανωμένους AH. correxit Stephanus.
23 αὐτῷ : αὐτῶν A.

5. ἐν ἅπασι] Stallbaum would have τούτοισιν ἅπασι. Plato frequently uses the old form of the dative plural: but there seems no real objection to the preposition.

7. μήτε χρυσόν] *Republic* 416 D, E.

15. περὶ γυναικῶν] Plato's regulations for the training of women will be found in *Republic* 451 C—457 B: he treats of παιδοποιία in the immediate sequel.

22. μηχανώμενοι] Hermann's defence of μηχανωμένους is vain; nor is Butt-

be able to show a due measure of mildness or sternness to friend or foe.

Timaeus. Yes.

Sokrates. And what of their training? were they not to have been trained in gymnastic and music and all studies which are connected with these?

Timaeus. Just so.

Sokrates. And those who had undergone this discipline, we said, must not consider that they have any private property in gold or silver or anything else whatsoever, but as auxiliaries drawing from those whom they preserved so much pay in return for their protection as was sufficient for temperate men, they were to spend it in common and pass their lives in company with one another, devoting themselves perpetually to the pursuit of virtue and relieved from all other occupations.

Timaeus. That also is the way it was put.

Sokrates. Moreover with regard to women we observed that their natures must be brought into harmonious similarity with those of men, and that the same employments must be assigned to them all both in war and in their general mode of life.

Timaeus. Yes, that was what we said.

Sokrates. And what were our rules concerning the procreation of children? This, I think, is easy of recollection because of the novelty of our scheme. We ordained that the rights of marriage and of children should be common to all, to the end that no one should ever know his own offspring, but that each should look upon all as his kindred, regarding as sisters and brethren all such as were between suitable limits of age, and those of the former and previous generations as parents and grandparents, and those after them as children and children's children.

Timaeus. Yes, it is very easy to remember this too as you describe it.

Sokrates. Next with a view to securing immediately the utmost possible perfection in their natures, do we not remember that it was incumbent on the rulers of both sexes to make

---

mann's μηχανωμένοις very satisfactory. I agree with Stallbaum in receiving the nominative.

31. εἰς τὴν τῶν γάμων σύνερξιν] *Republic* 459 D, E.

60   ΠΛΑΤΩΝΟΣ   [18 E—

κλήροις τισίν, ὅπως οἱ κακοὶ χωρὶς οἵ τ' ἀγαθοὶ ταῖς ὁμοίαις
ἑκάτεροι ξυλλήξονται, καὶ μή τις αὐτοῖς ἔχθρα διὰ ταῦτα γίγνηται,
τύχην ἡγουμένοις αἰτίαν τῆς ξυλλήξεως;
  ΤΙ. Μεμνήμεθα.
5 ΣΩ. Καὶ μὴν ὅτι γε τὰ μὲν τῶν ἀγαθῶν θρεπτέον ἔφαμεν 19 A
εἶναι, τὰ δὲ τῶν κακῶν εἰς τὴν ἄλλην λάθρᾳ διαδοτέον πόλιν·
ἐπαυξανομένων δὲ σκοποῦντας ἀεὶ τοὺς ἀξίους πάλιν ἀνάγειν δεῖν,
τοὺς δὲ παρὰ σφίσιν ἀναξίους εἰς τὴν τῶν ἐπανιόντων χώραν
μεταλλάττειν;
10 ΤΙ. Οὕτως.
  ΣΩ. Ἆρ' οὖν δὴ διεληλύθαμεν ἤδη καθάπερ χθές, ὡς ἐν
κεφαλαίοις πάλιν ἐπανελθεῖν, ἢ ποθοῦμεν ἔτι τι τῶν ῥηθέντων, ὦ
φίλε Τίμαιε, ὡς ἀπολειπόμενον;
  ΤΙ. Οὐδαμῶς, ἀλλὰ ταὐτὰ ταῦτ' ἦν τὰ λεχθέντα, ὦ Σώκρατες. B
15 ΙΙ. ΣΩ. Ἀκούοιτ' ἂν ἤδη τὰ μετὰ ταῦτα περὶ τῆς πολιτείας,
ἣν διήλθομεν, οἷόν τι πρὸς αὐτὴν πεπονθὼς τυγχάνω. προσέοικε
δὲ δή τινί μοι τοιῷδε τὸ πάθος, οἷον εἴ τις ζῷα καλά που θεασά-
μενος, εἴτε ὑπὸ γραφῆς εἰργασμένα εἴτε καὶ ζῶντα ἀληθινῶς, ἡσυχίαν δὲ ἄγοντα, εἰς ἐπιθυμίαν ἀφίκοιτο θεάσασθαι κινούμενά τε
20 αὐτὰ καί τι τῶν τοῖς σώμασι δοκούντων προσήκειν κατὰ τὴν ἀγω-
νίαν ἀθλοῦντα· ταὐτὸν καὶ ἐγὼ πέπονθα πρὸς τὴν πόλιν ἣν C
διήλθομεν. ἡδέως γὰρ ἄν του λόγῳ διεξιόντος ἀκούσαιμ' ἂν
ἄθλους, οὓς πόλις ἀθλεῖ, τούτους αὐτὴν ἀγωνιζομένην πρὸς πόλεις
ἄλλας πρεπόντως, εἴς τε πόλεμον ἀφικομένην καὶ ἐν τῷ πολεμεῖν
25 τὰ προσήκοντα ἀποδιδοῦσαν τῇ παιδείᾳ καὶ τροφῇ κατά τε τὰς

9 μεταλλάττειν: διαλλάττειν A.   14 ταὐτὰ: αὐτά S.
24 τε: γε A. omittit S.

6. **λάθρᾳ διαδοτέον**] Plato has here somewhat mitigated the rigour of his ordinance in the *Republic*: see 459 D τοὺς ἀρίστους ταῖς ἀρίσταις συγγίγνεσθαι ὡς πλειστάκις, τοὺς δὲ φαυλοτάτους ταῖς φαυλοτάταις τοὐναντίον, καὶ τῶν μὲν τὰ ἔκγονα τρέφειν, τῶν δὲ μή. Compare too 460 C τὰ δὲ τῶν χειρόνων, καὶ ἐάν τι τῶν ἄλλων ἀνάπηρον γίγνηται, ἐν ἀπορρήτῳ τε καὶ ἀδήλῳ κατακρύψουσιν ὡς πρέπει: and again, 461 C μάλιστα μὲν μηδ' εἰς φῶς ἐκφέρειν κύημα μηδέ γ' ἕν, ἐὰν γένηται, ἐὰν δέ τι βιάσηται, οὕτω τιθέναι ὡς οὐκ οὔσης τροφῆς τῷ τοιούτῳ. But in 415 B the milder course is enjoined: ἐὰν τε σφέτερος ἔκγονος ὑπόχαλκος ἢ ὑποσίδηρος γένηται, μηδενὶ τρόπῳ κατελεήσουσιν, ἀλλὰ τὴν τῇ φύσει προσήκουσαν τιμὴν ἀποδόντες ὤσουσιν εἰς δημιουργοὺς ἢ εἰς γεωργούς. Probably then, when Plato speaks of not rearing the inferior children, he merely means that they are not to be reared by the state as infant φύλακες.

7. **ἐπαυξανομένων δὲ σκοποῦντας**] Plato clearly recognises that the laws of heredity are only imperfectly understood by us, and that therefore the results may often baffle our expectation.

provision for the contraction of marriages by some secret mode of allotment, that to the good and bad separately might be allotted mates of their own kind, and so no ill-feeling should arise among them, supposing as they would that chance governed the allotment?

Timaeus. We remember that.

Sokrates. And the offspring of the good we said must be reared, while that of the bad was to be secretly dispersed among the other classes of the state; and continually observing them as they grew up, the rulers were to restore to their rank such as were worthy, and in the places of those so promoted substitute the unworthy in their own rank.

Timaeus. Quite so.

Sokrates. Have we now said enough for a summary recapitulation of yesterday's discourse? or do we feel that anything is lacking, my dear Timaeus, to our account?

Timaeus. Not at all : you have exactly described what was said, Sokrates.

II. *Sokrates.* Listen then and I will tell you in the next place what I feel about the constitution which we described. My feeling is something like this: suppose a man, on beholding beautiful creatures, whether the work of the painter or really alive but at rest, should conceive a desire to see them in motion and putting into active exercise the qualities which seemed to belong to their form—this is just what I feel about our city which we described: I would fain listen to one who depicted her engaged in a becoming manner with other countries in those struggles which cities must undergo, and going to war, and when at war showing a result worthy of her training and educa-

---

19 B—21 A, *c.* ii. Sokrates now expresses his desire to see his pictured city called as it were into life and action; he would have a representation of her actual doings and dealings with other cities. He distrusts his own power to do this worthily, nor has he any greater confidence in poets or sophists. But he declares that his three companions are of all men the best fitted by genius and training to accomplish it; and he therefore calls on them to gratify his wish. Hermokrates readily assents, but first begs Kritias to narrate a forgotten legend of ancient Athens, which he thinks is apposite to the matter in hand : to this Kritias consents.

17. οἷον εἴ τις] This passage is referred to by Athenaeus XI 507 D in support of the truly remarkable charge of φιλοδοξία which he brings against Plato.

ἐν τοῖς ἔργοις πράξεις καὶ κατὰ τὰς ἐν τοῖς λόγοις διερμηνεύσεις πρὸς ἑκάστας τῶν πόλεων. ταῦτ' οὖν, ὦ Κριτία καὶ Ἑρμόκρατες, ἐμαυτοῦ μὲν αὐτὸς κατέγνωκα μή ποτ' ἂν δυνατὸς γενέσθαι τοὺς D ἄνδρας καὶ τὴν πόλιν ἱκανῶς ἐγκωμιάσαι. καὶ τὸ μὲν ἐμὸν οὐδὲν 5 θαυμαστόν· ἀλλὰ τὴν αὐτὴν δόξαν εἴληφα καὶ περὶ τῶν πάλαι γεγονότων καὶ τῶν νῦν ὄντων ποιητῶν, οὔ τι τὸ ποιητικὸν ἀτιμάζων γένος, ἀλλὰ παντὶ δῆλον ὡς τὸ μιμητικὸν ἔθνος, οἷς ἂν ἐντραφῇ, ταῦτα μιμήσεται ῥᾷστα καὶ ἄριστα, τὸ δ' ἐκτὸς τῆς τροφῆς ἑκάστοις γιγνόμενον] χαλεπὸν μὲν ἔργοις, ἔτι δὲ χαλεπώτερον Ε 10 λόγοις εὖ μιμεῖσθαι. τὸ δὲ τῶν σοφιστῶν γένος αὖ πολλῶν μὲν λόγων καὶ καλῶν ἄλλων μάλ' ἔμπειρον ἥγημαι, φοβοῦμαι δέ, μή πως, ἅτε πλανητὸν ὂν κατὰ πόλεις οἰκήσεις τε ἰδίας οὐδαμῇ διῳκηκός, ἄστοχον ἅμα φιλοσόφων ἀνδρῶν ᾖ καὶ πολιτικῶν, ὅσ' ἂν οἷά τε ἐν πολέμῳ καὶ μάχαις πράττοντες ἔργῳ καὶ λόγῳ προσομι- 15 λοῦντες ἑκάστοις πράττοιεν καὶ λέγοιεν. καταλέλειπται δὴ τὸ τῆς ὑμετέρας ἕξεως γένος, ἅμα ἀμφοτέρων φύσει καὶ τροφῇ μετέχον. 20 A Τίμαιός τε γὰρ ὅδε, εὐνομωτάτης ὢν πόλεως τῆς ἐν Ἰταλίᾳ Λοκρίδος, οὐσίᾳ καὶ γένει οὐδενὸς ὕστερος ὢν τῶν ἐκεῖ, τὰς μεγίστας μὲν ἀρχάς τε καὶ τιμὰς τῶν ἐν τῇ πόλει μετακεχείρισται, φιλο- 20 σοφίας δ' αὖ κατ' ἐμὴν δόξαν ἐπ' ἄκρον ἁπάσης ἐλήλυθε· Κριτίαν δέ που πάντες οἱ τῇδ' ἴσμεν οὐδενὸς ἰδιώτην ὄντα ὧν λέγομεν· τῆς δὲ Ἑρμοκράτους αὖ περὶ φύσεως καὶ τροφῆς, πρὸς ἅπαντα ταῦτ'

6 καὶ τῶν: καὶ περὶ τῶν A.

7. τὸ μιμητικὸν ἔθνος] See *Republic* 392 D, 398 A, 597 E foll. Poetry, says Plato, is an imitative art; and poets cannot imitate what is outside of their experience. For the use of ἔθνος compare *Sophist* 242 D, *Gorgias* 455 B, *Politicus* 290 B.

9. ἔτι δὲ χαλεπώτερον λόγοις] Proklos raises needless difficulty about this. Plato simply means that to describe such things worthily requires a rare literary gift: it is far easier to find an Agamemnon than a Homer.

12. ἅτε πλανητὸν ὄν] cf. *Sophist* 224 B, where one kind of sophist is described as τὸν μαθήματα συνωνούμενον πόλιν τε ἐκ πόλεως νομίσματος ἀμείβοντα.

15. τὸ τῆς ὑμετέρας ἕξεως γένος] i.e. men of a philosophical habit. We have a very similar phrase below at 42 D τὸ τῆς πρώτης καὶ ἀρίστης ἀφίκοιτο εἶδος ἕξεως. ἕξις expresses a permanent habit of mind.

16. ἀμφοτέρων] sc. φιλοσόφου καὶ πολιτικοῦ.

17. τε γάρ] The τε is not answered: see Shilleto on Demosth. *fals. leg.* § 176.

εὐνομωτάτης ὢν πόλεως] The laws of the Epizephyrian Lokrians were ascribed to Zaleukos, 660 B.C. From Demosthenes κατὰ Τιμοκράτους p. 744 it appears that this people was so conservative as to pass no new law, with a single amusing exception, during a period of 200 years. In *Laws* 638 D they are said εὐνομώτατοι τῶν περὶ ἐκεῖνον τὸν τόπον γεγονέναι. Pindar adds his testimony, *Olymp.* XI (X) 17 νέμει γὰρ

tion, both when dealing in action and parleying in speech with other cities. Now, Kritias and Hermokrates, my own verdict upon myself is that I should never be capable of celebrating the city and her people according to their merit. So far as concerns me indeed, that is no marvel; but I have formed the same opinion about the poets, both past and present; not that I disparage the poetic race, but any one can see that the imitative tribe will most easily and perfectly imitate the surroundings amid which they have been brought up, but that which lies outside the range of each man's experience is hard to imitate correctly in actions and yet harder in words. As to the class of sophists on the other hand, I have always held them to be well furnished with many fine discourses on other subjects; yet I am afraid, seeing they wander from city to city and have never had dwellings of their own to manage, they may somehow fall short in their conception of philosophers and statesmen, as to what in time of war and battles they would do and say in their dealings and converse with divers people. One class then remains, those who share your habit of mind, having by nature and training a capacity for both philosophy and statecraft. Timaeus for instance, belonging to an admirably governed state, the Italian Lokris, and one of the foremost of its citizens in wealth and birth, has filled offices of the highest authority and honour in his native city, and has also in my judgment climbed to the topmost peak of all philosophy: while at Athens we all know that Kritias is no novice in any of the questions we are discussing: of Hermokrates too we must believe on the evidence of

'Ατρέκεια πόλιν Λοκρῶν Ζεφυρίων.

20. ἐπ' ἄκρον ἁπάσης] Plato's judgment of the historical Timaeus can hardly have gone so far as this: that however he must have set a high estimate on the Pythagorean's philosophical capacity he has proved by making him the mouthpiece of his own profoundest speculations.

21. οὐδενὸς ἰδιώτην] ἐκαλεῖτο ἰδιώτης μὲν ἐν φιλοσόφοις, φιλόσοφος δὲ ἐν ἰδιώταις, says Proklos. He seems to have been one of those who made a good show out of a little knowledge: cf. *Char-mides* 169 C κἀκεῖνος [sc. Κριτίας] ἔδοξέ μοι ὑπ' ἐμοῦ ἀποροῦντος ἀναγκασθῆναι καὶ αὐτὸς ἁλῶναι ὑπὸ ἀπορίας. ἅτε οὖν εὐδοκιμῶν ἑκάστοτε ᾐσχύνετο τοὺς παρόντας, καὶ οὔτε ξυγχωρῆσαί μοι ἤθελεν ἀδύνατος εἶναι διελέσθαι ἃ προυκαλούμην αὐτόν, ἔλεγέ τε οὐδὲν σαφές, ἐπικαλύπτων τὴν ἀπορίαν.

22. 'Ερμοκράτους] This was the celebrated Syracusan general and statesman, distinguished in the Peloponnesian war. A Hermokrates mentioned among the friends of Sokrates by Xenophon *memorabilia* I ii 48 is doubtless a different

64   ΠΛΑΤΩΝΟΣ   [20 A—

εἶναι ἱκανῆς πολλῶν μαρτυρούντων πιστευτέον δή. ὃ καὶ χθὲς ἐγὼ B
διανοούμενος ὑμῶν δεομένων τὰ περὶ τῆς πολιτείας διελθεῖν προθύμως ἐχαριζόμην, εἰδώς, ὅτι τὸν ἑξῆς λόγον οὐδένες ἂν ὑμῶν
ἐθελόντων ἱκανώτερον ἀποδοῖεν· εἰς γὰρ πόλεμον πρέποντα κατα-
5 στήσαντες τὴν πόλιν ἅπαντ' αὐτῇ τὰ προσήκοντα ἀποδοῖτ' ἂν
μόνοι τῶν νῦν. εἰπὼν δὴ τἀπιταχθέντα ἀντεπέταξα ὑμῖν ἃ καὶ
νῦν λέγω. ξυνωμολογήσατ' οὖν κοινῇ σκεψάμενοι πρὸς ὑμᾶς
αὐτοὺς εἰς νῦν ἀνταποδώσειν μοι τὰ τῶν λόγων ξένια, πάρειμί τε C
οὖν δὴ κεκοσμημένος ἐπ' αὐτὰ καὶ πάντων ἑτοιμότατος ὢν δέ-
10 χεσθαι.

ΕΡ. Καὶ μὲν δή, καθάπερ εἶπε Τίμαιος ὅδε, ὦ Σώκρατες, οὔτε
ἐλλείψομεν προθυμίας οὐδὲν οὔτε ἔστιν οὐδεμία πρόφασις ἡμῖν
τοῦ μὴ δρᾶν ταῦτα· ὥστε καὶ χθὲς εὐθὺς ἐνθένδε, ἐπειδὴ παρὰ
Κριτίαν πρὸς τὸν ξενῶνα, οὗ καὶ καταλύομεν, ἀφικόμεθα, καὶ ἔτι
15 πρότερον καθ' ὁδὸν αὐτὰ ταῦτ' ἐσκοποῦμεν. ὁ δ' οὖν ἡμῖν λόγον D
εἰσηγήσατο ἐκ παλαιᾶς ἀκοῆς· ὃν καὶ νῦν λέγε, ὦ Κριτία, τῷδε,
ἵνα ξυνδοκιμάσῃ πρὸς τὴν ἐπίταξιν εἴτ' ἐπιτήδειος εἴτ' ἀνεπιτήδειός
ἐστιν.

ΚΡ. Ταῦτα χρὴ δρᾶν, εἰ καὶ τῷ τρίτῳ κοινωνῷ Τιμαίῳ ξυν-
20 δοκεῖ.

ΤΙ. Δοκεῖ μήν.

ΚΡ. Ἄκουε δή, ὦ Σώκρατες, λόγου μάλα μὲν ἀτόπου, παντάπασί γε μὴν ἀληθοῦς, ὡς ὁ τῶν ἑπτὰ σοφώτατος Σόλων ποτ' ἔφη. E
ἦν μὲν οὖν οἰκεῖος καὶ σφόδρα φίλος ἡμῖν Δρωπίδου τοῦ προ-
25 πάππου, καθάπερ λέγει πολλαχοῦ καὶ αὐτὸς ἐν τῇ ποιήσει· πρὸς

1 ἱκανῆς: ἱκανήν H.   ὃ: διό ASZ.   9 ὢν omittit S.   13 τοῦ μή: τὸ μή S.
14 ἀφικόμεθα: ἀφικοίμεθα A.   19 χρή: δή A.

person: a friendship between Sokrates and the Syracusan leader is in itself improbable, if not impossible, and the language of Sokrates in the present passage seems inconsistent with the existence of any intimacy. That however the Syracusan is the interlocutor in this dialogue seems to me certain. Plato has assembled a company of the very highest distinction, among whom an obscure companion of Sokrates would be out of place.

4. **εἰς γὰρ πόλεμον πρέποντα**] The prominence given to war throughout the passage is notable: it is considered as a normal mode of a state's activity. And in fact, when Plato wrote, it could hardly be regarded otherwise.

9. **κεκοσμημένος**] i.e. with festal attire and garland.

11. **καὶ μὲν δή**] This is the only occasion throughout the dialogue on which Hermokrates opens his lips.

24. **Δρωπίδου**] Proklos makes out the genealogy thus:

many witnesses that his genius and acquirements qualify him to deal with all such matters. This was in my mind yesterday when I willingly complied with your request that I should repeat the conversation concerning the ideal polity; for I knew that no men were more competent than you, if you were willing, to supply the sequel: no one else indeed at the present day could, after engaging our city in an honourable war, render her conduct worthy of her in all respects. So after saying all that was enjoined on me I in my turn enjoined upon you the task of which I now remind you. Accordingly you consulted together and agreed to entertain me at this time with a return 'feast of reason'. I am here then ready for it in festal array, and never was there a more eager guest.

Hermokrates. Indeed, Sokrates, as Timaeus said, there will be no lack of zeal on our part, nor can we attempt to excuse ourselves from performing the task. In fact yesterday immediately on leaving this spot, when we reached the guest-chamber at the house of Kritias where we are staying, and even before that on our way thither, we were discussing this very matter. Kritias then told us a story from an old tradition, which you had better repeat now, Kritias, to Sokrates, that he may help us to judge whether it will answer the purpose for our present task or not.

Kritias. So be it, if our third partner Timaeus agrees.

Timaeus. I quite agree.

Kritias. Listen then, Sokrates, to a tale which, strange though it be, is yet perfectly true, as Solon, the wisest of the seven, once affirmed. He was a relation and dear friend of Dropides, my great-grandfather, as he says himself in many

```
            Exekestides
         ┌──────┴──────┐
       Solon        Dropides
                       │
                 Kritias (the elder)
         ┌─────────────┴─────────────┐
    Kallaischros              Glaukon (the elder)
         │                  ┌────────┴────────┐
    Kritias (the younger)  Periktione      Charmides
              ┌─────────────┼─────────────┐
            Plato        Glaukon       Adeimantos
```

He must however be mistaken in making Solon and Dropides brothers: Plato's words evidently do not imply so close a relationship. Moreover it would seem that Solon has been placed a generation too near to the elder Kritias.

δὲ Κριτίαν που τὸν ἡμέτερον πάππον εἶπεν, ὡς ἀπεμνημόνευεν αὖ
πρὸς ἡμᾶς ὁ γέρων, ὅτι μεγάλα καὶ θαυμαστὰ τῆσδ᾽ εἴη παλαιὰ
ἔργα τῆς πόλεως ὑπὸ χρόνου καὶ φθορᾶς ἀνθρώπων ἠφανισμένα,
πάντων δὲ ἓν μέγιστον, οὗ νῦν ἐπιμνησθεῖσι πρέπον ἂν ἡμῖν εἴη σοί 21 Α
τε ἀποδοῦναι χάριν καὶ τὴν θεὸν ἅμα ἐν τῇ πανηγύρει δικαίως τε
καὶ ἀληθῶς οἱόνπερ ὑμνοῦντας ἐγκωμιάζειν.

ΣΩ. Εὖ λέγεις. ἀλλὰ δὴ ποῖον ἔργον τοῦτο Κριτίας οὐ
λεγόμενον μέν, ὡς δὲ πραχθὲν ὄντως ὑπὸ τῆσδε τῆς πόλεως ἀρ-
χαῖον διηγεῖτο κατὰ τὴν Σόλωνος ἀκοήν;

III. ΚΡ. Ἐγὼ φράσω παλαιὸν ἀκηκοὼς λόγον οὐ νέου
ἀνδρός. ἦν μὲν γὰρ δὴ τότε Κριτίας, ὡς ἔφη, σχεδὸν ἐγγὺς ἤδη
τῶν ἐνενήκοντα ἐτῶν, ἐγὼ δέ πῃ μάλιστα δεκέτης· ἡ δὲ Κουρεῶτις Β
ἡμῖν οὖσα ἐτύγχανεν Ἀπατουρίων. τὸ δὴ τῆς ἑορτῆς σύνηθες
ἑκάστοτε καὶ τότε ξυνέβη τοῖς παισίν· ἆθλα γὰρ ἡμῖν οἱ πατέρες

1 του τόν: που omittunt SZ.   εἶπεν: εἰπεῖν A.

5. **ἐν τῇ πανηγύρει**] The goddess is of course Athena; and the festival would seem to be the lesser Panathenaia, as Proklos tells us. Considerable discussion has arisen as to the time of year in which this festival was held. The greater Panathenaia, which took place once in four years, lasted from the 17th to the 25th Hekatombaion. The lesser festival was annual. Demosthenes κατὰ Τιμοκράτους § 26 refers to a Panathenaic festival which took place in Hekatombaion; and it is affirmed by some scholars that he is speaking of the lesser Panathenaia. Were this so, it would follow that the greater and lesser festivals were held at the same time of year. But Proklos has an explicit statement to the contrary: ὅτι γε μὴν τὰ Παναθήναια (sc. τὰ μικρά) τοῖς Βενδιδείοις εἴπετο λέγουσιν οἱ ὑπομνηματισταί, καὶ Ἀριστοτέλης ὁ Ῥόδιος μαρτυρεῖ τὰ μὲν ἐν Πειραιεῖ Βενδίδεια τῇ εἰκάδι τοῦ Θαργηλιῶνος ἐπιτελεῖσθαι, ἕπεσθαι δὲ τὰς περὶ τὴν Ἀθηνᾶν ἑορτάς. It seems to me that this direct evidence is not to be outweighed by an uncertain argument based on the passage of Demosthenes. Clinton *Fasti Hellenici* II pp. 332—5 has a careful discussion of the question and decides in favour of placing the lesser Panathenaia in Thargelion.

7. **οὐ λεγόμενον μέν**] Stallbaum is ill advised in adopting the interpretation of Proklos μὴ πάνυ μὲν τεθρυλημένον, γενόμενον δὲ ὅμως. The meaning is beyond question 'not a mere figment of the imagination (like the commonwealth described in the *Republic*), but a history of facts that actually occurred'. Cf. 26 E τό τε μὴ πλασθέντα μῦθον ἀλλ᾽ ἀληθινὸν λόγον εἶναι πάμμεγά που.

21 A—25 D, c. iii. Kritias proceeds to tell a story which his grandfather once learned from Solon: that when Solon was travelling in Egypt he conversed with a priest at Sais; and beginning to recount to the priest some of the most ancient Hellenic legends he was interrupted by him with the exclamation 'Solon, ye are all children in Hellas, and no truly ancient history is to be found among you. For ever and anon there comes upon the earth a great destruction by fire or by water, and the people perish, and all their records and monuments are swept away. Only in the mountains survive a scattered remnant of shepherds and unlettered men,

passages of his poems: and Dropides told my grandfather Kritias, who when advanced in life repeated it to us, that there were great and marvellous exploits achieved by Athens in days of old, which through lapse of time and the perishing of men have vanished from memory: and the greatest of all is one which it were fitting for us to narrate, and so at once discharge our debt of gratitude to you and worthily and truly extol the goddess in this her festival by a kind of hymn in her honour.

Sokrates. A good proposal. But what was this deed which Kritias described on the authority of Solon as actually performed of old by this city, though unrecorded in history?

III. *Kritias.* I will tell an ancient story that I heard from a man no longer young. For Kritias was then, as he said, hard upon ninety years of age, while I was about ten. It happened to be the 'children's day' of the Apaturia; and then as usual the boys enjoyed their customary pastime, our fathers giving us

knowing nought of the past: and when again a civilisation has slowly grown up, presently there comes another visitation of fire or water and overwhelms it. So that in Greece and most other lands the records only go back to the last great cataclysm. But in Egypt we are preserved from fire by the inundation of the Nile, and from flood because no rain falls in our land: therefore our people has never been destroyed, and our records are far more ancient than in any other country on earth'. Then the priest goes on to tell Solon one of these histories: how that nine thousand years ago Athens was founded by Athena, and a thousand years later Sais was founded by the same goddess; how the ancient Athenians excelled all nations in good government and in the arts of war; and above all how they overthrew the power of Atlantis. For Atlantis was a vast island in the ocean, over against the pillars of Herakles, and her people were mighty men of valour and had brought much of Europe and Africa under their sway. And once the kings of Atlantis resolved at one blow to enslave all the countries that were not yet subject to them, and led forth a great host to subdue them. Then Athens put herself at the head of the nations that were fighting for freedom, and after passing through many a deadly peril, she smote the invaders and drove them back to their own country. Soon after there came dreadful earthquakes and floods; and the earth opened and swallowed up all the warriors of Athens; and Atlantis too sank beneath the sea and was never more seen.

13. ’Απατουρίων] Apaturia was the name of a festival in honour of Diónysos, held in the month Pyanepsion, which corresponded, roughly speaking, with our October. It lasted three days, of which the first was called δόρπεια, the second ἀνάρρυσις, the third κουρεῶτις. On this third day the names of children three or four years of age were enrolled on the register of their φρατρία. Proklos seems mistaken in making ἀνάρρυσις the first day; all other authorities place δόρπεια first.

68 ΠΛΑΤΩΝΟΣ [21 B—

ἔθεσαν ῥαψῳδίας. πολλῶν μὲν οὖν δὴ καὶ πολλὰ ἐλέχθη ποιητῶν
ποιήματα, ἅτε δὲ νέα κατ' ἐκεῖνον τὸν χρόνον ὄντα τὰ Σόλωνος
πολλοὶ τῶν παίδων ᾔσαμεν. εἶπεν οὖν δή τις τῶν φρατέρων, εἴτε
δὴ δοκοῦν αὐτῷ τότε εἴτε καὶ χάριν τινὰ τῷ Κριτίᾳ φέρων, δοκεῖν
5 οἱ τά τε ἄλλα σοφώτατον γεγονέναι Σόλωνα καὶ κατὰ τὴν ποίησιν C
αὖ τῶν ποιητῶν πάντων ἐλευθεριώτατον. ὁ δὴ γέρων, σφόδρα γὰρ
οὖν μέμνημαι, μάλα τε ἤσθη καὶ διαμειδιάσας εἶπεν· Εἴ γε, ὦ
Ἀμύνανδρε, μὴ παρέργῳ τῇ ποιήσει κατεχρήσατο, ἀλλ' ἐσπουδάκει
καθάπερ ἄλλοι, τόν τε λόγον, ὃν ἀπ' Αἰγύπτου δεῦρο ἠνέγκατο,
10 ἀπετέλεσε καὶ μὴ διὰ τὰς στάσεις ὑπὸ κακῶν τε ἄλλων, ὅσα εὗρεν
ἐνθάδε ἥκων, ἠναγκάσθη καταμελῆσαι, κατά γε ἐμὴν δόξαν οὔτε D
Ἡσίοδος οὔτε Ὅμηρος οὔτε ἄλλος οὐδεὶς ποιητὴς εὐδοκιμώτερος
ἐγένετο ἄν ποτε αὐτοῦ. Τίς δ' ἦν ὁ λόγος, ἦ δ' ὅς, ὦ Κριτία; Ἡ
περὶ μεγίστης, ἔφη, καὶ ὀνομαστοτάτης πασῶν δικαιότατ' ἂν πρά-
15 ξεως οὔσης, ἣν ἥδε ἡ πόλις ἔπραξε μέν, διὰ δὲ χρόνον καὶ φθορὰν
τῶν ἐργασαμένων οὐ διήρκεσε δεῦρο ὁ λόγος. Λέγε ἐξ ἀρχῆς, ἦ
δ' ὅς, τί τε καὶ πῶς καὶ παρὰ τίνων ὡς ἀληθῆ διακηκοὼς ἔλεγεν
ὁ Σόλων. Ἔστι τις κατ' Αἴγυπτον, ἦ δ' ὅς, ἐν τῷ Δέλτα, περὶ E
ὃ κατὰ κορυφὴν σχίζεται τὸ τοῦ Νείλου ῥεῦμα, Σαϊτικὸς ἐπικα-
20 λούμενος νομός, τούτου δὲ τοῦ νομοῦ μεγίστη πόλις Σάις, ὅθεν δὴ
καὶ Ἄμασις ἦν ὁ βασιλεύς· οἷς τῆς πόλεως θεὸς ἀρχηγός τίς
ἐστιν, Αἰγυπτιστὶ μὲν τοὔνομα Νηίθ, Ἑλληνιστὶ δέ, ὥς ὁ ἐκείνων
λόγος, Ἀθηνᾶ· μάλα δὲ φιλαθήναιοι καί τινα τρόπον οἰκεῖοι τῶνδ'
εἶναί φασιν. οἱ δὴ Σόλων ἔφη πορευθεὶς σφόδρα τε γενέσθαι
25 παρ' αὐτοῖς ἔντιμος, καὶ δὴ καὶ τὰ παλαιὰ ἀνερωτῶν τοὺς μάλιστα 22 A
περὶ ταῦτα τῶν ἱερέων ἐμπείρους σχεδὸν οὔτε αὐτὸν οὔτε ἄλλον
Ἕλληνα οὐδένα οὐδὲν ὡς ἔπος εἰπεῖν εἰδότα περὶ τῶν τοιούτων
ἀνευρεῖν. καί ποτε προαγαγεῖν βουληθεὶς αὐτοὺς περὶ τῶν ἀρ-

10 καὶ μή: καὶ εἰ μή A.   13 ἦ περί: ἦ omittit S.   25 ἀνερωτῶν: ἀνερωτῶντός ποτε A.

10. **διὰ τὰς στάσεις**] Plutarch *Solon* c. 31 says it was old age, not civil troubles, which prevented Solon from carrying out his designs.

14. **ἄν...οὔσης**] i.e. it would have been, had circumstances been less unfavourable.

21. **Ἄμασις ὁ βασιλεύς**] According to Herodotus II 172 the birthplace of Amasis was not Sais itself, but Siouph, another city in the Saitic nome. From Stallbaum's note it appears that this reference to Amasis placed in Solon's mouth has been regarded as an anachronism, and so Stallbaum himself seems to consider it. But since Amasis ascended the Egyptian throne in 569 B.C., according to Clinton, there is no obvious reason why Solon should not mention him, or why he may not even have visited him, as Herodotus affirms, I 30. For Solon was certainly alive after the usur-

prizes for reciting poetry. A great deal of poetry by various authors was recited, and since that of Solon was new at the time, many of us children sang his poems. So one of the clansmen said, whether he really thought so or whether he wished to please Kritias, he considered that Solon was not only in other respects the wisest of mankind but also the noblest of all poets The old man—how well I recollect it—was extremely pleased and said smiling, Yes, Amynandros, if he had not treated poetry merely as a by-work, but had made a serious business of it like the rest, and if he had finished the legend which he brought hither from Egypt, instead of being compelled to abandon it by the factions and other troubles which he found here on his return, my belief is that neither Hesiod nor Homer nor any other poet would have enjoyed greater fame than he. What was the legend, Kritias? asked Amynandros. It concerned a mighty achievement, he replied, and one that deserved to be the most famous in the world; a deed which our city actually performed, but owing to time and the destruction of the doers thereof the story has not lasted to our times. Tell us from the beginning, said the other, what was the tale that Solon told, and how and from whom he heard it as true.

There is in Egypt, said Kritias, in the Delta, at the apex of which the stream of the Nile divides, a province called the Saitic; and the chief city of this province is Sais, the birthplace of Amasis the king. The founder of their city is a goddess, whose name in the Egyptian tongue is Neith, and in Greek, as they aver, Athena: the people are great lovers of the Athenians and claim a certain kinship with our countrymen. Now when Solon travelled to this city he said he was most honourably entreated by the citizens; moreover when he questioned concerning ancient things such of the priests as were most versed therein, he found that neither he nor any other Grecian man, one might wellnigh say, knew aught about such matters. And once, when he wished to lead them on to talk of ancient times,

pation of Peisistratos, which occurred in 560.

22. Νη(θ)] This goddess is identified by Plutarch with Isis, *de Iside et Osiride* § 9 τὸ δ' ἐν Σάει τῆς Ἀθηνᾶς, ἣν καὶ Ἶσιν νομίζουσιν, ἕδος ἐπιγραφὴν εἶχε τοιαύτην, Ἐγώ εἰμι πᾶν τὸ γεγονὸς καὶ ὂν καὶ ἐσόμενον· καὶ τὸν ἐμὸν πέπλον οὐδείς πω θνητὸς ἀπεκάλυψεν.

ΠΛΑΤΩΝΟΣ [22 A—

χαίων εἰς λόγους τῶν τῇδε τὰ ἀρχαιότατα λέγειν ἐπιχειρεῖν, περὶ
Φορωνέως τε τοῦ πρώτου λεχθέντος καὶ Νιόβης, καὶ μετὰ τὸν κατακλυσμὸν αὖ περὶ Δευκαλίωνος καὶ Πύρρας ὡς διεγένοντο μυθολογεῖν,
καὶ τοὺς ἐξ αὐτῶν γενεαλογεῖν, καὶ τὰ τῶν ἐτῶν ὅσα ἦν οἷς ἔλεγε B
5 πειρᾶσθαι διαμνημονεύων τοὺς χρόνους ἀριθμεῖν· καί τινα εἰπεῖν τῶν
ἱερέων εὖ μάλα παλαιόν· Ὦ Σόλων, Σόλων, Ἕλληνες ἀεὶ παῖδές
ἐστε, γέρων δὲ Ἕλλην οὐκ ἔστιν. ἀκούσας οὖν, Πῶς τί τοῦτο λέγεις;
φάναι. Νέοι ἐστέ, εἰπεῖν, τὰς ψυχὰς πάντες· οὐδεμίαν γὰρ ἐν αὐταῖς
ἔχετε δι' ἀρχαίαν ἀκοὴν παλαιὰν δόξαν οὐδὲ μάθημα χρόνῳ πολιὸν
10 οὐδέν. τὸ δὲ τούτων αἴτιον τόδε. πολλαὶ καὶ κατὰ πολλὰ φθοραὶ C
γεγόνασιν ἀνθρώπων καὶ ἔσονται πυρὶ μὲν καὶ ὕδατι μέγισται,
μυρίοις δὲ ἄλλοις ἕτεραι βραχύτεραι. τὸ γὰρ οὖν καὶ παρ' ὑμῖν
λεγόμενον, ὥς ποτε Φαέθων Ἡλίου παῖς τὸ τοῦ πατρὸς ἅρμα
ζεύξας διὰ τὸ μὴ δυνατὸς εἶναι κατὰ τὴν τοῦ πατρὸς ὁδὸν ἐλαύνειν
15 τά τ' ἐπὶ γῆς ξυνέκαυσε καὶ αὐτὸς κεραυνωθεὶς διεφθάρη, τοῦτο
μύθου μὲν σχῆμα ἔχον λέγεται, τὸ δὲ ἀληθές ἐστι τῶν περὶ γῆν
καὶ κατ' οὐρανὸν ἰόντων παράλλαξις καὶ διὰ μακρῶν χρόνων D
γιγνομένη τῶν ἐπὶ γῆς πυρὶ πολλῷ φθορά. τότε οὖν ὅσοι κατ'
ὄρη καὶ ἐν ὑψηλοῖς τόποις καὶ ἐν ξηροῖς οἰκοῦσι, μᾶλλον διόλλυν-
20 ται τῶν ποταμοῖς καὶ θαλάττῃ προσοικούντων· ἡμῖν δὲ ὁ Νεῖλος
εἴς τε τὰ ἄλλα σωτὴρ καὶ τότε ἐκ ταύτης τῆς ἀπορίας σῴζει
λυόμενος. ὅταν δ' αὖ θεοὶ τὴν γῆν ὕδασι καθαίροντες κατακλύ-

22 θεοί: οἱ θεοί SZ.

2. **Φορωνέως**] Phoroneus is said in the legend to have been the son of Inachos : he was nevertheless the first man according to the explanation in Pausanias II xv λέγεται δὲ καὶ ὅδε λόγος· Φορωνέα ἐν τῇ γῇ ταύτῃ γενέσθαι πρῶτον, Ἴναχον δὲ οὐκ ἄνδρα ἀλλὰ τὸν ποταμὸν πατέρα εἶναι Φορωνεῖ...Φορωνεὺς δὲ ὁ Ἰνάχου τοὺς ἀνθρώπους συνήγαγε πρῶτον ἐς κοινόν, σποράδας τέως καὶ ἐφ' ἑαυτῶν ἑκάστοτε οἰκοῦν-τας· καὶ τὸ χωρίον ἐς ὃ πρῶτον ἠθροίσθησαν ἄστυ ὠνομάσθη Φορωνικόν. Proklos gives a list of several persons who enjoyed the distinction of being accounted 'first men' in various parts of Greece.

3. **ὡς διεγένοντο**] 'how they survived'. This seems clearly the meaning here ; but it is a rare use, which we find also in Hippokrates περὶ ἐπιδημιῶν I vol. III p.

384 Kühn καὶ τῶν κατακλιθέντων οὐκ οἶδ' εἴ τις καὶ μέτριον χρόνον διεγένετο.

16. **μύθου μὲν σχῆμα**] Compare *Politicus* 268 E, where another myth is similarly explained as a fragmentary reminiscence of the great convulsion that took place when the motion of the universe was reversed.

17. **παράλλαξις**] This does not signify a reverse motion, like the ἀνακύκλησις of *Politicus* 269 E, where the same word occurs, but some deviation from the wonted orbits, as in *Republic* 530 B γίγνεσθαί τε ταῦτα ἀεὶ ὡσαύτως καὶ οὐδαμῇ οὐδὲν παραλλάττειν. The παράλλαξις must not be regarded as due to accident, which Plato does not admit into his scheme : it is a phenomenon which, occurring at long but definite intervals, is strictly in the

D] ΤΙΜΑΙΟΣ. 71

he essayed to tell them of the oldest legends of Hellas, of Phoroneus who was called the first man, and of Niobe; and again he told the tale of Deukalion and Pyrrha, how they survived after the deluge, and he reckoned up their descendants, and tried, by calculating the periods, to count up the number of years that passed during the events he related. Then said one of the priests, a man well stricken in years, O Solon, Solon, ye Greeks are ever children, and old man that is a Grecian is there none. And when Solon heard it, he said, What meanest thou by this? And the priest said, Ye are all young in your souls; for ye have not in them because of old tradition any ancient belief nor knowledge that is hoary with eld. And the reason of it is this: many and manifold are the destructions of mankind that have been and shall be; the greatest are by fire and by water; but besides these there are lesser ones in countless other fashions. For indeed that tale that is also told among you, how that Phaethon, the child of the Sun, yoked his father's chariot, and for that he could not drive in his father's path, he burnt up all things upon earth and himself was smitten by a thunderbolt and slain—this story, as it is told, has the fashion of a fable; but the truth of it is a deviation of the bodies that move round the earth in the heavens, whereby comes at long intervals of time a destruction with much fire of the things that are upon earth. Thus do such as dwell on mountains and in high places and in dry perish more widely than they who live beside rivers and by the sea. Now the Nile, which is in all else our preserver, saves us then also from this distress by releasing his founts: but when the gods send a flood upon the earth, cleansing her with

regular course of nature.

22. λυόμενος] The explanation given of this word by Proklos is utterly worthless: λύεται γὰρ 'Αττικῶς ὅτι λύει τῆς ἀπορίας ἡμᾶς ὁ Νεῖλος. Even conceding the more than doubtful Atticism of λυόμενος = λύων (the only authority Stallbaum can quote is a very uncertain instance in Xenophon *de venatu* I 17), the clumsy tautology of the participle, thus understood, is glaring. It appears to me that the right interpretation has been suggested by Porphyrios, whom Proklos quotes with disapprobation. Πορφύριος μὲν δή φησιν, ὅτι δόξα ἦν παλαιὰ Αἰγυπτίων τὸ ὕδωρ κάτωθεν ἀναβλυστάνειν τῇ ἀναβάσει τοῦ Νείλου, διὸ καὶ ἱδρῶτα γῆς ἐκάλουν τὸν Νεῖλον, καὶ τὸ ἐπανιέναι κάτωθεν ταὐτὸ τῷ Αἰγυπτίῳ δηλοῦν καὶ τὸ σώζειν λυόμενον, οὐχ ὅτι ἡ χιὼν λυομένη τὸ πλῆθος τῶν ὑδάτων ποιεῖ, ἀλλ' ὅτι λύεται ἀπὸ τῶν ἑαυτοῦ πηγῶν καὶ πρόεισιν εἰς τὸ ἐμφανὲς ἐπεχόμενος πρότερον. Nothing can be more natural than that the Egyptians should have believed that the 'earth is full of secret springs', which by their

ζωσιν, οἱ μὲν ἐν τοῖς ὄρεσι διασῴζονται βουκόλοι νομεῖς τε, οἱ δ' ἐν
ταῖς παρ' ὑμῖν πόλεσιν εἰς τὴν θάλατταν ὑπὸ τῶν ποταμῶν φέ- Ε
ρονται, κατὰ δὲ τήνδε τὴν χώραν οὔτε τότε οὔτε ἄλλοτε ἄνωθεν
ἐπὶ τὰς ἀρούρας ὕδωρ ἐπιρρεῖ· τὸ δ' ἐναντίον κάτωθεν πᾶν ἐπα-
5 νιέναι πέφυκεν. ὅθεν καὶ δι' ἃς αἰτίας τἀνθάδε σῳζόμενα λέγεται
παλαιότατα. τὸ δὲ ἀληθὲς ἐν πᾶσι τοῖς τόποις, ὅπου μὴ χειμὼν
ἐξαίσιος ἢ καῦμα ἀπείργει, πλέον, τοτὲ δὲ ἔλαττον ἀεὶ γένος ἐστὶν
ἀνθρώπων. ὅσα δὲ ἢ παρ' ὑμῖν ἢ τῇδε ἢ καὶ κατ' ἄλλον τόπον ὧν 23 A
ἀκοὴν ἴσμεν, εἴ πού τι καλὸν ἢ μέγα γέγονεν ἢ καί τινα διαφορὰν
10 ἄλλην ἔχον, πάντα γεγραμμένα ἐκ παλαιοῦ τῇδ' ἐστὶν ἐν τοῖς
ἱεροῖς καὶ σεσῳσμένα. τὰ δὲ παρ' ὑμῖν καὶ τοῖς ἄλλοις ἄρτι
κατεσκευασμένα ἑκάστοτε τυγχάνει γράμμασι καὶ ἅπασιν, ὁπόσων
πόλεις δέονται, καὶ πάλιν δι' εἰωθότων ἐτῶν ὥσπερ νόσημα ἥκει
φερόμενον αὐτοῖς ῥεῦμα οὐράνιον καὶ τοὺς ἀγραμμάτους τε καὶ
15 ἀμούσους ἔλιπεν ὑμῶν, ὥστε πάλιν ἐξ ἀρχῆς οἷον νέοι γίγνεσθε, B
οὐδὲν εἰδότες οὔτε τῶν τῇδε οὔτε τῶν παρ' ὑμῖν, ὅσα ἦν ἐν τοῖς
παλαιοῖς χρόνοις. τὰ γοῦν νῦν δὴ γενεαλογηθέντα, ὦ Σόλων, περὶ
τῶν παρ' ὑμῖν ἃ διῆλθες, παίδων βραχύ τι διαφέρει μύθων, οἳ
πρῶτον μὲν ἕνα γῆς κατακλυσμὸν μέμνησθε πολλῶν ἔμπροσθεν

4 κάτωθεν πᾶν : πᾶν omittit Z.   9 ἀκοὴν dedi ex Λ. ἀκοῇ HSZ.

breaking forth gave rise to the inundation. It is true that there is still need of an explanation why the springs burst forth at a certain season: but the ancient Egyptians do not stand alone in supposing that they solve a difficulty by removing it a stage further back. λυόμενος will therefore mean 'being released' by the unsealing of its subterranean founts. This explanation also gives a good and natural sense to κάτωθεν ἐπανιέναι below. I hold it then undesirable to admit ῥυόμενος, which is the reading of some inferior mss.

3. κατὰ τήνδε τὴν χώραν] The priest's theory is as follows. The destruction of ancient records is due (1) to conflagrations, (2) to deluges. From the first the Egyptians are preserved by the inundation of the Nile, from the second by the total absence of rain in their country. Accordingly their population is continuous, and their monuments and other records escape destruction. But in Greece and elsewhere, when a deluge comes, the inhabitants of cities and the low countries are swept into the sea, and only the rude dwellers in the mountains escape: cf. *Critias* 109 D, *Laws* 677 B. Thus from time to time the more cultivated portion of the inhabitants, with all their memorials, are cut off, and civilisation has to make a fresh start : on which account all their history is of yesterday compared with that of the Egyptians. It would seem however that a conflagration which should occur in the winter or spring might take Egypt at a disadvantage.

6. τὸ δὲ ἀληθές] The application of this remark is not very obvious, but I take it to be this. We have seen that the history of the Egyptians, owing to their immunity from φθοραί, goes back to an extremely remote period, and consequently many φθοραὶ ἀνθρώπων are recorded. Elsewhere this immunity does

23 B]   ΤΙΜΑΙΟΣ.   73

waters, those in the mountains are saved, the neatherds and shepherds, but the inhabitants of the cities in your land are swept by the rivers into the sea. But in this country neither then nor at any time does water fall from on high upon the fields, but contrariwise all rises up by nature from below. Wherefore and for which causes the legends preserved here are the most ancient that are told: but the truth is that in all places, where exceeding cold or heat does not forbid, there are ever human beings, now more, now fewer. Now whether at Athens or in Egypt, or in any other place whereof we have tidings, anything noble or great or otherwise notable has occurred, we have all written down and preserved from ancient times in our temple here. But with you and other nations the commonwealth has only just been enriched with letters and all else that cities require: and again after the wonted term of years like a recurring sickness comes rushing on them the torrent from heaven; and it leaves only the unlettered and untaught among you, so that as it were ye become young again with a new birth, knowing nought of what happened in the ancient times either in our country or in yours. For instance the genealogies, Solon, which you just now recounted, concerning the people of your country, are little better than children's tales. For in the first place ye

not exist: tradition tells of but one φθορά; and people suppose that there has been but one, and that the existence of man in their country dates from a comparatively recent time. But the truth is, says the priest, that in all countries where the climate admits of human life there has been a human population of varying extent surviving a number of φθοραί, although no memorial of the earlier inhabitants remains. It was a common belief that as the North from cold, so the South from heat was uninhabitable by man: cf. Aristotle *meteorologica* II v 361ᵇ 26 ἔνθα μὲν γὰρ διὰ ψῦχος οὐκέτι κατοικοῦσιν, ἔνθα δὲ διὰ τὴν ἀλέαν. The difficulty about the sentence is that τὸ δ' ἀληθές has the air of correcting the statement in the preceding clause: whereas what is really corrected is the implied misconception; i.e. that the antiquity of man in other countries is no greater than that of the records.

12. κατεσκευασμένα...γράμμασι] 'literis mandata', says Stallbaum, a rendering which will surely find few friends: nor can we confine ἅπασιν ὁπόσων πόλεις δέονται to public monuments, as he would have us. κατεσκευασμένα means 'furnished' or 'enriched', a sense which it bears several times in Thucydides: see VI 91, VIII 24. The following words generally comprehend all the appurtenances of civilisation: amongst others, as Proklos says, τέχναι καὶ ἀγοραὶ καὶ λουτρά. τὰ παρ' ὑμῖν is also a general phrase, = your institutions or commonwealths. Compare *Critias* 110 A ὅταν ἴδητόν τισιν ἤδη τοῦ βίου τἀναγκαῖα κατεσκευασμένα.

13. δι' εἰωθότων ἐτῶν] These words show conclusively that the φθοραί were normal and regularly recurrent.

74   ΠΛΑΤΩΝΟΣ   [23 B—

γεγονότων, ἔτι δὲ τὸ κάλλιστον καὶ ἄριστον γένος ἐπ' ἀνθρώπους
ἐν τῇ χώρᾳ τῇ παρ' ὑμῖν οὐκ ἴστε γεγονός, ἐξ ὧν σύ τε καὶ πᾶσα
ἡ πόλις ἔστι τὰ νῦν ὑμῶν, περιλειφθέντος ποτὲ σπέρματος βραχέος, C
ἀλλ' ὑμᾶς λέληθε διὰ τὸ τοὺς περιγενομένους ἐπὶ πολλὰς γενεὰς
5 γράμμασι τελευτᾶν ἀφώνους. ἦν γὰρ δή ποτε, ὦ Σόλων, ὑπὲρ τὴν
μεγίστην φθορὰν ὕδασιν ἡ νῦν Ἀθηναίων οὖσα πόλις ἀρίστη πρός
τε τὸν πόλεμον καὶ κατὰ πάντα εὐνομωτάτη διαφερόντως· ᾗ κάλλιστα ἔργα καὶ πολιτεῖαι γενέσθαι λέγονται κάλλισται πασῶν,
ὁπόσων ὑπὸ τὸν οὐρανὸν ἡμεῖς ἀκοὴν παρεδεξάμεθα. ἀκούσας D
10 οὖν ὁ Σόλων ἔφη θαυμάσαι καὶ πᾶσαν προθυμίαν ἔχειν δεόμενος
τῶν ἱερέων πάντα δι' ἀκριβείας οἱ τὰ περὶ τῶν πάλαι πολιτῶν
ἑξῆς διελθεῖν. τὸν οὖν ἱερέα φάναι· Φθόνος οὐδείς, ὦ Σόλων, ἀλλὰ
σοῦ τε ἕνεκα ἐρῶ καὶ τῆς πόλεως ὑμῶν, μάλιστα δὲ τῆς θεοῦ χάριν,
ἣ τήν τε ὑμετέραν καὶ τήνδε ἔλαχε καὶ ἔθρεψε καὶ ἐπαίδευσε, προ-
15 τέραν μὲν τὴν παρ' ὑμῖν ἔτεσι χιλίοις, ἐκ Γῆς τε καὶ Ἡφαίστου τὸ E
σπέρμα παραλαβοῦσα ὑμῶν, τήνδε δὲ ὑστέραν. τῆς δὲ ἐνθάδε
διακοσμήσεως παρ' ἡμῖν ἐν τοῖς ἱεροῖς γράμμασιν ὀκτακισχιλίων
ἐτῶν ἀριθμὸς γέγραπται. περὶ δὴ τῶν ἐνακισχίλια γεγονότων ἔτη
πολιτῶν σοι δηλώσω διὰ βραχέων νόμους, καὶ τῶν ἔργων αὐτοῖς
20 ὃ κάλλιστον ἐπράχθη· τὸ δ' ἀκριβὲς περὶ πάντων ἐφεξῆς εἰσαῦθις 24 A
κατὰ σχολὴν αὐτὰ τὰ γράμματα λαβόντες διέξιμεν. τοὺς μὲν οὖν
νόμους σκόπει πρὸς τοὺς τῇδε. πολλὰ γὰρ παραδείγματα τῶν
τότε παρ' ὑμῖν ὄντων ἐνθάδε νῦν ἀνευρήσεις, πρῶτον μὲν τὸ τῶν
ἱερέων γένος ἀπὸ τῶν ἄλλων χωρὶς ἀφωρισμένον, μετὰ δὲ τοῦτο τὸ

9 ὁπόσων ὑπό: ὁπόσων νῦν ὑπό ΗΖ.   10 ἔχειν: σχεῖν SZ.
16 ἐνθάδε: ἐνθαδὶ S.   22 τῇδε: τῆσδε Α.

1. ἐπ' ἀνθρώπους] ἐπί signifies extension over: a use exceedingly rare in Attic prose, but occurring again in *Critias* 112 E ἐπὶ πᾶσαν Εὐρώπην καὶ Ἀσίαν κατά τε σωμάτων κάλλη καὶ κατὰ τὴν τῶν ψυχῶν παντοίαν ἀρετὴν ἐλλόγιμοί τε ἦσαν καὶ ὀνομαστότατοι πάντων τῶν τότε: and a similar, though not identical, use is to be found in *Protagoras* 322 D. It is not uncommon in Homer, e.g. *Iliad* X 213 μέγα κέν οἱ ὑπουράνιον κλέος εἴη | πάντας ἐπ' ἀνθρώποις.

5. ὑπὲρ τὴν μεγίστην φθοράν] ὑπὲρ = back beyond.

8. πολιτεῖαι] The plural is somewhat curious: it seems to stand for 'political institutions'.

15. Γῆς τε καὶ Ἡφαίστου] As we shall presently see, earth and fire are the two principal elements of which material nature is composed, air and water being means between them; cf. 31 C foll. Fire is the simplest combination of one of the two primary bases, while earth is the only form of the other, 51 D foll. These were the two ἀρχαί of Parmenides: Arist. *metaph.* I v 986[b] 33 δύο τὰς αἰτίας καὶ δύο τὰς ἀρχὰς πάλιν τίθησι, θερμὸν καὶ ψυχρόν, οἷον πῦρ καὶ γῆν λέγων. Cf. *physica* I v 188[a] 20. Plato's statement falls in with

remember but one deluge, whereas there had been many before it; and moreover ye know not that the fairest and noblest race among mankind lived once in your country, whence ye sprang and all your city which now is, from a very little seed that of old was left over. Ye however know it not, because the survivors lived and died for many generations without utterance in writing. For once upon a time, Solon, far back beyond the greatest destruction by waters, that which is now the city of the Athenians was foremost both in war and in all besides, and her laws were exceedingly righteous above all cities. Her deeds and her government are said to have been the noblest among all under heaven whereof the report has come to our ears. And Solon said that on hearing this he was astonished, and used all urgency in entreating the priests to relate to him from beginning to end all about those ancient citizens. So the priest said, I grudge thee not, O Solon, and I will tell it for thy sake and for the sake of thy city, and chiefly for the honour of the goddess who was the possessor and nurse and instructress both of your city and of ours; for she founded yours earlier by a thousand years, having taken the seed of you from Earth and Hephaistos; and ours in later time. And the date of our city's foundation is recorded in our sacred writings to be eight thousand years ago. But concerning the citizens of Athens nine thousand years ago I will inform you in brief of their laws and of the noblest of the deeds which they performed: the exact truth concerning everything we will examine in due order hereafter, taking the actual records at our leisure.

Consider now their laws in comparison with those of our country; for you will find here at the present day many examples of the laws which then existed among you:—first the separation of the priestly caste from the rest; next the distinc-

---

Athenian mythology: Erechtheus was the son of Earth and Hephaistos.

22. παραδείγματα is of course not put for εἰκόνας, as Proklos would have it, but signifies samples, specimens.

23. τὸ τῶν ἱερέων γένος] Plato's classification does not coincide with that given in Herodotus II 164. The latter makes seven castes: ἔστι δὲ Αἰγυπτίων ἑπτὰ γένεα, καὶ τούτων οἱ μὲν ἱρέες, οἱ δὲ μάχιμοι κεκλέαται, οἱ δὲ βουκόλοι, οἱ δὲ συβῶται, οἱ δὲ κάπηλοι, οἱ δὲ ἑρμηνέες, οἱ δὲ κυβερνῆται. The discrepancy arises from the fact that there were actually three castes, the two higher being priests and warriors, and the lowest comprising men following various occupations which are differently enumerated by different authorities.

76 ΠΛΑΤΩΝΟΣ [24 A—

τῶν δημιουργῶν, ὅτι καθ' αὑτὸ ἕκαστον ἄλλῳ δὲ οὐκ ἐπιμιγνύμενον δημιουργεῖ, τό τε τῶν νομέων καὶ τὸ τῶν θηρευτῶν τό τε τῶν γεωργῶν· καὶ δὴ καὶ τὸ μάχιμον γένος ᾔσθησαί που τῇδε ἀπὸ B πάντων τῶν γενῶν κεχωρισμένον, οἷς οὐδὲν ἄλλο πλὴν τὰ περὶ τὸν
5 πόλεμον ὑπὸ τοῦ νόμου προσετάχθη μέλειν· ἔτι δὲ ἡ τῆς ὁπλίσεως αὐτῶν σχέσις ἀσπίδων καὶ δοράτων, οἷς ἡμεῖς πρῶτοι τῶν περὶ τὴν Ἀσίαν ὡπλίσμεθα, τῆς θεοῦ καθάπερ ἐν ἐκείνοις τοῖς τόποις παρ' ὑμῖν πρώτοις ἐνδειξαμένης. τὸ δ' αὖ περὶ τῆς φρονήσεως, ὁρᾷς που τὸν νόμον τῇδε ὅσην ἐπιμέλειαν ἐποιήσατο εὐθὺς κατ' ἀρχὰς
10 περί τε τὸν κόσμον ἅπαντα, μέχρι μαντικῆς καὶ ἰατρικῆς πρὸς C ὑγίειαν, ἐκ τούτων θείων ὄντων εἰς τὰ ἀνθρώπινα ἀνευρών, ἴσα τε ἄλλα τούτοις ἕπεται μαθήματα πάντα κτησάμενος. ταύτην οὖν δὴ τότε ξύμπασαν τὴν διακόσμησιν καὶ σύνταξιν ἡ θεὸς προτέρους ὑμᾶς διακοσμήσασα κατῴκισεν, ἐκλεξαμένη τὸν τόπον ἐν ᾧ γε-
15 γένησθε, τὴν εὐκρασίαν τῶν ὡρῶν ἐν αὐτῷ κατιδοῦσα, ὅτι φρονιμωτάτους ἄνδρας οἴσοι· ἅτε οὖν φιλοπόλεμός τε καὶ φιλόσοφος ἡ θεὸς οὖσα τὸν προσφερεστάτους αὐτῇ μέλλοντα οἴσειν τόπον D ἄνδρας, τοῦτον ἐκλεξαμένη πρῶτον κατῴκισεν. ᾠκεῖτε δὴ οὖν νόμοις τε τοιούτοις χρώμενοι καὶ ἔτι μᾶλλον εὐνομούμενοι πάσῃ
20 τε πάντας ἀνθρώπους ὑπερβεβηκότες ἀρετῇ, καθάπερ εἰκὸς γεννήματα καὶ παιδεύματα θεῶν ὄντας. πολλὰ μὲν οὖν ὑμῶν καὶ μεγάλα ἔργα τῆς πόλεως τῇδε γεγραμμένα θαυμάζεται, πάντων γε

2 τὸ τῶν θηρευτῶν: τὸ omittit S. 20 πάντας: παρὰ πάντας A.
ὑπερβεβηκότες: ὑπερβεβληκότες H.

1. **οὐκ ἐπιμιγνύμενον**] i.e. each minded his own business, like the citizens of Plato's model republic.

6. **τῶν περὶ τὴν Ἀσίαν**] Egypt was commonly regarded in Plato's time as belonging to Asia rather than Africa. All Africa was indeed often regarded as part of Asia; but that Plato distinguished them is made clear below in 24 E.

8. **τὸ δ' αὖ περὶ τῆς φρονήσεως**] Having described the ordinances relating to externals he now proceeds to the training of the mind.

10. **περί τε τὸν κόσμον**] The meaning of this curiously involved and complex sentence seems to be this. The lawgiver, beginning with the study of the nature of the universe, which is divine, deduced from thence principles of practical use for human needs, applying them to divination and medicine and the other sciences therewith connected. The peculiarity of the law in fact consisted in basing its precepts concerning practical arts such as medicine (ἀνθρώπινα) upon universal truths of nature (θεῖα). μέχρι μαντικῆς, i.e. bringing its deductions down to divination. In the words ἐκ τούτων θείων ὄντων εἰς τὰ ἀνθρώπινα ἀνευρών we certainly have a difficulty of construction. I take the meaning to be 'from these divine studies (i.e. of the κόσμος) having invented them (μαντικὴ and ἰατρική) for human needs'. But the lack of an object to ἀνευρών and the construction of εἰς τὰ ἀνθρώπινα are alike unsatisfactory; and I

D] ΤΙΜΑΙΟΣ. 77

tion of the craftsmen, that each kind plies its own craft by itself and mingles not with another; and the class of shepherds and of hunters and of husbandmen are set apart; and that of the warriors too you have surely noticed is here sundered from all the other classes; for on them the law enjoins to study the art of war and nought else. Furthermore there is the fashion of their arming with spears and shields, wherewith we have been the first men in Asia to arm ourselves; for the goddess taught this to us, as she did first to you in that country of yours. Again as regards knowledge, you see how careful our law is in its first principles, investigating the laws of nature till it arrives at divination and medicine, the object of which is health, drawing from these divine studies lessons useful for human needs, and adding to these all the sciences that are connected therewithal. With all this constitution and order the goddess established you when she founded your nation first; choosing out the spot in which ye were born because she saw that the mild temperament of its seasons would produce the highest intelligence in its people. Seeing then that the goddess was a lover of war and of wisdom, she selected the spot that should bring forth men likest to herself, and therein she first founded your race. Thus then did ye dwell governed by such laws as I have described, ay and even better still, surpassing all men in excellence, as was meet for them that were offspring and nurslings of gods.

Many and mighty are the deeds of your city recorded here for the marvel of men; but one is there which for greatness and

much doubt whether the text is sound. The whole sentence reads strangely in a passage of such singular literary brilliance as this chapter. With regard to μαντικῆς καὶ ἰατρικῆς Proklos observes that the Egyptians combined these two professions.

15. **φρονιμωτάτους ἄνδρας**] Compare Laws 642 C, Menexenus 237 C foll. The Euripidean ἀεὶ διὰ λαμπροτάτου βαίνοντες ἀβρῶς αἰθέρος will occur to every one. How much importance was attached by Greek medical science to the influence of climate upon the nature of a people may be gathered

from the treatise of Hippokrates *de aere locis et aquis:* cf. especially εὑρήσεις γὰρ ἐπὶ τὸ πλῆθος τῆς χώρης τῇ φύσι ἀκολουθεῦντα καὶ εἴδεα τῶν ἀνθρώπων καὶ τοὺς τρόπους. Kühn vol. 1 p. 567. Compare too Plotinos *ennead* III i 5 ἀκολουθεῖν δὲ τοῖς τόποις οὐ μόνον τὰ ἄλλα φυτά τε καὶ ζῷα, ἀλλὰ καὶ ἀνθρώπων εἴδη τε καὶ μεγέθη καὶ χρόας καὶ θυμοὺς καὶ ἐπιθυμίας, ἐπιτηδεύματά τε καὶ ἤθη.

22. **πάντων γε μὴν ἕν**] The amount of speculation and misdirected ingenuity which Plato's story of Atlantis has awakened surpasses belief. Plato is our

78   ΠΛΑΤΩΝΟΣ   [24 D—

μὴν ἐν ὑπερέχει μεγέθει καὶ ἀρετῇ· λέγει γὰρ τὰ γεγραμμένα, ὅσην E
ἡ πόλις ὑμῶν ἔπαυσέ ποτε δύναμιν ὕβρει πορευομένην ἅμα ἐπὶ
πᾶσαν Εὐρώπην καὶ Ἀσίαν, ἔξωθεν ὁρμηθεῖσαν ἐκ τοῦ Ἀτλαντι-
κοῦ πελάγους. τότε γὰρ πορεύσιμον ἦν τὸ ἐκεῖ πέλαγος· νῆσον
5 γὰρ πρὸ τοῦ στόματος εἶχεν, ὃ καλεῖται, ὥς φατε ὑμεῖς, Ἡρακλέους
στῆλαι· ἡ δὲ νῆσος ἅμα Λιβύης ἦν καὶ Ἀσίας μείζων, ἐξ ἧς ἐπιβα-
τὸν ἐπὶ τὰς ἄλλας νήσους τοῖς τότε ἐγίγνετο πορευομένοις, ἐκ δὲ
τῶν νήσων ἐπὶ τὴν καταντικρὺ πᾶσαν ἤπειρον τὴν περὶ τὸν ἀληθι- 25 A
νὸν ἐκεῖνον πόντον. τάδε μὲν γάρ, ὅσα ἐντὸς τοῦ στόματος οὗ
10 λέγομεν, φαίνεται λιμὴν στενόν τινα ἔχων εἴσπλουν· ἐκεῖνο δὲ
πέλαγος ὄντως ἥ τε περιέχουσα αὐτὸ γῆ παντελῶς [ἀληθῶς]
ὀρθότατ᾽ ἂν λέγοιτο ἤπειρος. ἐν δὲ δὴ τῇ Ἀτλαντίδι νήσῳ ταύτῃ
μεγάλη συνέστη καὶ θαυμαστὴ δύναμις βασιλέων, κρατοῦσα μὲν
ἁπάσης τῆς νήσου, πολλῶν δὲ ἄλλων νήσων καὶ μερῶν τῆς ἠπείρου·
15 πρὸς δὲ τούτοις ἔτι τῶν ἐντὸς τῇδε Λιβύης μὲν ἦρχον μέχρι πρὸς B
Αἴγυπτον, τῆς δὲ Εὐρώπης μέχρι Τυρρηνίας. αὕτη δὴ πᾶσα ξυνα-
θροισθεῖσα εἰς ἓν ἡ δύναμις τόν τε παρ᾽ ὑμῖν καὶ τὸν παρ᾽ ἡμῖν καὶ
τὸν ἐντὸς τοῦ στόματος πάντα τόπον μιᾷ ποτὲ ἐπεχείρησεν ὁρμῇ
δουλοῦσθαι. τότε οὖν ὑμῶν, ὦ Σόλων, τῆς πόλεως ἡ δύναμις εἰς
20 ἅπαντας ἀνθρώπους διαφανὴς ἀρετῇ τε καὶ ῥώμῃ ἐγένετο· πάντων
γὰρ προστᾶσα εὐψυχίᾳ καὶ τέχναις ὅσαι κατὰ πόλεμον, τὰ μὲν
τῶν Ἑλλήνων ἡγουμένη, τὰ δ᾽ αὐτὴ μονωθεῖσα ἐξ ἀνάγκης τῶν C

5 καλεῖται...στῆλαι: καλεῖτε...στήλας AHSZ.   11 ἀληθῶς erasit A. ego inclusi.

only authority for the legend: there is no trace of confirmation from any independent source. It appears to me impossible to determine whether Plato has invented the story from beginning to end —ῥᾳδίως Αἰγυπτίους καὶ ὁποδαποὺς ἂν ἐθέλῃ λόγους ποιεῖ—or whether it really more or less represents some Egyptian legend brought home by Solon. Stallbaum supposes that the ancient Egyptians really had some information of the existence of America. But this is entirely incredible, considering the limited powers of navigation possessed by even the boldest seafarers of those times. The greatest voyage on record was the circumnavigation of Africa related by Herodotus IV 42: but that is mere child's play to crossing and recrossing the Atlantic without a compass. The explorers took over two years for their enterprise and went ashore each year to raise a crop. The view that Atlantis did actually exist and disappear, as Plato describes, receives, I believe, no countenance from geology. The wild absurdity of most of the theories on the subject may be gathered from Martin's learned and amusing dissertation. There is hardly a country on the face of the globe, not only from China to Peru, but from New Zealand to Spitzbergen, including such an eminently unpromising locality as Palestine, which has not been confidently identified with the Platonic Atlantis. It can only be said that such speculations are δεινοῦ καὶ ἐπιπόνου καὶ οὐ πάνυ εὐτυχοῦς ἀνδρός.

4. πορεύσιμον] Plato means that since

nobleness surpasses all the rest. For our chronicles tell what a power your city quelled of old, that marched in wanton insolence upon all Europe and Asia together, issuing yonder from the Atlantic ocean. For in those days the sea there could be crossed, since it had an island before the mouth of the strait which is called, as ye say, the pillars of Herakles. Now this island was greater than Libya and Asia together; and therefrom there was passage for the sea-farers of those times to the other islands, and from the islands to all the opposite continent which bounds that ocean truly named. For these regions that lie within the strait aforesaid seem to be but a bay having a narrow entrance; but the other is ocean verily, and the land surrounding it may with fullest truth and fitness be named a continent. In this island Atlantis arose a great and marvellous might of kings, ruling over all the island itself, and many other islands, and parts of the mainland; and besides these, of the lands east of the strait they governed Libya as far as Egypt, and Europe to the borders of Etruria. So all this power gathered itself together, and your country and ours and the whole region within the strait it sought with one single swoop to enslave. Then, O Solon, did the power of your city shine forth in all men's eyes glorious in valour and in strength. For being foremost upon earth in courage and the arts of war, sometimes she was leader of the Hellenes, sometimes she stood alone perforce,

the Atlantic was thickly studded with large islands, it was possible for mariners to pass from one to another by easy stages until they reached the transatlantic continent, without the necessity of a long sea voyage. We know from Thucydides that even the passage across the Ionian sea was regarded as formidable; we may readily conceive then that many halting places would be required to make the Atlantic ocean πορεύσιμον.

5. τοῦ στόματος] i.e. the strait of Gibraltar.

8 καλεῖται] The mss. give καλεῖται ...στήλας, which is usually corrected into καλεῖτε. But owing to the tautology thus produced, I prefer on Stallbaum's suggestion to retain καλεῖται and read στῆλαι.

6. Λιβύης ἦν καὶ 'Ασίας μείζων] In estimating the size of Atlantis allowance must be made for Plato's imperfect knowledge of the magnitude of Asia and Africa.

8. τὴν καταντικρὺ πᾶσαν ἤπειρον] Martin suggests that the notion of a transatlantic continent may have arisen from the early conception of Ocean as a river, implying a further shore.

20. πάντων γὰρ προστᾶσα] The unmistakable similarity between the position of the legendary Athens in the Atlantine war and that of the historical Athens in the Persian invasion indicates that if Plato is using an ancient legend, he has freely adapted it to his own ends: for the existence of such a coincidence in the original is highly improbable.

ἄλλων ἀποστάντων, ἐπὶ τοὺς ἐσχάτους ἀφικομένη κινδύνους, κρατήσασα μὲν τῶν ἐπιόντων τρόπαια ἔστησε, τοὺς δὲ μήπω δεδουλωμένους διεκώλυσε δουλωθῆναι, τοὺς δ' ἄλλους, ὅσοι κατοικοῦμεν ἐντὸς ὅρων Ἡρακλείων, ἀφθόνως ἅπαντας ἠλευθέρωσεν. ὑστέρῳ
5 δὲ χρόνῳ σεισμῶν ἐξαισίων καὶ κατακλυσμῶν γενομένων, μιᾶς ἡμέρας καὶ νυκτὸς χαλεπῆς ἐπελθούσης, τό τε παρ' ὑμῖν μάχιμον D πᾶν ἀθρόον ἔδυ κατὰ γῆς, ἥ τε Ἀτλαντὶς νῆσος ὡσαύτως κατὰ τῆς θαλάττης δῦσα ἠφανίσθη· διὸ καὶ νῦν ἄπορον καὶ ἀδιερεύνητον γέγονε τὸ ἐκεῖ πέλαγος, πηλοῦ κάρτα βραχέος ἐμποδὼν ὄντος, ὃν
10 ἡ νῆσος ἱζομένη παρέσχετο.

IV. Τὰ μὲν δὴ ῥηθέντα, ὦ Σώκρατες, ὑπὸ τοῦ παλαιοῦ Κριτίου κατ' ἀκοὴν τὴν Σόλωνος, ὡς συντόμως εἰπεῖν, ἀκήκοας· E λέγοντος δὲ δὴ χθὲς σοῦ περὶ πολιτείας καὶ τῶν ἀνδρῶν, οὓς ἔλεγες, ἐθαύμαζον ἀναμιμνησκόμενος αὐτὰ ἃ νῦν λέγω, κατανοῶν,
15 ὡς δαιμονίως ἔκ τινος τύχης οὐκ ἀπὸ σκοποῦ ξυνηνέχθης τὰ πολλὰ οἷς Σόλων εἶπεν. οὐ μὴν ἐβουλήθην παραχρῆμα εἰπεῖν· διὰ 26 A χρόνου γὰρ οὐχ ἱκανῶς ἐμεμνήμην· ἐνενόησα οὖν, ὅτι χρεὼν εἴη με πρὸς ἐμαυτὸν πρῶτον ἱκανῶς πάντα ἀναλαβόντα λέγειν οὕτως. ὅθεν ταχὺ ξυνωμολόγησά σοι τἀπιταχθέντα χθές, ἡγούμενος,
20 ὅπερ ἐν ἅπασι τοῖς τοιοῖσδε μέγιστον ἔργον, λόγον τινὰ πρέποντα τοῖς βουλήμασιν ὑποθέσθαι, τούτου μετρίως ἡμᾶς εὐπορήσειν. οὕτω δή, καθάπερ ὅδ' εἶπε, χθές τε εὐθὺς ἐνθένδε ἀπιὼν πρὸς

6 ἐπελθούσης: ἐλθούσης Ζ.   9 βραχέος: βαθέος ΑΖ.

6. **τό τε παρ' ὑμῖν μάχιμον**] We must suppose the chief fury of the earthquake was spent on Athens itself, so that all the more cultivated and intelligent citizens, who, as in Plato's own republic, included the fighting men, were destroyed; while the Attic race was continued by the rude inhabitants of country districts.

8. **ἄπορον καὶ ἀδιερεύνητον**] Aristotle agrees, though assigning a different reason, about the shallowness of the Atlantic near Gibraltar: cf. *meteorologica* II i 354ᵃ 22 τὰ δ' ἔξω στηλῶν βραχέα μὲν διὰ τὸν πηλόν, ἄπνοα δ' ἐστὶν ὡς ἐν κοίλῳ τῆς θαλάττης οὔσης. ὥσπερ οὖν καὶ κατὰ μέρος ἐκ τῶν ὑψηλῶν οἱ ποταμοὶ φαίνονται ῥέοντες, οὕτω καὶ τῆς ὅλης γῆς ἐκ τῶν ὑψηλοτέρων τῶν πρὸς ἄρκτον τὸ ῥεῦμα γίνεται. τὸ πλεῖστον, ὥστε τὰ μὲν διὰ τὴν ἔκχυσιν οὐ βαθέα, τὰ δ' ἔξω πελάγη βαθέα μᾶλλον. Aristotle's notion was that the more northerly parts of the globe were higher than the southern: hence the marine currents flowed southward carrying with them quantities of sand which, being deposited off the coasts of southern Europe, silted up the entrance to the Mediterranean.

9. **πηλοῦ κάρτα βραχέος**] I believe this reading to be perfectly correct, although I am unable to produce an exact parallel. βραχέα was the regular word for shoals: cf. Herodotus II 102 θάλασσαν οὐκέτι πλωτὴν ὑπὸ βραχέων: also IV 179, and Plutarch *de genio Socratis* § 22 ἀραιὰ τενάγη καὶ βραχέα. The peculiarity in our passage is of course that βραχέος is

when the rest fell away from her; and after being brought into the uttermost perils, she vanquished the invaders and triumphed over them: and the nations that were not yet enslaved she preserved from slavery; while the rest of us who dwell this side the pillars of Herakles, all did she set free with ungrudging hand. But in later time, after there had been exceeding great earthquakes and floods, there fell one day and night of destruction; and the warriors in your land all in one body were swallowed up by the earth, and in like manner did the island Atlantis sink beneath the sea and vanish away. Wherefore to this day the ocean there is impassable and unsearchable, being blocked by very shallow shoals, which the island caused as she settled down.

IV. You have heard this brief statement, Sokrates, of what the ancient Kritias reported that he heard from Solon: and when you were speaking yesterday about the polity and the men whom you described, I was amazed as I called to mind the story I have just told you, remarking how by some miraculous coincidence most of your account agreed unerringly with the description of Solon. I was unwilling however to say anything at the moment, for after so long a time my memory was at fault. I conceived therefore that I must not speak until I had thoroughly gone over the whole story by myself. Accordingly I was quick to accept the task you imposed on us yesterday, thinking that for the most arduous part of all such undertakings, I mean supplying a story fitly corresponding to our intentions,

---

an adjective agreeing with πηλοῦ. But though this use does not seem to occur elsewhere, I see no conclusive reason for rejecting it here; and certainly no tolerable substitute has been offered for it. A gives βαθέος, which is pointless: surely the question that would interest a sailor is how near the mud was to the surface; its depth he would regard with profound indifference. And there is little more to be said for Stallbaum's suggestion τραχέος. Accordingly I retain πηλοῦ κάρτα βραχέος in the sense of 'very shoaly mud'.

25 D—27 B, c. iv. Kritias proceeds to say that he was greatly struck by the resemblance between the ideal common-

wealth as painted by Sokrates and ancient Athens as described in Solon's legend. He therefore taxed his memory to recover every detail of the history, thinking it would serve to fulfil Sokrates' wish to see his imaginary citizens brought into life and action. Sokrates welcomes the suggestion; and it is agreed that Timaeus shall first expound the order of the universe down to the creation of man, and that Kritias shall follow with his account of the former Athenians and of their war with Atlantis.

18. **πάντα ἀναλαβόντα**] referring to the detailed account to be given in the *Critias*.

82   ΠΛΑΤΩΝΟΣ   [26 B—

τούσδε ἀνέφερον αὐτὰ ἀναμιμνησκόμενος, ἀπελθών τε σχεδόν τι B
πάντα ἐπισκοπῶν τῆς νυκτὸς ἀνέλαβον. ὡς δή τοι, τὸ λεγόμενον,
τὰ παίδων μαθήματα θαυμαστὸν ἔχει τι μνημεῖον. ἐγὼ γάρ, ἃ
μὲν χθὲς ἤκουσα, οὐκ ἂν οἶδ᾽ εἰ δυναίμην ἅπαντα ἐν μνήμῃ πάλιν
5 λαβεῖν· ταῦτα δέ, ἃ πάμπολυν χρόνον διακήκοα, παντάπασι
θαυμάσαιμ᾽ ἂν εἴ τί με αὐτῶν διαπέφευγεν. ἦν μὲν οὖν μετὰ
πολλῆς ἡδονῆς καὶ παιδικῆς τότε ἀκουόμενα, καὶ τοῦ πρεσβύτου C
προθύμως με διδάσκοντος, ἅτ᾽ ἐμοῦ πολλάκις ἐπανερωτῶντος,
ὥστε οἷον ἐγκαύματα ἀνεκπλύτου γραφῆς ἔμμονά μοι γέγονε·
10 καὶ δὴ καὶ τοῖσδε εὐθὺς ἔλεγον ἕωθεν αὐτὰ ταῦτα, ἵνα εὐποροῖεν
λόγων μετ᾽ ἐμοῦ. νῦν οὖν, οὗπερ ἕνεκα πάντα ταῦτα εἴρηται,
λέγειν εἰμὶ ἕτοιμος, ὦ Σώκρατες, μὴ μόνον ἐν κεφαλαίοις ἀλλ᾽
ὥσπερ ἤκουσα καθ᾽ ἕκαστον· τοὺς δὲ πολίτας καὶ τὴν πόλιν, ἣν
χθὲς ἡμῖν ὡς ἐν μύθῳ διῄεισθα σύ, νῦν μετενεγκόντες ἐπὶ τἀληθὲς D
15 δεῦρο θήσομεν ὡς ἐκείνην τήνδε οὖσαν, καὶ τοὺς πολίτας, οὓς
διενοοῦ, φήσομεν ἐκείνους τοὺς ἀληθινοὺς εἶναι προγόνους ἡμῶν,
οὓς ἔλεγεν ὁ ἱερεύς. πάντως ἁρμόσουσι καὶ οὐκ ἀπᾳσόμεθα λέγοντες αὐτοὺς εἶναι τοὺς ἐν τῷ τότε ὄντας χρόνῳ· κοινῇ δὲ δια-
λαμβάνοντες ἅπαντες πειρασόμεθα τὸ πρέπον εἰς δύναμιν οἷς
20 ἐπέταξας ἀποδοῦναι. σκοπεῖν οὖν δὴ χρή, ὦ Σώκρατες, εἰ κατὰ
νοῦν ὁ λόγος ἡμῖν οὗτος, ἤ τινα ἔτ᾽ ἄλλον ἀντ᾽ αὐτοῦ ζητητέον.   E

ΣΩ. Καὶ τίν᾽ ἄν, ὦ Κριτία, μᾶλλον ἀντὶ τούτου μεταλάβοι-
μεν, ὃς τῇ τε παρούσῃ τῆς θεοῦ θυσίᾳ διὰ τὴν οἰκειότητ᾽ ἂν πρέποι
μάλιστα, τό τε μὴ πλασθέντα μῦθον ἀλλ᾽ ἀληθινὸν λόγον εἶναι
25 πάμμεγά που. πῶς γὰρ καὶ πόθεν ἄλλους ἀνευρήσομεν ἀφέμενοι
τούτων; οὐκ ἔστιν, ἀλλ᾽ ἀγαθῇ τύχῃ χρὴ λέγειν μὲν ὑμᾶς, ἐμὲ
δὲ ἀντὶ τῶν χθὲς λόγων νῦν ἡσυχίαν ἄγοντα ἀντακούειν.   27 A

ΚΡ. Σκόπει δὴ τὴν τῶν ξενίων σοι διάθεσιν, ὦ Σώκρατες, ᾗ
διέθεμεν. ἔδοξε γὰρ ἡμῖν Τίμαιον μέν, ἅτε ὄντα ἀστρονομικώτατον

2 πάντα: ἅπαντα S.   7 παιδικῆς: παιδιᾶς S.   14 νῦν ante μετενεγκόντες omittunt SZ.   19 post ἅπαντες inserit A τοὺς ἀνθρώπους.

4. **οὐκ ἂν οἶδ᾽ εἰ δυναίμην**] For the construction and position of ἂν see Euripides *Alcestis* 48, *Medea* 941. I have not noted another instance in Plato.

7. **παιδικῆς**] Stallbaum with very slight ms. authority reads παιδιᾶς, without noticing any other reading: apparently he failed to perceive that παιδικῆς was in agreement with ἡδονῆς.

9. **ἐγκαύματα**] For the methods of encaustic painting see Pliny *Nat. Hist.* XXXV § 149.

24. **μὴ πλασθέντα μῦθον**] Cf. 21 A. We must not bind Plato down too strictly to this affirmation.

29. **ἀστρονομικώτατον**] Not in the popular sense merely, but in the sublimated Platonic manner.

we should be fairly well provided. So then, as Hermokrates said, as soon as ever I departed hence yesterday, I began to repeat the legend to our friends as I remembered it; and when I got home I recovered nearly the whole of it by thinking it over at night. How true is the saying that what we learn in childhood has a wonderful hold on the memory. Of what I heard yesterday I know not if I could call to mind the whole: but though it is so very long since I heard this tale, I should be surprised if a single point in it has escaped me. It was with much boyish delight that I listened at the time, and the old man was glad to instruct me, (for I asked a great many questions); so that it is indelibly fixed in my mind, like those encaustic pictures which cannot be effaced. And I narrated the story to the rest the first thing in the morning, that they might share my affluence of words. Now therefore, to return to the object of all our conversation, I am ready to speak, Sokrates, not only in general terms, but entering into details, as I heard it. The citizens and the city which you yesterday described to us as in a fable we will transfer to the sphere of reality and to our own country, and we will suppose that ancient Athens is your ideal commonwealth, and say that the citizens whom you imagined are those veritable forefathers of ours of whom the priest spoke. They will fit exactly, and there will be nothing discordant in saying that they were the men who lived in those days. And dividing the work between us we will all endeavour to render an appropriate fulfilment of your injunctions. So you must consider, Sokrates, whether this story of ours satisfies you, or whether we must look for another in its stead.

Sokrates. How could we change it for the better, Kritias? It is specially appropriate to this festival of the goddess, owing to its connexion with her; while the fact that it is no fictitious tale but a true history is surely a great point. How shall we find other such citizens if we relinquish these? It cannot be: so with Fortune's favour do you speak on, while I in requital for my discourse of yesterday have in my turn the privilege of listening in silence.

Kritias. Now consider, Sokrates, how we proposed to distribute your entertainment. We resolved that Timaeus, who is the best astronomer among us, and who has most of all made it

ἡμῶν καὶ περὶ φύσεως τοῦ παντὸς εἰδέναι μάλιστα ἔργον πεποιη-
μένον, πρῶτον λέγειν ἀρχόμενον ἀπὸ τῆς τοῦ κόσμου γενέσεως,
τελευτᾶν δὲ εἰς ἀνθρώπων φύσιν· ἐμὲ δὲ μετὰ τοῦτον, ὡς παρὰ
μὲν τούτου δεδεγμένον ἀνθρώπους τῷ λόγῳ γεγονότας, παρὰ σοῦ
5 δὲ πεπαιδευμένους διαφερόντως αὐτῶν τινάς, κατὰ δὴ τὸν Σόλωνος B
λόγον τε καὶ νόμον εἰσαγαγόντα αὐτοὺς ὡς εἰς δικαστὰς ἡμᾶς
ποιῆσαι πολίτας τῆς πόλεως τῆσδε ὡς ὄντας τοὺς τότε Ἀθηναίους,
οὓς ἐμήνυσεν ἀφανεῖς ὄντας ἡ τῶν ἱερῶν γραμμάτων φήμη, τὰ
λοιπὰ δὲ ὡς περὶ πολιτῶν καὶ Ἀθηναίων ὄντων ἤδη ποιεῖσθαι
10 τοὺς λόγους.

ΣΩ. Τελέως τε καὶ λαμπρῶς ἔοικα ἀνταπολήψεσθαι τὴν τῶν
λόγων ἑστίασιν. σὸν οὖν ἔργον λέγειν ἄν, ὦ Τίμαιε, εἴη τὸ μετὰ
τοῦτο, ὡς ἔοικεν, ἐπικαλέσαντα κατὰ νόμον θεούς.

V. ΤΙ. Ἀλλ᾽, ὦ Σώκρατες, τοῦτό γε δὴ πάντες, ὅσοι καὶ C
15 κατὰ βραχὺ σωφροσύνης μετέχουσιν, ἐπὶ παντὸς ὁρμῇ καὶ σμικροῦ
καὶ μεγάλου πράγματος θεὸν ἀεί που καλοῦσιν· ἡμᾶς δὲ τοὺς περὶ
τοῦ παντὸς λόγους ποιεῖσθαί πῃ μέλλοντας, ᾗ γέγονεν ἢ καὶ ἀγενές
ἐστιν, εἰ μὴ παντάπασι παραλλάττομεν, ἀνάγκη θεούς τε καὶ
θεὰς ἐπικαλουμένους εὔχεσθαι πάντα κατὰ νοῦν ἐκείνοις μὲν
20 μάλιστα, ἑπομένως δὲ ἡμῖν εἰπεῖν. καὶ τὰ μὲν περὶ θεῶν ταύτῃ
παρακεκλήσθω· τὸ δ᾽ ἡμέτερον παρακλητέον, ᾗ ῥᾷστ᾽ ἂν ὑμεῖς D
μὲν μάθοιτε, ἐγὼ δὲ ᾗ διανοοῦμαι μάλιστ᾽ ἂν περὶ τῶν προκει-
μένων ἐνδειξαίμην.

Ἔστιν οὖν δὴ κατ᾽ ἐμὴν δόξαν πρῶτον διαιρετέον τάδε· τί τὸ

3 μετὰ τοῦτον: τούτων Λ.   5 δὴ pro δὲ reposui suadente S.
6 ἡμᾶς: ὑμᾶς HZ.   12 εἴη omittit A. ante ὦ Τίμαιε ponit S.

3. φύσιν seems to have its old sense of 'generation'.

4. τῷ λόγῳ γεγονότας] cf. *Republic* 361 B τὸν δίκαιον παρ᾽ αὐτὸν ἱστῶμεν τῷ λόγῳ, ὥσπερ ἁπλοῦν καὶ γενναῖον, also 534 D παῖδας οὓς τῷ λόγῳ τρέφεις τε καὶ παιδεύεις, εἴ ποτε ἔργῳ τρέφοις.

5. κατὰ δή] Stallbaum's suggestion of reading δή for δέ appears to me to restore the true structure of the sentence.

6. λόγον τε καὶ νόμον] i.e. accepting the statement of Solon that they were Athenian citizens, we formally admit their claim to citizenship in the mode prescribed by his law.

27 C—29 D, c. v. Timaeus, after due invocation of heavenly aid, thus begins his exposition. The first step is to distinguish the eternally existing object of thought and reason from the continually fleeting object of opinion and sensation. To which class does the material universe belong, to Being or Becoming? To Becoming, because it is apprehensible by the senses. All that comes to be comes from some cause; so therefore does the universe. Also it must be a likeness of something. Now what is modelled on the eternal must needs be fair, but what is modelled on the created

D] ΤΙΜΑΙΟΣ. 85

his business to understand universal nature, should speak first, beginning with the origin of the universe, and should end with the birth of mankind: and that I should follow, receiving from him mankind brought to being in theory, and from you a portion of them exceptionally cultivated; and that in accordance with Solon's laws, no less than with his statement, I should introduce them before our tribunal and make them our fellow-citizens, as being the Athenians of bygone days, whom the declaration of the sacred writings has delivered from their oblivion; and thenceforward we shall speak as if their claim to Athenian citizenship were fairly established.

Sokrates. Ample and splendid indeed, it seems, will be the banquet of discourse which I am to receive in my turn. So it would seem to be your business to speak next, Timaeus, after you have duly invoked the gods.

V. *Timaeus.* Yes indeed, Sokrates, that is what all do who possess the slightest share of judgment; at the outset of every work, great or small, they always call upon a god: and seeing that we are going to enter on a discussion of the universe, how far it is created or perchance uncreate, unless we are altogether beside ourselves, we must needs invoke the gods and goddesses and pray above all that our discourse may be pleasing in their sight, next that it may be consistent with itself. Let it suffice then thus to have called upon the gods; but we must call upon ourselves likewise to conduct the discourse in such a way that you will most readily comprehend me, and I shall most fully carry out my intentions in expounding the subject that is before us.

First then in my judgment this distinction must be made.

---

is not fair. The universe is most fair, therefore it was modelled on the eternal. And in dealing with the eternal type and the created image, we must remember that the words we use of each must correspond to their several natures: those which deal with the eternally existent must be so far as possible sure and true and incontrovertible; while with those which treat of the likeness we must be content if they are likely. To this Sokrates assents.

The first eight chapters of Timaeus' discourse, extending to 40 D, deal with the universe as a whole; after which he proceeds to its several portions.

21. τὸ δ' ἡμέτερον παρακλητέον] i.e. after appealing to the gods for aid, we must appeal to ourselves to put forth all our energies: heaven helps those who help themselves.

22. ᾗ διανοοῦμαι] Stallbaum proposes to read ἅ.

86　　　　　　　　ΠΛΑΤΩΝΟΣ　　　　　　　[27 D—

ὂν ἀεί, γένεσιν δὲ οὐκ ἔχον, καὶ τί τὸ γιγνόμενον μὲν ἀεί, ὂν δὲ
οὐδέποτε. τὸ μὲν δὴ νοήσει μετὰ λόγου περιληπτόν, ἀεὶ κατὰ 28 A
ταὐτὰ ὄν, τὸ δ' αὖ δόξῃ μετ' αἰσθήσεως ἀλόγου δοξαστόν, γιγνόμενον καὶ ἀπολλύμενον, ὄντως δὲ οὐδέποτε ὄν. πᾶν δὲ αὖ τὸ
5 γιγνόμενον ὑπ' αἰτίου τινὸς ἐξ ἀνάγκης γίγνεσθαι· παντὶ γὰρ
ἀδύνατον χωρὶς αἰτίου γένεσιν σχεῖν. ὅτου μὲν οὖν ἂν ὁ δημιουργὸς πρὸς τὸ κατὰ ταὐτὰ ἔχον βλέπων ἀεί, τοιούτῳ τινὶ
προσχρώμενος παραδείγματι, τὴν ἰδέαν καὶ δύναμιν αὐτοῦ ἀπεργάζηται, καλὸν ἐξ ἀνάγκης οὕτως ἀποτελεῖσθαι πᾶν· οὗ δ' ἂν B
10 εἰς τὸ γεγονός, γεννητῷ παραδείγματι προσχρώμενος, οὐ καλόν.
ὁ δὴ πᾶς οὐρανός—ἢ κόσμος ἢ καὶ ἄλλο ὅ τί ποτε ὀνομαζόμενος
μάλιστ' ἂν δέχοιτο, τοῦθ' ἡμῖν ὠνομάσθω· σκεπτέον δ' οὖν περὶ
αὐτοῦ πρῶτον, ὅπερ ὑπόκειται περὶ παντὸς ἐν ἀρχῇ δεῖν σκοπεῖν,
πότερον ἦν ἀεί, γενέσεως ἀρχὴν ἔχων οὐδεμίαν, ἢ γέγονεν, ἀπ'
15 ἀρχῆς τινος ἀρξάμενος. γέγονεν· ὁρατὸς γὰρ ἁπτός τέ ἐστι καὶ
σῶμα ἔχων, πάντα δὲ τὰ τοιαῦτα αἰσθητά, τὰ δ' αἰσθητά, δόξῃ
περιληπτὰ μετ' αἰσθήσεως, γιγνόμενα καὶ γεννητὰ ἐφάνη. τῷ δ' C
αὖ γενομένῳ φαμὲν ὑπ' αἰτίου τινὸς ἀνάγκην εἶναι γενέσθαι. τὸν
μὲν οὖν ποιητὴν καὶ πατέρα τοῦδε τοῦ παντὸς εὑρεῖν τε ἔργον καὶ
20 εὑρόντα εἰς πάντας ἀδύνατον λέγειν· τόδε δ' οὖν πάλιν ἐπισκε-

2. τὸ μὲν δὴ νοήσει] νόησις and δόξα denote the faculties, λόγος and αἴσθησις the processes. The language of the present passage precisely agrees with the account given at the end of the fifth book of the *Republic*.

5. ὑπ' αἰτίου τινός] So *Philebus* 26 E ὅρα γὰρ εἴ σοι δοκεῖ ἀναγκαῖον εἶναι πάντα τὰ γιγνόμενα διά τινα αἰτίαν γίγνεσθαι. Only the ὄντως ὄν, the changeless and abiding, is a cause to itself and needs no αἰτία from without: the γιγνόμενον has no principle of causation in itself and must find the source of its becoming in some ulterior force.

8. τὴν ἰδέαν καὶ δύναμιν] Neither of these words has a technical meaning, though δύναμιν is here not so very far removed from the Aristotelian sense. ἰδέαν = the form and fashion of it, δύναμιν its function or quality.

11. ἢ καὶ ἄλλο] The universe is a living god: Plato therefore uses the customary reverent diffidence in naming the divine: cf. Aeschylus *Agamemnon* 160 Ζεύς, ὅστις ποτ' ἐστίν, εἰ τόδ' αὐτῷ φίλον κεκλημένῳ, τοῦτό νιν προσεννέπω.
The sentence becomes an anacoluthon owing to the parenthetical words ἢ καὶ ἄλλο...ὠνομάσθω.

14. πότερον ἦν ἀεί] i.e. whether it belongs to things eternal or to things temporal. It cannot be too carefully borne in mind that there is throughout no question whatsoever of the beginning of the universe in time. The creation in time is simply part of the figurative representation: it is κατ' ἐπίνοιαν only. In Plato's highly poetical and allegorical exposition a logical analysis is represented as a process taking place in time, and to reach his true meaning we must strip off the veil of imagery. He conceived the universe to be a certain evolution of absolute thought; and the several elements in this evolution he

28 C]   ΤΙΜΑΙΟΣ.   87

What is that which is eternally and has no becoming, and again what is that which comes to be but is never? The one is comprehensible by thought with the aid of reason, ever changeless; the other opinable by opinion with the aid of reasonless sensation, becoming and perishing, never truly existent. Now all that comes to be must needs be brought into being by some cause: for it is impossible for anything without a cause to attain to birth. Of whatsoever thing then the Artificer, looking ever to the changeless and using that as his model, works out the design and function, all that is so accomplished must needs be fair: but if he look to that which has come to be, using the created as his model, the work is not fair. Now as to the whole heaven or order of the universe—for whatsoever name is most acceptable to it, be it so named by us—we must first ask concerning it the question which lies at the outset of every inquiry, whether did it exist eternally, having no beginning of generation, or has it come into being, starting from some beginning? It has come into being: for it can be seen and felt and has body; and all such things are sensible, and sensible things, apprehensible by opinion with sensation, belong, as we saw, to becoming and creation. We say that what has come to be must be brought into being by some cause. Now the maker and father of this All it were a hard task to find, and having found him, it

represents as a succession of events. Such criticism then as that of Aristotle in *de caelo* I x is wholly irrelevant: he treats a metaphysical conception from a merely physical point of view. Stobaeus *ecl.* I 450 says Πυθαγόρας φησὶ γεννητὸν κατ' ἐπίνοιαν τὸν κόσμον, οὐ κατὰ χρόνον: and presently he ascribes the same view to Herakleitos. Whether these philosophers really held that opinion there seems no means of determining: but since in the immediate context Stobaeus assigns to Pythagoras some distinctively Platonic notions, we may pretty fairly infer that the creation of the world κατ' ἐπίνοιαν was one of the many Platonic doctrines which were foisted by the later doxographers upon Pythagoras, whose school served them as a πανδοκεῖον for all views they had a difficulty in otherwise bestowing. As to the past tense ἦν ἀεί, Proklos very justly observes εἰ δὲ τὸ ἦν οὔ φησι προελθὼν οἰκεῖον εἶναι τοῖς αἰωνίοις, οὐ δεῖ ταράττεσθαι· πρὸ γὰρ τῆς διαρθρώσεως ἕπεται τῇ συνηθείᾳ. The said διάρθρωσις is at 37 E—38 n.

19. εὑρεῖν τε ἔργον] Proklos says this is a warning against superficially seeking our ἀρχή in the physical forces which served the old φυσιολόγοι. It may be observed also that, were we to accept the δημιουργός literally, Plato would surely not have used such language in referring to so simple and familiar a conception as a personal creator of the universe; but if the δημιουργός is but a mythical representative of a metaphysical ἀρχή, the justice of the remark is evident.

88   ΠΛΑΤΩΝΟΣ   [28 C—

πτέον περὶ αὐτοῦ, πρὸς πότερον τῶν παραδειγμάτων ὁ τεκταινόμενος αὐτὸν ἀπειργάζετο, πότερον πρὸς τὸ κατὰ ταὐτὰ καὶ 29 A
ὡσαύτως ἔχον ἢ πρὸς τὸ γεγονός. εἰ μὲν δὴ καλός ἐστιν ὅδε ὁ
κόσμος ὅ τε δημιουργὸς ἀγαθός, δῆλον ὡς πρὸς τὸ ἀίδιον ἔβλεπεν·
5 εἰ δὲ ὃ μηδ᾽ εἰπεῖν τινι θέμις, πρὸς τὸ γεγονός. παντὶ δὴ σαφὲς
ὅτι πρὸς τὸ ἀίδιον· ὁ μὲν γὰρ κάλλιστος τῶν γεγονότων, ὁ δ᾽
ἄριστος τῶν αἰτίων. οὕτω δὴ γεγενημένος πρὸς τὸ λόγῳ καὶ
φρονήσει περιληπτὸν καὶ κατὰ ταὐτὰ ἔχον δεδημιούργηται· τούτων δὲ ὑπαρχόντων αὖ πᾶσα ἀνάγκη τόνδε τὸν κόσμον εἰκόνα B
10 τινὸς εἶναι. μέγιστον δὴ παντὸς ἄρξασθαι κατὰ φύσιν ἀρχήν.
ὧδε οὖν περί τε εἰκόνος καὶ περὶ τοῦ παραδείγματος αὐτῆς διοριστέον, ὡς ἄρα τοὺς λόγους, ὧνπέρ εἰσιν ἐξηγηταί, τούτων αὐτῶν
καὶ ξυγγενεῖς ὄντας. τοῦ μὲν οὖν μονίμου καὶ βεβαίου καὶ μετὰ
νοῦ καταφανοῦς μονίμους καὶ ἀμεταπτώτους, καθ᾽ ὅσον [οἷόν] τε
15 ἀνελέγκτοις προσήκει λόγοις εἶναι καὶ ἀκινήτοις, τούτου δεῖ μηδὲν
ἐλλείπειν· τοὺς δὲ τοῦ πρὸς μὲν ἐκεῖνο ἀπεικασθέντος, ὄντος δὲ C
εἰκόνος, εἰκότας ἀνὰ λόγον τε ἐκείνων ὄντας· ὅ τί περ πρὸς γένεσιν
οὐσία, τοῦτο πρὸς πίστιν ἀλήθεια. ἐὰν οὖν, ὦ Σώκρατες, πολλὰ

3 πρὸς τὸ γεγονός: τὸ omittit A.   8 καὶ ante κατὰ omittit A.   14 καθ᾽
ὅσον οἷόν τε AZ.   καθ᾽ ὅσον οἷόν τε καὶ H.   καὶ καθ᾽ ὅσον οἷόν τε S.   inclusi οἷον.
15 ἀνελέγκτοις: ἀνελέγκτους et mox λόγους et ἀκινήτους S.   δεῖ: δέ S.

1. πρὸς πότερον τῶν παραδειγμάτων] It may reasonably be asked, how could the creator look πρὸς τὸ γεγονός, since at that stage there was no γεγονός to look to? Plato's meaning, I take it, is this: the γεγονός at which the Artificer would look can of course only be the γεγονός that he was about to produce. Now if he looked at this, instead of fixing his eyes upon any eternal type, that would mean that he created arbitrarily and at random a universe that simply fulfilled his fancy at the moment and did not express any underlying thought: the universe would in fact be a collection of incoherent phenomena, a mere plaything of the creator. But, says Plato, this is not so: material nature is but the visible counterpart of a spiritual reality; all things have their meaning. Creation is no merely arbitrary exercise of will on the part of the creator; it is the working out of an inevitable law.

6. κάλλιστος τῶν γεγονότων] i.e. there is nothing in the universe which, taken by itself, is so fair as the universe as a whole.

9. εἰκόνα τινὸς εἶναι] This leads the way to the question raised in 30 C. Seeing that the creator looked to a pattern in framing the universe, it follows that the universe is a copy of something; and we have to inquire what that is whereof it is the copy. Cicero renders these words 'simulacrum aeternum esse alicuius aeterni'; whence it would appear that his ms. gave εἰκόνα ἀίδιόν τινος ἀιδίου, which it has been proposed to restore. This however it were rash to do against all existing mss. and Proklos. The phrase εἰκόνα ἀίδιον might perhaps be defended on the same principle as

were impossible to declare him to all men. However we must again inquire concerning him, after which of the models did the framer of it fashion the universe, after the changeless and abiding, or after that which has come into being? If now this universe is fair and its Artificer good, it is plain that he looked to the eternal; but if—nay it may not even be uttered without impiety,—then it was to that which has come into being. Now it is manifest to every one that he looked to the eternal: for the universe is fairest of all things that have come to be, and he is the most excellent of causes. And having come on this wise into being it has been created in the image of that which is comprehensible by reason and wisdom and changes never. Granting this, it must needs be that this universe is a likeness of something. Now it is all-important to make our beginning according to nature: and this affirmation must be laid down with regard to a likeness and its model, that the words must be akin to the subjects of which they are the interpreters: therefore of that which is abiding and sure and discoverable by the aid of reason the words too must be abiding and unchanging, and so far as it lies in words to be incontrovertible and immovable, they must in no wise fall short of this; but those which deal with that which is made in the image of the former and which is a likeness must be likely and duly corresponding with their subject: as being is to becoming, so is truth to belief. If then, Sokrates, after so many men have said divers things concerning

αἰώνιον εἰκόνα in 37 D: but there the expression has a pointedness which is lacking here. ἀίδιον properly means exempt from time, and cannot strictly be applied to the phenomenal world, though its duration be everlasting.

13. τοῦ μὲν οὖν μονίμου] Some corruption has clearly found its way into this sentence. It seems to me that the simplest remedy is to reject οἷον, which I think may have arisen from a duplication of ὅσον. By this omission the sentence becomes perfectly grammatical. Stallbaum, reading καὶ before καθ' ὅσον, alters ἀνελέγκτοις, λόγοις, ἀκινήτοις, to the accusative, and writes δὲ for δεῖ. This method does indeed produce a sentence that can be construed; but it involves larger alterations of the text, and the position of the word λόγους seems extremely unsatisfactory. I cannot therefore concede his claim to have restored Plato's words. According to my version of the sentence εἶναι must be supplied with μονίμους καὶ ἀμεταπτώτους.

17. ἀνὰ λόγον] i.e. they stand in the same relation to the λόγοι of the παράδειγμα as the εἰκὼν to the παράδειγμα: as becoming is to being so is probability to truth. We have here precisely the analogy of *Republic* 511 E.

ΠΛΑΤΩΝΟΣ [29 C—

πολλῶν εἰπόντων περὶ θεῶν καὶ τῆς τοῦ παντὸς γενέσεως, μὴ
δυνατοὶ γιγνώμεθα πάντη πάντως αὐτοὺς ἑαυτοῖς ὁμολογουμένους
λόγους καὶ ἀπηκριβωμένους ἀποδοῦναι, μὴ θαυμάσῃ τις· ἀλλ᾽
ἐὰν ἄρα μηδενὸς ἧττον παρεχώμεθα εἰκότας, ἀγαπᾶν χρή, μεμνη-
5 μένον, ὡς ὁ λέγων ἐγὼ ὑμεῖς τε οἱ κριταὶ φύσιν ἀνθρωπίνην D
ἔχομεν, ὥστε περὶ τούτων τὸν εἰκότα μῦθον ἀποδεχομένους πρέπει
τούτου μηδὲν ἔτι πέρα ζητεῖν.

ΣΩ. Ἄριστα, ὦ Τίμαιε, παντάπασί τε ὡς κελεύεις ἀποδεκτέον·
τὸ μὲν οὖν προοίμιον θαυμασίως ἀπεδεξάμεθά σου, τὸν δὲ δὴ νόμον
10 ἡμῖν ἐφεξῆς πέραινε.

VI. ΤΙ. Λέγωμεν δὴ δι᾽ ἥν τινα αἰτίαν γένεσιν καὶ τὸ πᾶν
τόδε ὁ ξυνιστὰς ξυνέστησεν. ἀγαθὸς ἦν, ἀγαθῷ δὲ οὐδεὶς περὶ E
οὐδενὸς οὐδέποτε ἐγγίγνεται φθόνος· τούτου δ᾽ ἐκτὸς ὢν πάντα ὅ
τι μάλιστα γενέσθαι ἐβουλήθη παραπλήσια ἑαυτῷ. ταύτην δὴ
15 γενέσεως καὶ κόσμου μάλιστ᾽ ἄν τις ἀρχὴν κυριωτάτην παρ᾽
ἀνδρῶν φρονίμων ἀποδεχόμενος ὀρθότατα ἀποδέχοιτ᾽ ἄν. βου- 30 A
ληθεὶς γὰρ ὁ θεὸς ἀγαθὰ μὲν πάντα, φλαῦρον δὲ μηδὲν εἶναι

1 εἰπόντων omittit A.   3 θαυμάσῃ τις: θαυμάσης HSZ.   4 μεμνημένον:
μεμνημένους H.   9 νόμον: λόγον Z.   14 ταύτην δή: δέ AHZ.

2. αὐτοὺς ἑαυτοῖς ὁμολογουμένους] The modesty of Timaeus leads him rather unduly to depreciate his physical theories: it would be hard, I think, to detect any inconsistencies in them, though there may be points which are not altogether ἀπηκριβωμένα. But Plato insists with much urgent iteration upon the impossibility of attaining certainty in any account of the objects of sense. They have no veritable existence, therefore no positive truth or secure knowledge concerning them is attainable. It is his desire to keep this constantly before the reader's mind that induces Plato to refer so frequently to the εἰκὼς μῦθος. The difference between the εἰκὼς μῦθος and ὁ δι᾽ ἀκριβείας ἀληθὴς λόγος is instructively displayed when each is invoked to decide the question of the unity of the universe. In 31 A the latter authoritatively declares the κόσμος to be one only, and gives the metaphysical reason: in 55 D all the former ventures to say is τὸ μὲν οὖν δὴ παρ᾽ ἡμῶν ἕνα αὐτὸν κατὰ τὸν εἰκότα λόγον πεφυκότα μηνύει, ἄλλος δὲ εἰς ἄλλα πῃ βλέψας ἕτερα δοξάσει.

9. τὸ μὲν οὖν προοίμιον] The metaphor is from harp-playing: προοίμιον is the prelude, νόμος the main body of the composition: cf. Republic 531 D ἦ οὐκ ἴσμεν ὅτι πάντα ταῦτα προοίμιά ἐστιν αὐτοῦ τοῦ νόμου ὃν δεῖ μαθεῖν.

29 D—31 B, c. vi. What then was the cause of creation? The creator was good and desired that all things should be so far as possible good like himself. So he took the world of matter, a chaos of disturbance and confusion, and brought it to order and gave it life and intelligence. And the type after which he ordered it was the eternal universal animal in the world of ideas; that, even as this comprehends within it all ideal animals, so the visible universe should include in it all animals that are material. And as the ideal animal is of its very essence one and alone, so he created not two or many

the gods and the generation of the universe, we should not prove able to render an account everywhere and in all respects consistent and accurate, let no one be surprised; but if we can produce one as probable as any other, we must be content, remembering that I who speak and you my judges are but men: so that on these subjects we should be satisfied with the probable story and seek nothing further.

Sokrates. Quite right, Timaeus; we must accept it exactly as you say. Your prelude is exceedingly welcome to us, so please proceed with the strain itself.

VI. *Timaeus.* Let us declare then for what cause nature and this All was framed by him that framed it. He was good, and in none that is good can there arise jealousy of aught at any time. So being far aloof from this, he desired that all things should be as like unto himself as possible. This is that most sovereign cause of nature and the universe which we shall most surely be right in accepting from men of understanding. For God desiring that all things should be good, and that, so far as

systems of material nature, but one universe only-begotten to exist for ever.

12. ἀγαθὸς ἦν] Consistently with all his previous teaching Plato here makes the αὐτὸ ἀγαθὸν the source and cause of all existence; this in the allegory is symbolized by a benevolent creator bringing order out of a preexisting chaos. Of course Plato's words are not to be interpreted with a crude literalness. The cause of the existence of visible nature is the supreme law by virtue of which the one absolute intelligence differentiates itself into the plurality of material objects: that is the reason why the world of matter exists at all: then, since intelligence must needs work on a fixed plan and with the best end in view, the universe thus evolved was made as perfect as anything material can be. It is necessary to insist on this distinction, although, when we remember that for Plato existence and goodness are one and the same, the distinction ultimately vanishes: all things exist just so far as they are good, and no more. Thus the conception of the αὐτὸ ἀγαθὸν as the supreme cause, which is affirmed in the *Republic* but not expounded, is here definitely set forth, though still invested with the form of a vividly poetical allegory.

13. οὐδέποτε ἐγγίγνεται φθόνος] The vulgar notion τὸ θεῖον φθονερὸν was extremely distasteful to Plato: cf. *Phaedrus* 247 A φθόνος γὰρ ἔξω θείου χοροῦ ἵσταται. So Aristotle *metaph.* A ii 983ᵃ 2 ἀλλ' οὔτε τὸ θεῖον φθονερὸν ἐνδέχεται εἶναι, ἀλλὰ καὶ κατὰ τὴν παροιμίαν πολλὰ ψεύδονται ἀοιδοί.

15. παρ' ἀνδρῶν φρονίμων] Who are the φρόνιμοι ἄνδρες? Probably some Pythagoreans. I have not traced the sentiment to any preplatonic thinker; but it is quite consonant with Pythagorean views: cf. Stobaeus *ecl.* ii 64 Σωκράτης Πλάτων ταὐτὰ τῷ Πυθαγόρᾳ· τέλος ὁμοίωσιν θεοῦ [? θεῷ]. Stallbaum cites the apophthegm attributed by Diogenes Laertius to Thales, κάλλιστον κόσμος, ποίημα γὰρ θεοῦ: but this does not seem specially apposite.

κατὰ δύναμιν, οὕτω δὴ πᾶν ὅσον ἦν ὁρατὸν παραλαβὼν οὐχ ἡσυχίαν ἄγον ἀλλὰ κινούμενον πλημμελῶς καὶ ἀτάκτως, εἰς τάξιν αὐτὸ ἤγαγεν ἐκ τῆς ἀταξίας, ἡγησάμενος ἐκεῖνο τούτου πάντως ἄμεινον. θέμις δὲ οὔτ' ἦν οὔτ' ἔστι τῷ ἀρίστῳ δρᾶν ἄλλο πλὴν τὸ κάλλιστον· λογισάμενος οὖν εὕρισκεν ἐκ τῶν κατὰ φύσιν ὁρατῶν οὐδὲν ἀνόητον τοῦ νοῦν ἔχοντος ὅλον ὅλου κάλλιον ἔσεσθαί ποτε B ἔργον, νοῦν δ' αὖ χωρὶς ψυχῆς ἀδύνατον παραγενέσθαι τῳ. διὰ δὴ τὸν λογισμὸν τόνδε νοῦν μὲν ἐν ψυχῇ, ψυχὴν δὲ ἐν σώματι ξυνιστὰς τὸ πᾶν ξυνετεκταίνετο, ὅπως ὅ τι κάλλιστον εἴη κατὰ φύσιν ἄριστόν τε ἔργον ἀπειργασμένος. οὕτως οὖν δὴ κατὰ λόγον τὸν εἰκότα δεῖ λέγειν, τόνδε τὸν κόσμον ζῷον ἔμψυχον ἔννουν τε τῇ ἀληθείᾳ διὰ τὴν τοῦ θεοῦ γενέσθαι πρόνοιαν.

Τούτου δ' ὑπάρχοντος αὖ τὰ τούτοις ἐφεξῆς ἡμῖν λεκτέον, τίνι C τῶν ζῴων αὐτὸν εἰς ὁμοιότητα ὁ ξυνιστὰς ξυνέστησε. τῶν μὲν

1. **κατὰ δύναμιν**] To make the material universe absolutely perfect was impossible, since evil, whatever it may be, is more or less inherent in the very nature of matter and can never be totally abolished: cf. *Theaetetus* 176 A ἀλλ' οὔτ' ἀπολέσθαι τὰ κακὰ δυνατόν, ὦ Θεόδωρε· ὑπεναντίον γάρ τι τῷ ἀγαθῷ ἀεὶ εἶναι ἀνάγκη· οὔτ' ἐν θεοῖς αὐτὰ ἱδρῦσθαι, τὴν δὲ θνητὴν φύσιν καὶ τόνδε τὸν τόπον περιπολεῖ ἐξ ἀνάγκης. See also *Politicus* 273 B, C. Evil is in fact, just as much as perception in space and time, an inevitable accompaniment of the differentiation of absolute intelligence into the multiplicity of finite intelligences. It is much to be regretted that Plato has not left us a dialogue dealing with the nature of evil and the cause of its necessary inherence in matter: as it is, we can only conjecture the line he would have taken.

πᾶν ὅσον ἦν ὁρατὸν παραλαβών] Martin finds in this passage a clear indication that chaos actually as a fact existed before the ordering of the κόσμος. But this is due to a misunderstanding of Plato's figurative exposition. Proklos says with perfect correctness κατ' ἐπίνοιαν θεωρεῖται πρὸ τῆς κοσμοποιίας. The statement that the δημιουργὸς found chaotic matter ready to his hand is one which πολὺ μετέχει τοῦ προστυχόντος. We learn in 34 C that soul is prior to matter, which can only mean that matter is evolved out of soul. What Plato expressed as a process taking place in time must be regarded as a logical conception only. When he speaks of matter as chaotic, he does not mean that there was a time when matter existed uninformed by mind and that afterwards νοῦς ἐλθὼν διεκόσμησεν: he means that matter, as conceived in itself, is without any formative principle of order: it is only when we think of it as the outcome of mind that it can have any system or meaning. Compare Appuleius *de dogm. Plat.* I viii 198 et hunc quidem mundum nunc sine initio esse dicit, alias originem habere natumque esse: nullum autem eius exordium atque initium esse ideo quod semper fuerit; nativum vero videri, quod ex his rebus substantia eius et natura constet, quae nascendi sortitae sunt qualitates.

οὐχ ἡσυχίαν ἄγον] The very fact that matter is described as in motion, though the motion be chaotic, is sufficient to prove conclusively that it is a phase of ψυχή, since for Plato ψυχή is the sole ἀρχὴ κινήσεως. **κινούμενον πλημμελῶς καὶ**

c]   ΤΙΜΑΙΟΣ.   93

this might be, there should be nought evil, having received all that is visible not in a state of rest, but moving without harmony or measure, brought it from its disorder into order, thinking that this was in all ways better than the other. Now it neither has been nor is permitted to the most perfect to do aught but what is most fair. Therefore he took thought and perceived that of all things which are by nature visible, no work that is without reason will ever be fairer than that which has reason, setting whole against whole, and that without soul reason cannot dwell in anything. Because then he argued thus, in forming the universe he created reason in soul and soul in body, that he might be the maker of a work that was by nature most fair and perfect. In this way then we ought to affirm according to the probable account that this universe is a living creature in very truth possessing soul and reason by the providence of God.

Having attained thus far, we must go on to tell what follows: after the similitude of what animal its framer fashioned it. To

ἀτάκτως describes the condition of matter as it would be were it not derived from an intelligent ἀρχή. Aristotle refers to this passage *de caelo* III ii 300^b 17, comparing Plato's chaotic motion to that attributed by Demokritos to his atoms. And this philosopheme of Demokritos is doubtless what Plato had in view: such a motion as the former conceives, not proceeding from intelligence, could not produce a κόσμος. It is impossible that Plato could have imagined that this disorderly motion ever actually existed: since all motion is of ψυχή, and ψυχή is intelligent.

3. ἡγησάμενος ἐκεῖνο τούτου πάντως ἄμεινον] sc. τάξιν ἀταξίας. Throughout this passage Plato is careful to remedy the defect he found in Anaxagoras. 'All was chaos', said Anaxagoras; 'then Mind came and brought it into order', 'because', Plato adds, 'Mind thought order better than disorder'. Thus the final cause is supplied which was wanting in the elder philosopher, and we now see Mind working ἐπὶ τὸ βέλτιστον.

7. νοῦν δ' αὖ χωρὶς ψυχῆς] Compare *Philebus* 30 C σοφία μὴν καὶ νοῦς ἄνευ ψυχῆς οὐκ ἄν ποτε γενοίσθην. Stallbaum, following the misty light of neoplatonic inspiration, says of ψυχή, 'media est inter corpora atque mentem'. But in truth νοῦς is simply the activity of ψυχή according to her own proper nature: it is soul undiluted, as it were; apprehending not through any bodily organs, but by the exercise of pure thought: it is not something distinct from ψυχή, but a particular function of ψυχή.

8. ψυχὴν δὲ ἐν σώματι] Plato is here employing popular language: accurately speaking, God constructed body within soul, as we see in 36 E. Plutarch *quaest. platon.* IV wrongly infers from this passage that, as νοῦς can only exist in ψυχή, so ψυχή can only exist in σῶμα. This of course is not so: the converse would be more correct, that σῶμα can only exist in ψυχή. The phrase νοῦν ἐν ψυχῇ is also an exoteric expression; for Plato is not here concerned to use technical language.

οὖν ἐν μέρους εἴδει πεφυκότων μηδενὶ καταξιώσωμεν· ἀτελεῖ γὰρ
ἐοικὸς οὐδέν ποτ' ἂν γένοιτο καλόν· οὗ δ' ἔστι τἆλλα ζῷα καθ' ἓν
καὶ κατὰ γένη μόρια, τούτῳ πάντων ὁμοιότατον αὐτὸν εἶναι τιθῶ-
μεν. τὰ γὰρ δὴ νοητὰ ζῷα πάντα ἐκεῖνο ἐν ἑαυτῷ περιλαβὸν
5 ἔχει, καθάπερ ὅδε ὁ κόσμος ἡμᾶς ὅσα τε ἄλλα θρέμματα ξυνέ-
στηκεν ὁρατά. τῷ γὰρ τῶν νοουμένων καλλίστῳ καὶ κατὰ πάντα D
τελέῳ μάλιστα αὐτὸν ὁ θεὸς ὁμοιῶσαι βουληθεὶς ζῷον ἓν ὁρατόν,
πάνθ' ὅσα αὐτοῦ κατὰ φύσιν ξυγγενῆ ζῷα ἐντὸς ἔχον ἑαυτοῦ,
ξυνέστησε. πότερον οὖν ὀρθῶς ἕνα οὐρανὸν προσειρήκαμεν, ἢ 31
10 πολλοὺς καὶ ἀπείρους λέγειν ἦν ὀρθότερον; ἕνα, εἴπερ κατὰ τὸ
παράδειγμα δεδημιουργημένος ἔσται. τὸ γὰρ περιέχον πάντα,
ὁπόσα νοητὰ ζῷα, μεθ' ἑτέρου δεύτερον οὐκ ἄν ποτ' εἴη· πάλιν
γὰρ ἂν ἕτερον εἶναι τὸ περὶ ἐκείνω δέοι ζῷον, οὗ μέρος ἂν εἴτην
ἐκείνω, καὶ οὐκ ἂν ἔτι ἐκείνοιν ἀλλ' ἐκείνῳ τῷ περιέχοντι τόδ' ἂν
15 ἀφωμοιωμένον λέγοιτο ὀρθότερον. ἵνα οὖν τόδε κατὰ τὴν μόνωσιν B

13 ἐκείνω: ἐκείνῳ A.

1. ἐν μέρους εἴδει] Stallbaum cites *Cratylus* 394 D ἐν τέρατος εἴδει, *Phaedo* 91 D ἐν ἁρμονίας εἴδει, *Republic* 389 B ὡς ἐν φαρμάκου εἴδει, *Hippias maior* 297 B ἐν πατρὸς τινος ἰδέᾳ.

2. καθ' ἓν καὶ κατὰ γένη] The neoplatonic commentators are at variance whether ἕν or γένη is to be regarded as the more universal expression. I think Plato's usage is pretty conclusive in favour of taking ἕν as the more special. ἕν will thus signify the separate species, such as horse or tree; while γένη, I am disposed to think, refers to the four classes mentioned in 40 A, corresponding to the four elements to which they severally belong. In any case the αὐτὸ ὅ ἐστι ζῷον comprehends in it all the scale of inferior ideas from the four highest to the lowest species.

6. τῶν νοουμένων καλλίστῳ] As we saw that the material universe is fairer than any of its parts, so the universal idea is fairer than any of the ideas which it comprehends: cf. 39 E ἵνα τόδε ὡς ὁμοιότατον ᾖ τῷ τελέῳ καὶ νοητῷ ζῴῳ.

8. αὐτοῦ κατὰ φύσιν ξυγγενῆ] For the construction αὐτοῦ ξυγγενῆ compare 29 B, 77 A, *Philebus* 11 B.

10. ἕνα, εἴπερ κατὰ τὸ παράδειγμα] The objection might occur that every other idea, just as much as the αὐτὸ ζῷον, is necessarily one and unique. That is true; but the difference lies in this: the αὐτὸ ζῷον is ἕν as being πᾶν; there cannot be a second αὐτὸ ζῷον, else it would not contain within it all νοητὰ ζῷα. Therefore while the other particulars may be satisfactory μιμήματα of their ideas, although they are many, the ὁρατὸς κόσμος must be one only, else it would not copy the νοητὸς κόσμος in the essential attribute of all-comprehensiveness.

It is noticeable that in this case we have an idea with only one particular corresponding. This would have been impossible in the earlier phase of Plato's metaphysic. He says in *Republic* 596 A εἶδος γάρ πού τι ἓν ἕκαστον εἰώθαμεν τίθεσθαι περὶ ἕκαστα τὰ πολλά, οἷς ταὐτὸν ὄνομα ἐπιφέρομεν. But now that the ideas are restricted to ὁπόσα φύσει, now that they are naturally determined and their existence is no longer inferred from a group of particulars, there is for Plato no reason why a natural genus should not exist containing but a single particular.

none of these which naturally belong to the class of the partial must we deign to liken it: for nothing that is like to the imperfect could ever become fair; but that of which the other animals severally and in their kinds are portions, to this above all things we must declare that the universe is most like. For that comprehends and contains in itself all ideal animals, even as this universe contains us and all other creatures that have been formed to be visible. For since God desired to liken it most nearly to what is fairest of the objects of reason and in all respects perfect, he made it a single visible living being, containing within itself all animals that are by nature akin to it. Are we right then in affirming the universe to be one, or had it been more true to speak of a great and boundless number? One it must be, if it is to be created according to its pattern. For that which comprehends all ideal animals that are could never be a second in company of another: for there must again exist another animal comprehending them, whereof the two would be parts, and no longer to them but to that which comprehended them should we more truly affirm the universe to have been likened. To the end then that in its solitude this universe might be like

But what is this αὐτὸ ζῷον? Surely not an essence existing outside the κόσμος, else we should have something over and above the All, and the All would not be all. It is then (to keep up Plato's metaphor) the idea of the κόσμος existing in the mind of the δημιουργός: or, translating poetry into prose, it is the primal ἕν which finds its realisation and ultimate unity through its manifestation as πολλά: there will be more to say about this on 92 C. Proklos has for once expressed the truth with some aptness: τὸ μὲν γὰρ [παράδειγμα] ἦν νοητῶς πᾶν, αὐτὸς δὲ [ὁ δημιουργός] νοερῶς πᾶν, ὁ δὲ κόσμος αἰσθητῶς πᾶν: i.e. the παράδειγμα is universal thought regarded as the supreme intelligible, the δημιουργός represents the same regarded as the supreme intelligence, and the κόσμος is the same in material manifestation. See introduction § 38.

Aristotle deduces the unity of the οὐρανός thus: metaph. Λ viii 1074ª 31 ὅτι δὲ εἷς οὐρανὸς φανερόν. εἰ γὰρ πλείους οὐρανοὶ ὥσπερ ἄνθρωποι, ἔσται εἴδει μία ἡ περὶ ἕκαστον ἀρχή, ἀριθμῷ δέ γε πολλαί. ἀλλ' ὅσα ἀριθμῷ πολλά, ὕλην ἔχει...τὸ δὲ τί ἦν εἶναι οὐκ ἔχει ὕλην τὸ πρῶτον. ἐντελέχεια γάρ. ἓν ἄρα καὶ λόγῳ καὶ ἀριθμῷ τὸ πρῶτον κινοῦν ἀκίνητον ὄν. καὶ τὸ κινούμενον ἄρα ἀεὶ καὶ συνεχῶς ἓν μόνον· εἷς ἄρα οὐρανὸς μόνος.

12. πάλιν γὰρ ἄν] Compare *Republic* 597 C εἰ δύο μόνας ποιήσειε, πάλιν ἂν μία καταφανείη, ἧς ἐκεῖναι ἂν αὖ ἀμφότεραι τὸ εἶδος ἔχοιεν, καὶ εἴη ἂν ὁ ἔστι κλίνη ἐκείνη, ἀλλ' οὐχ αἱ δύο.

13. μέρος] i.e. a subdivision, a lower generalisation.

15. κατὰ τὴν μόνωσιν] i.e. respect of its isolation, of being the only one of its kind. This would not have called for explanation, but for Stallbaum's strange remark 'mox κατὰ τὴν μόνωσιν i. q. μόνον'.

96   ΠΛΑΤΩΝΟΣ   [31 B—

ὅμοιον ᾖ τῷ παντελεῖ ζῴῳ, διὰ ταῦτα οὔτε δύο οὔτ' ἀπείρους
ἐποίησεν ὁ ποιῶν κόσμους, ἀλλ' εἷς ὅδε μονογενὴς οὐρανὸς γεγονὼς
ἔστι τε καὶ ἔτ' ἔσται.

VII. Σωματοειδὲς δὲ δὴ καὶ ὁρατὸν ἁπτόν τε δεῖ τὸ γενόμενον
5 εἶναι· χωρισθὲν δὲ πυρὸς οὐδὲν ἄν ποτε ὁρατὸν γένοιτο, οὐδὲ
ἁπτὸν ἄνευ τινὸς στερεοῦ, στερεὸν δὲ οὐκ ἄνευ γῆς· ὅθεν ἐκ
πυρὸς καὶ γῆς τὸ τοῦ παντὸς ἀρχόμενος ξυνιστάναι σῶμα ὁ θεὸς
ἐποίει. δύο δὲ μόνω καλῶς ξυνίστασθαι τρίτου χωρὶς οὐ δυνατόν·
δεσμὸν γὰρ ἐν μέσῳ δεῖ τινὰ ἀμφοῖν ξυναγωγὸν γίγνεσθαι· δεσ- C
10 μῶν δὲ κάλλιστος ὃς ἂν αὐτόν τε καὶ τὰ ξυνδούμενα ὅ τι μάλιστα ἓν ποιῇ. τοῦτο δὲ πέφυκεν ἀναλογία κάλλιστα ἀποτελεῖν·
ὁπόταν γὰρ ἀριθμῶν τριῶν εἴτε ὄγκων εἴτε δυνάμεων ὡντινωνοῦν
ᾖ τὸ μέσον, ὅ τί περ τὸ πρῶτον πρὸς αὐτό, τοῦτο αὐτὸ πρὸς τὸ 32 A
ἔσχατον, καὶ πάλιν αὖθις, ὅ τι τὸ ἔσχατον πρὸς τὸ μέσον, τὸ μέσον
15 πρὸς τὸ πρῶτον, τότε τὸ μέσον μὲν πρῶτον καὶ ἔσχατον γιγνόμενον,
τὸ δ' ἔσχατον καὶ τὸ πρῶτον αὖ μέσα ἀμφότερα, πάνθ' οὕτως ἐξ
ἀνάγκης τὰ αὐτὰ εἶναι ξυμβήσεται, τὰ αὐτὰ δὲ γενόμενα ἀλλήλοις ἓν

10 τε omittunt SZ.   14 τοῦτο ante alterum τὸ μέσον habent SZ.

1. **οὔτε δύο οὔτ' ἀπείρους**] This is directed against the theory of Demokritos, that there were an infinite number of κόσμοι: a theory which is of course a perfectly just inference from Demokritean principles.

2. **εἷς ὅδε μονογενὴς οὐρανός**] Compare 92 C εἷς οὐρανὸς ὅδε μονογενὴς ὤν. The words that follow must be understood as an affirmation of the everlasting continuance of the κόσμος, and γεγονώς, as I have already done my best to show, does not imply its beginning in time.

31 B—34 A, c. vii. Now the world must be visible and tangible, therefore God constructed it of fire and earth. But two things cannot be harmoniously blended without a third as a mean: therefore he set proportionals between them. Between plane surfaces one proportional suffices; but seeing that the bodies of fire and earth are solid, two proportionals were required. Therefore he created air and water, in such wise that as fire is to air, so is air to water, and so is water to earth: thus the four became one harmony. And of these substances God used the whole in constructing the universe, so that nothing was left outside it which might be a source of danger to it. And he gave it a spherical form, because that shape comprehends within it all other shapes whatsoever: and he gave it the motion therewith conformable, namely rotation on its own axis. And he bestowed on it neither eyes nor ears nor hands nor feet nor any organs of respiration or nutrition; for as nothing existed outside it, nor had it requirement of aught, it was sufficient to itself and needed none of these things.

4. **ὁρατὸν ἁπτόν τε**] Visibility and tangibility are the two most conspicuous characteristics of matter: therefore the fundamental constituents of the universe are fire and earth. This agrees with the view of Parmenides: cf. Aristotle *physica* I v 188ᵃ 20 καὶ γὰρ Παρμενίδης θερμὸν καὶ ψυχρὸν ἀρχὰς ποιεῖ, ταῦτα δὲ προσαγορεύει πῦρ καὶ γῆν: and Parmenides 112 foll. (Karsten): see too Aristotle *de gen. et corr.* II ix 336ᵃ 3. The four elements

32 A]   ΤΙΜΑΙΟΣ.   97

the all-perfect animal, the maker made neither two universes nor an infinite number; but as it has come into being, this universe one and only-begotten, so it is and shall be for ever.

VII. Now that which came into being must be material and such as can be seen and touched. Apart from fire nothing could ever become visible, nor without something solid could it be tangible, and solid cannot exist without earth: therefore did God when he set about to frame the body of the universe form it of fire and of earth. But it is not possible for two things to be fairly united without a third; for they need a bond between them which shall join them both. The best of bonds is that which makes itself and those which it binds as complete a unity as possible; and the nature of proportion is to accomplish this most perfectly. For when of any three numbers, whether expressing three or two dimensions, one is a mean term, so that as the first is to the middle, so is the middle to the last; and conversely as the last is to the middle, so is the middle to the first; then since the middle becomes first and last, and the last and the first both become middle, of necessity all will come to be the same, and being the same with one another all will be a unity. Now if the

of Empedokles likewise reduced themselves to two: cf. Aristotle *metaph.* A iv 985ᵃ 33 οὐ μὴν χρῆταί γε τέτταρσιν, ἀλλ' ὡς δυοῖν οὖσι μόνοις, πυρὶ μὲν καθ' αὑτό, τοῖς δ' ἀντικειμένοις ὡς μιᾷ φύσει, γῇ τε καὶ ἀέρι καὶ ὕδατι: and *de gen. et corr.* II iii 330ᵇ 20. His division however does not agree with that of Plato, who classes fire air and water as forms of the same base, and places earth alone by itself.

8. δύο δὲ μόνω] Two things alone cannot be formed into a perfect harmony because they cannot constitute an ἀναλογία.

12. εἴτε ὄγκων εἴτε δυνάμεων] 'whether cubic or square.' The Greek mathematician in the time of Plato looked upon number from a geometrical standpoint, as the expression of geometrical figures. ὄγκος is a solid body, here a number representing a solid body, i.e. composed of three factors, so as to represent three dimensions. δύναμις is the technical term for a square, or sometimes

a square root; cf. *Theaetetus* 148 A; and here stands for a number composed of two factors and representing two dimensions. This interpretation of the terms seems to me the only one at all apposite to the present passage. Another explanation is that they represent the distinction made by Aristotle in *Categories* I vi 4ᵇ 20 between continuous and discrete number; the former being a geometrical figure, the latter a number in the strict sense. But as our present passage is not concerned with pure numbers at all, this does not seem to the purpose.

13. ὅ τί περ τὸ πρῶτον πρὸς αὐτό] e.g. the continuous proportion 4 : 6 :: 6 : 9 may either be reversed so that ἔσχατον becomes πρῶτον, 9 : 6 :: 6 : 4: or alternated so that the μέσον becomes ἔσχατον and πρῶτον, as 6 : 9 :: 4 : 6, or 6 : 4 :: 9 : 6. Thus, says Plato, the ἀναλογία forms a coherent whole, in which the members may freely interchange their positions.

P. T.   7

98   ΠΛΑΤΩΝΟΣ   [32 A—

πάντα ἔσται. εἰ μὲν οὖν ἐπίπεδον μέν, βάθος δὲ μηδὲν ἔχον
ἔδει γίγνεσθαι τὸ τοῦ παντὸς σῶμα, μία μεσότης ἂν ἐξήρκει
τά τε μεθ' ἑαυτῆς ξυνδεῖν καὶ ἑαυτήν· νῦν δέ—στερεοειδῆ γὰρ B
αὐτὸν προσῆκεν εἶναι, τὰ δὲ στερεὰ μία μὲν οὐδέποτε, δύο δὲ ἀεὶ
5 μεσότητες ξυναρμόττουσιν· οὕτω δὴ πυρός τε καὶ γῆς ὕδωρ ἀέρα
τε ὁ θεὸς ἐν μέσῳ θείς, καὶ πρὸς ἄλληλα καθ' ὅσον ἦν δυνατὸν
ἀνὰ τὸν αὐτὸν λόγον ἀπεργασάμενος, ὅ τί περ πῦρ πρὸς ἀέρα,
τοῦτο ἀέρα πρὸς ὕδωρ, καὶ ὅ τι ἀὴρ πρὸς ὕδωρ, ὕδωρ πρὸς γῆν,
ξυνέδησε καὶ ξυνεστήσατο οὐρανὸν ὁρατὸν καὶ ἁπτόν. καὶ διὰ
10 ταῦτα ἔκ τε δὴ τούτων τοιούτων καὶ τὸν ἀριθμὸν τεττάρων τὸ C
τοῦ κόσμου σῶμα ἐγεννήθη δι' ἀναλογίας ὁμολογῆσαν, φιλίαν τε
ἔσχεν ἐκ τούτων, ὥστ' εἰς ταὐτὸν αὐτῷ ξυνελθὸν ἄλυτον ὑπό του

3 στερεοειδῆ: στεροείδη (sic) A.   8 τοῦτο ante ὕδωρ dedit S.
10 τούτων τοιούτων: τούτων [καὶ] τοιούτων H.   12 ξυνελθόν: ξυνελθεῖν A.

2. **μία μεσότης ἂν ἐξήρκει.**) Plato lays down the law that between two plane numbers one rational and integral mean can be obtained, while between solid numbers two are required. But here we are met by a difficulty. For there are certain solid numbers between which one mean can be found; and this certainly was not unknown to Plato, who was one of the first mathematicians of his day. For instance, between 8 (2^3) and 512 (8^3) we have the proportion 8 : 64 :: 64 : 512. A second point, regarded by both Böckh and Martin as a difficulty, is really no difficulty at all, viz. the fact that there are plane numbers between which two means can be found, e.g. between 4 (2^2) and 256 (16^2) we have 4 : 16 :: 64 : 256. This is immaterial; for Plato does not say that two means can never be found between two planes, but merely that one is sufficient. The other point however does require elucidation. Böckh, who has written two able essays on the subject, offers the following explanation: 'Philosophus noster non universe planorum et solidorum magnitudinem spectavit, sed solum eam comparabilium figurarum rationem, quae fit, ubi alterum alteri inscribas, ut supra fecimus, et ibi notatas lineas exares: idque etiam quadratis et cubis accommodari potest.' This he supports by a geometrical demonstration. Martin's explanation however (with some modifications), despite Böckh's criticism of it, appears to me simpler and better. He points out that Plato's statement is true, if we suppose him to be using the words ἐπίπεδον and στερεόν in their strictest sense, so that a plane number consists of two factors only, and the solid only of three; all the factors being primes. Now it is *a priori* in the highest degree probable that Plato is using these terms in their strictest possible sense. Martin is not indeed correct in saying that between two such strictly plane numbers two means can never be intercalated: for, given that a, b, c are prime numbers, we may have this proportion: $ab : ac :: bc : c^2$, where ac, bc are integral. But this, as we have seen, is of no importance, since Plato does not deny the possibility of such a series, and since his extremes must be squares. On the other hand, provided that both the extremes are squares, we can always interpose a single mean between them, e.g. $a^2 : ab :: ab : b^2$. Again between solids formed of prime numbers we can never (with one exception) find one rational mean: for if $a^3 : x :: x : b^3$,

body of the universe were to have been made a plane surface having no thickness, one mean would have sufficed to unify itself and the extremes; but now since it behoved it to be solid, and since solids can never be united by one mean, but require two —God accordingly set air and water betwixt fire and earth, and making them as far as possible exactly proportional, so that fire is to air as air to water, and as air is to water water is to earth, thus he compacted and constructed a universe visible and tangible. For these reasons and out of elements of this kind, four in number, the body of the universe was created, being brought into concord through proportion; and from these it derived friendship, so that coming to unity with itself it became indissoluble by any force save the will of him who joined it.

then $x = ab\sqrt{ab}$; and similarly if the extremes are of the form a^2b or abc. The exception is the case $a^2b : abc :: abc : bc^2$. We can however obtain two rational and integral means, whether the extremes be cubes or compounded of unequal factors. Howbeit for Plato's purpose the extremes must be cubes, since a continuous proportion is required corresponding to fire : air :: air : water :: water : earth. This we represent by $a^3 : a^2b :: a^2b : ab^2 :: ab^2 : b^3$. The necessity of this proviso Martin has overlooked. Thus the exceptional case of a single mean is excluded. This limitation of the extremes to actual cubes is urged by Böckh as an objection to Martin's theory: but surely the cube would naturally commend itself to Plato's love of symmetry in representing his extremes, more especially as his plane extremes are necessarily squares. It is clear to my mind that, in formulating his law, Plato had in view two squares and two cubes as extremes: in the first case it is obviously possible to extract the square root of their product and so obtain a single mean; in the second it is as obviously impossible. Böckh's defence of his own explanation is to be found in vol. III of his *Kleine Schriften* pp. 253—265. The Neoplatonists attempted to extend this proportion to the physical qualities which they assigned to the four elements in groups of three; but as these belong to them in various degrees, the analogy will not hold: *e.g.* mobility is shared by fire air and water, but not to the same extent in each; and similarly with the rest. As to Stallbaum's attempt at explanation I can only echo the comment of Martin: 'je ne sais vraiment comment M. Stallbaum a pu se faire illusion au point de s'imaginer qu'il se comprenait lui-même'.

9. διὰ ταῦτα ἔκ τε δὴ τούτων] 'on this principle and out of these materials': ταῦτα signifies the ἀναλογία, τούτων the στοιχεῖα. Plato is accounting for the fact that the so-called elements are four in number by representing this as the expression of a mathematical law; and thus he shows how number acts as a formative principle in nature. In φιλίαν we have an obvious allusion to Empedokles. It is noteworthy that as Plato's application of number in his cosmogony is incomparably more intelligent than that of the Pythagoreans, so too he excels Empedokles in this matter of φιλία: he is not content with the vague assertion that φιλία keeps the universe together; he must show how φιλία comes about.

7—2

ΠΛΑΤΩΝΟΣ [32 C—

ἄλλου πλὴν ὑπὸ τοῦ ξυνδήσαντος γενέσθαι. τῶν δὲ δὴ τεττάρων
ἓν ὅλον ἕκαστον εἴληφεν ἡ τοῦ κόσμου ξύστασις. ἐκ γὰρ πυρὸς
παντὸς ὕδατός τε καὶ ἀέρος καὶ γῆς ξυνέστησεν αὐτὸν ὁ ξυνιστάς,
μέρος οὐδὲν οὐδενὸς οὐδὲ δύναμιν ἔξωθεν ὑπολιπών, τάδε διανοη-
5 θείς, πρῶτον μὲν ἵνα ὅλον ὅ τι μάλιστα ζῷον τέλεον ἐκ τελέων D
τῶν μερῶν εἴη, πρὸς δὲ τούτοις ἕν, ἅτε οὐχ ὑπολελειμμένων ἐξ 33 A
ὧν ἄλλο τοιοῦτον γένοιτ᾿ ἄν, ἔτι δὲ ἵνα ἀγήρων καὶ ἄνοσον ᾖ,
κατανοῶν, ὡς ξυστάτῳ σώματι θερμὰ καὶ ψυχρὰ καὶ πάνθ᾽ ὅσα
δυνάμεις ἰσχυρὰς ἔχει περιιστάμενα ἔξωθεν καὶ προσπίπτοντα
10 ἀκαίρως, λύει καὶ νόσους γήράς τε ἐπάγοντα φθίνειν ποιεῖ. διὰ
δὴ τὴν αἰτίαν καὶ τὸν λογισμὸν τόνδε ἓν ὅλον ὅλων ἐξ ἁπάντων
τέλεον καὶ ἀγήρων καὶ ἄνοσον αὐτὸν ἐτεκτήνατο. σχῆμα δὲ B
ἔδωκεν αὐτῷ τὸ πρέπον καὶ τὸ ξυγγενές. τῷ δὲ τὰ πάντ᾿ ἐν
αὑτῷ ζῷα περιέχειν μέλλοντι ζῴῳ πρέπον ἂν εἴη σχῆμα τὸ
15 περιειληφὸς ἐν αὑτῷ πάντα ὁπόσα σχήματα· διὸ καὶ σφαιροειδές,
ἐκ μέσου πάντῃ πρὸς τὰς τελευτὰς ἴσον ἀπέχον, κυκλοτερὲς αὐτὸ
ἐτορνεύσατο, πάντων τελεώτατον ὁμοιότατόν τε αὐτὸ ἑαυτῷ σχημάτων, νομίσας μυρίῳ κάλλιον ὅμοιον ἀνομοίου. λεῖον δὲ δὴ
κύκλῳ πᾶν ἔξωθεν αὐτὸ ἀπηκριβοῦτο πολλῶν χάριν. ὀμμάτων
20 τε γὰρ ἐπεδεῖτο οὐδέν, ὁρατὸν γὰρ οὐδὲν ὑπελείπετο ἔξωθεν· οὐδ᾿ C
ἀκοῆς, οὐδὲ γὰρ ἀκουστόν· πνεῦμά τε οὐκ ἦν περιεστὸς δεόμενον
ἀναπνοῆς· οὐδ᾿ αὖ τινὸς ἐπιδεὲς ἦν ὀργάνου σχεῖν, ᾧ τὴν μὲν εἰς
ἑαυτὸ τροφὴν δέξοιτο, τὴν δὲ πρότερον ἐξικμασμένην ἀποπέμψοι
πάλιν. ἀπῄει τε γὰρ οὐδὲν οὐδὲ προσῄειν αὐτῷ ποθέν· οὐδὲ γὰρ

8 ξυστάτῳ σώματι dedi cum H e W. Wagneri coniectura. ξυνιστὰς τῷ σώματι A.
ἃ ξυνιστᾷ τὰ σώματα SZ. 20 ὑπελείπετο: ὑπέλειπτο A.

4. **οὐδὲ δύναμιν**] δύναμιν is not to be understood as 'potentiality', but as 'power' or 'faculty'.

5. **τέλεον**] 'complete' and so perfect: cf. Aristotle *metaph*. Δ xvi 1021ᵇ 12 τέλειον λέγεται ἐν μὲν οὗ μὴ ἔστιν ἔξω τι λαβεῖν μηδὲ ἐν μόριον: and from this sense Aristotle derives all the other meanings of this word.

8. **ὡς ξυστάτῳ σώματι**] I have adopted the correction of W. Wagner. The reading of Stallbaum and the Zürich edition ἃ ξυνιστᾷ τὰ σώματα has poor ms. authority and is weak in sense; moreover the form ξυνιστᾷ is extremely doubtful

Attic. The mss. for the most part have ξυνιστὰς or ξυνιστὰν τῷ σώματι. ξυστάτῳ σώματι is supported by Cicero's rendering 'coagmentatio corporis'.

9. **περιιστάμενα ἔξωθεν καὶ προσπίπτοντα**] Compare the statement in 81 D as to the cause of disease and decay.

11. **ἓν ὅλον**] It is needless either with Stallbaum to read ἵνα or to change αὐτὸν into αὐτό: the meaning is 'he made it (the κόσμος) one single whole'.

14. **τὸ περιειληφὸς ἐν αὑτῷ**] The sphere is said to contain within it all other shapes, because of all figures having an equal periphery it is the great-

Now the making of the universe took up the whole bulk of each of these four elements. Of all fire and all water and air and earth its framer fashioned it, leaving over no part nor power without. Therein he had this intent: first that it might be a creature perfect to the utmost with all its parts perfect; next that it might be one, seeing that nothing was left over whereof another should be formed; furthermore that it might be free from age and sickness; for he reflected that when hot things and cold and all such as have strong powers gather round a composite body from without and fall unseasonably upon it, they undermine it, and bringing upon it sickness and age cause its decay. For such motives and reasons he fashioned it as one whole, with each of its parts whole in itself, so as to be perfect and free from age and sickness. And he assigned to it its proper and natural shape. To that which is to comprehend all animals in itself that shape seems proper which comprehends in itself all shapes that are. Wherefore he turned it of a rounded and spherical shape, having its bounding surface in all points at an equal distance from the centre: this being the most perfect and regular shape; for he thought that a regular shape was infinitely fairer than an irregular. And all round about he finished off the outer surface perfectly smooth, for many reasons. It needed not eyes, for naught visible was left outside; nor hearing, for there was nothing to hear; and there was no surrounding air which made breathing needful. Nor must it have any organ whereby it should receive into itself its sustenance, and again reject that which was already digested; for nothing went forth of it nor entered in from anywhere; for

18. λεῖον δὲ δή] This might be supposed to be involved in what has been said: but Plato is insisting that not only is the general shape of the κόσμος spherical, but that it is a sphere without any appendages.

21. πνεῦμά τε οὐκ ἦν περιεστός] This is directed against a Pythagorean fancy, that outside the universe there existed κενόν, or ἄπειρον πνεῦμα, which passed into the cavities in the universe, as though the latter were respiring it: cf.

Aristotle *physica* IV vi 213ᵇ 22 εἶναι δ' ἔφασαν καὶ οἱ Πυθαγόρειοι κενόν, καὶ ἐπεισιέναι αὐτὸ τῷ οὐρανῷ ἐκ τοῦ ἀπείρου πνεύματος ὡς ἀναπνέοντι καὶ τὸ κενόν, ὃ διορίζει τὰς φύσεις, ὡς ὄντος τοῦ κενοῦ χωρισμοῦ τινὸς τῶν ἐφεξῆς καὶ τῆς διορίσεως· καὶ τοῦτ' εἶναι πρῶτον ἐν τοῖς ἀριθμοῖς· τὸ γὰρ κενὸν διορίζειν τὴν φύσιν αὐτῶν: and *physica* III iv 203ᵃ 6 οἱ μὲν Πυθαγόρειοι ἐν τοῖς αἰσθητοῖς [sc. τιθέασι τὸ ἄπειρον]· οὐ γὰρ χωριστὸν ποιοῦσι τὸν ἀριθμόν· καὶ εἶναι τὸ ἔξω τοῦ οὐρανοῦ ἄπειρον. See too Stobaeus *ecl.* I 382.

ἦν· αὐτὸ γὰρ ἑαυτῷ τροφὴν τὴν ἑαυτοῦ φθίσιν παρέχον καὶ
πάντα ἐν ἑαυτῷ καὶ ὑφ' ἑαυτοῦ πάσχον καὶ δρῶν ἐκ τέχνης D
γέγονεν· ἡγήσατο γὰρ αὐτὸ ὁ ξυνθεὶς αὔταρκες ὂν ἄμεινον ἔσεσθαι
μᾶλλον ἢ προσδεὲς ἄλλων. χειρῶν δέ, αἷς οὔτε λαβεῖν οὔτε αὖ
5 τινὰ ἀμύνασθαι χρεία τις ἦν, μάτην οὐκ ᾤετο δεῖν αὐτῷ προσάπτειν, οὐδὲ ποδῶν οὐδὲ ὅλως τῆς περὶ τὴν βάσιν ὑπηρεσίας.
κίνησιν γὰρ ἀπένειμεν αὐτῷ τὴν τοῦ σώματος οἰκείαν, τῶν ἑπτὰ 34 A
τὴν περὶ νοῦν καὶ φρόνησιν μάλιστα οὖσαν· διὸ δὴ κατὰ ταὐτὰ
ἐν τῷ αὐτῷ καὶ ἐν ἑαυτῷ περιαγαγὼν αὐτὸ ἐποίησε κύκλῳ κι-
10 νεῖσθαι στρεφόμενον, τὰς δὲ ἓξ ἁπάσας κινήσεις ἀφεῖλε καὶ
ἀπλανὲς ἀπειργάσατο ἐκείνων· ἐπὶ δὲ τὴν περίοδον ταύτην ἅτ'
οὐδὲν ποδῶν δέον ἀσκελὲς καὶ ἄπουν αὐτὸ ἐγέννησεν.

VIII. Οὗτος δὴ πᾶς ὄντος ἀεὶ λογισμὸς θεοῦ περὶ τὸν ποτὲ

1. **τροφὴν τὴν ἑαυτοῦ φθίσιν παρέχον**] By this striking phrase Plato means that the nutrition of one thing is effected by the decomposition of another: all the elements of which the universe is composed feed upon each other and are fed upon in turn. The idea is still more boldly expressed by Herakleitos fr. 25 (Bywater) ζῇ πῦρ τὸν γῆς θάνατον καὶ ἀὴρ ζῇ τὸν πυρὸς θάνατον, ὕδωρ ζῇ τὸν ἀέρος θάνατον, γῆ τὸν ὕδατος.

4. **χειρῶν δέ**] There is an anacoluthon: the genitive is written as though χρεία τις ἦν belonged to the main clause.

7. **τὴν τοῦ σώματος οἰκείαν**] Plato does not of course mean that the motion belongs to the body in the sense of being its own attribute, because all motion is of soul; but simply that the most perfect motion suits the most perfect form. For τῶν ἑπτὰ see 43 B: the seven are up and down, forwards and backwards, to right and to left, and finally rotation upon an axis.

8. **τὴν περὶ νοῦν καὶ φρόνησιν**] Compare *Laws* 898 A τὸ κατὰ ταὐτὰ δήπου καὶ ὡσαύτως καὶ ἐν τῷ αὐτῷ καὶ περὶ τὰ αὐτὰ καὶ πρὸς τὰ αὐτὰ καὶ καθ' ἕνα λόγον καὶ τάξιν μίαν ἄμφω κινεῖσθαι λέγοντες νοῦν τήν τε ἐν ἑνὶ φερομένην κίνησιν, σφαίρας εὐτόρνου ἀπεικασμένα φοραῖς, οὐκ ἄν ποτε φανεῖμεν φαῦλοι δημιουργοὶ λόγῳ καλῶν εἰκόνων. Aristotle states his objections (which are not very cogent) to the comparison in *de anima* I iii § 15.

9. **κύκλῳ κινεῖσθαι στρεφόμενον**] If we compare the account given in the *Timaeus* concerning the motion of the κόσμος with that in the myth of the *Politicus*, we shall observe a peculiar and very significant discrepancy. In a passage of the latter dialogue, 269 A foll., we are told that for a fixed period God turns the universe in a given direction, making it revolve upon its axis; at the end of this period he lets go of it and suffers it to rotate by itself for a like period in a reverse direction: its motion being the recoil from that which had been imparted by God. And this alternation recurs *ad infinitum*. Now the reason for this singular arrangement is thus stated by Plato: τὸ κατὰ ταὐτὰ καὶ ὡσαύτως ἔχειν ἀεὶ καὶ ταὐτὸν εἶναι τοῖς πάντων θειοτάτοις προσήκει μόνοις, σώματος δὲ φύσις οὐ ταύτης τῆς τάξεως. ὃν δὲ οὐρανὸν καὶ κόσμον ἐπωνομάκαμεν, πολλῶν μὲν καὶ μακαρίων παρὰ τοῦ γεννήσαντος μετείληφεν, ἀτὰρ οὖν δὴ κεκοινώνηκε καὶ σώματος. For this cause it was impossible to give it the same motion unchanged for ever; so God devised this ἀνακύκλησις as the slightest παράλ-

there was nothing. For by design was it created to supply its own sustenance by its own wasting, and to have all its action and passion in itself and by itself: for its framer deemed that were it self-sufficing it would be far better than if it required aught else. And hands, wherewith it had no need to grasp aught nor to defend itself against another, he thought not fit idly to bestow upon it, nor yet feet, nor in a word anything to serve as the means of movement. For he assigned it that motion which was proper to its bodily form, of all the seven that which most belongs to reason and intelligence. Wherefore turning it about uniformly in the same spot on its own axis, he made it to revolve round and round; but all the six motions he took away from it and left it without part in their wanderings. And since for this revolution there was no need of feet he made it without legs and without feet.

VIII. So the universal design of the ever-living God, that

λαξις from a perpetually constant motion. But in the *Timaeus* the movement of the universe is changeless and everlastingly in the same direction. Now the interpretation of this difference is in my judgment indubitably this. The passage in the *Politicus* belongs to a different class of myth to the allegory of the *Timaeus*. Plato is not there expounding his metaphysical theories under a similitude; he is telling a tale with a moral to it. Therefore it suited his convenience to adopt the popular distinction between spirit and matter; and since the κόσμος was material, he was forced to deny it the motion peculiar to τὸ θειότατον. In the *Timaeus*, on the contrary, when the entire universe is the self-evolution of νοῦς, the distinction between spirit and matter is finally eliminated; and there is now no reason for refusing, or rather there is a necessity for assigning to the κόσμος the unchanging motion of the Same. I do not mean to imply that Plato's view on this subject was different when he wrote the *Politicus*; merely that the circumstances and object of his writing were other.

34 A—36 D, c. viii. So God made the universe a sphere, even and smooth and perfect, quickened through and through with soul, alone and sufficient to itself. But he made not soul later than body, as we idly speak of it: but rather, as soul was to be mistress and queen over body, he framed her first, of three elements blended, of Same and of Other and of Essence. And when the blending was finished, he ordered and apportioned her according to the intervals of a musical scale, so that the harmony thereof pervaded all her substance. And then he divided the whole soul into two portions, which he formed into two intersecting circles; and he called them the circle of the Same and the circle of the Other: and he gave the circle of the Same dominion over the circle of the Other. And the outer circle, which is of the Same, he left undivided, but the circle of the Other he cleft into seven circles, four one way revolving and three the other; and their distances one from another were ordained according to the proportion of the seven harmonic numbers of the soul.

ἐσόμενον θεὸν λογισθεὶς λεῖον καὶ ὁμαλὸν πανταχῇ τε ἐκ μέσου B
ἴσον καὶ ὅλον καὶ τέλεον ἐκ τελέων σωμάτων σῶμα ἐποίησε·
ψυχὴν δὲ εἰς τὸ μέσον αὐτοῦ θεὶς διὰ παντός τε ἔτεινε καὶ ἔτι
ἔξωθεν τὸ σῶμα αὐτῇ περιεκάλυψε ταύτῃ, καὶ κύκλῳ δὴ κύκλον
5 στρεφόμενον οὐρανὸν ἕνα μόνον ἔρημον κατέστησε, δι' ἀρετὴν δὲ
αὐτὸν αὑτῷ δυνάμενον ξυγγίγνεσθαι καὶ οὐδενὸς ἑτέρου προσδεό-
μενον, γνώριμον δὲ καὶ φίλον ἱκανῶς αὐτὸν αὑτῷ. διὰ πάντα
δὴ ταῦτα εὐδαίμονα θεὸν αὐτὸν ἐγεννήσατο.

Τὴν δὲ δὴ ψυχὴν οὐχ ὡς νῦν ὑστέραν ἐπιχειροῦμεν λέγειν,
10 οὕτως ἐμηχανήσατο καὶ ὁ θεὸς νεωτέραν· οὐ γὰρ ἂν ἄρχεσθαι C
πρεσβύτερον ὑπὸ νεωτέρου ξυνέρξας εἴασεν· ἀλλά πως ἡμεῖς
πολὺ μετέχοντες τοῦ προστυχόντος τε καὶ εἰκῇ ταύτῃ πῃ καὶ
λέγομεν· ὁ δὲ καὶ γενέσει καὶ ἀρετῇ προτέραν καὶ πρεσβυτέραν
ψυχὴν σώματος ὡς δεσπότιν καὶ ἄρξουσαν ἀρξομένου ξυνεστή-
15 σατο ἐκ τῶνδέ τε καὶ τοιῷδε τρόπῳ. τῆς ἀμερίστου καὶ ἀεὶ 35 A

2 καὶ ante ἐκ habet A.

2. **τέλεον ἐκ τελέων σωμάτων**] i.e. it was a complete whole constructed out of the whole quantity that existed of its constituent elements, as stated in 32 C.

3. **ψυχὴν δὲ εἰς τὸ μέσον**] Soul being unextended, this is of course metaphorical, signifying that every part of the material universe from centre to circumference is informed and instinct with soul. In the words that follow, ἔξωθεν τὸ σῶμα αὐτῇ περιεκάλυψε ταύτῃ, Stallbaum (who seems throughout to regard Plato as incapable of originating any idea for himself) will have it that he is following Philolaos. Now the Pythagorean πνεῦμα ἄπειρον, the existence of which is peremptorily denied by Plato in 33 C, has not a trace of community with the Platonic world-soul: nor is there any reasonable evidence that Philolaos or any other Pythagorean conceived such a soul. Plato seems by this phrase simply to assert the absolute domination of soul over body. The old physicists regarded soul or life as a function of material things, but for Plato matter is but an accident of soul: neither will he allow that soul is contained in body, as the Epicureans later held—corpus quod vas quasi constitit eius, Lucr. III 440—rather she comprehends it. The same figure recurs 36 E. Aristotle's criticism in *metaph.* Λ vi 1071[b] 37 is based on a confusion between κατὰ χρόνον and κατ' ἐπίνοιαν.

9. **οὐχ ὡς νῦν ὑστέραν**] This passage ought surely to be warning enough to those who will not allow Plato the ordinary licence of a story-teller. A similar rectification of an inexact statement is to be found at 54 B.

12. **τοῦ προστυχόντος τε καὶ εἰκῇ**] Cf. *Philebus* 28 D τὴν τοῦ ἀλόγου καὶ εἰκῇ δύναμιν. Stallbaum has the following curious remark: 'egregie convenit cum iis quae Legum libro X. 904 A disputantur, ubi animam indelebilem quidem esse docetur, nec vero aeternam'. This were 'inconstantia Platonis' with a vengeance: fortunately nothing of the kind is taught in the passage cited. The words are ἀνώλεθρον δὲ ὂν γενόμενον [τὸ γενόμενον Herm.] ἀλλ' οὐκ αἰώνιον, ὥσπερ οἱ κατὰ νόμον ὄντες θεοί. Plato here plainly denies eternity, not to soul, but to the ξύστασις of soul and body, which

he planned for the God that was some time to be, made its surface smooth and even, everywhere equally distant from the centre, a body whole and perfect out of perfect bodies. And God set soul in the midst thereof and spread her through all its body and even wrapped the body about with her from without, and he made it a sphere in a circle revolving, a universe one and alone; but for its excellence it was able to be company to itself and needed no other, being sufficient for itself as acquaintance and friend. For all these things then he created it a happy god.

But the soul was not made by God younger than the body, even as she comes later in this account we are essaying to give; for he would not when he had joined them together have suffered the elder to be governed by the younger: but we are far too prone to a casual and random habit of mind which shows itself in our speech. God made soul in birth and in excellence earlier and elder than body, to be its mistress and governor; and he framed her out of the following elements and in the following

is ἀνώλεθρος, since such a mode of existence must subsist perpetually, but not αἰώνιος, since it belongs to γένεσις.

13. γενέσει καὶ ἀρετῇ προτέραν] The statement that soul is prior to matter in order of generation can mean nothing else but that matter is evolved out of soul: for had matter an independent ἀρχή, it would not be ὕστερον γενέσει. Again the priority is logical not temporal.

15. ἐκ τῶνδε] Aristotle de anima I ii 404ᵇ 16 says τὸν αὐτὸν δὲ τρόπον καὶ Πλάτων ἐν τῷ Τιμαίῳ τὴν ψυχὴν ἐκ τῶν στοιχείων ποιεῖ· γινώσκεσθαι γὰρ τῷ ὁμοίῳ τὸ ὅμοιον, καὶ τὰ πράγματα ἐκ τῶν ἀρχῶν εἶναι. This statement is in more than one respect gravely misleading. First, although it is impossible to suppose that Aristotle really meant to classify Plato's στοιχεῖα along with the material στοιχεῖα of Empedokles and the rest, yet, after stating the theories of the materialists, to proceed τὸν αὐτὸν δὲ τρόπον καὶ Πλάτων is, to say the least, a singularly infelicitous mode of exposition. Next, while it is true that in Plato's scheme like is known by like, yet that is not the fundamental principle. The antithesis Same and Other, One and Many, is the very basis of his whole metaphysic, and must inevitably be the basis of his psychogony. γινώσκεσθαι τῷ ὁμοίῳ τὸ ὅμοιον is consequent, not antecedent.

τῆς ἀμερίστου] First a word concerning the Greek. The genitives τῆς ἀμερίστου...μεριστῆς might well enough be taken with Proklos as dependent on ἐν μέσῳ. I think however they are rather to be considered as in a somewhat loose anticipative apposition to ἐξ ἀμφοῖν, with which words the construction first becomes determinate. Stallbaum is certainly wrong in connecting them with εἶδος. Presently the words αὖ περὶ after τῆς τε ταὐτοῦ φύσεως are unquestionably spurious—repeated no doubt from τῆς αὖ περὶ τὰ σώματα. In the phrase ἀεὶ κατὰ ταὐτὰ ἐχούσης οὐσίας Dr Jackson has with some probability suggested that for οὐσίας we should read φύσεως: there is certainly an awkwardness in this use of οὐσίας, when we have the word directly afterwards in so very peculiar and technical a sense.

κατὰ ταὐτὰ ἐχούσης οὐσίας καὶ τῆς αὖ περὶ τὰ σώματα γιγνο-
μένης μεριστῆς τρίτον ἐξ ἀμφοῖν ἐν μέσῳ ξυνεκεράσατο οὐσίας
εἶδος, τῆς τε ταὐτοῦ φύσεως καὶ τῆς θατέρου, καὶ κατὰ ταῦτα
ξυνέστησεν ἐν μέσῳ τοῦ τε ἀμεροῦς αὐτῶν καὶ τοῦ κατὰ τὰ
5 σώματα μεριστοῦ· καὶ τρία λαβὼν αὐτὰ ὄντα συνεκεράσατο εἰς
μίαν πάντα ἰδέαν, τὴν θατέρου φύσιν δύσμικτον οὖσαν εἰς ταὐτὸν
ξυναρμόττων βίᾳ· μιγνὺς δὲ μετὰ τῆς οὐσίας καὶ ἐκ τριῶν ποιη- B
σάμενος ἕν, πάλιν ὅλον τοῦτο μοίρας ὅσας προσῆκε διένειμεν,
ἑκάστην δὲ ἔκ τε ταὐτοῦ καὶ θατέρου καὶ τῆς οὐσίας μεμιγμένην.
10 ἤρχετο δὲ διαιρεῖν ὧδε. μίαν ἀφεῖλε τὸ πρῶτον ἀπὸ παντὸς
μοῖραν, μετὰ δὲ ταύτην ἀφῄρει διπλασίαν ταύτης, τὴν δ' αὖ τρίτην
ἡμιολίαν μὲν τῆς δευτέρας, τριπλασίαν δὲ τῆς πρώτης, τετάρτην

3 Post φύσεως delevi αὖ πέρι, quae cum consensu codicum retinent SZ: inclusit H.

This passage is obviously one of the most important in the dialogue; and it is necessary to use the utmost care in interpreting the terms. ταὐτὸν and θάτερον are in their widest and most radical sense respectively the principle of unity and identity and the principle of multiplicity and difference: but they are likewise used in special applications of these significations. Such applications are ἡ ἀμέριστος οὐσία and ἡ περὶ τὰ σώματα γιγνομένη μεριστή, which are identical but not coextensive with ταὐτὸν and θάτερον. Regarded objectively, ταὐτὸν is the element of changeless unity in the κόσμος, the intelligible ἀρχή, θάτερον is the plurality of variable phenomena, in which the primal unity is materially and visibly manifested. The first is ἡ ἀμέριστος οὐσία, pure mind as it is in its own nature, the second is mind as it becomes differentiated into material existence. Regarded subjectively, ταὐτὸν is that faculty in the world-soul which deals with the intelligible unity, θάτερον that which deals with sensible multiplicity. One is the simple activity of thought as such, the other the operation of thought as subjected to the conditions of time and space.

But what is οὐσία? This is stated by Plato to be τρίτον ἐξ ἀμφοῖν ἐν μέσῳ τῆς τε ταὐτοῦ φύσεως καὶ τῆς τοῦ ἑτέρου—a third term arising from the other two and intermediate between them. I think the nature of οὐσία will be made clearest if we take the case of an individual soul. Every one has (1) the faculty of pure thought, of reasoning apart from sensation, (2) the faculty of perceiving sensible impressions. Now if we hold that these two faculties are simply processes which go on in the brain, so that thought and perception are merely affections of the substance of the brain and nothing more—there is an end: there is no οὐσία: the two faculties have no bond of union further than they are affections of the same brain. But if we consider, as Plato did, that the physical action of the brain which accompanies thought and sensation does not constitute these, but that there is a thinking and sentient substance which acts by means of these brain-processes, at once we have a unity: the two faculties are no longer independent physical processes but diverse activities of one and the same intelligence: the subject is no more a series of consciousnesses but a conscious personality. Just so the κόσμος, being a sentient intelligence, must be conscious of itself as a whole: by ταὐτὸν it apprehends itself as unity, by θάτερον it apprehends itself as multi-

B] ΤΙΜΑΙΟΣ. 107

way. From the undivided and ever changeless substance and that which becomes divided in material bodies, of both these he mingled in the third place the form of Essence, in the midst between the Same and the Other; and this he composed on such wise between the undivided and that which is in material bodies divided; and taking them, three in number, he blended them into one form, forcing the nature of the Other, hard as it was to mingle, into union with the Same. And mingling them with Essence and of the three making one, again he divided this into as many parts as was meet, each part mingled of Same and of Other and of Essence. And he began his dividing thus: first he took one portion from the whole; then he went on to take a portion double of this; and the third half as much again as the

plicity: and as these are not apart, but are activities of the same thinking subject, we have οὐσία, their union as modes of one and the same consciousness. οὐσία then is neither identical with ταὐτὸν or θάτερον nor a substance apart from both: it is the identification of the two as one substance. And as in the particular soul the reasoning and perceptive faculties have no independent existence of their own, but, if they are to exist, must coexist in a soul and thus obtain οὐσία, so it is in the cosmic soul. Taken apart, both ταὐτὸν and θάτερον are mere logical abstractions, they have no existence. Combined they instantly unite into a single οὐσία, they are no longer abstract, but concrete. Thus οὐσία is said to be τρίτον ἐξ ἀμφοῖν, because it arises from their union. So again we see that the One and the Many cannot exist but in combination.

2. ἐν μέσῳ] i.e. it is a bond of union and connecting link between them. I would draw special attention to the fact that according as they are regarded objectively or subjectively, ἀμερής and μεριστὴ οὐσία have a distinct significance: they are (α) ψυχή as the primal and eternal ἐν ὄν, and ψυχή as evolved into a plurality of γιγνόμενα, (β) ψυχή as dealing

directly by pure thought with absolute unity, and ψυχή as dealing sensually with the multitude of material phenomena.

6. δύσμικτον οὖσαν] The element of difference and divergency was naturally refractory and hard to force into union with the rest. Plato, while convinced of the necessity of conciliating the opposites ἕν and πολλά, is fully alive to the magnitude of the undertaking.

10. ἤρχετο δὲ διαιρεῖν ὧδε] Here Plato is really pythagorising. The numbers which follow are those which compose the geometrical τετρακτὺς of the Pythagoreans. This τετρακτὺς is double, proceeding in one branch from 1 to 2^3, in the other from 1 to 3^3, thus:

It will be observed that the sum of the first six numbers, 1, 2, 3, 4, 8, 9 equals the last, 27. This τετρακτὺς was significant of many things to the Pythagoreans: of these it will suffice to mention one, which Plato may have had in view in selecting these numbers: 1 denotes the point; then in the διπλάσια διαστή-

δὲ τῆς δευτέρας διπλῆν, πέμπτην δὲ τριπλῆν τῆς τρίτης, τὴν δ' C
ἕκτην τῆς πρώτης ὀκταπλασίαν, ἑβδόμην δὲ ἑπτακαιεικοσιπλα-
σίαν τῆς πρώτης· μετὰ δὲ ταῦτα συνεπληροῦτο τά τε διπλάσια
καὶ τριπλάσια διαστήματα, μοίρας ἔτι ἐκεῖθεν ἀποτέμνων καὶ 36 A
5 τιθεὶς εἰς τὸ μεταξὺ τούτων, ὥστε ἐν ἑκάστῳ διαστήματι δύο
εἶναι μεσότητας, τὴν μὲν ταὐτῷ μέρει τῶν ἄκρων αὐτῶν ὑπερέ-
χουσαν καὶ ὑπερεχομένην, τὴν δὲ ἴσῳ μὲν κατ' ἀριθμὸν ὑπερέ-
χουσαν, ἴσῳ δὲ ὑπερεχομένην· ἡμιολίων δὲ διαστάσεων καὶ
ἐπιτρίτων καὶ ἐπογδόων γενομένων ἐκ τούτων τῶν δεσμῶν ἐν
10 ταῖς πρόσθεν διαστάσεσι, τῷ τοῦ ἐπογδόου διαστήματι τὰ ἐπί- B

μαта 2 stands for the straight line, 4 for the rectilinear plane, 8 for the rectilinear solid. In the τριπλάσια διαστήματα 3 is the curved line, 9 the curvilinear superficies, 27 the curvilinear solid. These numbers also, as we presently see, form the basis of a musical scale. The simple Pythagorean τετρακτύς, $1+2+3+4=10$ is not employed by Plato.

1. **πέμπτην δὲ τριπλῆν τῆς τρίτης**] Note that 9 is prior in the enumeration to 8: this is because 9 is a lower power, being the square of 3, while 8 is the cube of 2.

3. **μετὰ δὲ ταῦτα συνεπληροῦτο**] Next between every two members of the double and triple intervals severally he inserted two means, the harmonical and the arithmetical. The harmonical mean is such that it exceeds the lesser extreme and is exceeded by the greater in the same fraction of each extreme respectively: i.e. if x and y be the extremes and m the mean, $x + \frac{x}{n} = y - \frac{y}{n} = m$. The arithmetical mean exceeds the lesser extreme by the same number whereby it is exceeded by the greater extreme, $x + n = y - n = m$. Thus between 6 and 12 we have 8 as the harmonical mean, 9 as the arithmetical. Now inserting these means in the two series above, we get

In the διπλάσια διαστήματα 1, $\frac{4}{3}$, $\frac{3}{2}$, 2, $\frac{8}{3}$, 3, 4, $\frac{16}{3}$, 6, 8:
In the τριπλάσια διαστήματα 1, $\frac{3}{2}$, 2, 3, $\frac{9}{2}$, 6, 9, $\frac{27}{2}$, 18, 27.

8. **ἡμιολίων δέ**] It will be seen that the first of the two series given in the preceding note proceeds regularly in the ratios $\frac{4}{3}$, $\frac{9}{8}$, $\frac{4}{3}$ &c; while the second proceeds in the ratios $\frac{3}{2}$, $\frac{4}{3}$, $\frac{3}{2}$ &c : there being in the first series three sets of $\frac{4}{3}$, $\frac{9}{8}$, $\frac{4}{3}$, in the second three sets of $\frac{3}{2}$, $\frac{4}{3}$, $\frac{3}{2}$.

10. **τῷ τοῦ ἐπογδόου διαστήματι**] In order to understand this passage it is only necessary to bear in mind one or two simple acoustical facts. The pitch of a musical note depends upon the rapidity with which the sounding body vibrates. To take for example two vibrating strings: if one string be twice the length of the other, the shorter string will, other things being equal, produce twice as many vibrations in a given time as the longer and will give a note an octave above the first. Another string $\frac{2}{3}$ the length of the first will give the fifth above the second string, or the twelfth above the first. Therefore we express the octave by the ratio 1 : 2 and the fifth by 2 : 3. The other ratios with which we are here concerned are 3 : 4, which gives the fourth; 8 : 9, which gives a whole tone; 16 : 27, which gives the (Pythagorean) major sixth; and 243 : 256, which will be treated of presently, but which is very nearly a semitone. Now in reckoning these ratios we may either take as our basis the num-

[36 B] ΤΙΜΑΙΟΣ.

second and triple of the first; the fourth double of the second; the fifth three times the third; the sixth eight times the first, the seventh twenty-seven times the first. After that, he filled up the interval between the powers of two and of three by severing yet more from the original mass and placing it between them in such a manner that within each interval were two means, the first exceeding one extreme in the same proportion as it was exceeded by the other, the second by the same number exceeding the one as it was exceeded by the other. And whereas by these links there were formed in the original intervals new intervals of $\frac{3}{2}$ and $\frac{4}{3}$ and $\frac{9}{8}$, he went on to fill up all the intervals of $\frac{4}{3}$ with that of $\frac{9}{8}$, leaving in each a fraction over; and the

ber of vibrations executed in a given time—as is the practice of modern musicians—or the relative lengths of string required to produce the several notes, as was usual among the Greeks. In the first case it is obvious that the ratio $\frac{1}{2}$ expresses the octave upwards, in the second downwards. As Plato doubtless followed the latter plan, I shall follow it too—that is, we shall reckon the scale from top to bottom. Now taking the διπλάσια διαστήματα with their harmonical and arithmetical means, and filling up the intervals as Plato directs, we shall have:

$$1 \quad \begin{array}{cccccccc} 8:9 & 8:9 & \frac{243}{256} & 8:9 & 8:9 & 8:9 & \frac{243}{256} \\ \frac{9}{8} & \frac{81}{64} & \frac{4}{3} & \frac{3}{2} & \frac{27}{16} & \frac{243}{128} \end{array} \quad 2$$

$$2 \quad \begin{array}{cccccccc} 8:9 & 8:9 & \frac{243}{256} & 8:9 & 8:9 & 8:9 & \frac{243}{256} \\ \frac{9}{4} & \frac{81}{32} & \frac{8}{3} & 3 & \frac{27}{8} & \frac{243}{64} \end{array} \quad 4$$

$$4 \quad \begin{array}{cccccccc} 8:9 & 8:9 & \frac{243}{256} & 8:9 & 8:9 & 8:9 & \frac{243}{256} \\ \frac{9}{2} & \frac{81}{16} & \frac{16}{3} & 6 & \frac{27}{4} & \frac{243}{32} \end{array} \quad 8$$

The small figures denote the ratio between each term and its successor.

Now giving these intervals their musical value, we get the following scale:

The original notes of the τετρακτὺς are marked as semibreves, the means as minims, and the insertions of the ἐπόγδοα and λείμματα as crotchets. Thus we get a system of three octaves in the Dorian mode, which was identical with one form of our modern minor scale.

So far all is simple. But it is not so easy to determine how the scale of τριπλάσια διαστήματα should be constructed. The most obvious method is to continue the system of ἐπίτριτα or tetrachords in the lower octaves by supplying the octaves of the means belonging to the binary system. Thus we shall have one continuous scale formed of the two sets of intervals: we shall add two more lines to our series of numbers,

ΠΛΑΤΩΝΟΣ [36 B—

τρίτα πάντα ξυνεπληροῦτο, λείπων αὐτῶν ἑκάστου μόριον, τῆς
τοῦ μορίου ταύτης διαστάσεως λειφθείσης ἀριθμοῦ πρὸς ἀριθμὸν
ἐχούσης τοὺς ὅρους ἓξ καὶ πεντήκοντα καὶ διακοσίων πρὸς τρία
καὶ τετταράκοντα καὶ διακόσια. καὶ δὴ καὶ τὸ μιχθέν, ἐξ οὗ
5 ταῦτα κατέτεμνεν, οὕτως ἤδη πᾶν ἀναλώκει. ταύτην οὖν τὴν
ξύστασιν πᾶσαν διπλῆν κατὰ μῆκος σχίσας μέσην πρὸς μέσην
ἑκατέραν ἀλλήλαις οἷον χῖ προσβαλὼν κατέκαμψεν, εἰς ἓν κύκλῳ C

1 τῆς τοῦ: τῆς δὲ τοῦ H cum rc. A.   4 καὶ δὴ καί: alterum καὶ omittunt SZ.
5 πᾶν: πάντ' A.   ἀναλώκει dedi cum A.   ἀνηλώκει H.   καταναλώκει SZ.

8, 9, $\frac{81}{8}$, $\frac{32}{3}$, 12, $\frac{27}{2}$, $\frac{243}{16}$, 16,

16, 18, $\frac{81}{4}$, $\frac{64}{3}$, 24, 27,

where 12, 16, 24 are derived from the octaves of the former series: and we shall continue the scale thus from where it left off:

But a serious, if not fatal, objection to this scale is that it does not constitute a perfect system or systems in any one of the Greek modes. It would seem then as if we must, with Westphal (*Musik d. gr. Alterthums*), construct the triple scale quite independently of the other. Then for each of the intervals 1 : 3, 3 : 9, 9 : 27 we shall have three dodecachords:

Here we have three conjunct dodecachords in the Dorian or Aeolian mode, passing from A minor to D minor and G minor. This scale, which is identical with that given by Westphal, does not seem free from objection; but it is more

ΤΙΜΑΙΟΣ.

terms of the interval forming this fraction are in the numerical proportion of 256 to 243. By this time the mixture, whence he cut off these portions, was all used up. Next he cleft the structure so formed lengthwise into two halves, and laying the two so as to meet in the centre in the shape of the letter X, he bent them into a circle and joined them, causing them to meet

satisfactory than any other I can suggest. The scale given by Proklos is not suitable; nor yet one which he attributes to Severus, who, supposing him to start from A minor, modulates as far as C minor. The extent of Plato's scale, four octaves and a major sixth, is far greater than any that actually occurred in Greek music, which employed at most but two octaves. It has been suggested by Proklos that Plato's reason for using so extensive a scale is that ψυχή has to apprehend not only spirit but matter, which has three dimensions; hence in the symbol the cubes 8 and 27 were required.

1. **λείπων αὐτῶν ἑκάστου μόριον**] Taking the first tetrachord of our scale, E to B, if we proceed to insert as many ἐπόγδοα as we can, we find we can introduce two, viz. E to D, D to C : a third would take us to B♭ instead of B. This interval then, C to B, is the μόριον which remains over. This is called the λεῖμμα and has the ratio 243 : 256. The Pythagoreans held that the tone cannot be divided into two equal parts, because there is not a rational mean between 8 and 9: they accordingly distributed it into a minor semitone or λεῖμμα, $\frac{243}{256}$, and a major semitone or ἀποτομή, $\frac{2048}{2187}$; of which two the product = $\frac{8}{9}$. The Pythagorean λεῖμμα is slightly less than the 'natural' semitone, which is $\frac{15}{16}$ or $\frac{240}{256}$.

The pseudo-Timaeus Locrus in his abstract of this passage (96 B) says the number of terms in the series is 36: a similar view is held, according to Proklos, by some of the old Platonists; apparently for no other reason than that 36 is the sum of another double τετρακτὺς given by Plutarch, consisting of the first four odd and the first four even numbers. This number of terms is gained by forming the two scales separately and then combining them so that the apotome twice occurs; e.g. C, B, B♭, A: the interval C—B being a λεῖμμα, the interval B—B♭ is an ἀποτομή. But the apotome is totally foreign to Plato's scale, which is διάτονον σύντονον of the strictest kind. Nor is there any Greek scale which would tolerate three half-tones successively: even in the χρῶμα τονιαῖον only two occur in succession. Nor do I see on what plan the apotome could be made to occur twice and no more. Therefore, although this view is supported by no less an authority than Böckh, we must refuse to attribute to Plato a scale which is altogether barbarous.

τῆς τοῦ μορίου] τῆς δὲ has been retained by Hermann, who defends it as coordinating λείπων and ἐχούσης. But it seems to me rather clumsy.

7. **οἷον χῖ προσβαλών**] We are to conceive the soul, after having been duly blended and having received her mathematical ratios, as extended like a horizontal band: then the creator cleaves it lengthwise and lays the two strips across each other in the shape of the letter X (i.e. at an acute angle), and so that the two centres coincide: next he bends them both round till the ends meet, so that each becomes a circle touching the other at a point in their circumferences opposite to the original point of contact. Thus we have two circles bisecting each other

112 ΠΛΑΤΩΝΟΣ [36 C—

ξυνάψας αὐταῖς τε καὶ ἀλλήλαις ἐν τῷ καταντικρὺ τῆς προσβολῆς, καὶ τῇ κατὰ ταὐτὰ καὶ ἐν ταὐτῷ περιαγομένῃ κινήσει πέριξ
αὐτὰς ἔλαβε, καὶ τὸν μὲν ἔξω, τὸν δ' ἐντὸς ἐποιεῖτο τῶν κύκλων.
τὴν μὲν οὖν ἔξω φορὰν ἐπεφήμισεν εἶναι τῆς ταὐτοῦ φύσεως, τὴν
5 δ' ἐντὸς τῆς θατέρου. τὴν μὲν δὴ ταὐτοῦ κατὰ πλευρὰν ἐπὶ δεξιὰ
περιήγαγε, τὴν δὲ θατέρου κατὰ διάμετρον ἐπ' ἀριστερά, κράτος
δ' ἔδωκε τῇ ταὐτοῦ καὶ ὁμοίου περιφορᾷ· μίαν γὰρ αὐτὴν ἄσχι- D
στον εἴασε, τὴν δ' ἐντὸς σχίσας ἑξαχῇ ἑπτὰ κύκλους ἀνίσους κατὰ

3 αὐτάς: αὐτῆς A.

and inclined at an acute angle. The obliquity of the inclination is insisted on, because, as we shall presently see, the two circles represent respectively (amongst other things) the equator and the ecliptic.

2. **πέριξ αὐτὰς ἔλαβε**] As the soul was interfused throughout the whole sphere of the universe, we must regard the two circles simply as a framework, so to speak, denoting the directions of the two movements. These two circles are encompassed by a moving spherical envelope, being the circumference of the entire sphere of soul, revolving κατὰ ταὐτὰ καὶ ἐν ταὐτῷ.

3. **τὸν μὲν ἔξω**] The circle of the Same is made exterior, because it was to control the circle of the Other, and also because it symbolises the sphere of the fixed stars.

5. **κατὰ πλευράν**] This expression will be readily understood by means of the accompanying diagram. ACE, CDG

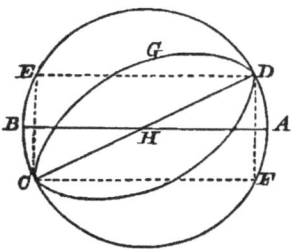

are two circles in different planes, cutting each other at the points C, D. AB and CD are their respective diameters, bisecting one another in H. The dotted lines are a parallelogram inscribed in the circle ACE, having its sides ED, CF parallel to AB and having CD for its diagonal. The rotation of the circle ACE, which is the circle of the Same, is κατὰ πλευράν, in the direction of DE; that is, its axis is perpendicular to DE or AB, and it revolves from east to west. CDG, the circle of the Other, rotates κατὰ διάμετρον, i.e. in the direction of the diagonal CD, from WSW to ENE. The Greek term ἡ διάμετρος generally means diagonal, not diameter. Proklos sees a special significance in the circle of the Other moving κατὰ διάμετρον, inasmuch as (the sides of the rectangle being expressed by integral numbers) the diagonal is irrational. It is quite possible that Plato may have thought of this: but, as Böckh has remarked, unless the rectangle is a square, the diagonal is not necessarily a surd: e.g. if the sides are 3 and 4, the diagonal will be 5.

ἐπὶ δεξιά.. ἐπ' ἀριστερά] This has given rise to much discussion, because according to the usual Greek nomenclature the east was the right side of the heavens and the west the left: and so we have it in Laws 760 D τὸ δ' ἐπὶ δεξιὰ γιγνέσθω τὸ πρὸς ἕω; cf. Epinomis 987 B. This mode of reckoning seems to have arisen from the fact that the Greek diviners stood facing the north in taking the omens. I think the explanation of Plato's present departure from ordinary custom is simple enough. The diurnal motion

D]   ΤΙΜΑΙΟΣ.   113

themselves and each other at a point opposite to that of their original contact : and he comprehended them in the motion that revolves uniformly on the same axis, and one of the circles he made exterior and one interior. The exterior motion he named the motion of the Same, the interior that of the Other. And the circle of the Same he made revolve to the right by way of the side, that of the Other to the left by way of the diagonal. And he gave the supremacy to the motion of the same and uniform, for he left that single and undivided; but the inner circle he cleft into seven unequal circles in the proportion of the

of the universe is visible only by the daily motion of the heavenly bodies, especially the sun. An observer in Europe can only see the sun's motions by looking towards the south, when of course the west is on his right hand : compare Pliny *natur. hist.* VI § 24 (of some visitors from the tropics) sed maxume mirum iis erat umbras suas in nostrum caelum cadere, non in suum, solemque a laeva oriri et in dextram occidere potius quam e diverso. Plato's use of the terms right and left seems then perfectly natural. The universe being a sphere, Plato knew that the right and left, like up and down, are perfectly arbitrary terms (see 62 C foll.) and he therefore did not hesitate to apply them just as suited his purpose. Those who are curious on the subject may find (to put it mildly) some very singular arguing in the opposite sense in Aristotle *de caelo* II ii 284ᵇ 6 foll.

6. **κράτος δ' ἔδωκε τῇ ταὐτοῦ**] That is, while the circle of the Other retains its independent rotation round its own centre, it is also carried round by the revolution of the Same.

ἄσχιστον ἔασε] Note that though the circle of the Same is one and undivided, it contains the same mathematical ratios as the Other: this clearly signifies that the multiplicity of the Other is only a different form of the unity of the Same— there exists in immaterial soul a law or principle which, when expressed in terms of matter (or here rather of the apprehen-

sion of matter), assumes the form of these mathematical ratios. Note also that the portion of the soul which constitutes the circle of the Same is composed both of Same and of Other, as also is the circle of the Same. The antithesis Same and Other pervades all οὐσία from highest to lowest.

8. **σχίσας ἑξαχῇ**] The circle of the Other is subdivided into seven concentric circles corresponding to the seven planets which were reckoned in Plato's day. These are ordered at distances from the earth corresponding to the seven numbers of the τετρακτύς: 1 represents the distance of the moon, 2 the sun, 3 Venus, 4 Mercury, 8 Mars, 9 Jupiter, 27 Saturn.

The question might suggest itself, how would Plato have been affected, had he become aware that the real position of the heavenly bodies is widely different from his supposition? In my judgment he would have been absolutely unconcerned. How these bodies are situated is to him a matter of profound indifference: what does concern him is that wherever they are and whatever they do should be the result of the orderly evolution of νοῦς. For it should be borne in mind that, strange and fantastic as this ψυχογονία may seem at first sight, Plato has but one aim steadily in view throughout. Whatever exists and happens in material nature is simply the material symbol of immaterial truth : it is the inevitable result of the regular evolution of spirit, according to the eternal law of its nature,

P. T.   8

τὴν τοῦ διπλασίου καὶ τριπλασίου διάστασιν ἑκάστην, οὐσῶν
ἑκατέρων τριῶν, κατὰ τἀναντία μὲν ἀλλήλοις προσέταξεν ἰέναι
τοὺς κύκλους, τάχει δὲ τρεῖς μὲν ὁμοίως, τοὺς δὲ τέτταρας ἀλλήλοις καὶ τοῖς τρισὶν ἀνομοίως, ἐν λόγῳ δὲ φερομένους.
5 IX. Ἐπεὶ δὲ κατὰ νοῦν τῷ ξυνιστάντι πᾶσα ἡ τῆς ψυχῆς
ξύστασις ἐγεγένητο, μετὰ τοῦτο πᾶν τὸ σωματοειδὲς ἐντὸς αὐτῆς
ἐτεκταίνετο καὶ μέσον μέσῃ ξυναγαγὼν προσήρμοττεν· ἡ δ' ἐκ E
μέσου πρὸς τὸν ἔσχατον οὐρανὸν πάντῃ διαπλακεῖσα κύκλῳ τε
αὐτὸν ἔξωθεν περικαλύψασα, αὐτὴ ἐν αὑτῇ στρεφομένη, θείαν
10 ἀρχὴν ἤρξατο ἀπαύστου καὶ ἔμφρονος βίου πρὸς τὸν ξύμπαντα
χρόνον. καὶ τὸ μὲν δὴ σῶμα ὁρατὸν οὐρανοῦ γέγονεν, αὐτὴ δὲ
ἀόρατος. μέν, λογισμοῦ δὲ μετέχουσα καὶ ἁρμονίας ψυχή, τῶν 37 A
νοητῶν ἀεί τε ὄντων ὑπὸ τοῦ ἀρίστου ἀρίστη γενομένη τῶν

3 ἀλλήλοις : ἀλλήλοις τε S.   · 8 διαπλακεῖσα : διαπλεκεῖσα Λ.

in corporeal manifestation. Plato does not of course mean that the immaterial and indivisible essence of soul is composed of circles and distributed in mathematical proportions. The circle is with him a common symbol of the activity of thought : and by assigning the harmonic numbers to soul he declares that whatever relations or harmonies, mathematical or otherwise, are found in the world of space and time, these are the natural expression in material terms of some eternal law of soul. It is perhaps advisable to notice this, because of the amusing literalness with which Aristotle has treated the subject in *de anima* I iii 407ᵃ 2 foll.—a piece of criticism which at first it is hard to believe was intended seriously.

2. κατὰ τἀναντία] As seven circles cannot all be contrary each to each, we are to suppose that the three planets having the same period revolve in one direction, and the four others in the opposite. It is usually supposed that Mercury and Venus alone have the contrary motion ; but if Plato's theory is to be anything like an explanation of the facts, the sun must have the same direction as these two : see note on 38 D τὴν δ' ἐναντίαν εἰληχότας αὐτῷ δύναμιν, where the motive

for this arrangement is discussed. In the parallel passage of the *Republic*, 616 D—617 C, it is not said that any of the planets have a contrary motion, though it is stated that Venus, Mercury and the Sun complete their orbits in the same period. The harmonic numbers of the *Timaeus* seem to be represented by the eight Sirens, who stood on the σφόνδυλοι, each singing one tone. In the *Republic* there are eight spheres, because the fixed stars are included, which here are assigned to the circle of the Same. For Aristotle's views about the music of the spheres see *de caelo* II ix 290ᵇ 12 foll.: he thinks the idea κομψόν, ἐμμελές, and μουσικόν, but cannot believe it.

36 D—37 C, c. ix. So when God had ended the framing of the soul to his mind, next he formed within her all the visible body of the universe: but she herself is invisible, the noblest creation of the most perfect creator. And seeing that she is composed of Same and Other and Essence, whenever she comes in contact with aught that has being, be it divided or indivisible, she discerns sameness in it and difference and all else that is predicable of it. And her verdict is true both concerning material and immaterial

37 A] ΤΙΜΑΙΟΣ. 115

double and triple intervals severally, each being three in number; and he appointed that the circles should move in opposite directions, three at the same speed, the other four differing in speed from the three and among themselves, yet moving in a due ratio.

IX. Now after that the framing of the soul was finished to the mind of him that framed her, next he fashioned within her all that is bodily, and he drew them together and fitted them middle to middle. And from the midst even unto the ends of heaven she was woven in everywhere and encompassed it around from without, and having her movement in herself she began a divine beginning of endless and reasonable life for ever and evermore. Now the body of the universe has been created visible; but she is invisible, and she, even soul, has part in reason and in harmony. And whereas she is made by the best of all whereunto belong reason and eternal being, so she is

existence: for when, by the circle of the Other she deals with sensibles, she forms sure opinions and beliefs; but when by the circle of the Same she apprehends intelligible being, then knowledge and reason, which soul alone possesses, are made perfect in her.

5. κατὰ νοῦν] Probably, as in *Phaedo* 97 D, there is a double meaning in these words 'to his mind', and 'according to reason'.

6. μετὰ τοῦτο] τὸ δὲ μετὰ τοῦτο μὴ χρονικὸν ὑπολάβῃς, ἀλλὰ τάξεως σημαντικόν, says Proklos very rightly.

7. μέσον μέσῃ] Soul, being immaterial, has of course no centre. The phrase simply means that the whole sphere of material nature from centre to circumference was instinct with the indwelling vital force. πάντῃ διαπλακεῖσα, i.e. she interpenetrated its every particle, being everywhere present in her two modes of Same and of Other.

9. ἔξωθεν περικαλύψασα] See note on 34 B. Plutarch *de anim. procr.* § 21 says Palto is ὥσπερ ἀπωθούμενος τῆς ψυχῆς τὴν ἐκ σώματος γένεσιν. Compare Plotinos *ennead* II ix 7 ἐν γὰρ τῇ πάσῃ ψυχῇ ἡ τοῦ

σώματος φύσις δεδεμένη ἤδη συνδεῖ ὃ ἂν περιλάβῃ, αὐτὴ δὲ ἡ τοῦ παντὸς ψυχὴ οὐκ ἂν δέοιτο ὑπὸ τῶν ὑπ' αὐτῆς δεδεμένων.

10. ἤρξατο] Again of course a beginning κατ' ἐπίνοιαν only.

11. καὶ τὸ μὲν δὴ σῶμα] So *Laws* 898 D ἡλίου πᾶς ἄνθρωπος σῶμα μὲν ὁρᾷ, ψυχὴν δὲ οὐδείς.

12. λογισμοῦ δὲ μετέχουσα καὶ ἀρμονίας ψυχή] Notwithstanding Stallbaum's defence of ψυχή, I feel strong misgivings as to its genuineness: its position is strange and disturbs the connexion.

τῶν νοητῶν ἀεί τε ὄντων] It is very significant that the δημιουργὸς is identified with the object of reason, νοῦς with νοητόν. Here then we have another token that the δημιουργὸς is merely a mythological representative of universal νοῦς which evolves itself in the form of the κόσμος. Still more remarkable is the use of λογιστικὸν below in 37 C. There is no other passage in Plato where λογιστικὸν is contrasted with αἰσθητόν: the regular term is of course νοητόν. It is surely impossible that Plato could have substituted λογιστικὸν for νοη-

8—2

γεννηθέντων. ἄτε οὖν ἐκ τῆς ταὐτοῦ καὶ τῆς θατέρου φύσεως ἔκ τε οὐσίας τριῶν τούτων συγκραθεῖσα μοιρῶν, καὶ ἀνὰ λόγον μερισθεῖσα καὶ ξυνδεθεῖσα, αὐτή τε ἀνακυκλουμένη πρὸς αὑτήν, ὅταν οὐσίαν σκεδαστὴν ἔχοντός τινος ἐφάπτηται καὶ ὅταν ἀμέ-
5 ριστον, λέγει κινουμένη διὰ πάσης ἑαυτῆς, ὅτῳ τ' ἄν τι ταὐτὸν ᾖ καὶ ὅτου ἂν ἕτερον, πρὸς ὅ τί τε μάλιστα καὶ ὅπῃ καὶ ὅπως καὶ B ὁπότε ξυμβαίνει κατὰ τὰ γιγνόμενά τε πρὸς ἕκαστον ἕκαστα εἶναι καὶ πάσχειν καὶ πρὸς τὰ κατὰ ταὐτὰ ἔχοντα ἀεί· λόγος δὲ ὁ κατὰ ταὐτὸν ἀληθὴς γιγνόμενος περί τε θάτερον ὢν καὶ περὶ
10 τὸ ταὐτόν, ἐν τῷ κινουμένῳ ὑφ' αὑτοῦ φερόμενος ἄνευ φθόγγου καὶ ἠχῆς, ὅταν μὲν περὶ τὸ αἰσθητὸν γίγνηται καὶ ὁ τοῦ θατέρου κύκλος ὀρθὸς ὢν εἰς πᾶσαν αὐτὰ τὴν ψυχὴν διαγγείλῃ, δόξαι καὶ πίστεις γίγνονται βέβαιοι καὶ ἀληθεῖς· ὅταν δὲ αὖ περὶ τὸ λογιστικὸν ᾖ καὶ ὁ τοῦ ταὐτοῦ κύκλος εὔτροχος ὢν αὐτὰ μηνύσῃ, C
15 νοῦς ἐπιστήμη τε ἐξ ἀνάγκης ἀποτελεῖται· τούτω δὲ ἐν ᾧ τῶν

7 ξυμβαίνει: ξυμβαίνηι Λ.   9 ὢν: ὂν ΛΗ.   12 αὐτὰ scripsi: αὐτοῦ ΛΗΣΖ.

τὸν until he had reached a period in his metaphysic where he deliberately affirmed the identity of thought and its object. I believe also his present use both of νοητῶν and of λογιστικὸν is purposely designed to draw attention to this.

3. μερισθεῖσα καὶ ξυνδεθεῖσα] μερισθεῖσα refers to the original distribution of the soul according to the seven numbers of the τετρακτύς, ξυνδεθεῖσα to the introduction of the δεσμοί, the arithmetical and harmonical means which mediated between them.

αὐτή τε ἀνακυκλουμένη πρὸς αὑτή[ν] This is merely Plato's favourite metaphor describing the activity of thought, which is complete and perfect in itself.

4. οὐσίαν σκεδαστήν] Formerly called ἡ κατὰ τὰ σώματα μεριστή: i.e. οὐσία which appears in the form of plurality, sensible phenomena, opposed to ἀμέριστον, which is νοητόν.

5. κινουμένη διὰ πάσης ἑαυτῆς] This is the consequence of the soul being composed not only of ταὐτὸν and θάτερον but of οὐσία. Had the circles of Same and Other been the only possession of the soul, the experiences of each circle might have been confined to it: but now, since the elements of ταὐτὸν and θάτερον are unified in οὐσία, the reports received from either circle are the property of the whole soul.

ὅτῳ τ' ἄν τι ταὐτὸν ᾖ] Stallbaum, affirming that no one has hitherto understood this passage, takes the antecedent of ὅτῳ as the subject of ξυμβαίνει: 'she declares of that wherewith anything is the same and wherefrom it is different, in relation to what &c'. It may well be doubted whether he has thus improved upon his predecessors. Surely the discernment of sameness and difference is a function necessarily belonging to soul and necessarily included in the catalogue of her functions: yet Stallbaum's rendering excludes it from that catalogue. The fact that we have ὅτῳ ἄν ᾖ, not ὅτῳ ἐστί, does not really favour his view—'with whatsoever a thing may be the same, she declares it the same'. I coincide then with the other interpreters in regarding the whole sentence from ὅτῳ τ' ἄν as indirect interrogation subordinate to λέγει.

6. πρὸς ὅ τί τε μάλιστα] Lindau has justly remarked that all or nearly all

c]   ΤΙΜΑΙΟΣ.   117

the best of all that is brought into being. Therefore since she is
formed of the nature of Same and of Other and of Being, of these
three portions blended, in due proportion divided and bound
together, and turns about and returns into herself, whenever she
touches aught that has manifold existence or aught that has
undivided, she is stirred through all her substance, and she tells
that wherewith the thing is same and that wherefrom it is
different, and in what relation or place or manner or time
it comes to pass both in the region of the changing and in the
region of the changeless that each thing affects another and
is affected. This word of hers is true alike, whether it deal with
Same or with Other, without voice or sound in the Self-moved
arising; and when she is busied with the sensible, and the circle
of the Other, being true, announces it throughout all the soul,
then are formed sure opinions and true beliefs; and when she is
busy with the rational, and the circle of the Same declares
it, running smoothly, then reason and knowledge cannot but be
made perfect. And in whatsoever existing thing these two are

Aristotle's ten categories are to be found in this sentence.

8. πρὸς τὰ κατὰ ταὐτά] This phrase is exactly parallel to κατὰ τὰ γιγνόμενα above. The only reason for the change of preposition is the obvious lack of euphony in κατὰ τὰ κατὰ ταὐτά.

λόγος] 'her verdict'. λόγος = ὁ λέγει, what she pronounces concerning that which is submitted to her judgment. Stallbaum aptly refers to *Sophist* 263 E οὐκοῦν διάνοια μὲν καὶ λόγος ταὐτόν· πλὴν ὁ μὲν ἐντὸς τῆς ψυχῆς πρὸς αὐτὴν διάλογος ἄνευ φωνῆς γιγνόμενος τοῦτ' αὐτὸ ἡμῖν ἐπωνομάσθη, διάνοια. See too *Philebus* 39 A, and *Theaetetus* 189 E, where Sokrates defines διανοεῖσθαι as λόγον ὃν αὐτὴ πρὸς αὑτὴν ἡ ψυχὴ διεξέρχεται περὶ ὧν ἂν σκοπῇ.

9. κατὰ ταὐτόν is adverbial, 'equally': there is nothing in it of the technical sense of ταὐτόν.

10. ἐν τῷ κινουμένῳ ὑφ' αὑτοῦ] i.e. ἐν ψυχῇ, ψυχὴ being αὐτοκίνητος.

12. ὀρθὸς ὤν] Proklos draws attention to the difference of the language applied to the two circles; of the circle of the Same it is said εὔτροχος ὤν. The change of expression is readily understood if we turn to 43 D foll. where Plato is speaking of the disturbance of the circles by the continuous influx. of bodily nutriment: the circle of the Other is distorted and displaced, but the circle of the Same is only blocked (ἐπέδησαν).

εἰς πᾶσαν αὐτὰ τὴν ψυχὴν διαγγείλῃ] The ms. reading αὐτοῦ is clearly wrong, though Martin defends it. Stallbaum proposes αὐτό: but as we presently have αὐτὰ referring to λογιστικόν, that is perhaps more likely to be right here.

13. βέβαιοι καὶ ἀληθεῖς] There is a slight chiasmus: βέβαιοι is appropriate to πίστεις and ἀληθεῖς to δόξαι.

περὶ τὸ λογιστικὸν ᾖ] Of the peculiar use of λογιστικόν I have already spoken. Note however that the verb is changed from γίγνηται to ᾖ and for διαγγείλῃ we have the more authoritative word μηνύσῃ.

15. τούτῳ δέ] There has been much

ὄντων ἐγγίγνεσθον, ἄν ποτέ τις αὐτὸ ἄλλο πλὴν ψυχὴν εἴπῃ, πᾶν μᾶλλον ἢ τἀληθὲς ἐρεῖ.

X. Ὡς δὲ κινηθὲν αὐτὸ καὶ ζῶν ἐνόησε τῶν ἀιδίων θεῶν γεγονὸς ἄγαλμα ὁ γεννήσας πατήρ, ἠγάσθη τε καὶ εὐφρανθεὶς ἔτι
5 δὴ μᾶλλον ὅμοιον πρὸς τὸ παράδειγμα ἐπενόησεν ἀπεργάσασθαι. καθάπερ οὖν αὐτὸ τυγχάνει ζῷον ἀίδιον ὄν, καὶ τόδε τὸ πᾶν οὕτως D εἰς δύναμιν ἐπεχείρησε τοιοῦτον ἀποτελεῖν. ἡ μὲν οὖν τοῦ ζῴου φύσις ἐτύγχανεν οὖσα αἰώνιος. καὶ τοῦτο μὲν δὴ τῷ γεννητῷ παντελῶς προσάπτειν οὐκ ἦν δυνατόν· εἰκὼ δ' ἐπινοεῖ κινητόν
10 τινα αἰῶνος ποιῆσαι, καὶ διακοσμῶν ἅμα οὐρανὸν ποιεῖ μένοντος αἰῶνος ἐν ἑνὶ κατ' ἀριθμὸν ἰοῦσαν αἰώνιον εἰκόνα, τοῦτον ὃν δὴ

3 ἐνόησε : ἐνενόησε SZ. 6 ὄν omittunt AS. 9 ἐπινοεῖ : ἐπενόει A.

discussion as to the exact reference of τούτω. One interpretation, mentioned by Proklos, is to refer it to the two pairs, δόξαι πίστεις, νοῦς ἐπιστήμη: and this is practically the view of Stallbaum, who understands δόξα and ἐπιστήμη. The natural grammatical reference however is to νοῦς ἐπιστήμη τε, and so I believe we should understand it: cf. 30 B νοῦν δ' αὖ χωρὶς ψυχῆς ἀδύνατον παραγενέσθαι τῳ. No doubt it is true that δόξα and πίστις are equally impossible χωρὶς ψυχῆς: but these are functions of soul in her material relations, whereas the other two are characteristic of soul qua soul, in the activity of pure thought. The distinction between νοῦς and ἐπιστήμη is that between the faculty of reason and the possession of knowledge.

37 C—38 B, c. x. So when the universe was quickened with soul, God was well pleased; and he bethought him to make it yet more like its type. And whereas the type is eternal and nought that is created can be eternal, he devised for it a moving image of abiding eternity, which we call time. And he made days and months and years, which are portions of time; and past and future are forms of time, though we wrongly attribute them also to eternity. For of eternal Being we ought not to say 'it was', 'it shall be', but 'it is' alone: and in like manner we are wrong in saying 'it is' of sensible things which become and perish; for these are ever fleeting and changing, having their existence in time.

3. κινηθὲν αὐτὸ καὶ ζῶν] Motion is always for Plato the inalienable characteristic of life: cf. *Phaedrus* 245 E and *Theaetetus* 153 A τὸ μὲν εἶναι δοκοῦν καὶ τὸ γίγνεσθαι κίνησις παρέχει, τὸ δὲ μὴ εἶναι καὶ τὸ ἀπόλλυσθαι ἡσυχία.

τῶν ἀιδίων θεῶν γεγονὸς ἄγαλμα] This is a very singular phrase. The κόσμος we know is the image of the αὐτὸ ζῷον, and the creatures in it are images of the νοητὰ ζῷα. Therefore the ἀίδιοι θεοί can be nothing else than the ideas. But nowhere else does Plato call the ideas 'gods', and the significance of so calling them is very hard to see. If however Plato wrote θεῶν (which I cannot help regarding as doubtful), I am convinced that he used this strange phrase with some deliberate purpose in view; but what that purpose was, I confess myself unable to divine. The interpretation of Proklos is naught.

6. αὐτό] sc. τὸ παράδειγμα.

8. ἐτύγχανεν οὖσα αἰώνιος] Presently Plato tells us that the past tense is not applicable to eternal existence: the use of it is however necessitated by the narrative form into which he has thrown his theory. This use of ἐτύγχανεν, in

D]   ΤΙΜΑΙΟΣ.   119

found, if a man affirm it is aught but soul, what he says will be anything rather than the truth.

X. And when the father who begat it perceived the created image of the eternal gods, that it had motion and life, he rejoiced and was well pleased; and he bethought him to make it yet more nearly like its pattern. Now whereas that is a living being eternally existent, even so he essayed to make this All the like to the best of his power. Now so it was that the nature of the ideal was eternal. But to bestow this attribute altogether upon a created thing was impossible; so he bethought him to make a moving image of eternity, and while he was ordering the universe he made of eternity that abides in unity an eternal image moving according to number, even that which we have

the face of the explicit declaration a few lines later, is an additional proof, if more were wanted, that the creation of the κόσμος is pure allegory. For if Plato meant to be understood literally, he is flagrantly violating his own law.

9. εἰκὼ δ' ἐπινοεῖ] Plato's meaning in terming time the εἰκών of eternity may thus be stated. As extension is to the immaterial, so is succession to the eternal. The material existence is the εἰκών of pure being or thought: that is to say, it is the mode in which the One manifests itself in the form of multiplicity. Now the two main characteristics of material existence are (α) extension, (β) succession. The universe then regarded as extended is the εἰκών of νοῦς regarded as unextended: the same universe regarded as a succession of phenomena is the εἰκών of νοῦς regarded as eternal. As then space is the image of the immaterial, so is time the image of the eternal: space and time being the conditions under which the spaceless and timeless ἕν evolves itself in the apprehension of finite intelligences.

11. κατ' ἀριθμὸν ἰοῦσαν] i.e. moving by measurable periods: the ἀριθμὸς is the temporal reflection of the changeless ἕν of eternity.

αἰώνιον εἰκόνα] This phrase surely deserves more notice than it has hitherto obtained. In the present passage we have time and eternity most sharply contrasted; time being explained as a condition belonging to that which is not eternal. And notwithstanding this, time is itself declared to be eternal. Plato's careful definition of the word αἰώνιος entirely precludes the supposition that it here denotes merely the everlasting duration of time. In what sense then is it eternal? I think but one answer is possible. The universal mind has of necessity not only existence in the form of unity, but also existence in the form of multiplicity. It is to the existence in multiplicity that time appertains. But although time is a condition of the phenomena contained in this manifold existence, that existence is itself eternal; for mind is eternal whether existing as one or as many: its self-evolution is eternal, not in time. Temporality then is the attribute of the particular things comprised in μεριστὴ οὐσία, but the mode of mind's existence which takes that form is eternal. It is in fact part of the eternal essence of mind that it should exist in the form of things which are subject to time. Thus there is a sense in which time may be termed eternal, as one element in the eternal

χρόνον ὠνομάκαμεν. ἡμέρας γὰρ καὶ νύκτας καὶ μῆνας καὶ
ἐνιαυτούς, οὐκ ὄντας πρὶν οὐρανὸν γενέσθαι, τότε ἅμα ἐκείνῳ E
ξυνισταμένῳ τὴν γένεσιν αὐτῶν μηχανᾶται· ταῦτα δὲ πάντα
μέρη χρόνου, καὶ τό τ' ἦν τό τ' ἔσται χρόνου γεγονότα εἴδη, ἃ δὴ
5 φέροντες λανθάνομεν ἐπὶ τὴν ἀίδιον οὐσίαν οὐκ ὀρθῶς. λέγομεν
γὰρ δὴ ὡς ἦν ἔστι τε καὶ ἔσται, τῇ δὲ τὸ ἔστι μόνον κατὰ τὸν
ἀληθῆ λόγον προσήκει, τὸ δ' ἦν τό τ' ἔσται περὶ τὴν ἐν χρόνῳ 38 A
γένεσιν ἰοῦσαν πρέπει λέγεσθαι· κινήσεις γάρ ἐστον· τὸ δὲ ἀεὶ
κατὰ ταὐτὰ ἔχον ἀκινήτως οὔτε πρεσβύτερον οὔτε νεώτερον προσ-
10 ήκει γίγνεσθαι διὰ χρόνου οὐδὲ γενέσθαι ποτὲ οὐδὲ γεγονέναι
νῦν οὐδ' εἰσαῦθις ἔσεσθαι, τὸ παράπαν τε οὐδὲν ὅσα γένεσις τοῖς
ἐν αἰσθήσει φερομένοις προσῆψεν, ἀλλὰ χρόνου ταῦτα αἰῶνα
μιμουμένου καὶ κατ' ἀριθμὸν κυκλουμένου γέγονεν εἴδη. καὶ
πρὸς τούτοις ἔτι τὰ τοιάδε, τό τε γεγονὸς εἶναι γεγονὸς καὶ τὸ B
15 γιγνόμενον εἶναι γιγνόμενον, ἔτι δὲ τὸ γενησόμενον εἶναι γενησόμενον καὶ τὸ μὴ ὂν μὴ ὂν εἶναι, ὧν οὐδὲν ἀκριβὲς λέγομεν.
περὶ μὲν οὖν τούτων τάχ' ἂν οὐκ εἴη καιρὸς πρέπων ἐν τῷ
παρόντι διακριβολογεῖσθαι.

4 καὶ post ἦν inserit A.   12 αἰῶνα : αἰῶνά τε SZ.   15 ἔτι δέ : ἔτι τε A.

evolution of thought. It is eternal, not as an aggregate, but as a whole.

1. ἡμέρας...ἐνιαυτούς] There is a slight anacoluthon, τὴν γένεσιν αὐτῶν being substituted for the original object.

2. οὐκ ὄντας πρὶν οὐρανὸν γενέσθαι] That is to say, time and its divisions are not logically conceivable without the existence of a world of phenomena: if there is to be succession, there must be things to succeed each other. But as there is no beginning of the κόσμος in time, there is no beginning of time itself. Aristotle, with his usual confusion between metaphor and substance, accuses Plato of generating time in time: *physica* VIII i 251ᵇ 17 Πλάτων δ' αὐτὸν γεννᾷ μόνος. In Plato's narrative no other mode of expressing it would be admissible. Proklos well says χρόνος γὰρ μετ' οὐρανοῦ γέγονεν, οὐ χρόνου μόριον, ἀλλ' ὁ πᾶς χρόνος, ὥστε ἐν τῷ ἀπείρῳ χρόνῳ γίνεται ὁ οὐρανὸς καὶ ἀνέκλειπτός ἐστιν ἐφ' ἑκάτερα καθάπερ ὁ χρόνος.

4. γεγονότα εἴδη] i.e. forms or modes of time, and therefore belonging to γένεσις.

6. τῇ δὲ τὸ ἔστι] This passage leaves no doubt about the perfect clearness of Plato's conception of eternity as distinguished from time. Eternity is quite another thing from everlasting duration: it is that which μένει ἐν ἑνί, it is apart from time and has nothing to do with succession. Time has been and shall be for everlasting; but the infinity of its duration has nothing in common with eternity, for it is a succession. Plato, as he was certainly the first to form a real conception of immateriality, was probably the first who firmly grasped the notion of eternity. Parmenides indeed uses similar language, verse 64 (Karsten), οὔποτ' ἔην οὐδ' ἔσται, ἐπεὶ νῦν ἔστιν ὁμοῦ πᾶν | ἐν ξυνεχές. But the materiality attaching to his conception of ἕν renders it very doubtful whether he actually realised the full meaning of

named time. For whereas days and nights and months and years were not before the universe was created, he then devised the generation of them along with the fashioning of the universe. Now all these are portions of time, and *was* and *shall be* are forms of time that have come to be, although we wrongly ascribe them unawares to the eternal essence. For we say that it was and is and shall be, but in verity *is* alone belongs to it: and *was* and *shall be* it is meet should be applied only to Becoming which moves in time; for these are motions. But that which is ever changeless without motion must not become elder or younger in time, neither must it have become so in the past nor be so in the future; nor has it to do with any attributes that Becoming attaches to the moving objects of sense: these have come into being as forms of time, which is the image of eternity and revolves according to number. Moreover we say that the become *is* the become, and the becoming *is* the becoming, and that which shall become *is* that which shall become, and not-being *is* not-being. In all this we speak incorrectly. But concerning these things the present were perchance not the right season to inquire particularly.

this. It may even be doubted whether Aristotle, though Plato had preceded him, held an equally clear view: see for instance *de caelo* I ix 279ᵃ 23 foll. With the present passage may be compared the minute discussion in *Parmenides* 140 E—142 A.

8. κινήσεις γάρ ἐστον] i.e. they imply succession.

13. κατ' ἀριθμὸν κυκλουμένου] i.e. fulfilling regular periodic cycles, such as years months and days.

14. πρὸς τούτοις ἔτι τὰ τοιάδε] sc. οὐκ ὀρθῶς λέγομεν.

τὸ γεγονὸς εἶναι γεγονός] One inaccuracy of which we are guilty is to apply the terms ἦν and ἔσται to eternity: a second is to apply ἔστι to phenomena and to non-existence. To say that γεγονός *is* γεγονός is incorrect; for even as we say 'is', it has changed from what it was: it is ever moving and we can find no stable point where we can say it *is*. Compare Plutarch *de ei apud Delphos* § 19. Again to say μὴ ὂν *is* μὴ ὄν is absurd and contradictory. It might be rejoined that Plato has himself proved that μὴ ὄν does in a certain sense exist: *Sophist* 259 A ἔστι σαφέστατα ἐξ ἀνάγκης εἶναι τὸ μὴ ὄν. And in *Parmenides* 162 A he shows that δεῖ αὐτὸ δεσμὸν ἔχειν τοῦ μὴ εἶναι τὸ εἶναι μὴ ὄν, εἰ μέλλει μὴ εἶναι. In the *Sophist* however Plato, by elucidating the true nature of μὴ ὄν, is controverting the logical and metaphysical errors which arose from assuming that μὴ ὄν was an absolute contradictory of ὄν, and from ignoring the copulative force of ἐστί. Here he is complaining of that very use of ἐστίν as a copula: it is wrong, he says, that the word should have been employed for that purpose: it is the inaccuracy of human thought represented in language.

38 B—39 E, c. xi. So time is created

XI. Χρόνος δ' οὖν μετ' οὐρανοῦ γέγονεν, ἵνα ἅμα γεννηθέντες ἅμα καὶ λυθῶσιν, ἄν ποτε λύσις τις αὐτῶν γίγνηται, καὶ κατὰ τὸ παράδειγμα τῆς διαιωνίας φύσεως, ἵν' ὡς ὁμοιότατος αὐτῷ κατὰ δύναμιν ᾖ· τὸ μὲν γὰρ δὴ παράδειγμα πάντα αἰῶνά ἐστιν C ὄν, ὁ δ' αὖ διὰ τέλους τὸν ἅπαντα χρόνον γεγονώς τε καὶ ὢν καὶ ἐσόμενος. ἐξ οὖν λόγου καὶ διανοίας θεοῦ τοιαύτης πρὸς χρόνου γένεσιν, [ἵνα γεννηθῇ χρόνος,] ἥλιος καὶ σελήνη καὶ πέντε ἄλλα ἄστρα, ἐπίκλην ἔχοντα πλανητά, εἰς διορισμὸν καὶ φυλακὴν ἀριθμῶν χρόνου γέγονε. σώματα δὲ αὐτῶν ἑκάστων ποιήσας ὁ θεὸς ἔθηκεν εἰς τὰς περιφοράς, ἃς ἡ θατέρου περίοδος ᾔειν, ἑπτὰ D οὔσας ὄντα ἑπτά, σελήνην μὲν εἰς τὸν περὶ γῆν πρῶτον, ἥλιον δ' εἰς τὸν δεύτερον ὑπὲρ γῆς, ἑωσφόρον δὲ καὶ τὸν ἱερὸν Ἑρμοῦ λεγόμενον εἰς τοὺς τάχει μὲν ἰσόδρομον ἡλίῳ κύκλον ἰόντας,

3 διαιωνίας: αἰωνίας S.   7 ἵνα γεννηθῇ χρόνος inclusi.
8 πλανητά: πλανῆται S.   13 τούς: τόν AHZ.

along with the material universe and coeval therewithal, to complete its similitude to the eternal type. And for the measuring of time God made the sun and the moon and five other planets; and he set them in the seven orbits into which the circle of the Other was sundered, and gave each of them its fitting period: and being instinct with living soul every planet learnt and understood its appointed task. And those that revolved in smaller orbits fulfilled their revolutions more speedily than those which moved in larger. And whereas their orbits were inclined at an angle to the direction wherein the universe moves, the motion of the Same in its diurnal round converted all their circles into spirals: and since their motion was opposed to the rotation of the universe, whereby they were carried round, the slower, as making less way against this rotation, seemed more swift than the swifter and to overtake those by which they were in truth overtaken. And God kindled a light, even the sun, in the second orbit, that it should shine to the ends of the universe, and men might learn number from the heavenly periods.

For night and day are measured by the revolution of the universe, and months and years by the moon and the sun; and all the other planets give measures of time, diverse and manifold, though they are not accounted such by the multitude: and the perfect year is fulfilled when all the revolutions come round at the same time to the same point. For these causes were the heavenly bodies created.

1. **μετ' οὐρανοῦ γέγονεν**] 'has come into being in our story', as the tense denotes. Time and the material universe are of necessity strictly coeval, since each implies the other nor can exist apart from it.

2. **ἄν ποτε λύσις**] Proklos has some sensible remarks on this passage, saying σαφῶς ἀγέννητον καὶ ἄφθαρτον δείκνυσι τὸν οὐρανόν. εἰ γὰρ γέγονεν, ἐν χρόνῳ γέγονεν. εἰ δὲ μετὰ χρόνου γέγονεν, οὐκ ἐν χρόνῳ γέγονεν· οὐδὲ γὰρ ὁ χρόνος ἐν χρόνῳ γέγονεν, ἵνα μὴ πρὸ χρόνου χρόνος ᾖ. δεῖ γὰρ πᾶν τὸ γιγνόμενον μεταγενέστερον εἶναι χρόνου· ὁ δ' οὐρανὸς οὐδαμῶς ἐστι χρόνου μεταγενέστερος...ὅμοιον οὖν ὡς εἰ τις περιττὰς εἶναι βουλόμενος τὰς θατέρου περιφορὰς ἑπτάδα λέγοι συνυπάρχειν αὐταῖς,

ΤΙΜΑΙΟΣ.

XI. Time then has come into being along with the universe, that being generated together, together they may be dissolved, should a dissolution of them ever come to pass; and it was made after the pattern of the eternal nature, that it might be as like to it as was possible. For the pattern is existent for all eternity; but the copy has been and is and shall be throughout all time continually. So then this was the plan and intent of God for the generation of time; the sun and the moon and five other stars which have the name of planets have been created for defining and preserving the numbers of time. And when God had made their several bodies, he set them in the orbits wherein the revolution of the Other was moving, in seven circles seven stars. The moon he placed in that nearest the earth, and in the second above the earth he set the sun; and the morning-star and that which is held sacred to Hermes he assigned to those that moved in an orbit having equal speed with the sun,

ἵνα ἐάν ποτε ἡ ἑπτὰς ἀρτία γίγνηται, καὶ αὗται ἄρτιαι γίγνωνται, σημαίνων μὴ μεταπεσεῖσθαι τὰς περιφορὰς ἐπὶ τὸ ἄρτιον, οὕτω δὴ καὶ νῦν ἡγεῖσθαι νομιστέον περὶ τῆς ἀλυσίας τῆς τοῦ κόσμου τε καὶ τοῦ χρόνου.

5. ὁ δ' αὖ] Lindau understands χρόνος: but this produces tautology; evidently οὐρανός is to be supplied.

7. [ἵνα γεννηθῇ χρόνος] Although these words are in all mss. and in Proklos, they appear to me so unmistakably a mere gloss on πρὸς χρόνου γένεσιν that I have bracketed them. They are not represented in Cicero's translation.

8. ἐπίκλην ἔχοντα πλανητά] I have retained the reading of A, though Stallbaum's πλανῆται is perfectly good grammar; ἐπίκλην ἔχοντα being equivalent to ἐπικαλούμενα: compare *Symposium* 205 D τὸ τοῦ ὅλου ὄνομα ἴσχουσιν, ἐρωτᾶ τε καὶ ἐρᾶν καὶ ἐρασταί. In *Laws* 821 B Plato condemns the term πλανητά on the score of irreverence, as implying that these bodies wandered at random without law.

10. εἰς τὰς περιφοράς] sc. the zodiac.

11. ἥλιον δ' εἰς τὸν δεύτερον] This was the usual arrangement in Plato's time and down to Eudoxos and Aristotle: later astronomers placed the sun in the fourth or middle circle, above Venus and Mercury.

12. ἑωσφόρον] i.e. Venus. Plato was aware of the identity of ἑωσφόρος and ἕσπερος. It is somewhat strange that he gives none of the planets their usual appellations except Mercury; for these names must have been current in his day: they are all given in *Epinomis* 987 B, C. Other Greek names were for Saturn φαίνων, for Jupiter φαέθων, for Mars πυρόεις, for Mercury στίλβων, while Venus was φωσφόρος, ἑωσφόρος, or ἕσπερος: see Cicero *de natura deorum* II §§ 52, 53; pseudo-Aristotle *de mundo* 392ª 23.

13. εἰς τοὺς τάχει μὲν ἰσόδρομον] I have with Stallbaum adopted τούς. The reading τόν, which has best authority, can nevertheless hardly be right, since it would imply that Venus and Mercury had one and the same orbit. It may be objected that, if κύκλους is to be supplied, we have an awkward tautology in κύκλους κύκλον ἰόντας. But may we not understand πλανήτας? As to the equality of the periods assigned to the Sun, Venus, and Mercury, compare *Republic* 617 B

124   ΠΛΑΤΩΝΟΣ   [38 D—

τὴν δ' ἐναντίαν εἰληχότας αὐτῷ δύναμιν· ὅθεν καταλαμβάνουσί τε
καὶ καταλαμβάνονται κατὰ ταὐτὰ ὑπ' ἀλλήλων ἥλιός τε καὶ ὁ
τοῦ Ἑρμοῦ καὶ ἑωσφόρος. τὰ δ' ἄλλα οἱ δὴ καὶ δι' ἃς αἰτίας
ἱδρύσατο, εἴ τις ἐπεξίοι πάσας, ὁ λόγος πάρεργος ὢν πλέον ἂν
5 ἔργον ὧν ἕνεκα λέγεται παράσχοι. ταῦτα μὲν οὖν ἴσως τάχ' ἂν E
κατὰ σχολὴν ὕστερον τῆς ἀξίας τύχοι διηγήσεως. ἐπειδὴ δὲ οὖν

δευτέρους τε καὶ ἅμα ἀλλήλοις τόν τε ἕβ-
δομον καὶ ἕκτον and πέμπτον. The author
of the *Epinomis*, though in rather indefinite language, gives the same account,
986 E ἡ τετάρτη δὲ φορὰ καὶ διέξοδος ἅμα
καὶ πέμπτη τάχει μὲν ἡλίῳ σχεδὸν ἴση, καὶ
οὔτε βραδυτέρα οὔτε θάττων: cf. 990 B.
Probably, as Martin suggests, Plato was
led to this hypothesis by the observation
that at the end of the sun's annual revolution the two planets are in close proximity to him.

1. **τὴν δ' ἐναντίαν εἰληχότας αὐτῷ
δύναμιν**] These words are usually understood to mean that Venus and Mercury
revolve in a direction contrary to that of
the sun. This view I believe to be untenable. Aristotle indeed says, *metaph.*
Δ xii 1019ᵃ 15, δύναμις λέγεται ἡ μὲν ἀρχὴ
κινήσεως ἢ μεταβολῆς ἡ ἐν ἑτέρῳ ἢ ᾗ ἕτερον.
But still δύναμις ἐναντία cannot amount
in itself to contrary motion, only to a
contrary tendency, whatever that may be.
Moreover the facts which fell under

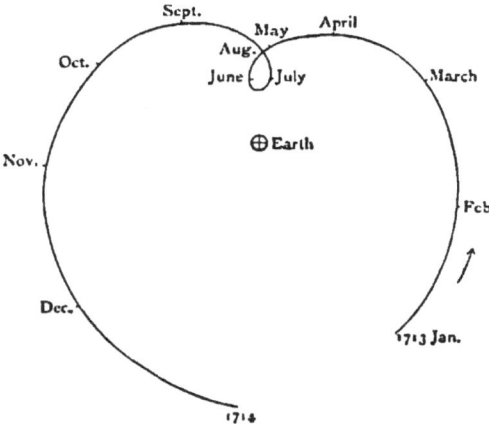

Plato's observation do not in the slightest
degree lend themselves to such a hypothesis. Martin gives the following statement of the facts which it is supposed
the contrary motion is intended to explain. After the conjunction of either
Venus or Mercury with the sun at perigee,
for some time the planet gains upon the
sun; then for several days it is nearly
stationary in relation to him; after which
it begins to lose ground, comes into conjunction with the sun at apogee, continues
for some time longer to lose ground, and
then again appears stationary: once more
it begins to gain on the sun, comes into
conjunction at perigee, and so forth *ad
infinitum*.

Now, as Martin observes, the theory of

but having a contrary tendency: wherefore the sun and Hermes and the morning star in like manner overtake and are overtaken one by another. And as to the rest, were we to set forth all the orbits wherein he put them and the causes wherefore he did so, the account, though only by the way, would lay on us a heavier task than that which was our chief object in giving it. These things perhaps may hereafter, when we have leisure, find a fitting exposition.

contrary motion is flagrantly inadequate to account for these facts; for since the motion of the planets will thus be approximately in the same direction as the motion of the Same, they would regularly and rapidly gain upon the sun. The truth is, as I believe, that Plato meant the sun to share the contrary motion of Venus and Mercury in relation to the other four planets. It is quite natural, seeing that the sun and the orbits of Venus and Mercury are encircled by the orbit of the earth, while Plato supposed them all to revolve about the earth, that he should class them together apart from the four whose orbits really do encircle that of the earth: his observations would very readily lead him to attributing to these three a motion contrary to the rest; but there seems nothing which could possibly have induced him to class the sun apart from the two inferior planets. But if this is so, what is the ἐναντία δύναμις? What I believe it to be may be understood from the accompanying figure, which is copied from part of a diagram in Arago's *Popular Astronomy*. This represents the motion of Venus relative to the earth during one year, as observed in 1713. It will be seen that the planet pursues her path among the stars pretty steadily from January to May; after that she wavers, begins a retrograde movement, and then once more resumes her old course, thus forming a loop, which is traversed from May to August. After that she proceeds unfaltering on her way for the rest of the year. This process is repeated so that five such loops are formed in eight years. Mercury behaves in precisely the same way, except that his curve is very much more complex and the loops occur at far shorter intervals. Now this is just what I believe is the ἐναντία δύναμις, this tendency on the part of Venus, as viewed from the earth, periodically to retrace her steps. These retrogressions of the planets were well known to the Greek astronomers, who invented a complex theory of revolving spheres to account for them. Probably Plato meant to put forward no very definite astronomical theory: for instance he gives no hint of the revolving spheres: he merely records the fact of this retrogressive tendency being observable.

If the contrary motion of the two planets is insisted on, the result follows that we have here the one theory in the whole dialogue which is manifestly and flagrantly inadequate. Plato's physical theories, however far they may differ from the conclusions of modern science, usually offer a fair and reasonable explanation of such facts as were known to him: they are sometimes singularly felicitous, and never absurd. I cannot then believe that he has here presented us with a hypothesis so obviously futile. And if he had, how did it escape the vigilance of Aristotle, who would have been ready enough to seize the occasion of making a telling point against Plato?

It is remarkable that neither in *Republic* 617 A, nor in *Epinomis* 986 E (the author of which must have been well acquainted with Plato's astronomy), nor

εἰς τὴν ἑαυτῷ πρέπουσαν ἕκαστον ἀφίκετο φορὰν τῶν ὅσα ἔδει
ξυναπεργάζεσθαι χρόνον, δεσμοῖς τε ἐμψύχοις σώματα δεθέντα
ζῷα ἐγεννήθη τό τε προσταχθὲν ἔμαθε, κατὰ δὴ τὴν θατέρου
φορὰν πλαγίαν οὖσαν, διὰ τῆς ταὐτοῦ φορᾶς ἰοῦσάν τε καὶ κρα- 39 A
5 τουμένην, τὸ μὲν μείζονα αὐτῶν, τὸ δ' ἐλάττω κύκλον ἰόν, θᾶττον
μὲν τὰ τὸν ἐλάττω, τὰ δὲ τὸν μείζω βραδύτερον περιῄειν. τῇ δὴ
ταὐτοῦ φορᾷ τὰ τάχιστα περιιόντα ὑπὸ τῶν βραδύτερον ἰόντων
ἐφαίνετο καταλαμβάνοντα καταλαμβάνεσθαι· πάντας γὰρ τοὺς
κύκλους αὐτῶν στρέφουσα ἕλικα, διὰ τὸ διχῇ κατὰ τὰ ἐναντία

4 ἰοῦσαν: ἰούσης et mox κρατουμένης AHZ.   7 τὰ τάχιστα: τὰ omittit A.
βραδύτερον: βραδυτέρων A.

yet in the pseudo-Timaeus Locrus, who has a rather minute paraphrase of the present passage, is there mention of a contrary motion as belonging to any of the planets.

4. **ἰοῦσάν τε καὶ κρατουμένην**] This correction is absolutely necessary. The circle of the Other passes διὰ τῆς ταὐτοῦ φορᾶς, that is, traverses it at the angle which the ecliptic makes with the equator, and is controlled by it, that is, it is carried round as a whole by the rotation of the Same. The relative motion of the Same and the Other are precisely exemplified, if we suppose an ordinary terrestrial globe to be revolving on its own axis, and a point upon its surface traversing it along the circle of the ecliptic in a direction approximately contrary to the globe's rotation: thus the point, while retaining its own independent motion on the surface of the globe, shares the rotary motion of the whole. Lindau would justify ἰούσης καὶ κρατουμένης by treating it as a genitive absolute referring to τὴν θατέρου φοράν: but this is hopeless.

5. **θᾶττον μὲν τὰ τὸν ἐλάττω**] Thus the periods of revolution continuously increase from the Moon to Saturn. Böckh has sufficiently demonstrated that the words θᾶττον and βραδύτερον do not refer to the absolute velocity of the planets through space, but to the celerity with which they accomplish their revolutions:

thus the moon, having the smallest orbit to traverse, completes it in by far the shortest period; although her actual velocity may be much less than that of Saturn who has the largest orbit and the longest period. Thus the Sun, Venus, and Mercury, having the same period for ἀποκατάστασις, differ in actual velocity in the proportion 2, 3, 4.

6. **τῇ δὴ ταὐτοῦ φορᾷ**] The difficult passage which follows has been very lucidly expounded by Böckh in his invaluable essay 'Ueber das kosmische System des Platon' pp. 38—48. Martin's note also is excellent: of Stallbaum's the less said the better. The two chief points requiring explanation are the apparent overtaking of the swifter planets by the slower, and the formation of the spirals. To take the former first, the sentence τῇ δὲ ταὐτοῦ...καταλαμβάνεσθαι is explained by the following πάντας γὰρ...ἀπέφαινεν.

Let the circle ACBD represent the universe, diurnally rotating from east to west on its own axis, which is perpendicular to the plane of the equator AB. The representation being in two dimensions, the straight lines AB, CD must be taken to indicate great circles of a sphere. Thus the motion of the Same is in the direction AB. The motion of the Other, or of the planets, is in the direction CD. Let us suppose two planets to be at a given time at the point E. Now had these planets, which we

But when each of the beings which were to join in creating time had arrived in its proper orbit, and had been generated as animate creatures, their bodies secured with living bonds, and had learnt their appointed task; then in the motion of the Other, which was slanting and crossed the motion of the Same and was thereby controlled, whereas one of these planets had a larger, another a smaller circuit, the lesser orbit was completed more swiftly, the larger more slowly: but because of the motion of the Same those which revolved most swiftly seemed to be overtaken by those that went more slowly, though really they overtook them. For the motion of the Same, twisting all their circles into spirals, because they have a separate and simul-

will call P^1 and P^2, no independent motion of their own, but were stationary relatively to the universe, it is obvious that in twenty-four hours the revolution of the Same would bring them both round to the same point E. But suppose that P^2

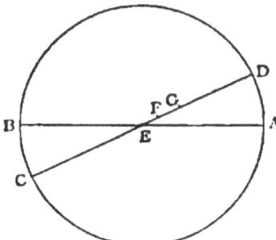

travels twice as fast as P^1 (that is accomplishes twice as great a fraction of its own orbit in the same time): then, while during the day P^1 has arrived at F, P^2 has got as far as G. Thus, since the course of the planets is approximately opposite to the rotation of the whole, P^2 has counteracted that motion to twice as great an extent as P^1, and accordingly is proportionally longer in being carried back opposite E. Thus P^1, departing more slowly from the revolution of the Same (βραδύτατα ἀπιὸν ἀπ' αὐτῆς), arrives at the same region of the heavens earlier than P^2, and so seems to the popular eye to have outstripped it. The revolution of the Same being immeasurably the swiftest, it is the motion imparted by this which attracts the eye from day to day; and when the leeway due to the planet's own motion is made up, the slower planet appears faster because it accomplishes the rotation of the Same in the shorter time. Supposing for instance on a given day the moon rises as the sun sets, on the following day the moon will not rise for perhaps an hour after sunset, thus appearing to have lost an hour on the sun.

9. **στρέφουσα ἕλικα**] The motion of the Same produces the spirals as follows. In the foregoing diagram we will suppose a planet at a given time to be at the point E. Now, as before said, were the planet itself stationary, this diurnal revolution would in twenty-four hours bring it round again to the point E; and the figure described by the planet would be a perfect circle. But, as it is, while the motion of the Same is whirling it round, the planet is travelling along its own path towards G. At the end of twenty-four hours then the planet is not at E but at G; and the figure it has described under the influence of the motion of the Same is accordingly not a circle but a spiral. Similarly the next diurnal revolution brings it back not to G, but to a point between G and D; so that each daily journey of the planet caused by the revolution of the Same is

128 ΠΛΑΤΩΝΟΣ [39 A—

ἅμα προϊέναι, τὸ βραδύτατα ἀπιὸν ἀφ' αὑτῆς οὔσης ταχίστης Β
ἐγγύτατα ἀπέφαινεν. ἵνα δ' εἴη μέτρον ἐναργές τι πρὸς ἄλληλα
βραδυτῆτι καὶ τάχει, καθ' ἃ περὶ τὰς ὀκτὼ φορὰς πορεύοιτο, φῶς
ὁ θεὸς ἀνῆψεν ἐν τῇ πρὸς γῆν δευτέρᾳ τῶν περιόδων, ὃ δὴ νῦν
5 κεκλήκαμεν ἥλιον, ἵνα ὅ τι μάλιστα εἰς ἅπαντα φαίνοι τὸν οὐρανὸν μετάσχοι τε ἀριθμοῦ τὰ ζῷα, ὅσοις ἦν προσῆκον, μαθόντα
παρὰ τῆς ταὐτοῦ καὶ ὁμοίου περιφορᾶς. νὺξ μὲν οὖν ἡμέρα τε
γέγονεν οὕτως καὶ διὰ ταῦτα, ἡ τῆς μιᾶς καὶ φρονιμωτάτης κυ- C
κλήσεως περίοδος· μεὶς δὲ ἐπειδὰν σελήνη περιελθοῦσα τὸν
10 ἑαυτῆς κύκλον ἥλιον ἐπικαταλάβῃ, ἐνιαυτὸς δὲ ὁπόταν ἥλιος
τὸν ἑαυτοῦ περιέλθῃ κύκλον. τῶν δ' ἄλλων τὰς περιόδους οὐκ
ἐννενοηκότες ἄνθρωποι, πλὴν ὀλίγοι τῶν πολλῶν, οὔτε ὀνομά-
ζουσιν οὔτε πρὸς ἄλληλα ξυμμετροῦνται σκοποῦντες ἀριθμοῖς,
ὥστε ὡς ἔπος εἰπεῖν οὐκ ἴσασι χρόνον ὄντα τὰς τούτων πλάνας,
15 πλήθει μὲν ἀμηχάνῳ χρωμένας, πεποικιλμένας δὲ θαυμαστῶς· D
ἔστι δ' ὅμως οὐδὲν ἧττον κατανοῆσαι δυνατόν, ὡς ὅ γε τέλεος
ἀριθμὸς χρόνου τὸν τέλεον ἐνιαυτὸν πληροῖ τότε, ὅταν ἁπασῶν

3 καθ' ἃ scripsi. καὶ τά ASZ. ὥς τά H.

a spiral. This of course in no wise affects its own proper movement along the circle of the Other.

It is necessary to bear clearly in mind that the apparent overtaking of the swifter by the slower planets has nothing to do with the spirals. The spirals are due solely to the obliquity of the ecliptic. But if there were no such obliquity, if the motion of the Other were directly opposed to that of the Same, the illusion concerning the swifter and slower planets would be unaltered. In that case P^1 and P^2, instead of travelling to F and G, would travel to points on EA equidistant with F and G from E. In this case no spirals would arise; the planets would all in good time get back to E; but P^1 would equally appear to have outstripped P^2.

A few words must be said concerning the construction, which is not quite free from obscurity. I agree with Böckh in joining διὰ τὸ διχῇ...προϊέναι with the preceding clause, but not in taking πάντας τοὺς κύκλους as the subject; for then it is hardly possible to give a suitable sense to διχῇ. But if we regard τὴν θατέρου φορὰν and τὴν ταὐτοῦ jointly as the subject of προϊέναι we are enabled to do so. The spirals are formed because the circles move διχῇ, that is, separately, asunder: i.e. they are not two contrary motions in the same circle, but two approximately contrary motions in two separate intersecting circles. κατὰ τἀναντία does not constitute any part of the cause why the spirals are formed; they would arise equally were the motion of the Other from D to C; but Plato is in fact condensing into this one clause a statement of how the spirals are formed and how the slower planets seem to overtake the swifter: the first is given by διχῇ, the second by κατὰ τἀναντία. The difficulty of the passage mainly arises from this extreme brevity.

3. καθ' ἅ] I have ventured upon this correction of the ms. reading καὶ τά, which certainly cannot stand, involving as it does the absurd conception that the hea-

D] ΤΙΜΑΙΟΣ. 129

taneous motion in the opposite way, being of all the swiftest displays closest to itself that which departs most slowly from it. And that there might be some clear measure of the relative swiftness and slowness with which they moved in their eight revolutions, God kindled a light in the second orbit from the earth, which we now have named the sun, in order that it might shine most brightly to the ends of heaven, and that living things, so many as was meet, should possess number, learning it from the motion of the same and uniform. Night then and day have been created in this manner and for these causes; and this is one revolution of the undivided and most intelligent circuit; and a month is fulfilled when the moon, after completing her own orbit, overtakes the sun; a year, when the sun has completed his own course. But the courses of the others men have not taken into account, save a few out of many; and they neither give them names nor measure them against one another, comparing them by means of numbers—nay I may say they do not know that time arises from the wanderings of these, which are incalculable in multitude and marvellously intricate. None the less however can we observe that the perfect number of time fulfils the perfect year at the moment

venly bodies could not see their way until their orbits were illumined by the Sun.

6. μαθόντα παρὰ τῆς ταὐτοῦ] Day and night are caused by the diurnal rotation of the universe, which is the motion of the Same, round the earth: and these, being smaller than any other divisions of time produced by the celestial bodies, are taken as the unit of measurement. Hence man derived the conception of number: compare 47 A, and *Epinomis* 978 C foll.

8. ἡ τῆς μιᾶς] The circle of the Same, it will be remembered, was left ἄσχιστος. The περίοδος is here put for the time consumed in completing the περίοδος, the νυχθήμερον, as Proklos calls it.

10. ἥλιον ἐπικαταλάβῃ] i.e. the synodic month of 29½ days; the sidereal month, or period in which the moon completes her own circuit, being about 27⅓.

14. οὐκ ἴσασι χρόνον ὄντα] Plato means that men have not generalised con-

cerning time: they do not reflect that the revolutions of the other celestial bodies equally afford measurements of time.

17. τὸν τέλεον ἐνιαυτόν] The perfect year is when all the planets return to one and the same region of the heavens at the same time. See Stobaeus *ecl.* I 264. σχῇ κεφαλήν, 'attain their starting-point'; as Stobaeus *l.l.* puts it, ὅταν ἐπὶ τοὺς ἀφ' ὧν ἤρξαντο τῆς κινήσεως ἀφικνῶνται τόπους. Alkinoos also says that the perfect number is complete when all the planets arrive in the same sign of the zodiac and are so situate that a radius drawn from the earth to the sphere of the fixed stars passes through the centres of all. The phrase σχῇ κεφαλήν seems like a technical term of astronomy, but I have found no other example of it, though Stobaeus speaks of a κεφαλὴ Κρόνου. As to the duration of the μέγας ἐνιαυτός there is no agreement among the ancients. Tacitus *dial. de orat.* 16 gives

P. T. 9

τῶν ὀκτὼ περιόδων τὰ πρὸς ἄλληλα ξυμπερανθέντα τάχη σχῇ
κεφαλὴν τῷ τοῦ ταὐτοῦ καὶ ὁμοίως ἰόντος ἀναμετρηθέντα κύκλῳ.
κατὰ ταῦτα δὴ καὶ τούτων ἕνεκα ἐγεννήθη τῶν ἄστρων ὅσα δι᾽
οὐρανοῦ πορευόμενα ἔσχε τροπάς, ἵνα τόδ᾽ ὡς ὁμοιότατον ᾖ τῷ E
5 τελέῳ καὶ νοητῷ ζῴῳ πρὸς τὴν τῆς διαιωνίας μίμησιν φύσεως.
 XII. Καὶ τὰ μὲν ἄλλα ἤδη μέχρι χρόνου γενέσεως ἀπείρ-
γαστο εἰς ὁμοιότητα ᾧπερ ἀπεικάζετο, τῷ δὲ μήπω τὰ πάντα
ζῷα ἐντὸς αὐτοῦ γεγενημένα περιειληφέναι, ταύτῃ ἔτι εἶχεν ἀνομοίως. τοῦτο δὴ τὸ κατάλοιπον ἀπειργάζετο αὐτοῦ πρὸς τὴν
10 τοῦ παραδείγματος ἀποτυπούμενος φύσιν. ᾗπερ οὖν νοῦς ἐνούσας
ἰδέας τῷ ὃ ἔστι ζῷον, οἷαί τε ἔνεισι καὶ ὅσαι, καθορᾷ, τοιαύτας
καὶ τοσαύτας διενοήθη δεῖν καὶ τόδε σχεῖν. εἰσὶ δὴ τέτταρες,
μία μὲν οὐράνιον θεῶν γένος, ἄλλη δὲ πτηνὸν καὶ ἀεροπόρον, 40 A
τρίτη δὲ ἔνυδρον εἶδος, πεζὸν δὲ καὶ χερσαῖον τέταρτον. τοῦ μὲν
15 οὖν θείου τὴν πλείστην ἰδέαν ἐκ πυρὸς ἀπειργάζετο, ὅπως ὅ τι
λαμπρότατον ἰδεῖν τε κάλλιστον εἴη, τῷ δὲ παντὶ προσεικάζων
εὔκυκλον ἐποίει, τίθησί τε εἰς τὴν τοῦ κρατίστου φρόνησιν ἐκείνῳ
ξυνεπόμενον, νείμας περὶ πάντα κύκλῳ τὸν οὐρανόν, κόσμον
ἀληθινὸν αὐτῷ πεποικιλμένον εἶναι καθ᾽ ὅλον. κινήσεις δὲ δύο

3 ἐγεννήθη : ἐγενήθη A.   9 ἀπειργάζετο : ἀπήρξατο AZ.   12 δή : δὲ S.

it on the authority of Cicero at 12954 years; but Cicero himself, *de natura deorum* II § 52, expresses no opinion.

1. τὰ πρὸς ἄλληλα ξυμπερανθέντα τάχη] i.e. when their several periods are accomplished simultaneously: τάχη of course refers to the period of ἀποκατάστασις, not to the actual velocity.

2. τῷ ταὐτοῦ] Because the periods are measured by the number of days and nights they contain.

39 E—40 D, *c.* xii. Next God created four kinds of living creatures in the universe, so many forms as he saw there were in the type. One, the race of the heavenly gods, he fashioned for the most part of fire; the second soared in the air; the third dwelt in the waters; and the fourth went upon dry land. The gods, who are the stars of heaven, he placed in the sphere of the Same to follow its revolution, so many of them as are fixed stars; and he gave them two motions, one a uniform rotation on their own axis, the other a forward revolution about the centre of the universe; but in the other five motions they had no part. The planets he set, as aforesaid, in the sphere of the Other. But the earth he made motionless at the centre, fast about the axis of the universe, to be the measure of day and night, first and most august of divine beings. Now all the motions of these stars and their crossings and conjunctions and occultations it were vain to describe without an orrery: let this account of them then suffice.

11. τοιαύτας καὶ τοσαύτας] The influence of οἷαί τε καὶ ὅσαι preceding has caused these words to be substituted for ταύτῃ, which would regularly correspond to ᾗπερ.

13. οὐράνιον θεῶν γένος] i.e. the stars and planets. The γένη are four in number

when the relative swiftnesses of all the eight revolutions accomplish their course together and reach their starting-point, being measured by the circle of the same and uniformly moving. In this way then and for these causes were created all such of the stars as wander through the heavens and turn about therein, in order that this universe may be most like to the perfect and ideal animal by its assimilation to the eternal being.

XII. Now up to the generation of time all else had been accomplished in the likeness of that whereunto it was likened: but in that it did not yet contain all living creatures created within it, herein it was still unlike. So he went on to complete this that remained unfinished, moulding it after the nature of the pattern. So many forms then as Mind perceived to exist in the ideal animal, according to their variety and multitude, such kinds and such a number did he think fit that this universe should possess. These are fourfold: first the race of the heavenly gods, next the winged tribe whose path is in the air, third whatso dwells in the water, and fourth that which goes upon dry land. The visible form of the deities he created chiefly of fire, that it might be most radiant and most fair to behold; and likening it to the All he shaped it like a sphere and assigned it to the intelligence of the supreme to follow after it; and he disposed it throughout all the firmament of heaven, to be an adornment of it in very truth, wrought cunningly over the whole expanse. And he bestowed two movements upon

to correspond with the four elements. It is to be observed that only in the first class does the correspondence depend upon the structure: the remaining three are classed according to their place of abode.

15. τὴν πλείστην ἰδέαν] cf. *Epinomis* 981 D τὸ γὰρ πλεῖστον πυρὸς ἔχει, ἔχει μὴν γῆς τε καὶ ἀέρος, ἔχει δὲ καὶ ἁπάντων τῶν ἄλλων βραχέα μέρη. The reason for the qualification is doubtless that were they constituted solely of fire, they would be ὁρατά, but not ἁπτά: some admixture of earth was necessary to give them the second distinctive property of bodily existence; cf. 31 D.

ἀπειργάζετο] This reading, which is that of all mss. except A, seems certainly preferable to ἀπήρξατο—an entirely inappropriate word. I cannot think that the authority of A ought to prevail to the exclusion of sense.

17. εἰς τὴν τοῦ κρατίστου φρόνησιν] A very bold substitute for εἰς τὴν τοῦ κρατίστου περιφορὰν φρονιμωτάτην οὖσαν. τὸ κράτιστον evidently signifies the Same, cf. 36 C; and the phrase means that the fixed stars, situate in the outermost sphere, follow the diurnal rotation of the universe, but do not change their positions relative to it.

18. κόσμον ἀληθινόν] The play on the word κόσμος is obvious, though hardly capable of being retained in translation.

προσῆψεν ἑκάστῳ, τὴν μὲν ἐν ταὐτῷ κατὰ ταὐτὰ περὶ τῶν αὐτῶν
ἀεὶ τὰ αὐτὰ ἑαυτῷ διανοουμένῳ, τὴν δὲ εἰς τὸ πρόσθεν ὑπὸ τῆς B
ταὐτοῦ καὶ ὁμοίου περιφορᾶς κρατουμένῳ· τὰς δὲ πέντε κινήσεις
ἀκίνητον καὶ ἑστός, ἵν' ὅ τι μάλιστα αὐτῶν ἕκαστον γένοιτο ὡς
5 ἄριστον. ἐξ ἧς δὴ τῆς αἰτίας γέγονεν ὅσ' ἀπλανῆ τῶν ἄστρων
ζῷα θεῖα ὄντα καὶ ἀίδια καὶ κατὰ ταὐτὰ ἐν ταὐτῷ στρεφόμενα
ἀεὶ μένει· τὰ δὲ τρεπόμενα καὶ πλάνην τοιαύτην ἴσχοντα, καθάπερ ἐν τοῖς πρόσθεν ἐρρήθη, κατ' ἐκεῖνα γέγονε. γῆν δὲ τροφὸν
μὲν ἡμετέραν, εἰλλομένην δὲ περὶ τὸν διὰ παντὸς πόλον τετα-

3 κρατουμένῳ : κρατουμένων A.   9 τὴν ante περὶ habet A.

1. **τὴν μὲν ἐν ταὐτῷ**] No more is meant than that rotation upon an axis, being of all motions the most uniform, is the best symbol of the unerring uniformity pertaining to the activity of pure reason. The stars then, being the highest of finite intelligences, naturally have this motion. A curious instance of false conclusion from a true premiss is to be found in Aristotle *de caelo* II viii 290ᵃ 25, where the rotation of the heavenly bodies is denied on the ground that the same side of the moon is always turned towards us.

2. **ὑπὸ τῆς ταὐτοῦ**] i.e. the motion εἰς τὸ πρόσθεν is not an advance in a straight line, but by the revolution of the Same is formed into a circular orbit.

7. **τρεπόμενα**] sc. τροπὰς ἔχοντα, as above, 39 D.

8. **ἐν τοῖς πρόσθεν**] 38 C foll.: κατ' ἐκεῖνα is merely antecedent to καθάπερ.

9. **εἰλλομένην δὲ περὶ τὸν διὰ παντὸς πόλον**] For an exhaustive and very masterly examination of this passage see Böckh's essay 'Ueber das kosmische System des Platon'. Böckh has proved beyond all controversy that Plato does not here affirm the rotation of the earth upon her axis. Grote has indeed attempted to reply to his arguments, but only to meet with a crushing refutation: see Böckh's 'Kleine Schriften' vol. III p. 294 foll. It is indeed evident from one consideration alone that Plato cannot have intended the earth to move. The universe, he says, revolves diurnally on its axis, and thus, by carrying the sun round with its revolution, causes the alternation of day and night on any given region of the earth once in 24 hours. Now if the earth had an independent revolution of her own, whether in the same or in a contrary direction, it is self-evident that this whole arrangement would be overthrown: if the theory is to account for the phenomena, the earth must be absolutely motionless.

The word εἴλλεσθαι, εἰλεῖσθαι, or ἴλ-λεσθαι, though it does not necessarily exclude the idea of motion, in itself in no wise implies it. Its signification is forcible compression or conglobation: the earth is packed or balled round the centre. Cicero's translation is 'quae traiecto axe sustinetur'. Various forms of the word are extremely common in Homer to express the dense packing of a crowd of men: e.g. *Iliad* VIII 215. In passages where the meaning is extended to include motion, such as Sophokles *Antigone* 340 ἰλλομένων ἀρότρων ἔτος εἰς ἔτος, the real force of the word lies, not in the motion, but in the confinement of the motion within certain restricted limits, as is justly pointed out by Prof. Campbell, who says 'the force of ἴλλειν is "limited motion"'.

It is indeed safe to affirm that no controversy would ever have arisen on the subject, but for a passage in Aristotle, *de caelo* II xiii 293ᵇ 30. In the Berlin text this reads as follows: ἔνιοι δὲ καὶ κειμένην

B] ΤΙΜΑΙΟΣ. 133

each, one in the same spot and uniform, whereby it should be ever constant to its own thoughts concerning the same thing; the other forward, but controlled by the revolution of the same and uniform: but for the other five movements he made it motionless and still, that each star might attain the highest completeness of perfection. From which cause have been created all the stars that wander not but abide fast for ever, living beings divine and eternal and in one spot revolving: while those that move in a circle and wander as aforesaid have come into being on those principles which in the foregoing we have declared.

And the earth our foster-mother, that is globed round the axis stretched from pole to pole of the universe, her he fashioned

ἐπὶ τοῦ κέντρου φασὶν αὐτὴν Ἴλλεσθαι περὶ τὸν διὰ παντὸς τεταμένον πόλον, ὥσπερ ἐν τῷ Τιμαίῳ γέγραπται. This (except that for Ἴλλεσθαι they give εἰλεῖσθαι) is the reading of two mss.; three others add καὶ κινεῖσθαι. Thus there arise three ἀπορίαι: (1) are the words καὶ κινεῖσθαι, which Simplicius had in his text, genuine? (2) has Aristotle misstated Plato's view? (3) if we admit καὶ κινεῖσθαι, can the passage be so understood as to harmonise with Plato's statement? Böckh, adopting the third hypothesis, interprets Aristotle thus: φασὶν αὐτὴν "Ἴλλεσθαι" καὶ κινεῖσθαι "περὶ τὸν διὰ παντὸς τεταμένον πόλον". That is, he supposes Aristotle to be stating, not Plato's view, but that of some who conceived the earth to rotate, quoting the words of the *Timaeus*, but adding καὶ κινεῖσθαι to adapt them to his present purpose. This however is perhaps too ingenious. As for the second alternative, we have seen and have yet to see that Aristotle has repeatedly misrepresented Plato; and if he was here citing the *Timaeus* from memory, it is impossible to say that he may not have done so in the present instance. On the whole however I am disposed to believe that the words καὶ κινεῖσθαι were added by some unwise annotator, who had in his mind the sentence which occurs soon afterwards,

296ᵃ 26 οἱ δ᾽ ἐπὶ τοῦ μέσου θέντες Ἴλλεσθαι καὶ κινεῖσθαι φασι: where the added words distinguish the theory there stated from Plato's.

One argument of Grote's may briefly be noticed. The inconsistency, he says, between the rotation of the earth on her axis and the diurnal rotation of the universe escaped Aristotle (since he does not advert to it), why then should it not have escaped Plato? But Aristotle is not criticising the cosmogony of the *Timaeus*, but discussing the mobility of the earth; therefore he is not concerned to notice such an inconsistency: moreover Grote is herein guilty of *petitio principii* respecting Aristotle's text. But it is really supererogatory to expose the weakness of a hypothesis which has reduced so able a reasoner as Grote, in his eagerness to convict Plato of an irrationality, to insist on importing the ἄτρακτος from the mythical imagery of *Republic* x into the serious cosmology of the *Timaeus*, to serve as a solid axis of the universe. Plato was never guilty of such an absurdity as to conceive the axis as other than a mathematical line. If we are to find a place in the *Timaeus* for the ἄτρακτος, why not also for the σφόνδυλοι, for the knees of Necessity, in short for the whole apparatus of the myth?

μένον, φύλακα καὶ δημιουργὸν νυκτός τε καὶ ἡμέρας ἐμηχανήσατο, c
πρώτην καὶ πρεσβυτάτην θεῶν ὅσοι ἐντὸς οὐρανοῦ γεγόνασι.
χορείας δὲ τούτων αὐτῶν καὶ παραβολὰς ἀλλήλων, καὶ <τὰ>
περὶ τὰς τῶν κύκλων πρὸς ἑαυτοὺς ἐπανακυκλήσεις καὶ προσ-
5 χωρήσεις, ἔν τε ταῖς ξυνάψεσιν ὁποῖοι τῶν θεῶν κατ' ἀλλήλους
γιγνόμενοι καὶ ὅσοι καταντικρύ, μεθ' οὕστινάς τε ἐπίπροσθεν
ἀλλήλοις ἡμῖν τε κατὰ χρόνους οὕστινας ἕκαστοι κατακαλύπτον-
ται καὶ πάλιν ἀναφαινόμενοι φόβους καὶ σημεῖα τῶν μετὰ ταῦτα D
γενησομένων τοῖς οὐ δυναμένοις λογίζεσθαι πέμπουσι, τὸ λέγειν

3 τὰ addidi.   9 οὐ δυναμένοις: οὐ omittunt SZ.

It may be asked, must not the earth, having a soul, possess motion, seeing that all the other heavenly bodies move because they are ἔμψυχοι? To this Martin acutely replies that, had she not a soul of her own, she must rotate on her own axis (which is part of the axis of the universe), following the rotation of the whole. But her vital force enables her to resist this rotation, and by remaining fixed to measure day and night: her rest in fact is equivalent to a motion countervailing the motion of the whole.

1. **φύλακα καὶ δημιουργόν**] Earth is the 'guardian' of day and night inasmuch as without her they could not be measured; the 'creatress', because it is her shadow which causes night to be distinct from day. Proklos says μᾶλλον μὴν ὁ μὲν ἥλιος ἡμέρας, ἡ δὲ νυκτὸς αἰτία. But day, regarded as the light portion of the νυχθήμερον, cannot exist unless night exists wherewith to contrast it; therefore in that sense earth is its δημιουργός: without her there would be light, but not day. Martin puts it thus: '[elle] est ainsi la productrice du jour par sa résistance au mouvement, en même temps qu'elle en est la gardienne par son immobilité'.

2. **ὅσοι ἐντὸς οὐρανοῦ**] i.e. she is inferior only to the οὐρανὸς as a whole.

3. **χορείας**] This is an astronomical term signifying the revolution of the planets around a common centre, as it were in a round dance: see *Epinomis* 982 E πορείαν δὲ καὶ χορείαν πάντων χορῶν καλλίστην καὶ μεγαλοπρεπεστάτην χορεύοντα. παραβολὴ is explained by Proklos to denote the position of two planets in the same longitude, though different latitude, or their rising or setting simultaneously: παραβολὰς δὲ τὰς κατὰ μῆκος αὐτῶν συντάξεις, ὅταν κατὰ πλάτος διαφέρωσιν, ἢ κατὰ βάθος, τὰς συναντολὰς λέγω καὶ συγκαταδύσεις.

καὶ <τὰ> περὶ τάς] The vulgate καὶ περὶ τάς cannot be right, nor is the conjecture of Stephanus, περιττάς, much more satisfactory than Stallbaum's ποικίλας. Acting on a suggestion of the Engelmann translator I have inserted τά, which at least gives a good sense. From *Republic* 617 B τρίτον δὲ φορᾷ ἰέναι, ὡς σφίσι φαίνεσθαι, ἐπανακυκλούμενον τὸν τέταρτον we might infer that ἐπανακύκλησις simply means the planet's ἀποκατάστασις 'the 'return of the circle upon itself' denoting the revolution of the περιφορά again to a given point. If Proklos is to be trusted however, it means the retardation of one heavenly body in relation to another, as προσχώρησις means the gaining by one upon another. For προσχωρήσεις it is probable that we ought to read προχωρήσεις, which is given by one ms.

5. **ἔν τε ταῖς ξυνάψεσιν**] This sentence is certainly complex and involved, but I see no sufficient reason for meddling

ΤΙΜΑΙΟΣ.

to be the guardian and creator of night and day, the first and most august of the gods that have been created within the heavens. But the circlings of them and their crossings one of another, and the manner of the returning of their orbits upon themselves and their approximations, and which of the deities meet in their conjunctions and which are in opposition, and how they pass before and behind each other, and at what times they are hidden from us and again reappearing send to them who cannot calculate their motions panics and portents of things to come—to declare all this without visible illustrations of their

with the text. The chief causes of offence are (1) the repeated interrogative μεθ' οὕστινας—οὕστινας, (2) the position of τε after ἡμῖν. Stallbaum would read κατὰ χρόνους τινάς. I think however that the ms. reading may be defended as a double indirect interrogative: a construction which, though by far less common than the double direct interrogation, is yet quite a good one: cf. Sophokles *Antigone* 1341 οὐδ' ἔχω ὅπα πρὸς πότερον ἴδω. The literal rendering of the clause will then be 'behind what stars at what times they pass before one another and are now severally hidden from us, now again reappearing &c.' The τε after ἡμῖν really belongs to κατακρύπτονται and is answered by the following καί, *quasi* ἡμῖν...κατακρύπτονταί τε καὶ ἀναφαινόμενοι...πέμπουσι. For the irregular position of τε compare Thukydides IV 115 οἱ δὲ Ἀθηναῖοι ἠμύναντό τε ἐκ φαύλου τειχίσματος καὶ ἀπ' οἰκιῶν ἐπάλξεις ἐχουσῶν. And instances might be multiplied. So much for the main difficulties: there remain a few lesser points. ἔν τε ταῖς ξυνάψεσιν (ξύναψις is in technical language 'conjunction') must of course be taken with κατ' ἀλλήλους γιγνόμενοι alone: ὅσοι καταντικρύ denotes the contrary situation, 'opposition'. γιγνόμενοι must be supplied with ὅσοι καταντικρύ, and again with μεθ' οὕστινάς τε ἐπίπροσθεν ἀλλήλοις: i.e. when a given star passes behind a second and before a third. The whole sentence, as I read it,

is undeniably a very complicated piece of syntax; and it is possible enough that some mischief may have befallen the text; but I have seen no emendation convincing enough to warrant me in deserting the mss. And it should be remembered that the *Timaeus* contains much more of involved construction than the earlier dialogues in general do. With μεθ' οὕστινας is to be understood τῶν θεῶν.

9. **τοῖς οὐ δυναμένοις**] Although the negative rests on the authority of A alone, I have yet retained it, understanding the sense to be that the celestial movements are held for signs and portents by those who do not comprehend the natural laws which govern them. The οὐ would very readily be omitted by a copyist living at a time when astrology had become prevalent, and recourse was had to the professional astrologer for interpretation of the signs of the heavens. If it be objected that the negative ought to be μή, I should reply that this is one of many cases where the negative coheres so closely with the participle as practically to form one word: cf. Isokrates *de pace* § 13 νομίζετε δημοτικωτέρους εἶναι τοὺς μεθύοντας τῶν νηφόντων καὶ τοὺς νοῦν οὐκ ἔχοντας τῶν εὖ φρονούντων. There νοῦν οὐκ ἔχοντας = ἀνοήτους, as here οὐ δυναμένοις = ἀδυνατοῦσιν.

136   ΠΛΑΤΩΝΟΣ   [40 D—

ἄνευ <τῶν> δι' ὄψεως τούτων αὐτῶν μιμημάτων μάταιος ἂν εἴη
πόνος· ἀλλὰ ταῦτά τε ἱκανῶς ἡμῖν ταύτῃ καὶ τὰ περὶ θεῶν
ὁρατῶν καὶ γεννητῶν εἰρημένα φύσεως ἐχέτω τέλος.
    XIII. Περὶ δὲ τῶν ἄλλων δαιμόνων εἰπεῖν καὶ γνῶναι τὴν
5 γένεσιν μεῖζον ἢ καθ' ἡμᾶς, πειστέον δὲ τοῖς εἰρηκόσιν ἔμπροσθεν,
ἐκγόνοις μὲν θεῶν οὖσιν, ὡς ἔφασαν, σαφῶς δέ που τούς γε αὐτῶν
προγόνους εἰδόσιν· ἀδύνατον οὖν θεῶν παισὶν ἀπιστεῖν, καίπερ E
ἄνευ τε εἰκότων καὶ ἀναγκαίων ἀποδείξεων λέγουσιν, ἀλλ' ὡς
οἰκεῖα φασκόντων ἀπαγγέλλειν ἑπομένους τῷ νόμῳ πιστευτέον.
10 οὕτως οὖν κατ' ἐκείνους ἡμῖν ἡ γένεσις περὶ τούτων τῶν θεῶν
ἐχέτω καὶ λεγέσθω. Γῆς τε καὶ Οὐρανοῦ παῖδες Ὠκεανός τε
καὶ Τηθὺς ἐγενέσθην, τούτων δὲ Φόρκυς Κρόνος τε καὶ Ῥέα καὶ
ὅσοι μετὰ τούτων, ἐκ δὲ Κρόνου καὶ Ῥέας Ζεὺς Ἥρα τε καὶ 41 A
πάντες ὅσους ἴσμεν ἀδελφοὺς λεγομένους αὐτῶν, ἔτι τε τούτων
15 ἄλλους ἐκγόνους.
    Ἐπεὶ δ' οὖν πάντες, ὅσοι τε περιπολοῦσι φανερῶς καὶ ὅσοι
φαίνονται καθ' ὅσον ἂν ἐθέλωσιν, οἱ θεοὶ γένεσιν ἔσχον, λέγει
πρὸς αὐτοὺς ὁ τόδε τὸ πᾶν γεννήσας τάδε· Θεοὶ θεῶν, ὧν ἐγὼ

1 ἄνευ τῶν δι' ὄψεως scripsi auctore Proclo. ἄνευ διόψεως AHSZ.   αὐτῶν
scripsi.   αὖ τῶν AHSZ.   4 δαιμόνων : δαιμονίων A.   9 φασκόντων : φάσ-
κουσιν SZ.   17 οἱ θεοί : οἱ omittunt SZ.

1. ἄνευ < τῶν > δι' ὄψεως] Proklos, in first citing this passage, gives ἄνευ δι' ὄψεως αὐτῶν τούτων μιμημάτων: presently, quoting it again, he says ἄνευ τῶν δι' ὄψεως, and this I believe to be what Plato wrote. The vulgate ἄνευ διόψεως τούτων αὖ τῶν μιμημάτων is so uncouth a phrase that it surely cannot have proceeded from him: even the word διόψις itself seems suspicious; it occurs nowhere else before Plutarch. Following the text of Proklos then I construe ἄνευ τῶν δι' ὄψεως μιμημάτων αὐτῶν τούτων—without ocular representations of precisely these things: i.e. without a planetarium to illustrate the movements. Ficinus seems to have read αὐτῶν, to judge from the word 'ipsorum' in his rendering.

6. σαφῶς δέ που] The irony of this passage, though it seems to have generally escaped the commentators, is evident; more especially in the opening sentence of the next chapter. Plato had no cause for embroiling himself with popular religion. To his metaphysical scheme it is quite immaterial whether mankind is the highest order of finite intelligences beneath the stars, or whether there exist anthropomorphic beings of superior rank, such as the gods and daemons of the old mythology.

40 D—41 D, c. xiii. Let us then acquiesce in the account given by children of the gods concerning their own lineage and accept the deities of the national mythology. When therefore all the gods of whatsoever nature had come into being, the Artificer addressed the work of his hands, and showed them how that, since they had a beginning, they were not in their own nature immortal altogether, yet should they never suffer dissolution, seeing that the sovereign will of their creator was a firmer

very movements were labour lost. So let thus much suffice on this head and let our exposition concerning the nature of the gods visible and created be brought to an end.

XIII. But concerning the other divinities, to declare and determine their generation were a task too mighty for us: therefore we must trust in those who have revealed it heretofore, seeing that they are offspring, as they said, of gods, and without doubt know their own forefathers. We cannot then mistrust the children of gods, though they speak without probable or inevitable demonstrations; but since they profess to announce what pertains to their own kindred, we must conform to usage and believe them. Let us then accept on their word this account of the generation of these gods. Of Earth and Heaven were born children, Okeanos and Tethys; of these Phorkys and Kronos and Rhea and all their brethren: and of Kronos and Rhea, Zeus and Hera and all whom we know to be called their brothers; and they in their turn had children after them.

Now when all the gods had come to birth, both those who revolve before our eyes and those who reveal themselves in so far as they will, he who begat this universe spake to them these words: Gods of gods, whose creator am I and father of works, which

surely for their endurance than the vital bonds wherewith their being was bound together. But the universe was not yet complete: three kinds of creatures must yet be born, which are mortal. Now if the Artificer created these himself, they must needs be immortal, since he could not will the dissolution of his own work; they must therefore derive their birth from the created gods. Receiving then from him the immortal essence, the gods should implant it in a mortal frame and so generate mortal living creatures, that the universe may be a perfect copy of its type.

9. ἑπομένους τῷ νόμῳ πιστευτέον] cf. *Laws* 904 A οἱ κατὰ νόμον ὄντες θεοί. Plato indifferently acquiesces in the established custom. His theogony is said by Proklos to be Orphic; it differs from that of Hesiod. For the construction compare *Phaedrus* 272 E πάντως λέγοντα τὸ δὴ εἰκὸς διωκτέον: the idiom is common enough.

16. ὅσοι τε περιπολοῦσι φανερῶς] Those who 'revolve visibly' are of course Plato's own gods, the stars of heaven; the others are the deities of popular belief, who ξείνοισιν ἐοικότες ἀλλοδαποῖσιν, παντοῖοι τελέθοντες, ἐπιστρωφῶσι πόληας. There seems again to be a quiet irony in the words φαίνονται καθ' ὅσον ἂν ἐθέλωσιν.

18. θεοὶ θεῶν] The exact sense of these words has been much disputed. Setting aside neoplatonic mystifications, which the curious may find in the commentary of Proklos, the interpretations which seem to deserve notice are as follows. (1) 'Gods born of gods'. This, though

138   ΠΛΑΤΩΝΟΣ   [41 A—

δημιουργὸς πατήρ τε ἔργων, ἃ δι' ἐμοῦ γενόμενα ἄλυτα ἐμοῦ γε
μὴ ἐθέλοντος· τὸ μὲν οὖν δὴ δεθὲν πᾶν λυτόν, τό γε μὴν καλῶς
ἁρμοσθὲν καὶ ἔχον εὖ λύειν ἐθέλειν κακοῦ· δι' ἃ καὶ ἐπείπερ B
γεγένησθε, ἀθάνατοι μὲν οὐκ ἐστὲ οὐδ' ἄλυτοι τὸ πάμπαν, οὔ τι
5 μὲν δὴ λυθήσεσθέ γε οὐδὲ τεύξεσθε θανάτου μοίρας, τῆς ἐμῆς
βουλήσεως μείζονος ἔτι δεσμοῦ καὶ κυριωτέρου λαχόντες ἐκείνων,
οἷς ὅτ' ἐγίγνεσθε ξυνεδεῖσθε. νῦν οὖν ὃ λέγω πρὸς ὑμᾶς ἐνδει-
κνύμενος, μάθετε. θνητὰ ἔτι γένη λοιπὰ τρί' ἀγέννητα· τούτων
δὲ μὴ γενομένων οὐρανὸς ἀτελὴς ἔσται· τὰ γὰρ ἅπαντ' ἐν αὐτῷ
10 γένη ζῴων οὐχ ἕξει, δεῖ δέ, εἰ μέλλει τέλεος ἱκανῶς εἶναι.   δι' C
ἐμοῦ δὲ ταῦτα γενόμενα καὶ βίου μετασχόντα θεοῖς ἰσάζοιτ' ἄν·

1 ἐμοῦ γε μὴ ἐθέλοντος: ἐμοῦ γ' ἐθέλοντος SZ.   8 ἀγέννητα: ἀγένητα A.
τούτων δέ: τούτων οὖν S.

supported by Martin as well as Stallbaum, seems to me inadmissible, for the plain reason that the only source whence they derived their birth was the δημιουργὸς himself; the plural θεῶν then is without propriety or meaning. (2) 'Gods, images of gods', cf. τῶν ἀιδίων θεῶν γεγονὸς ἄγαλμα. But 'images' is not in the Greek, nor can be got out of it: and even granting that it could, the obscure words just quoted are far too unstable a basis for such an interpretation. (3) In my own judgment the phrase is simply an instance of rhetorical ὄγκος, well suited to the stately pomp characterising the whole passage. 'Gods of gods' comes nearest, I believe, to the sense of the original, signifying solely the transcendent dignity of the οὐράνιοι θεοί, the first-fruits of creation. Superlatives of this kind, though not perhaps common in Greek, certainly exist: compare Sophokles *Oed. Col.* 1237 ἵνα πρόπαντα κακὰ κακῶν ξυνοικεῖ: also *Oed. Tyr.* 465 ἄρρητ' ἀρρήτων τελέσαντα φοινίαισι χερσίν: Aeschylus *Persae* 681 ὦ πιστὰ πιστῶν. Plato may have in his mind a comparison between the highest gods and δαίμονες of a lower rank, such as those of *Phaedrus* 247 A or *Epinomis* 984 E: but this is not necessary.

1. ὧν ἐγὼ δημιουργὸς πατήρ τε ἔργων] These words are almost as much debated as the preceding. (1) The clause may be taken in apposition with θεοί: sc. ἔργα, ὧν ἐγὼ δημιουργὸς πατήρ τε: (2) ὧν may be governed by ἔργων, as Stallbaum takes it: (3) or by δημιουργός. It can hardly be doubted that the interpretation is to be preferred which best lends itself to the majestic flow of Plato's rhythm; and on that ground I should give the preference to the last, making ὧν masculine: 'whose maker am I and father of works which through me coming into being &c.' The construction will thus really follow the same principle as the familiar idiom whereby a demonstrative is substituted for the relative in the second member of a relative clause: as for instance in *Euthydemus* 301 E ταῦτα ἡγεῖ σὰ εἶναι, ὧν ἂν ἄρξῃς καὶ ἐξῇ σοι αὐτοῖς χρῆσθαι ὅ τι ἂν βούλῃ.

Badham (on *Philebus* 30 D) proposes to read the opening clauses thus: θεοί, ὅσων ἐγὼ δημιουργὸς πατήρ τε ἔργων, ἅτε δι' ἐμοῦ γενόμενα, ἄλυτα ἐμοῦ γ' ἐθέλοντος. This is grammatically faultless, but, it is to be feared, sorely inadequate to the 'large utterance' of the Artificer. The omission of μὴ before ἐθέλοντος has the support of most mss. and gives an equally good sense: I retain however the reading of A, which is confirmed by Cicero's 'me invito'.

ΤΙΜΑΙΟΣ.

by me coming into being are indissoluble save by my will: Behold, all which hath been fastened may be loosed, yet to loose that which is well fitted and in good case were the will of an evil one. Wherefore, forasmuch as ye have come into being, immortal ye are not, nor indissoluble altogether; nevertheless shall ye not be loosed nor meet with the doom of death, having found in my will a bond yet mightier and more sovereign than those that ye were bound withal when ye came into being. Now therefore hearken to the word that I declare unto you. Three kinds of mortal beings are yet uncreate. And if these be not created, the heaven will be imperfect; for it will not have within it all kinds of living things; yet these it must have, if it is to be perfect. But if these were created by my hands and from me received their life, they would be equal to gods.

2. τὸ μὲν οὖν δή] It is impossible not to admire the serenity with which all the editors set a full stop after ἐθέλοντος, and then make a fresh start, as though the words from θεοὶ to ἐθέλοντος were a sentence; as though γίγνεται stood in place of γενόμενα. It were easy to convert this into a sentence through milder means than Badham employed, by substituting τὰ for ἅ. But a certain unpleasing curtness is thereby introduced, which leads me to shrink from tampering with the text. I regard then all the words down to ἐθέλοντος as constituting an appellation. The difficulty then arises however, that the particles μὲν οὖν δή seem to indicate the commencement of a fresh sentence. Yet the objection is not, I think, fatal: for although the words θεοὶ...ἐθέλοντος are not in form a sentence containing a statement, they do practically convey a statement; and the προσηγορία being somewhat extended, Plato proceeds as if the information implied in a description were given in the form of a direct assertion. The massive form of the opening address seems to justify a stronger combination of particles at the commencement of the main sentence than could ordinarily be used.

4. οὔ τι μὲν δή] For this strong adversative formula compare *Theaetetus* 187 A, *Philebus* 46 B, *Phaedrus* 259 B; and, without γε, *Theaetetus* 148 E.

7. οἷς ὅτ' ἐγίγνεσθε ξυνεδεῖσθε] Compare 43 A ξυνεκόλλων οὐ τοῖς ἀλύτοις, οἷς αὐτοὶ ξυνείχοντο, δεσμοῖς: and 73 D καθάπερ ἐξ ἀγκυρῶν βαλλόμενος ἐκ τούτων πάσης ψυχῆς δεσμούς.

8. γένη λοιπὰ τρία] i.e. those which made their habitation in air, in water, and on land.

11. θεοῖς ἰσάζοιτ' ἄν] This assertion of the δημιουργὸς that whatsoever immediately proceeds from him must be immortal is, I think, not without its metaphysical significance. The creation of the universe by the δημιουργός, we take it, symbolises the evolution of absolute intelligence into material nature, i.e. into the perceptions of finite intelligences. Now this evolution, the manifestation of supreme thought in the material world, is *per se* eternal—it is an essential element in the being of eternal thought. But, the evolution once given, the things that belong to it as such are all transitory. Considered as making up the sum total of phenomenal nature, the infinite series of phenomena is eternal: but the phenomena themselves belong not

140 ΠΛΑΤΩΝΟΣ [41 C—

ἵνα οὖν θνητά τε ᾖ τό τε πᾶν τόδε ὄντως ἅπαν ᾖ, τρέπεσθε κατὰ
φύσιν ὑμεῖς ἐπὶ τὴν τῶν ζῴων δημιουργίαν, μιμούμενοι τὴν ἐμὴν
δύναμιν περὶ τὴν ὑμετέραν γένεσιν. καὶ καθ᾽ ὅσον μὲν αὐτῶν
ἀθανάτοις ὁμώνυμον εἶναι προσήκει, θεῖον λεγόμενον ἡγεμονοῦν
5 τε ἐν αὐτοῖς τῶν ἀεὶ δίκῃ καὶ ὑμῖν ἐθελόντων ἕπεσθαι, σπείρας
καὶ ὑπαρξάμενος ἐγὼ παραδώσω· τὸ δὲ λοιπὸν ὑμεῖς, ἀθανάτῳ
θνητὸν προσυφαίνοντες, ἀπεργάζεσθε ζῷα καὶ γεννᾶτε τροφήν τε D
διδόντες αὐξάνετε καὶ φθίνοντα πάλιν δέχεσθε.

XIV. Ταῦτ᾽ εἶπε, καὶ πάλιν ἐπὶ τὸν πρότερον κρατῆρα, ἐν ᾧ
10 τὴν τοῦ παντὸς ψυχὴν κεραννὺς ἔμισγε, τὰ τῶν πρόσθεν ὑπό-
λοιπα κατεχεῖτο μίσγων τρόπον μέν τινα τὸν αὐτόν, ἀκήρατα δ᾽
οὐκέτι κατὰ ταὐτὰ ὡσαύτως, ἀλλὰ δεύτερα καὶ τρίτα. ξυστήσας
δὲ τὸ πᾶν διεῖλε ψυχὰς ἰσαρίθμους τοῖς ἄστροις, ἔνειμέ θ᾽ ἑκάστην

to eternity, but to γένεσις. In other words, the *existence* of time and space is part of the being of absolute intelligence: the apprehension of things in time and space pertains to finite intelligences. Therefore, as phenomena apprehended in time and space do not directly pertain to absolute intelligence, so in the allegory mortal things are not directly the work of the δημιουργός.

1. ἵνα οὖν θνητά τε ᾖ] Mortality is necessary in this way. The scheme of existence involves a material counterpart of the ideal world. To materiality belongs becoming and perishing: accordingly αἰσθητὰ ζῷα, the copies of the νοητὰ ζῷα, must, so far as material, be mortal. Mortality must correspond to immortality as inevitably as multiplicity to unity. Even the stars, which, being the handiwork of the Artificer himself, are immortal, contain within them the processes of γένεσις and φθορά.

κατὰ φύσιν] In the way of nature: i.e. βλέποντες πρὸς τὸ ἀίδιον.

3. καθ᾽ ὅσον] It has been proposed to omit καθ᾽: but I think the text is sufficiently defended by Stallbaum.

4. ἀθανάτοις ὁμώνυμον] The αἰσθητὰ ζῷα are ἀθάνατα, in so far as they possess the indestructible vital essence supplied by the creator; but only ὁμωνύμως, since their present mode of existence as individuals is transitory.

ἡγεμονοῦν] Here seems to be the first suggestion of a word which afterwards became a technical term common in the Stoic philosophy—τὸ ἡγεμονικόν, the reason. We have it again similarly used in 70 C: cf. *Laws* 963 A νοῦν δέ γε πάντων τούτων ἡγεμόνα. The genitive τῶν ἐθελόντων is governed by ἡγεμονοῦν.

6. ὑπαρξάμενος] This transitive use of the middle of this verb is not quoted in Liddell and Scott.

7. τροφήν τε διδόντες] How they did this we learn in 77 A. The gods of course had no need of sustenance; for, like the κόσμος, they αὐτοὶ ἑαυτοῖς τροφὴν τὴν ἑαυτῶν φθίσιν παρεῖχον. With φθίνοντα πάλιν δέχεσθε compare 42 E δανειζόμενοι μόρια ὡς ἀποδοθησόμενα πάλιν: they created mortals out of the substance of the universe, and at their dissolution restored the elements of them thither whence they were borrowed.

41 D—42 E, c. xiv. Thus having spoken, the Artificer prepared a second blending of soul, having its proportions like to the former, but less pure. And of the soul so formed he separated as many portions as there were stars in heaven, and set a portion in each star, and declared to them the laws of nature: how

D] ΤΙΜΑΙΟΣ. 141

Therefore in order that they may be mortal, and that this All may be truly all, turn ye according to nature unto the creation of living things, imitating my power that was put forth in the generation of you. Now such part of them as is worthy to share the name of the immortals, which is called divine and governs in the souls of those that are willing ever to follow after justice and after you, this I, having sown and provided it, will deliver unto you: and ye for the rest, weaving the mortal with the immortal, shall create living beings and bring them to birth, and giving them sustenance shall ye increase them, and when they perish receive them back again.

XIV. Thus spake he; and again into the same bowl wherein he mingled and blended the universal soul he poured what was left of the former, mingling it somewhat after the same manner, yet no longer so pure as before but second and third in pureness. And when he had compounded the whole, he portioned off souls equal in number to the stars and distributed a soul to

that every single soul should be first embodied in human form, clothed in a frame subject to vehement affections and passions. And whoso should conquer these and live righteously, after fulfilling his allotted span, he should return to the star of his affinity and dwell in blessedness; but if he failed thereof, he should pass at death into the form of some lower being, and cease not from such transmigrations until, obeying the reason rather than the passions, he should gradually raise himself again to the first and best form. Then God sowed the souls severally in the different planets, and gave the task of their incarnation to the gods he had created, to make them as fair and perfect as mortal nature may admit.

10. τὰ τῶν πρόσθεν ὑπόλοιπα] Not the remnants of the universal soul, as Stallbaum supposes; for that, we are told in 36 B, was all used up; but of the elements composing soul, ταυτὸν θάτερον and οὐσία.

11. ἀκήρατα δ' οὐκέτι] That is to say, the harmonical proportions are less accurate, and the Other is less fully subordinated to the Same: in other words, these souls are a stage further removed from pure thought, a degree more deeply immersed in the material. Compare *Philebus* 29 B foll. Plato's scheme includes a regular gradation of finite existences, from the glorious intelligence of a star down to the humblest herb of the field: all these are manifestations of the same eternal essence through forms more and more remote.

13. διεῖλε ψυχὰς ἰσαρίθμους τοῖς ἄστροις] There is a certain obscurity attending this part of the allegory, which has given rise to much misunderstanding. It is necessary to distinguish clearly between the νομή of the present passage and the σπόρος of 42 D. What the δημιουργὸς did, I conceive to be this. Having completed the admixture of soul he divided the whole into portions, assigning one portion to each star. These portions, be it understood, are not particular souls nor aggregates of particular souls: they are divisions of the whole quantity of soul, which is not as yet differentiated into particular souls.

πρὸς ἕκαστον, καὶ ἐμβιβάσας ὡς ἐς ὄχημα τὴν τοῦ παντὸς φύσιν E
ἔδειξε, νόμους τε τοὺς εἱμαρμένους εἶπεν αὐταῖς, ὅτι γένεσις πρώτη
μὲν ἔσοιτο τεταγμένη μία πᾶσιν, ἵνα μή τις ἐλαττοῖτο ὑπ' αὐτοῦ,
δέοι δὲ σπαρείσας αὐτὰς εἰς τὰ προσήκοντα ἑκάσταις ἕκαστα
5 ὄργανα χρόνου φῦναι ζῴων τὸ θεοσεβέστατον, διπλῆς δὲ οὔσης 42 A
τῆς ἀνθρωπίνης φύσεως τὸ κρεῖττον τοιοῦτον εἴη γένος, ὃ καὶ
ἔπειτα κεκλήσοιτο ἀνήρ. ὁπότε δὴ σώμασιν ἐμφυτευθεῖεν ἐξ
ἀνάγκης, καὶ τὸ μὲν προσίοι, τὸ δ' ἀπίοι τοῦ σώματος αὐτῶν,
πρῶτον μὲν αἴσθησιν ἀναγκαῖον εἴη μίαν πᾶσιν ἐκ βιαίων πα-
10 θημάτων ξύμφυτον γίγνεσθαι, δεύτερον δὲ ἡδονῇ καὶ λύπῃ μεμιγμένον ἔρωτα, πρὸς δὲ τούτοις φόβον καὶ θυμὸν ὅσα τε ἑπό-
μενα αὐτοῖς καὶ ὁπόσα ἐναντίως πέφυκε διεστηκότα· ὧν εἰ μὲν B

1 ἐς: εἰς S.   5 χρόνου: χρόνων AHSZ.

It is hardly necessary to observe that these ψυχαὶ ἰσάριθμοι τοῖς ἄστροις are quite distinct from the souls of the stars themselves. Next the δημιουργὸς explains to these still undifferentiated souls the laws of nature; after which he redistributes the whole quantity of soul among the planets (ὄργανα χρόνου, 42 D) for incarnation in mortal bodies. From the language of 42 D, τοὺς μέν...τοὺς δέ, it would seem that the differentiating of the souls into individual beings was done by the δημιουργὸς himself, before they were handed over to the created gods: in fact this is metaphysically necessary.

Martin's interpretation appears to me wholly unplatonic, indeed unintelligible. He regards the ψυχαὶ ἰσάριθμοι as distinct from the soul that was afterwards to inform mortal bodies. 'C'est à ces grandes âmes confiées aux astres, c'est à ces vastes dépôts de substance incorporelle et intelligente, que Dieu révèle ses desseins.' This he himself most justly terms an 'étrange doctrine', and certainly it is not Plato's. It is surely indubitable that what the δημιουργὸς mixed in the κρατὴρ was the whole substance of soul intended to be differentiated into particular souls; that this whole substance was first distributed in large portions among the fixed stars, to learn the laws of existence; and that finally it was redistributed among the planets for division into separate souls incorporated in bodies.

But what is the purpose and meaning of this distribution among the fixed stars? I think the explanation is suggested by *Phaedrus* 252 C, D, where different gods are assigned as patrons for persons of various temperament. The apportionment to diverse stars is thus a fanciful way of accounting for innate diversity of character and disposition; each individual being influenced by the star to which the division was assigned of which what was afterwards his soul formed a part.

1. **ὡς ἐς ὄχημα**] The same word is used in 69 D to express the relation of body to soul in the human being, although the relation is different to that here indicated; for these ψυχαὶ do not inform and vitalise the body of the star, which is to them solely a 'vehicle'.

τὴν τοῦ παντὸς φύσιν ἔδειξε] It is interesting to observe that here in Plato's maturest period we have something closely resembling the ἀνάμνησις of the *Phaedo* and *Phaedrus*. To say that the laws of the universe were declared to soul before it became differentiated into individual souls is very much the same thing as to say that the soul beheld the ideas in a previous existence. At the same

each star, and setting them in the stars as though in a chariot, he shewed them the nature of the universe and declared to them its fated laws; how that the first incarnation should be ordained to be the same for all, that none might suffer disadvantage at his hands; and how they must be sown into the instruments of time, each into that which was meet for it, and be born as the most god-fearing of all living creatures; and whereas human nature was twofold, the stronger was that race which should hereafter be called man. When therefore they should be of necessity implanted in bodily forms, and of their bodies something should ever be coming in and other passing away, in the first place they must needs all have innate one and the same faculty of sense, arising from forcible affections; next love mingled with pleasure and pain; and besides these fear and wrath and all the feelings that accompany these and such

time the tendency to merge the individual existence of the soul is characteristic of the *Timaeus* and of Plato's later thought.

2. **γένεσις πρώτη**] i.e. their first embodiment in human form. Stallbaum is obviously wrong in understanding by πρώτη γένεσις the distribution among the stars, since the δευτέρα γένεσις is the incarnation εἰς γυναικὸς φύσιν, 42 B. Here however a point presents itself in which the allegory appears *prima facie* inconsistent. At 39 E Plato says there are four εἴδη of νοητὰ ζῷα in the αὐτὸ ζῷον: yet of αἰσθητὰ ζῷα we only have two εἴδη at the outset: how then is the sensible world a faithful image of the intelligible world? The answer would seem to be that the δημιουργὸς foresaw that many souls must necessarily degenerate from the πρώτη καὶ ἀρίστη ἕξις, and therefore left the perfect assimilation of the image to the type to be worked out and completed in the course of nature, with which he did not choose arbitrarily to interfere, in order that no soul might start at a disadvantage through his doing: ἵνα μή τις ἐλαττοῖτο ὑπ' αὐτοῦ. It is remarkable however that the perfection of the copy should be accomplished through a process of degeneration.

4. **δέοι δὲ σπαρείσας**] Stallbaum for some incomprehensible reason would insert μετὰ before σπαρείσας. The δημιουργὸς is referring to the σπόρος of 42 D, which must take place before the incarnation in mortal bodies can be accomplished. ὄργανα χρόνου, a phrase recurring in 42 D, = the planets: the vulgate χρόνων is clearly a copyist's error. The reason why one planet was more suitable for some souls than another does not appear.

5. **ζῴων τὸ θεοσεβέστατον**] i.e. mankind: cf. *Laws* 902 B ἅμα καὶ θεοσεβέστατον αὐτό ἐστι πάντων ζῴων ἄνθρωπος.

7. **ἐξ ἀνάγκης**] This phrase expresses the unwilling conjunction of spirit and matter, the reluctance of soul to accept corporeal conditions: cf. 69 D συγκερασάμενοι ταῦτα ἀναγκαίως, and a little above δεινὰ καὶ ἀναγκαῖα παθήματα ἐν ἑαυτῷ ἔχον. The whole account in 69 C, D is full of echoes of the present passage.

8. **τὸ μὲν προσίοι τὸ δ' ἀπίοι**] i.e. the body is undergoing a perpetual process of waste and reparation: cf. 43 A ἐνέδουν εἰς ἐπίρρυτον σῶμα καὶ ἀπόρρυτον.

9. **βιαίων παθημάτων**] I take βιαίων to mean vehement and masterful, though it might be understood like ἀναγκαῖα in 69 C.

144   ΠΛΑΤΩΝΟΣ   [42 B—

κρατήσοιεν, δίκῃ βιώσοιντο, κρατηθέντες δὲ ἀδικίᾳ. καὶ ὁ μὲν εὖ τὸν
προσήκοντα χρόνον βιούς, πάλιν εἰς τὴν τοῦ ξυννόμου πορευθεὶς
οἴκησιν ἄστρου, βίον εὐδαίμονα καὶ συνήθη ἕξοι· σφαλεὶς δὲ
τούτων εἰς γυναικὸς φύσιν ἐν τῇ δευτέρᾳ γενέσει μεταβαλοῖ· μὴ
5 παυόμενος δὲ ἐν τούτοις ἔτι κακίας, τρόπον ὃν κακύνοιτο, κατὰ
τὴν ὁμοιότητα τῆς τοῦ τρόπου γενέσεως εἴς τινα τοιαύτην ἀεὶ
μεταβαλοῖ θήρειον φύσιν, ἀλλάττων τε οὐ πρότερον πόνων λήξοι,
πρὶν τῇ ταὐτοῦ καὶ ὁμοίου περιόδῳ τῇ ἐν αὑτῷ ξυνεπισπόμενος τὸν
πολὺν ὄχλον καὶ ὕστερον προσφύντα ἐκ πυρὸς καὶ ὕδατος καὶ
10 ἀέρος καὶ γῆς, θορυβώδη καὶ ἄλογον ὄντα, λόγῳ κρατήσας εἰς τὸ τῆς D
πρώτης καὶ ἀρίστης ἀφίκοιτο εἶδος ἕξεως. διαθεσμοθετήσας δὲ
πάντα αὐτοῖς ταῦτα, ἵνα τῆς ἔπειτα εἴη κακίας ἑκάστων ἀναίτιος,

1 κρατήσοιεν: κρατήσειαν S, qui mox ἐν δίκῃ dedit.   2 χρόνον βιούς: βιοὺς
χρόνον S, nescio an recte.   5 παυόμενος δέ: παυόμενός τε AHZ.   8 ξυνεπισπόμενος: ξυνεπισπώμενος AHZ.

1. **τὸν προσήκοντα χρόνον**] No definite period is ordained in the *Timaeus*, as is the case in the myths of the *Phaedrus* and *Republic*.

2. **τοῦ ξυννόμου**] i.e. the star to which was distributed the portion of soul whence his individual soul afterwards proceeded. συνήθη = congenial: the conditions of life in the σύννομον ἄστρον would be familiar from the soul's former residence in it, though she was not then differentiated.

4. **εἰς γυναικὸς φύσιν**] Here, it must be confessed, we have a piece of questionable metaphysic. For the distinction of sex cannot possibly stand on the same logical footing as the generic differences between various animals; and in the other forms of animal life the distinction is ignored. It is somewhat curious that Plato, who in his views about woman's position was immeasurably in advance of his age, has here yielded to Athenian prejudice so far as to introduce a dissonant element into his theory.

μεταβαλοῖ] After this word the old editions insert χιλιοστῷ δὲ ἔτει ἀμφότεραι ἀφικνούμεναι ἐπὶ κλήρωσιν καὶ αἵρεσιν τοῦ δευτέρου βίου αἱροῦνται ὃν ἂν ἐθέλῃ βίον ἑκάστη· ἔνθα καὶ εἰς θηρίου βίον ἀνθρωπίνη ψυχὴ ἀφικνεῖται. These words, which stand in the margin of two mss., are simply quoted from *Phaedrus* 249 B.

5. **κατὰ τὴν ὁμοιότητα**] That is to say, they assumed the form of those animals to whose natural character they had most assimilated themselves by their special mode of misbehaviour; cf. *Phaedo* 81 E ἐνδοῦνται δέ, ὥσπερ εἰκός, εἰς τοιαῦτα ἤθη ὁποῖ' ἄττ' ἂν καὶ μεμελετηκυῖαι τύχωσιν ἐν τῷ βίῳ: and presently we see that the sensual take the form of asses, the cruel and rapacious that of hawks and kites.

8. **τῇ ταὐτοῦ καὶ ὁμοίου περιόδῳ**] Even in the lower forms the principle of reason is present, only more or less in abeyance. But once let the soul listen to its dictates, so far as in that condition it can make itself heard, and she may retrieve one step of the lost ground at the next incarnation.

12. **ἵνα τῆς ἔπειτα**] Here as in the *Republic* Plato absolves God from all responsibility for evil: cf. *Republic* 379 C οὐδ' ἄρα ὁ θεός, ἐπειδὴ ἀγαθός, πάντων ἂν εἴη αἴτιος, ὡς οἱ πολλοὶ λέγουσιν, ἀλλ' ὀλίγων μὲν τοῖς ἀνθρώποις αἴτιος, πολλῶν δὲ ἀναίτιος· πολὺ γὰρ ἐλάττω τἀγαθὰ τῶν κακῶν ἡμῖν. καὶ τῶν μὲν ἀγαθῶν οὐδένα

as are of a contrary nature: and should they master these passions, they would live in righteousness; if otherwise, in unrighteousness. And he who lived well throughout his allotted time should be conveyed once more to a habitation in his kindred star, and there should enjoy a blissful and congenial life: but failing of this, he should pass in the second incarnation into the nature of a woman; and if in this condition he still would not turn from the evil of his ways, then, according to the manner of his wickedness, he should ever be changed into the nature of some beast in such form of incarnation as fitted his disposition, and should not rest from the weariness of these transformations, until by following the revolution that is within him of the same and uniform, he should overcome by reason all that burden that afterwards clung around him of fire and water and air and earth, a troublous and senseless mass, and should return once more to the form of his first and best nature.

And when he had ordained all these things for them, to the end that he might be guiltless of all the evil that should be in

ἄλλον αἰτιατέον, τῶν δὲ κακῶν ἄλλ' ἄττα δεῖ ζητεῖν τὰ αἴτια, ἀλλ' οὐ τὸν θεόν. See too *Republic* 617 C, *Laws* 900 E, 904 A—C, and especially *Theaetetus* 176 A ἀλλ' οὔτ' ἀπολέσθαι τὰ κακὰ δυνατόν, ὦ Θεόδωρε· ὑπεναντίον γάρ τι τῷ ἀγαθῷ ἀεὶ εἶναι ἀνάγκη· οὔτ' ἐν θεοῖς αὐτὰ ἱδρῦσθαι, τὴν δὲ θνητὴν φύσιν καὶ τόνδε τὸν τόπον περιπολεῖ ἐξ ἀνάγκης. In other words, to soul, as such, no evil can attach in any form whatsoever. Absolute spirit then in itself has no part in evil nor can he be the cause of any. With the evolution of absolute spirit into finite souls arises evil; it is one of the conditions of limitation as much as space and time are. Evil then attaches to finite souls, not *qua* souls, which were impossible, but *qua* finite. Yet, seeing that in the Platonic system the evolution of the infinite into the finite is a necessary law of being, can it be said that God, or absolute spirit, is irresponsible for evil, since that spirit necessarily must manifest itself in a mode of existence to which

Plato declares that evil must inevitably attach? and why is it that evil *must* arise together with limited existence? To these questions Plato has returned no explicit reply: only we may deduce thus much from his ontological scheme—since the realm of absolute essence is a stable unity, the realm of finite existence is a moving plurality, a process. And if a process, we can only conceive, on Plato's principles, that it is a process towards good. Therefore imperfection must always attach to it, since it is ever approaching but never reaches the good. Were perfection predicable of it, it would be the good—the eternal changeless unity: the two sides of the Platonic antithesis would coalesce; motion and plurality would vanish, and we should relapse into the Eleatic ἕν which has been proved unworkable. In this sense Plato may say that evil is necessary and that it belongs to matter, not to God. At the same time since the absolute cannot exist without

ἔσπειρε τοὺς μὲν εἰς γῆν, τοὺς δ' εἰς σελήνην, τοὺς δ' εἰς τἆλλα
ὅσα ὄργανα χρόνου· τὸ δὲ μετὰ τὸν σπόρον τοῖς νέοις παρέδωκε
θεοῖς σώματα πλάττειν θνητά, τό τε ἐπίλοιπον, ὅσον ἔτι ἦν
ψυχῆς ἀνθρωπίνης δέον προσγενέσθαι, τοῦτο καὶ πάνθ' ὅσα E
5 ἀκόλουθα ἐκείνοις ἀπεργασαμένους ἄρχειν, καὶ κατὰ δύναμιν ὅ τι
κάλλιστα καὶ ἄριστα τὸ θνητὸν διακυβερνᾶν ζῷον, ὅ τι μὴ κακῶν
αὐτὸ ἑαυτῷ γίγνοιτο αἴτιον.

XV. Καὶ ὁ μὲν δὴ ἅπαντα ταῦτα διατάξας ἔμενεν ἐν τῷ
ἑαυτοῦ κατὰ τρόπον ἤθει· μένοντος δὲ νοήσαντες οἱ παῖδες τὴν
10 τοῦ πατρὸς διάταξιν ἐπείθοντο αὐτῇ, καὶ λαβόντες ἀθάνατον
ἀρχὴν θνητοῦ ζῴου, μιμούμενοι τὸν σφέτερον δημιουργόν, πυρὸς
καὶ γῆς ὕδατός τε καὶ ἀέρος ἀπὸ τοῦ κόσμου δανειζόμενοι μόρια,
ὡς ἀποδοθησόμενα πάλιν, εἰς ταὐτὸν τὰ λαμβανόμενα συνεκόλλων, 43 A
οὐ τοῖς ἀλύτοις οἷς αὐτοὶ ξυνείχοντο δεσμοῖς, ἀλλὰ διὰ σμικρό-
15 τητα ἀοράτοις πυκνοῖς γόμφοις ξυντήκοντες, ἓν ἐξ ἁπάντων
ἀπεργαζόμενοι σῶμα ἕκαστον, τὰς τῆς ἀθανάτου ψυχῆς περιόδους

10 διάταξιν: τάξιν A pr. m. S.

manifesting itself as the finite, and since to the finite belongs evil, the ultimate cause of evil is really carried back to the absolute, though not *qua* absolute.

2. ὄργανα χρόνου] This sowing seems to have been confined to the earth and the seven planets; for these alone appear to be recognised as instruments of time in 39 C, D. It would presumably follow then that to these gods only was committed the formation of the mortal races.

3. τό τε ἐπίλοιπον] This clearly refers to the θνητὸν εἶδος ψυχῆς of 69 D: i.e. those functions and activities of the soul which are called into being by her conjunction with matter.

7. αὐτὸ ἑαυτῷ] Evil in some shape or other is, as we have seen, an inevitable concomitant of material existence. But if we follow after pure reason, this evil is kept at the lowest minimum; if we perversely forsake her, it is needlessly aggravated. So that while we are not answerable for whatsoever of evil is inseparable from limitation, for all that is the result of our own folly we are answerable. Compare *Laws* 904 B τῆς δὲ γενέσεως τοῦ ποίου τινὸς ἀφῆκε ταῖς βουλήσεσιν ἑκάστων τὰς αἰτίας· ὅπῃ γὰρ ἂν ἐπιθυμῇ καὶ ὁποῖός τις ὢν τὴν ψυχήν, ταύτῃ σχεδὸν ἑκάστοτε καὶ τοιοῦτος γίγνεται ἅπας ἡμῶν ὡς τὸ πολύ. A further discussion of Plato's position as regards the problem of free will is to be found in note on 86 D.

42 E—44 D, *c*. xv. And the eternal God was abiding in his own unity. But the created gods, following the example of their creator, fashioned mortal creatures, fettering the motions of the soul in a material body, whereof they borrowed the substance from that of the universe. And the soul, being imprisoned in a body subject to ceaseless inflowing and outflowing, is at first confounded and distracted. For the perpetual stream of nourishment that enters in, together with the bewildering effect of external sensations, throws her into disorder and tumult: the revolution of the Same in her is brought to a stand,

43 A]   ΤΙΜΑΙΟΣ.   147

each of them, God sowed some in the earth, some in the moon, and some in the other instruments of time. And what came after the sowing he gave into the hands of the young gods, to mould mortal bodies, and having wrought all the residue of human soul that needed yet to be added, to govern and guide as nobly and perfectly as they could the mortal creature, in so far as it brought not evil upon its own head.

XV. So when he had made all these ordinances for them God was abiding after the manner of his own nature: and as he so abode, the children thinking on the command of their father were obedient to it, and having received the immortal principle of a mortal creature, imitating their own artificer, they borrowed from the universe portions of fire and of earth and of water and of air, on condition that they should be returned again, and they cemented together what they took, not with the indissoluble bonds wherewithal they themselves were held together, but welding it with many rivets, invisible for smallness, and making of all the elements one body for each creature, they confined the revolutions of the immortal soul in a body in-

while that of the Other is distorted or reversed: its harmonic proportions cannot indeed be destroyed, save by the creator alone, but they are in every way strained and perturbed. Accordingly, when she has to judge concerning anything, that it is same or other, her judgment is wrong, and she is filled with falsehood and folly: and reason, which seems to rule, is really enslaved by sensation. For all these causes the soul, at her first entrance into a body, is devoid of reason. But presently, as the disturbance caused by the requirements of nutrition and growth diminishes, the circles of the Same and the Other gradually resume their proper functions, and reason regains her sway. But careful and rational training is requisite in order that a man may enjoy his full intellectual liberty: lacking this, his life will be maimed, and imperfect and unreasonable he will pass beneath the shades.

This chapter supplies a theory to account for the abeyance of reason in infants and young children.

8. ἔμενεν ἐν τῷ ἑαυτοῦ] This phrase is significant. Plato does not say that the δημιουργὸς *returned* to his own ἦθος, but that he 'was abiding' therein. The imperfect expresses that not only after he had given these instructions, but previously also, he was abiding. The eternal essence, while manifesting itself in multiplicity, still abides in unity. The process of thought-evolution does not affect the nature of thought as it is in itself: thought, while many and manifold, is one and simple still.

13. ὡς ἀποδοθησόμενα] Plato always insists that the sum of all things, whether spiritual or material, is a constant quantity. Accordingly the gods had to borrow from the store of materials already existing; there could be no addition.

15. πυκνοῖς γόμφοις] i.e. the law of cohesion in matter. The word γόμφοι, as contrasted with δεσμοί, gives the notion of inferior durability.

10—2

148   ΠΛΑΤΩΝΟΣ   [43 A—

ἐνέδουν εἰς ἐπίρρυτον σῶμα καὶ ἀπόρρυτον. αἱ δ᾽ εἰς ποταμὸν
ἐνδεθεῖσαι πολὺν οὔτ᾽ ἐκράτουν οὔτ᾽ ἐκρατοῦντο, βίᾳ δ᾽ ἐφέροντο
καὶ ἔφερον, ὥστε τὸ μὲν ὅλον κινεῖσθαι ζῷον, ἀτάκτως μὴν ὅπῃ Β
τύχοι προϊέναι καὶ ἀλόγως, τὰς ἐξ ἁπάσας κινήσεις ἔχον· εἴς τε
5 γὰρ τὸ πρόσθε καὶ ὄπισθεν καὶ πάλιν εἰς δεξιὰ καὶ ἀριστερὰ κάτω
τε καὶ ἄνω καὶ πάντῃ κατὰ τοὺς ἓξ τόπους πλανώμενα προῄειν.
πολλοῦ γὰρ ὄντος τοῦ κατακλύζοντος καὶ ἀπορρέοντος κύματος, ὃ
τὴν τροφὴν παρεῖχεν, ἔτι μείζω θόρυβον ἀπειργάζετο τὰ τῶν
προσπιπτόντων παθήματα ἑκάστοις, ὅτε πυρὶ προσκρούσειε τὸ C
10 σῶμά τινος ἔξωθεν ἀλλοτρίῳ περιτυχὸν ἢ καὶ στερεῷ γῆς ὑγροῖς
τε ὀλισθήμασιν ὑδάτων, εἴτε ζάλῃ πνευμάτων ὑπὸ ἀέρος φερομένων
καταληφθείη, καὶ ὑπὸ πάντων τούτων διὰ τοῦ σώματος αἱ κινήσεις ἐπὶ τὴν ψυχὴν φερόμεναι προσπίπτοιεν· αἳ δὴ καὶ ἔπειτα
διὰ ταῦτα ἐκλήθησάν τε καὶ νῦν ἔτι αἰσθήσεις ξυνάπασαι κέ-
15 κληνται. καὶ δὴ καὶ τότε ἐν τῷ παρόντι πλείστην καὶ μεγίστην
παρεχόμεναι κίνησιν, μετὰ τοῦ ῥέοντος ἐνδελεχῶς ὀχετοῦ κινοῦσαι D
καὶ σφοδρῶς σείουσαι τὰς τῆς ψυχῆς περιόδους, τὴν μὲν ταὐτοῦ
παντάπασιν ἐπέδησαν ἐναντία αὐτῇ ῥέουσαι καὶ ἐπέσχον ἄρ-
χουσαν καὶ ἰοῦσαν, τὴν δ᾽ αὖ θατέρου διέσεισαν, ὥστε τὰς τοῦ

4 προϊέναι : προσιέναι Α.   5 πρόσθε : πρόσθεν S.   11 φερομένων : φερομένου Α.

1. **ἐπίρρυτον σῶμα καὶ ἀπόρρυτον**] Plato's Herakleitean theory of matter could hardly find stronger expression than this. Fresh particles are being perpetually added to the body's substance to supply the place of others which are for ever flying off. Compare *Theaetetus* 159 B foll.

αἱ δ᾽ εἰς ποταμόν] It may be this expression was suggested by the well-known words of Herakleitos (fr. 41 Bywater) ποταμοῖσι δὶς τοῖσι αὐτοῖσι οὐκ ἂν ἐμβαίης· ἕτερα γὰρ καὶ ἕτερα ἐπιρρέει ὕδατα : cf. *Cratylus* 402 A. According to Aristotle *metaph*. Γ v 1010ᵃ 13, Kratylos found this statement not thorough-going enough : Ἡρακλείτῳ ἐπετίμα εἰπόντι ὅτι δὶς τῷ αὐτῷ ποταμῷ οὐκ ἔστιν ἐμβῆναι· αὐτὸς γὰρ ᾤετο οὐδ᾽ ἅπαξ. Proklos is perhaps right in supposing Plato's ποταμὸς to include not the body only in which the soul resides, but generally the region of γένεσις in which she is placed : ὁ μὲν δὴ ποταμὸς οὗ τὸ ἀνθρώπινον δὴ σῶμα σημαίνει μόνον, ἀλλὰ καὶ πᾶσαν τὴν περικειμένην ἔξωθεν ἡμῖν γένεσιν, διὰ τὴν ὀξύρροπον αὐτῆς καὶ ἀστάθμητον ῥοήν.

2. **ἐφέροντο καὶ ἔφερον**] The περίοδοι could not be altogether passive, that being impossible for an animate being ; the external impressions and the subjective consciousness mutually interacted and conditioned each other.

4. **τὰς ἐξ ἁπάσας**] These six are reckoned as all for the present purpose, since the seventh, or rotary motion, belongs only to beings of a higher order. It may be noted that a completely different classification of κινήσεις is given in *Laws* 893 C foll., where 10 kinds are enumerated.

7. **πολλοῦ γὰρ ὄντος**] Two chief causes are assigned by Plato for the dormant state of the intellect in the case of

D] ΤΙΜΑΙΟΣ. 149

flowing and out-flowing continually. And they, being confined in a great river, neither controlled it nor were controlled, but bore and were borne violently to and fro; so that the whole creature moved, but advanced at random without order or method, having all the six motions: for they moved forward and backward and again to right and to left and downward and upward, and in every way went straying in the six directions. For great as was the tide sweeping over them and flowing off which brought them sustenance, a yet greater tumult was caused by the effects of the bodies that struck against them; as when the body of any one came in contact with some alien fire that met it from without, or with solid earth, or with liquid glidings of water, or if he were caught in a tempest of winds borne on the air, and so the motions from all these elements rushing through the body penetrated to the soul. This is in fact the reason why these have all alike been called and still are called sensations (αἰσθήσεις). Then too did they produce the most wide and vehement agitation for the time being, joining with the perpetually streaming current in stirring and violently shaking the revolutions of the soul, so that they altogether hindered the circle of the Same by flowing contrary to it, and they stopped it from governing and from going; while the circle of the Other

---

infants: the first is the continual influx of nutriment, which the growing child requires; the second and yet more potent cause is the violent effect produced by outward sensations, which bewilder and overwhelm the soul but newly arrived in the world of becoming and inexperienced in its conditions.

10. ἀλλοτρίῳ περιτυχόν] Plato says 'alien' fire, because, as we learn in 45 B, there is a fire, viz. daylight, which is akin to the fire within our bodies and therefore harmless to us. All the four elements are described, each in its own way, as conspiring to the soul's confusion. The poetical tone of this passage is very noticeable.

13. ἐπὶ τὴν ψυχήν] This theory is fully set forth in 64 B foll.: see also *Philebus* 33 D.

14. διὰ ταῦτα ἐκλήθησαν] What is the etymology intended is not very obvious from the context; but probably, as Martin says, Plato meant to connect αἴσθησις with ἀΐσσω. Proklos also proposes the Homeric word ἀΐσθω: cf. *Iliad* XVI 468 ὁ δὲ βράχε θυμὸν ἀΐσθων: but this suggestion has not very much to recommend it.

16. μετὰ τοῦ ῥέοντος ἐνδελεχῶς ὀχετοῦ] i.e. combined with the κῦμα τῆς τροφῆς.

18. παντάπασιν ἐπέδησαν] It should be observed that the effect on the two circles is different: that of the Same is stopped; i.e. the reason does not act: that of the Other is dislocated and distorted; i.e. the reports of the senses are confused and inaccurate.

150   ΠΛΑΤΩΝΟΣ   [43 D—

διπλασίου καὶ τριπλασίου τρεῖς ἑκατέρας ἀποστάσεις καὶ τὰς τῶν
ἡμιολίων καὶ ἐπιτρίτων καὶ ἐπογδόων μεσότητας καὶ ξυνδέσεις,
ἐπειδὴ παντελῶς λυταὶ οὐκ ἦσαν πλὴν ὑπὸ τοῦ ξυνδήσαντος,
πάσας μὲν στρέψαι στροφάς, πάσας δὲ κλάσεις καὶ διαφορὰς E
5 τῶν κύκλων ἐμποιεῖν, ὁσαχῇπερ ἦν δυνατόν, ὥστε μετ' ἀλλήλων
μόγις ξυνεχομένας φέρεσθαι μέν, ἀλόγως δὲ φέρεσθαι, τοτὲ μὲν
ἀντίας, ἄλλοτε δὲ πλαγίας, τοτὲ δὲ ὑπτίας· οἷον ὅταν τις ὕπτιος
ἐρείσας τὴν κεφαλὴν μὲν ἐπὶ γῆς, τοὺς δὲ πόδας ἄνω προσβαλὼν
ἔχῃ πρός τινι, τότε ἐν τούτῳ τῷ πάθει τοῦ τε πάσχοντος καὶ τῶν
10 ὁρώντων τά τε δεξιὰ ἀριστερὰ καὶ τὰ ἀριστερὰ δεξιὰ ἑκατέροις
τὰ ἑκατέρων φαντάζεται. ταὐτὸν δὴ τοῦτο καὶ τοιαῦτα ἕτερα αἱ
περιφοραὶ πάσχουσαι σφοδρῶς, ὅταν γέ τῳ τῶν ἔξωθεν τοῦ ταὐτοῦ 44 A
γένους ἢ τοῦ θατέρου περιτύχωσι, τότε ταὐτόν τῳ καὶ θάτερόν
του τἀναντία τῶν ἀληθῶν προσαγορεύουσαι ψευδεῖς καὶ ἀνόητοι
15 γεγόνασιν, οὐδεμία τε ἐν αὐταῖς τότε περίοδος ἄρχουσα οὐδ'
ἡγεμών ἐστιν· αἷς δ' ἂν ἔξωθεν αἰσθήσεις τινὲς φερόμεναι καὶ
προσπεσοῦσαι ξυνεπισπάσωνται καὶ τὸ τῆς ψυχῆς ἅπαν κύτος,
τόθ' αὗται κρατούμεναι κρατεῖν δοκοῦσι. καὶ διὰ δὴ ταῦτα πάντα
τὰ παθήματα νῦν κατ' ἀρχάς τε ἄνους ψυχὴ γίγνεται τὸ πρῶτον, B
20 ὅταν εἰς σῶμα ἐνδεθῇ θνητόν. ὅταν δὲ τὸ τῆς αὔξης καὶ τροφῆς

12 ὅταν γε: ὅταν τε AH.   15 ἐν αὐταῖς: ἐν ἑαυταῖς Λ.   16 αἷς δ' ἂν: ἂν δ' αὖ S.

2. **μεσότητας καὶ συνδέσεις**] These words merely signify 'means and connecting links'; they contain no special reference to the λεῖμμα, as Stallbaum imagines.

3. **λυταὶ οὐκ ἦσαν**] The dissolution of the μεσότητες καὶ συνδέσεις would of course involve the destruction of the soul.

7. **ἀντίας...πλαγίας...ὑπτίας**] It is not very clear what is the precise import of these terms. Perhaps we may understand the meaning to be that the false report of the senses may be either a negation of the truth, or diverse from it, or contrary to it: e.g. fire is not hot, fire is smoke, fire is cold. So far as the figure is concerned, it would seem impossible to draw any distinction between ἀντίας and ὑπτίας.

10. **τά τε δεξιὰ ἀριστερά**] The nature of this inversion is thus expounded by Proklos. Suppose a man to stand facing the north; then he will of course have the east on his right hand, the west on his left: then let him lie down on his back, still keeping the east on his right, and then raise his feet in the air, so that he stands on his head: he will now be looking south, while east and west will still be to right and to left as before. But a person looking south in the natural way has east to the left, and west to the right. Therefore our inverted one, knowing that he is looking south, will feel as if the east were on his left, though it is not so. Thus along with his inverted position his notion of right and left is inverted. It seems to me however that such a display of athletic skill is unnecessary. All that Plato's meaning requires is this: if A and B stand face to face, B's right is of course opposite A's left. But if A stand on his head, still facing B, then

ΤΙΜΑΙΟΣ.

they displaced, so that the double and triple intervals, being three of each sort, and the means and junctures of $\frac{3}{2}$ and $\frac{4}{3}$ and $\frac{9}{8}$, since they could not be utterly undone save by him that joined them, were forced by them to turn in all kinds of ways and to admit all manner of breaking and twisting of the circles, in every possible form, so that they can barely hold to one another, and though they are in motion, it is motion without law, sometimes reversed, now slanting, and now inverted. It is as though a man should stand on his head, resting it on the earth and supporting his feet against something aloft; in this case the man in such condition and the spectators would reciprocally see right and left reversed in the persons of each other. The same and similar effects are produced with great intensity in the soul's revolutions: and when from external objects there meets them anything that belongs to the class of the Same or to that of the Other, then they declare its relative sameness or difference quite contrariwise to the truth, and show themselves false and irrational; and no circuit is governor or leader in them at that time. And whenever sensations from without rushing up and falling upon them drag along with them the whole vessel of the soul, then the circuits seem to govern though they really are governed. On account then of all these experiences the soul is at first bereft of reason, now as in the beginning, when she is confined in a mortal body. But when the stream of growth and nutriment

B's right will be opposite A's right; the normal relation being inverted.

17. ἅπαν κύτος] The soul is, as it were, an envelope containing the περιφοραί. Stallbaum compares *Laws* 964 E, where the city is compared to a κύτος.

18. αὗται κρατούμεναι κρατεῖν δοκοῦσι] Stallbaum, after Proklos, refers αὗται to αἰσθήσεις, interpreting 'they (the sensations) seem to rule the soul, which by rights rules them'. But this cannot be admitted, because the important addition 'by rights' is not in the Greek and cannot be dispensed with. Moreover the sensations do really and not only in appearance govern the soul under these circumstances. Martin's interpretation seems to me unquestionably right. αὗται refers to περίοδοι, and is the antecedent to αἷς. When, Plato says, any sensations rush upon the περίοδοι and carry the whole soul along with them, then the περίοδοι seem to govern, though really they are governed. That is to say, the motion of the circles which is imparted to them by the impulse of the αἰσθήσεις is mistaken for their own proper motion: their report of the perception is received as true, though in fact it is untrustworthy. The notion in ἅπαν κύτος seems to be that when the sensations are very overpowering, they give an impulse to the *whole* soul: there is no hesitation nor conflict of opinion. Since then the soul ratifies without question the report of the senses, she seems to be acting regularly and rightly

152   ΠΛΑΤΩΝΟΣ   [44 B—

ἔλαττον ἐπίῃ ῥεῦμα, πάλιν δὲ αἱ περίοδοι λαμβανόμεναι γαλήνης
τὴν ἑαυτῶν ὁδὸν ἴωσι καὶ καθιστῶνται μᾶλλον ἐπιόντος τοῦ
χρόνου, τότε ἤδη πρὸς τὸ κατὰ φύσιν ἰόντων σχῆμα ἑκάστων τῶν
κύκλων αἱ περιφοραὶ κατευθυνόμεναι, τό τε θάτερον καὶ τὸ ταὐτὸν
5 προσαγορεύουσαι κατ' ὀρθόν, ἔμφρονα τὸν ἔχοντα αὐτὰς γιγνόμενον ἀποτελοῦσιν. ἂν μὲν οὖν δὴ καὶ ξυνεπιλαμβάνηταί τις
ὀρθὴ τροφὴ παιδεύσεως, ὁλόκληρος ὑγιής τε παντελῶς, τὴν με- C
γίστην ἀποφυγὼν νόσον, γίγνεται, καταμελήσας δέ, χωλὴν τοῦ
βίου διαπορευθεὶς ζωήν, ἀτελὴς καὶ ἀνόητος εἰς Ἅιδου πάλιν
10 ἔρχεται. ταῦτα μὲν οὖν ὕστερά ποτε γίγνεται· περὶ δὲ τῶν νῦν
προτεθέντων δεῖ διελθεῖν ἀκριβέστερον, τὰ δὲ πρὸ τούτων περὶ
σωμάτων κατὰ μέρη τῆς γενέσεως καὶ περὶ ψυχῆς, δι' ἅς τε αἰτίας
καὶ προνοίας γέγονε θεῶν, τοῦ μάλιστα εἰκότος ἀντεχομένοις
οὕτω καὶ κατὰ ταῦτα πορευομένοις διεξιτέον.   D

9 ἀνόητος : ἀνόνητος A pr. m. S.

apprehending the phenomena, whereas really she is obeying an external impulse.

1. **ἔλαττον ἐπίῃ ῥεῦμα**] That is to say, as the child grows older the imperious necessities of nutrition become less predominant; also the sensations from without grow less distracting. Accordingly the intellect has freer play to exercise its functions.

5. **ἔμφρονα...γιγνόμενον**] Note that he is only put in the way to become rational.

7. **ὀρθὴ τροφὴ παιδεύσεως**] These words must be taken together, the genitive depending upon τροφή. Stallbaum, governing παιδεύσεως by ἐπιλαμβάνηται, wrongly understands ὀρθὴ τροφὴ to refer to the diminished influx of nutriment.

ὁλόκληρος] This is a technical term of the Eleusinian ritual. Plato is fond of borrowing such terms: cf. *Phaedrus* 250 C ὁλόκληρα δὲ καὶ ἁπλᾶ καὶ ἀτρεμῆ καὶ εὐδαίμονα φάσματα μυούμενοί τε καὶ ἐποπτεύοντες ἐν αὐγῇ καθαρᾷ, καθαροὶ ὄντες καὶ ἀσήμαντοι τούτου ὃ νῦν σῶμα περιφέροντες ὀνομάζομεν, ὀστρέου τρόπον δεδεσμευμένοι. See too *Laws* 759 C. Similarly ἀτελής is a ritual term. It is also possible that in τὴν μεγίστην ἀπο-φυγὼν νόσον we have an echo of the ejaculation of the initiates, ἔφυγον κακόν, εὗρον ἄμεινον : cf. Demosthenes *de corona* p. 312 § 259.

8. **χωλήν**] Compare 87 D, where it is said that if a disproportion exists between soul and body, the ὅλον ζῷον is ἀξύμμετρον ταῖς μεγίσταις ξυμμετρίαις.

τοῦ βίου διαπορευθεὶς ζωήν] βίου ζωή = 'the conscious existence of his lifetime', ζωή being a more subjective term than βίος. Compare on the other hand Euripides *Hercules furens* 664 ἁ δυσγένεια δ' ἁπλᾶν ἂν | εἶχε ζωᾶς βιοτάν.

10. **ὕστερά ποτε γίγνεται**] i.e. belong to a later part of our exposition: the subject is in fact dealt with in chapters 41—43.

τῶν νῦν προτεθέντων] I concur with Stallbaum in referring τὰ νῦν προτεθέντα to the inquiry into the operation of the several senses, while τὰ πρὸ τούτων signifies the investigation περὶ σωμάτων κατὰ μέρη γενέσεως καὶ περὶ ψυχῆς.

13. **τοῦ μάλιστα εἰκότος**] We are now fairly in the region of the physical, where we must be content with the 'probable account'.

ΤΙΜΑΙΟΣ.

flows in with smaller volume, and the revolutions calming down go their own way and become settled as time passes on, then the orbits are reduced to the form that belongs to the several circles in their natural motion, and declaring accurately the Other and the Same, they set their possessor in the way to become rational. And if any just discipline of education help this process, he becomes whole altogether without a blemish, having made his escape from the most grievous of plagues; but if he neglect it, he passes the days of his life halt and maimed, and unhallowed and unreasonable he comes again to Hades. These things however belong to a later time: we must discuss more exactly the subject immediately before us. And as to the matters which are previous to this, concerning the generation of the body in all its parts and concerning soul, and the reasons and designs of the gods whereby they have come into being, we must cling to the most probable theory, and by proceeding in this way so give an account of all.

44 D—47 E, *c.* xvi. The two revolutions of the soul were enclosed in a spherical case which we call the head: and all the rest of the body was framed that it might minister to the head, aiding it to move from place to place and preserving it from harm. And to man the gods assigned a forward progress as his most natural motion; for this was more dignified than the contrary. To distinguish front from rear they set the face with its organs of sense in one part of the head; and this they made the forward and leading side. The first organs they fashioned therein were the eyes that lighten the body. Now vision comes to pass on this wise. From the eyes issues forth a stream of clear and subtle fire, of the same substance as the sunlight in the air; with which it mingles, and the two combined meet the fire proceeding from the object which is in the line of vision; and so the united fires, becoming one body, transmit the vibrations from the object to the eye. But at night, when there is no more light in the air, the visual fire on passing forth into the darkness is quenched; and when the eyelids are closed, the flow of it is turned inwards, and calming the motions that are within, it produces sleep, more or less dreamless according as the calm is complete.

Then it is shown how images in mirrors arise through the reflection of the combined fires when they meet upon a smooth shining surface; how in plane mirrors right and left are reversed in the reflection; and how in a concave mirror, when it is held in one position, right and left are not transposed, but if it be held in another, the image is inverted.

But we must remember that all these physical laws are but a means to an end; we must learn to distinguish between spiritual causes, which are primary, and material causes, which are only subsidiary: and though both must be explained, the first alone is the true object of the wise man's search. Now the true motive of the gods in bestowing sight upon man was the attainment of philosophy by him: for had we never seen the celestial motions and from them

154   ΠΛΑΤΩΝΟΣ   [44 D—

XVI. Τὰς μὲν δὴ θείας περιόδους δύο οὔσας τὸ τοῦ παντὸς σχῆμα ἀπομιμησάμενοι περιφερὲς ὂν εἰς σφαιροειδὲς σῶμα ἐνέδησαν, τοῦτο ὃ νῦν κεφαλὴν ἐπονομάζομεν, ὃ θειότατόν τ' ἐστὶ καὶ τῶν ἐν ἡμῖν πάντων δεσποτοῦν· ᾧ καὶ πᾶν τὸ σῶμα παρέδοσαν 5 ὑπηρεσίαν αὐτῷ ξυναθροίσαντες θεοί, κατανοήσαντες, ὅτι πασῶν ὅσαι κινήσεις ἔσοιντο μετέχοι. ἵν' οὖν μὴ κυλινδούμενον ἐπὶ γῆς ὕψη τε καὶ βάθη παντοδαπὰ ἐχούσης ἀποροῖ τὰ μὲν ὑπερβαίνειν, E ἔνθεν δὲ ἐκβαίνειν, ὄχημ' αὐτῷ τοῦτο καὶ εὐπορίαν ἔδοσαν· ὅθεν δὴ μῆκος τὸ σῶμα ἔσχεν, ἐκτατά τε κῶλα καὶ καμπτὰ ἔφυσε τέτταρα 10 θεοῦ μηχανησαμένου πορείαν, οἷς ἀντιλαμβανόμενον καὶ ἀπερειδόμενον διὰ πάντων τόπων πορεύεσθαι δυνατὸν γέγονε, τὴν τοῦ θειοτάτου καὶ ἱερωτάτου φέρον οἴκησιν ἐπάνωθεν ἡμῶν. σκέλη 45 A μὲν οὖν χεῖρές τε ταύτῃ καὶ διὰ ταῦτα προσέφυ πᾶσι· τοῦ δ' ὄπισθεν τὸ πρόσθεν τιμιώτερον καὶ ἀρχικώτερον νομίζοντες θεοὶ 15 ταύτῃ τὸ πολὺ τῆς πορείας ἡμῖν ἔδοσαν. ἔδει δὴ διωρισμένον ἔχειν καὶ ἀνόμοιον τοῦ σώματος τὸ πρόσθεν ἄνθρωπον. διὸ πρῶτον μὲν περὶ τὸ τῆς κεφαλῆς κύτος, ὑποθέντες αὐτόσε τὸ πρόσωπον, ὄργανα ἐνέδησαν τούτῳ πάσῃ τῇ τῆς ψυχῆς προνοίᾳ, καὶ διέταξαν B τὸ μετέχον ἡγεμονίας τοῦτ' εἶναι τὸ κατὰ φύσιν πρόσθεν. τῶν 20 δὲ ὀργάνων πρῶτον μὲν φωσφόρα ξυνετεκτήναντο ὄμματα, τοιᾷδε ἐνδήσαντες αἰτίᾳ. τοῦ πυρὸς ὅσον τὸ μὲν καίειν οὐκ ἔσχε, τὸ δὲ παρέχειν φῶς ἥμερον, οἰκεῖον ἑκάστης ἡμέρας σῶμα ἐμηχανήσαντο

10 πορείαν: πορεῖα SZ.   18 τῇ omittit Λ.   διέταξαν τὸ μετέχον: διετάξαντο μέτοχον SZ.   22 post ἡμέρας commate vulgo interpungitur.

learnt number, philosophy could never have been ours. But now we are able to rule and correct the errant movements of our soul by contemplating the serene unswerving revolutions of the skies. And to the same end too they gave sound and music and harmony and rhythm, that we might bring order from disorder in our souls.

1. τὸ τοῦ παντὸς σχῆμα ἀπομιμησάμενοι] Cf. 73 C: see too 81 Α, where the whole human frame is regarded as a microcosm working on the same principles as the universe.

3. ὃ νῦν κεφαλήν] Plato, in placing the ἀρχή of consciousness in the head, agrees with Hippokrates: cf. de morbo sacro vol. I p. 614 Kühn διότι ἡ καρδίη αἰσθάνεταί τε μάλιστα καὶ αἱ φρένες. τῆς μέντοι φρονήσιος οὐδετέρῳ μέτεστιν, ἀλλὰ πάντων τουτέων ὁ ἐγκέφαλος αἴτιός ἐστιν. This view was afterwards upheld by Galen against the Peripatetics and Stoics, who made the heart the sole ἀρχή. With δεσποτοῦν compare a phrase in one of the Hippokratean epistles, III 824 Kühn: δεσπότην φύλακα διανοίης καλύπτουσιν ἐγκέφαλον.

5. πασῶν] i.e. all the six, excluding rotation: cf. 43 Β.

10. πορείαν] This reading has overwhelming ms. support, and may very well signify 'as means of locomotion': there seems no sufficient ground for changing it to πορεῖα.

13. προσέφυ] With this remarkable

XVI. Imitating the shape of the universe, which was spherical, they confined the two divine revolutions in a globe-shaped body, the same that we now call the head, which is the divinest part of us and has dominion over all our members. To this the gods gave the whole body, when they had put it together, to minister to it, reflecting that it possessed all the motions that should be. In order then that it might not have to roll upon the earth, which has hills and hollows of all kinds, nor be at a loss to surmount the one and climb out of the other, they gave it the body for a conveyance and for ease of going: whence the body was endowed with length and grew four limbs that could be stretched and bent, which the god devised for it to go withal, and by means of which clinging and supporting itself it is enabled to pass through every place, bearing at the top of us the habitation of the most divine and sacred element. In this way then and for these reasons were legs and hands added to all mankind; and the gods, deeming that the front was more honourable than the back and more fit to lead, made us to move for the most part in this direction. So it behoved man to have the front part distinguished and unlike the back. Therefore having set the face upon the globe of the head on that side, they attached to it organs for all the forethought of the soul, and they ordained that this which had the faculty of guidance should be by nature the front. And first of the organs they wrought light-giving eyes, which they fixed there on the plan I shall explain. Such sort of fire as had the property of yielding a gentle light

use of the singular compare the still stronger case in *Symposium* 188 B καὶ γὰρ πάχναι καὶ χάλαζαι καὶ ἐρυσίβαι ἐκ πλεονεξίας καὶ ἀκοσμίας περὶ ἄλληλα τῶν τοιούτων γίγνεται ἐρωτικῶν. The construction is of course distinct from the so-called 'schema Pindaricum', in which the verb precedes its subject, and which is not so very uncommon in Attic writers.

15. ἴδει δή] Forward motion is more dignified than retrograde; and man is to have the more dignified. But to attain this there must be something to distinguish front from rear; therefore the gods placed the sensory organs, eyes

nose and mouth, on the same side of the head, forming the face; and this side they called the front.

18. διέταξαν τὸ μετέχον] This reading is distinctly preferable to διετάξαντο μέτοχον. For μέτοχον ἡγεμονίας must be the predicate: the meaning however plainly is that the gods, to distinguish front from back, ordered that the face, which held the leading position (because it contained the ὄργανα τῇ τῆς ψυχῆς προνοίᾳ), should be τὸ κατὰ φύσιν πρόσθεν.

22. οἰκεῖον ἑκάστης ἡμέρας σῶμα] This punctuation is due to Madvig, who by

ΠΛΑΤΩΝΟΣ [45 B—

γίγνεσθαι. τὸ γὰρ ἐντὸς ἡμῶν ἀδελφὸν ὂν τούτου πῦρ εἰλικρινὲς ἐποίησαν διὰ τῶν ὀμμάτων ῥεῖν λεῖον καὶ πυκνόν, ὅλον μέν, μάλιστα δὲ τὸ μέσον ξυμπιλήσαντες τῶν ὀμμάτων, ὥστε τὸ μὲν C ἄλλο ὅσον παχύτερον στέγειν πᾶν, τὸ τοιοῦτον δὲ μόνον αὐτὸ καθαρὸν διηθεῖν. ὅταν οὖν μεθημερινὸν ᾖ φῶς περὶ τὸ τῆς ὄψεως ῥεῦμα, τότ' ἐκπῖπτον ὅμοιον πρὸς ὅμοιον, ξυμπαγὲς γενόμενον, ἓν σῶμα οἰκειωθὲν συνέστη κατὰ τὴν τῶν ὀμμάτων εὐθυωρίαν, ὅπῃπερ ἂν ἀντερείδῃ τὸ προσπῖπτον ἔνδοθεν πρὸς ὃ τῶν ἔξω συνέπεσεν. ὁμοιοπαθὲς δὴ δι' ὁμοιότητα πᾶν γενόμενον, ὅτου τε ἂν αὐτό ποτε ἐφάπτηται καὶ ὃ ἂν ἄλλο ἐκείνου, τούτων τὰς κινήσεις διαδιδὸν D εἰς ἅπαν τὸ σῶμα μέχρι τῆς ψυχῆς αἴσθησιν παρέσχετο ταύτην, ᾗ δὴ ὁρᾶν φαμέν. ἀπελθόντος δὲ εἰς νύκτα τοῦ ξυγγενοῦς πυρὸς ἀποτέτμηται· πρὸς γὰρ ἀνόμοιον ἐξιὸν ἀλλοιοῦταί τε αὐτὸ καὶ κατασβέννυται, ξυμφυὲς οὐκέτι τῷ πλησίον ἀέρι γιγνόμενον, ἅτε

7 ὅπῃπερ ἄν: ἄν omittit A.   9 ὅτου τε ἄν: ὅτου τε ἐὰν A.

merely expunging a comma has restored sense to the passage. Ordinarily a comma is placed after ἡμέρας, leaving us to face the inconvenient problem, how could the gods make into body that which was body already? For Martin's attempt to specialise the use of σῶμα in the sense of 'definitely formed matter' is hopeless. Eschewing the comma however, we get quite the right sense—they made it into a substance similar to the daylight, which is a subtle fire pervading the atmosphere. Thus too the γάρ immediately following, to which Stallbaum takes exception, is justified; it introduces the explanation how the gods made the fire within us similar to the fire without. There is an obvious play between ἥμερον—ἡμέρας. For Plato's etymology of ἡμέρα see *Cratylus* 418 C.

4. τὸ τοιοῦτον] sc. τὸ εἰλικρινὲς καὶ λεῖον καὶ πυκνόν.

6. ἓν σῶμα οἰκειωθέν] That is to say, wherever the eye is directed, the stream of fire from the eye and the fire in the atmosphere, which is of one and the same substance with it, combine and form a ray of homogeneous fire all along the line of vision.

10. τούτων τὰς κινήσεις διαδιδόν] Plato's theory may thus be briefly explained. There are three fires concerned : the fire that streams from the eye, the fire of daylight in the air, and the fire in the object seen, which is the cause of its visibility. The first two are absolutely homogeneous one with the other and combine into a perfectly uniform substance. This substance, on meeting the rays from the object, receives their vibrations and transmits them to the eye, whence they are delivered to the seat of consciousness, at which point of the process perception takes place. The problem with which Plato has to deal is, how is action at a distance effected? This he ingeniously attempts to explain by the hypothesis of an extension of the substance of the percipient in the direction of the object : for the ὄψεως ῥεῦμα is just as much part of ourselves as the brain or hand : this is clear from 64 D. If this passage be compared with the statements in *Theaetetus* 156 A foll. or 182 A, it will be seen that the physical theory of the *Timaeus* fits in perfectly well with the metaphysical doctrine of perception in the *Theaetetus*.

It is plain too that Plato's theory is

ΤΙΜΑΙΟΣ.

but not of burning, they contrived to form into a substance akin to the light of every day. The fire within us, which is akin to the daylight, they made to flow pure smooth and dense through the eyes, having made close the whole fabric of the eyes and especially the pupils, so that they kept back all that was coarser and suffered only this to filter through unmixed and pure. Whenever then there is daylight surrounding the current of vision, then this issues forth as like into like, and coalescing with the light is formed into one uniform substance in the direct line of vision, wherever the stream issuing from within strikes upon some external object that falls in its way. So the whole from its uniformity becomes sympathetic; and whenever it comes in contact with anything else, or anything with it, it passes on the motions thereof over the whole body until they reach the soul, and thus causes that sensation which we call seeing. But when its kindred fire departs into night, the visual current is cut off: for issuing into an alien element it is itself changed and quenched, having no longer a common nature with the surrounding air,

peculiar to himself and quite diverse from the Empedoklean (or Demokritean) doctrine of effluences, with which Stallbaum confuses it; although the two theories have some points in common, as appears from the statement of Aristotle *de sensu* 437^b 11 foll. Empedokles, as Aristotle informs us, wavered in his explanation, sometimes adopting the ἀπορροαί aforesaid, sometimes comparing the eye to a lantern, sending forth its visual ray through the humours and membranes which correspond to the frame of the lantern. But as propounded in the passage quoted by Aristotle (302—310 Karsten), this notion amounts merely to a metaphor or analogy and is not worked up into a physical theory: it agrees however with Plato in taking fire for the active force of the eye. The doctrine of effluences from the object corresponding to πόροι in the percipient is attributed to Empedokles in *Meno* 76 C: see too Aristotle *de gen. et corr.* I viii 324^b 25 foll. Plato himself assumes an effluence of rays

from the object, but this has little resemblance to the Empedoklean ἀπορροαί. An exposition of the peculiar theory of Demokritos will be found in Theophrastos *de sensu* § 49 foll. Aristotle's theory of vision is expounded in *de anima* II vii and *de sensu* ii, iii.

11. μέχρι τῆς ψυχῆς] See note on 43 C.

12. ᾗ δή] 'whereby' we see. The physical process is the soul's instrument: cf. *Theaetetus* 184 C.

14. κατασβέννυται] Plato explains quite clearly what he means by 'extinguished'. The visual fire, issuing into air destitute of light, finds no kindred substance with which to coalesce: it is thus modified, and losing its proper nature becomes unable to carry on the process of vision. Aristotle however, catching at the word κατασβέννυται, asks τίς γὰρ ἀπόσβεσις φωτός ἐστιν; σβέννυται γὰρ ἢ ὑγρῷ ἢ ψυχρῷ τὸ θερμὸν καὶ ξηρόν, οἷον δοκεῖ τό τ' ἐν τοῖς ἀνθρακώδεσιν εἶναι πῦρ καὶ ἡ φλόξ, ὧν τῷ φωτὶ οὐδέτερον φαίνεται ὑπάρχον. It is

158   ΠΛΑΤΩΝΟΣ   [45 D—

πῦρ οὐκ ἔχοντι. παύεταί τε οὖν ὁρῶν, ἔτι τε ἐπαγωγὸν ὕπνου
γίγνεται· σωτηρίαν γὰρ ἦν οἱ θεοὶ τῆς ὄψεως ἐμηχανήσαντο, τὴν
τῶν βλεφάρων φύσιν, ὅταν ταῦτα ξυμμύσῃ, καθείργνυσι τὴν E
τοῦ πυρὸς ἐντὸς δύναμιν, ἡ δὲ διαχεῖ τε καὶ ὁμαλύνει τὰς ἐντὸς
5 κινήσεις, ὁμαλυνθεισῶν δὲ ἡσυχία γίγνεται, γενομένης δὲ πολλῆς
μὲν ἡσυχίας βραχυόνειρος ὕπνος ἐμπίπτει, καταλειφθεισῶν δέ
τινων κινήσεων μειζόνων, οἷαι καὶ ἐν οἵοις ἂν τόποις λείπωνται, 46 A
τοιαῦτα καὶ τοσαῦτα παρέσχοντο ἀφομοιωθέντα ἐντὸς ἔξω τε
ἐγερθεῖσιν ἀπομνημονευόμενα φαντάσματα. τὸ δὲ περὶ τὴν τῶν
10 κατόπτρων εἰδωλοποιίαν, καὶ πάντα ὅσα ἐμφανῆ καὶ λεῖα, κατιδεῖν
οὐδὲν ἔτι χαλεπόν. ἐκ γὰρ τῆς ἐντὸς ἐκτός τε τοῦ πυρὸς ἑκατέρου
κοινωνίας ἀλλήλοις, ἑνός τε αὖ περὶ τὴν λειότητα ἑκάστοτε γενομένου καὶ πολλαχῇ μεταρρυθμισθέντος, πάντα τὰ τοιαῦτα ἐξ B
ἀνάγκης ἐμφαίνεται, τοῦ περὶ τὸ πρόσωπον πυρὸς τῷ περὶ τὴν
15 ὄψιν πυρὶ περὶ τὸ λεῖον καὶ λαμπρὸν ξυμπαγοῦς γιγνομένου.
δεξιὰ δὲ φαντάζεται τὰ ἀριστερά, ὅτι τοῖς ἐναντίοις μέρεσι τῆς

1 ὕπνου γίγνεται: γίγνεται ὕπνου S.   16 κατὰ post φαντάζεται habet A.

impossible to exonerate criticism of this kind from the charge of ὀνομάτων θήρευσις. The reference in *de anima* III xii 435ᵃ 5 is apparently to Empedokles, not to Plato.

4. ἡ δὲ διαχεῖ] sc. ἡ τοῦ πυρὸς δύναμις, not, as Stallbaum has it, ἡ τῶν βλεφάρων φύσις: to say nothing of the sense, the ἡ δὲ is sufficient to show that the subject of διαχεῖ is different from that of καθείργνυσι. Plato's view is that when the eyes are closed, the visual stream, unable to find an outlet, is directed inwards, and the smooth and subtle flow of fire mollifies and calms all the motions within, thus inducing sleep.

8. ἀφομοιωθέντα ἐντός] Dreams are the result of motions which are not thoroughly calmed down, whereby semblances of external things are presented to the mind from within: the κίνησις corresponding to any particular external impression producing a likeness of that impression in the sleeping consciousness. The sense is plain enough; but some difficulty attaches to the words ἐντὸς ἔξω τε. Martin, construing them with ἀφομοιωθέντα, trans-

lates 'images semblables à des objets soit intérieurs, soit extérieurs'. But what can be meant by 'objets intérieurs'? I had thought of substituting ἔξωθεν for ἔξω τε, 'copied within from without': in which case ἐγερθεῖσί τ' must be read. But though this gives a good sense, it overthrows the balance of the sentence. And the text may, I think, be explained as it stands: the images are copied within—that is, in the dream-world, and recalled to mind without—that is, when we have emerged from the dream-world. For Aristotle's theory of dreams see the treatise περὶ ἐνυπνίων.

11. ἐκ γὰρ τῆς ἐντός] Plato proceeds to explain the phenomena of reflection in mirrors. The rays from the object reflected are arrested by the smooth shining surface of the mirror, which they cannot penetrate: the combined ὄψεως ῥεῦμα and μεθημερινὸν φῶς are arrested on the same surface and thus come into conjunction with the rays from the object. Thus the mirror is the cause of contact between the fire of the subject and the fire of the object, and so an indirect vision is

which has in it no fire. Therefore it ceases from seeing and moreover becomes an allurement to sleep. For the gods had devised as a safeguard of the sight the structure of the eyelids; and when these are closed, they shut up the force of fire within; and this smoothes and calms the motions within; and when these are calmed, quiet ensues. And if the quiet is profound, sleep with few dreams falls on us; but if some of the stronger motions are left, according to their nature and the places where they remain, they engender visions corresponding in kind and in number; which are images within us, and when we awake are remembered as outside us. Now the explanation of reflections in mirrors and all bright smooth surfaces is no longer hard to discern. For because of the communion of the internal and external fire, which again is united on the smooth surface and in manifold ways deflected, all these reflections take place; the fire that belongs to the face coalescing with the fire of the visual current upon the surface of the smooth bright object. And left appears right and right left, because mutually opposite particles

effected. τοῦ πυρὸς ἑκατέρου signifies not the visual stream and the daylight, but the visual stream (combined with the daylight) and the rays from the object. These two fires combine upon the surface of the mirror (ἐκτός), and the κινήσεις of this combination are transmitted along the visual stream and impressed upon the retina (ἐντός). The foregoing interpretation gives the best meaning I can put upon the curious phrase τῆς ἐντὸς ἐκτός τε κοινωνίας, unless we may suppose that Plato rather loosely said 'the internal and external combination of the two fires' for 'the combination of the internal and external fires'. But I have strong suspicions that ἐντὸς ἐκτός τε is a marginal gloss upon ἑκατέρου. Seneca *natur. quaest.* I v I clearly expresses the distinctive characteristic of Plato's theory of reflections: 'de speculis duae opiniones sunt; alii enim in illis simulacra cerni putant, id est corporum nostrorum figuras a nostris corporibus emissas ac separatas, *alii non imagines in speculo, sed ipsa adspici cor-*

pora, retorta oculorum acie et in se rursus reflexa'. The italicised words express Plato's opinion. πολλαχῇ μεταρρυθμισθέντος refers, I conceive, to the various angles at which the rays are reflected, corresponding to the different angles of incidence.

14. ἐμφαίνεται] 'are reflected'. ἐμφαίνεσθαι is the technical term. The word ἔμφασις, 'reflection', does not occur in Plato but is frequent in Aristotle and Theophrastos.

τοῦ περὶ τὸ πρόσωπον πυρός] i.e. the fire belonging to the face, which is the object reflected. We must suppose the case of a person looking at his own face in a mirror: what happens is that the ray from the face, τὸ περὶ τὸ πρόσωπον, is checked on the surface of the mirror and is then amalgamated with the visual stream, τὸ περὶ τὴν ὄψιν, which meets it at that spot. Plato's theory of course applies to all reflections, although in this sentence he is speaking as of a particular case.

160   ΠΛΑΤΩΝΟΣ   [46 B—

ὄψεως περὶ τἀναντία μέρη γίγνεται ἐπαφὴ παρὰ τὸ καθεστὸς
ἔθος τῆς προσβολῆς· δεξιὰ δὲ τὰ δεξιὰ καὶ τὰ ἀριστερὰ ἀριστερὰ
τοὐναντίον, ὅταν μεταπέσῃ συμπηγνύμενον ᾧ συμπήγνυται φῶς·
τοῦτο δέ, ὅταν ἡ τῶν κατόπτρων λειότης, ἔνθεν καὶ ἔνθεν ὕψη C
5 λαβοῦσα, τὸ δεξιὸν εἰς τὸ ἀριστερὸν μέρος ἀπώσῃ τῆς ὄψεως καὶ
θάτερον ἐπὶ θάτερον. κατὰ δὲ τὸ μῆκος στραφὲν τοῦ προσώπου
ταὐτὸν τοῦτο ὕπτιον ἐποίησε πᾶν φαίνεσθαι, τὸ κάτω πρὸς τὸ ἄνω
τῆς αὐγῆς τό τ' ἄνω πρὸς τὸ κάτω πάλιν ἀπῶσαν.
   Ταῦτ' οὖν πάντα ἔστι τῶν ξυναιτίων, οἷς θεὸς ὑπηρετοῦσι
10 χρῆται τὴν τοῦ ἀρίστου κατὰ τὸ δυνατὸν ἰδέαν ἀποτελῶν· δοξά-
ζεται δὲ ὑπὸ τῶν πλείστων οὐ ξυναίτια ἀλλ' αἴτια εἶναι τῶν πάν- D
των, ψύχοντα καὶ θερμαίνοντα πηγνύντα τε καὶ διαχέοντα καὶ ὅσα
τοιαῦτα ἀπεργαζόμενα· λόγον δὲ οὐδένα οὐδὲ νοῦν εἰς οὐδὲν δυνατὰ
ἔχειν ἐστί. τῶν γὰρ ὄντων ᾧ νοῦν μόνῳ κτᾶσθαι προσήκει, λεκτέον

1. **περὶ τἀναντία μέρη**] Plato's meaning will be readily understood by means of a diagram, which, together with the explanation, is borrowed from Martin.

AB is a line in the mirror where it is cut by a plane which also passes through the eye of the observer and through the object reflected. *CD* is the line where the plane cuts the eye, *EF* the line where it cuts the object. *DH*, *CG* are two rays of the visual fire impinging upon the mirror in the points *G*, *H*: *EG*, *FH* are two rays from the objects impinging upon the mirror and meeting *DH*, *CG* in the

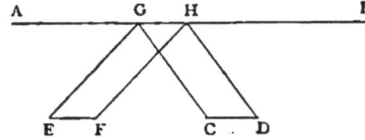

same two points. Then it will be seen that the ray *DH*, which proceeds from the right side of the eye, meets the ray *FH*, proceeding from the right side of the object: therefore (the angle of reflection being equal to the angle of incidence) the ray from *F* is reflected along *HD* to the right side of the eye. Similarly the ray *EG*, issuing from the left of the object, is reflected along *GC* to the left side of the eye. This is a reversal of what happens in the case of direct vision (παρὰ τὸ καθεστὸς ἔθος τῆς προσβολῆς). For if *A* and *B* look each other in the face, *A*'s right eye will be opposite *B*'s left, and so forth: but if *A* look at his own face in the glass, the eye in the reflection, which should be the left relatively to the reflection, will be the reflection of the right eye: for if *A* close his right eye, the eye in the mirror opposite his right will be closed. Plato's theory then is designed to explain why it is that in a reflection the right side of the visual current comes in contact with the rays from the right side of the object, whereas in direct vision it meets the rays from the left of the object. Compare *Sophist* 266 C διπλοῦν δὲ ἡνίκ' ἂν φῶς. οἰκεῖόν τε καὶ ἀλλότριον περὶ τὰ λαμπρὰ καὶ λεῖα εἰς ἓν ξυνελθὸν τῆς ἔμπροσθεν εἰωθυίας ὄψεως ἐναντίαν αἴσθησιν παρέχον εἶδος ἀπεργάζηται.

4. **ἔνθεν καὶ ἔνθεν ὕψη λαβοῦσα**] i.e. a concave mirror. Plato conceives the reversal of the phenomena of reflection as appearing in a plane mirror to be due to the concavity deflecting the rays at the

ΤΙΜΑΙΟΣ.

of the visual current and of the object seen come into contact, contrary to the wonted mode of collision. On the other hand right appears as right and left as left, when in the act of combination with that wherewith it combines the ray changes sides. This happens when the smooth surface of the mirror is curved upwards on each side and so throws the right portion of the visual current to the left side and the converse. But if it is turned lengthwise to the face, it makes this same reflection appear completely upside down, thrusting the lower portion of the ray to the upper end and the upper to the lower.

All these things are among the secondary causes which God uses to serve him in carrying out the idea of the best so far as is possible. But the multitude regard them not as secondary but as primary causes, which act by cooling and heating, condensing and rarefying, and all such processes. Yet they are incapable of all reason or thought for any purpose. For the only existing thing to which belongs the possession of reason

---

moment of impact. In the case of a concave mirror the section AB would be a curved line instead of straight; and thereby a ray from the right side, just at the moment of impact, while it is in act of amalgamating with the ray from the object, is shifted to the left side, and *vice versa*. It must be remembered that the concave mirrors of which Plato speaks are not of the sort with which we are most familiar, namely hemispherical mirrors: they are hemicylindrical: therefore when the mirror is held laterally, so that the curvature is from right to left, the position of right and left as compared with a reflection in a plane mirror is inverted; if it is held vertically (κατὰ μῆκος στραφὲν τοῦ προσώπου), so that the curvature is from top to bottom, the reflection is upside down. See Munro's note on Lucretius IV 317. If the mirror were hemispherical, or one which is concave all round from centre to circumference, both right and left and top and bottom would be inverted, as may be seen by simply looking into the bowl of a silver spoon. This case is not noticed by Plato, nor by Lucretius *l. l.* Martin gives a mathematical explanation of the phenomena.

9. τῶν ξυναιτίων] Plato now proceeds to guard against being supposed to mean that the physical principles which he has just laid down are the real cause: they are merely the means through which the true cause works, viz., νοῦς operating ἐπὶ τὸ βέλτιστον. Compare *Phaedo* 99 B. The whole of this latter part of the chapter contains a polemic partly against Anaxagoras, partly against Demokritos. Anaxagoras did indeed postulate νοῦς as his prime force, but he used it simply as a mechanical agent, without attributing to it a conscious effort to produce the best result. Demokritos conceives a blind unconscious force, ἀνάγκη, to be the motive power of the universe. Thus whereas the opposition between Demokritos and Plato is fundamental and essential, Plato's controversy with Anaxagoras is due rather to inconsequence or incompleteness on the part of the latter.

ψυχήν· τοῦτο δὲ ἀόρατον, πῦρ δὲ καὶ ὕδωρ καὶ γῆ καὶ ἀὴρ σώματα πάντα ὁρατὰ γέγονε· τὸν δὲ νοῦ καὶ ἐπιστήμης ἐραστὴν
ἀνάγκη τὰς τῆς ἔμφρονος φύσεως αἰτίας πρώτας μεταδιώκειν, ὅσαι
δὲ ὑπ' ἄλλων μὲν κινουμένων, ἕτερα δὲ ἐξ ἀνάγκης κινούντων E
5 γίγνονται, δευτέρας. ποιητέον δὴ κατὰ ταῦτα καὶ ἡμῖν· λεκτέα
μὲν ἀμφότερα τὰ τῶν αἰτιῶν γένη, χωρὶς δὲ ὅσαι μετὰ νοῦ καλῶν
καὶ ἀγαθῶν δημιουργοὶ καὶ ὅσαι μονωθεῖσαι φρονήσεως τὸ τυχὸν
ἄτακτον ἑκάστοτε ἐξεργάζονται. τὰ μὲν οὖν τῶν ὀμμάτων ξυμμετ-
αίτια πρὸς τὸ ἔχειν τὴν δύναμιν ἣν νῦν εἴληχεν εἰρήσθω· τὸ δὲ
10 μέγιστον αὐτῶν εἰς ὠφέλειαν ἔργον, δι' ὃ θεὸς αὔθ' ἡμῖν δεδώρηται, 47 A
μετὰ τοῦτο ῥητέον. ὄψις δὴ κατὰ τὸν ἐμὸν λόγον αἰτία τῆς μεγίστης ὠφελείας γέγονεν ἡμῖν, ὅτι τῶν νῦν λόγων περὶ τοῦ παντὸς
λεγομένων οὐδεὶς ἄν ποτε ἐρρήθη μήτε ἄστρα μήτε ἥλιον μήτε
οὐρανὸν ἰδόντων· νῦν δ' ἡμέρα τε καὶ νὺξ ὀφθεῖσαι μῆνές τε καὶ
15 ἐνιαυτῶν περίοδοι μεμηχάνηνται μὲν ἀριθμόν, χρόνου δὲ ἔννοιαν
περί τε τῆς τοῦ παντὸς φύσεως ζήτησιν ἔδοσαν· ἐξ ὧν ἐπορισάμεθα
φιλοσοφίας γένος, οὗ μεῖζον ἀγαθὸν οὔτ' ἦλθεν οὔτε ἥξει ποτὲ τῷ B
θνητῷ γένει δωρηθὲν ἐκ θεῶν. λέγω δὴ τοῦτο ὀμμάτων μέγιστον
ἀγαθόν· τἆλλα δέ, ὅσα ἐλάττω, τί ἂν ὑμνοῖμεν; ὧν ὁ μὴ φιλόσοφος

4 ἄλλων μέν : ἀλλήλων A.   9 ἔχειν : σχεῖν SZ.

3. **τὰς τῆς ἔμφρονος φύσεως αἰτίας**]
That is to say the final causes, the design
of Intelligence, as distinguished from the
physical means used to carry out the
design. Thus in the case of vision the
δεύτεραι αἰτίαι are the physical laws which
Plato has set forth, the πρώτη αἰτία is
what he is presently about to state. Both
classes of cause are to be investigated by
the lover of truth, but the secondary only
for the sake of the primary: compare
68 E.

ὅσαι δὲ ὑπ' ἄλλων κινουμένων] κινουμένων, κινούντων are partitive genitives
'such as are among things which are
moved by others'. ἐξ ἀνάγκης, i.e. without an intelligent purpose (since these
ξυναίτια have λόγον οὐδένα οὐδὲ νοῦν εἰς
οὐδέν), and not of their own free will.

7. **ὅσαι μονωθεῖσαι φρονήσεως**] The
nature of the two causes is dealt with in
the note on ἀνάγκη at the beginning of

the following chapter. Plato does not
mean that there is a blind force existing
in nature, acting at random and producing
hap-hazard effects. Such a conception is
totally foreign to his system, in which the
one cause, the one ἀρχὴ κινήσεως, is
ψυχή. What he does mean is this. It
is idle to treat the physical forces of
nature as causes, since in themselves they
have no intelligence or purpose. They
are indeed designed and set in motion by
Intelligence for the best ends; but the
conditions of their action may be such
that sometimes their immediate results
are not good, and they have no power in
themselves to avoid such results; they
must operate inevitably according to the
law of their nature. The point is well
put by Mr D. D. Heath in an able essay
in the *Journal of Philology*, vol. vii p.
111, where he is dealing with Aristotle's
views of causation. 'Any agent', he

we must affirm to be soul: and this is invisible, whereas fire and water and air and earth are all visible creations. Now the lover of reason and knowledge must first seek for the causes which belong to the rational order; and only in the second place those which belong to the class of things which are moved by others and move others in turn. This then is what we must also do: we must declare both classes of causes, distinguishing between those which with the aid of reason are the creators of fair things and good, and those which being destitute of reason produce from time to time chance effects without design. Enough then of the auxiliary causes which combine in giving the eyes the power they now possess; but the great result, for the sake of which God bestowed them on us, must be our next theme. Sight, according to my judgment, has been the cause of the greatest blessing to us, inasmuch as of our present discourse concerning the universe not one word would have been uttered had we never seen the stars and the sun and the heavens. But now day and night, being seen of us, and months and the revolution of the years have created number, and they gave us the notion of time and the power of searching into the nature of the All; whence we have derived philosophy, than which no greater good has come neither shall come hereafter as the gift of heaven to mortal man. This I declare to be the chiefest blessing due to the eyes: on the rest that are meaner why should we descant? let him who loves not

says, 'natural or artificial, may produce effects which do not naturally or necessarily flow from those qualities which give it its name or constitute its kind, but which result from properties common to it and other kinds, or from circumstances which bring it into casual relation with the thing it acts upon: a coal may break your head as well as warm you'. See Aristotle *physica* II iv 195^b 31 foll. In this sense only is an effect produced which is τὸ τυχὸν ἄτακτον. The falling of the coal is the natural effect of its gravity, a property bestowed upon it by νοῦς: and if your head happens to be in the line of the coal's descent, it is broken in con-sequence of the 'casual relation' which is thus established between it and the coal. But this is in complete conformity with the natural laws which arise solely from the evolution of νοῦς.

16. ἐξ ὧν ἐπορισάμεθα] The true final cause of sight then is the attainment of philosophy, which is the ultimate result of the knowledge of number, acquired by observation of the celestial bodies. The sciences of number and astronomy were for Plato a propaedeutic to philosophy, as we learn from *Republic* 525 A foll.: and it is well known that he regarded geometry as an indispensable part of a liberal education.

τυφλωθεὶς ὀδυρόμενος ἂν θρηνοῖ μάτην. ἀλλὰ τούτου λεγέσθω παρ' ἡμῶν αὕτη ἐπὶ ταῦτα αἰτία, θεὸν ἡμῖν ἀνευρεῖν δωρήσασθαί τε ὄψιν, ἵνα τὰς ἐν οὐρανῷ κατιδόντες τοῦ νοῦ περιόδους χρησαίμεθα ἐπὶ τὰς περιφορὰς τὰς τῆς παρ' ἡμῖν διανοήσεως, ξυγγενεῖς
5 ἐκείναις οὔσας, ἀταράκτοις τεταραγμένας, ἐκμαθόντες δὲ καὶ λο- C γισμῶν κατὰ φύσιν ὀρθότητος μετασχόντες, μιμούμενοι τὰς τοῦ θεοῦ πάντως ἀπλανεῖς οὔσας, τὰς ἐν ἡμῖν πεπλανημένας καταστησαίμεθα. φωνῆς τε δὴ καὶ ἀκοῆς πέρι πάλιν ὁ αὐτὸς λόγος, ἐπὶ ταὐτὰ τῶν αὐτῶν ἕνεκα παρὰ θεῶν δεδωρῆσθαι. λόγος τε γὰρ ἐπ'
10 αὐτὰ ταῦτα τέτακται, τὴν μεγίστην ξυμβαλλόμενος εἰς αὐτὰ μοῖραν, ὅσον τ' αὖ μουσικῆς φωνῇ χρήσιμον [πρὸς ἀκοήν], ἕνεκα ἁρμονίας ἐστὶ δοθέν· ἡ δὲ ἁρμονία, ξυγγενεῖς ἔχουσα φορὰς ταῖς ἐν D ἡμῖν τῆς ψυχῆς περιόδοις, τῷ μετὰ νοῦ προσχρωμένῳ Μούσαις οὐκ ἐφ' ἡδονὴν ἄλογον, καθάπερ νῦν εἶναι δοκεῖ χρήσιμος, ἀλλ'
15 ἐπὶ τὴν γεγονυῖαν ἐν ἡμῖν ἀνάρμοστον ψυχῆς περίοδον εἰς κατακόσμησιν καὶ συμφωνίαν ἑαυτῇ σύμμαχος ὑπὸ Μουσῶν δέδοται· καὶ ῥυθμὸς αὖ διὰ τὴν ἄμετρον ἐν ἡμῖν καὶ χαρίτων ἐπιδεᾶ γιγνο- E μένην ἐν τοῖς πλείστοις ἕξιν ἐπίκουρος ἐπὶ ταὐτὰ ὑπὸ τῶν αὐτῶν ἐδόθη.
20 XVII. Τὰ μὲν οὖν παρεληλυθότα τῶν εἰρημένων πλὴν βρα-

1 τούτου: τοῦτο SZ.   2 αὕτη ἐπὶ ταῦτα αἰτία: αὐτῇ ἐπὶ ταύτῃ αἰτίᾳ S.
2 ἀνευρεῖν: εὑρεῖν A.   10 τὴν ante μεγίστην omittunt SZ.   11 φωνῇ: φωνὴ A pr. m. φωνῆς HSZ.   mox inclusi πρὸς ἀκοήν.   18 ἐπὶ ταὐτά: ἐπὶ ταῦτα Z.

1. **θρηνοῖ μάτην**] This, as Lindau and Stallbaum have pointed out, is an echo of Euripides *Phoenissae* 1762 ἀλλὰ γὰρ τί ταῦτα θρηνῶ καὶ μάτην ὀδύρομαι;

3. **ἵνα τὰς ἐν οὐρανῷ**] Compare *Republic* 500 C, where we read of the philosophers εἰς τεταγμένα ἄττα καὶ κατὰ ταὐτὰ ἀεὶ ἔχοντα ὁρῶντας καὶ θεωμένους οὔτ' ἀδικοῦντα οὔτ' ἀδικούμενα ὑπ' ἀλλήλων, κόσμῳ δὲ πάντα καὶ κατὰ λόγον ἔχοντα, ταῦτα μιμεῖσθαί τε καὶ ὅτι μάλιστα ἀφομοιοῦσθαι.

11. **ὅσον τ' αὖ μουσικῆς**] The reading of the text, although I cannot consider it altogether satisfactory, affords a fairly good sense. μουσική is a comprehensive term, including much more than 'music' in the modern sense. Plato is therefore limiting the signification in the present case to such μουσική as consists of musical and vocal sounds, which he says were given us for the sake of harmony. The high educational value which Plato set upon music and harmony is again and again emphasised in his writings: see for instance *Republic* 401 D, *Laws* 666 D. Stallbaum's reading and punctuation are alike unsatisfactory. The words πρὸς ἀκοήν appear to me superfluous and unmeaning: I conceive them to have been a marginal gloss on φωνῇ.

12. **ξυγγενεῖς ἔχουσα φοράς**] Thus is brought out the significance of the harmonic ratios in 35 B: the laws of harmony and the laws of being are the same; the former being just one special aspect of the latter.

47 E—48 E, *c.* xvii. Hitherto our dis-

wisdom, if he be blinded of these, lament with idle moan. But on our part let this be affirmed to be the cause of vision, for these ends: God discovered and bestowed sight upon us in order that we might observe the orbits of reason which are in heaven and make use of them for the revolutions of thought in our own souls, which are akin to them, the troubled to the serene; and that learning them and acquiring natural truth of reasoning we might imitate the divine movements that are ever unerring and bring into order those within us which are all astray. And of sound and hearing again the same account must be given: to the same ends and with the same intent they have been bestowed on us by the gods. For not only has speech been appointed for this same purpose, whereto it contributes the largest share, but all such music as is expressed in sound has been granted, for the sake of harmony: and harmony, having her motions akin to the revolutions in our own souls, has been bestowed by the Muses on him who with reason seeks their help, not for any senseless pleasure, such as is now supposed to be its chiefest use, but as an ally against the discord which has grown up in the revolution of our soul, to bring her into order and into unison with herself: and rhythm too, because our habit of mind is mostly so faulty of measure and lacking in grace, is a succour bestowed on us by the same givers for the same ends.

XVII. Now in our foregoing discourse, with few exceptions,

course has been entirely or mainly concerned with the works of Intelligence; but now we must likewise take account of the operations of Necessity. For all the fabric of this universe is the effect of Intelligence acting upon Necessity and influencing it to produce the best possible result. Therefore in our account of creation we must find room for the Errant Cause. And first we must set forth the origin of fire and the other elements, which no man has yet declared. But in dealing with things material we cannot find any infallible first principle whereupon to base our discourse; we must be content, as we have always said, with the probable account. And so with heaven's blessing let us set forth on a new and strange journey of discovery.

20. τὰ μὲν οὖν παρεληλυθότα] Up to this point Plato has been treating of the general design and plan of creation, πλὴν βραχέων, with some small exceptions, e.g. the account of the συμμεταίτια which contribute to the process of vision. The inquiry into the effects of necessity, to which a great part of the remainder of the dialogue is devoted, consists of physical and physiological speculations concerning the various properties and forms of matter and their interaction one on another. This inquiry is however introduced by a metaphysical theory of the first importance, without which it

χέων ἐπιδέδεικται τὰ διὰ νοῦ δεδημιουργημένα· δεῖ δὲ καὶ τὰ δι'
ἀνάγκης γιγνόμενα τῷ λόγῳ παραθέσθαι. μεμιγμένη γὰρ οὖν ἡ
τοῦδε τοῦ κόσμου γένεσις ἐξ ἀνάγκης τε καὶ νοῦ συστάσεως ἐγεν- 48 A
νήθη· νοῦ δὲ ἀνάγκης ἄρχοντος τῷ πείθειν αὐτὴν τῶν γιγνομένων
5 τὰ πλεῖστα ἐπὶ τὸ βέλτιστον ἄγειν, ταύτῃ κατὰ ταὐτά τε δι'
ἀνάγκης ἡττωμένης ὑπὸ πειθοῦς ἔμφρονος οὕτω κατ' ἀρχὰς ξυνί-
στατο τόδε τὸ πᾶν. εἴ τις οὖν ᾗ γέγονε κατὰ ταῦτα ὄντως ἐρεῖ,
μικτέον καὶ τὸ τῆς πλανωμένης εἶδος αἰτίας, ᾗ φέρειν πέφυκεν.

is not too much to say that no conception of Platonism as a coherent whole could be formed. A thorough study of the eighteenth chapter of the *Timaeus* is absolutely essential before we can even think of beginning to understand Plato. To this theory the present chapter is prefatory.

3. **ἐξ ἀνάγκης τε καὶ νοῦ συστάσεως**] The first point which it is indispensable precisely to determine is the meaning of ἀνάγκη and ἡ πλανωμένη αἰτία, which clearly signify one and the same thing. I have already in the note on 46 E to some extent indicated what I conceive to be Plato's meaning. In the first place it is necessary once for all to discard the notion that ἀνάγκη is in any sense whatsoever an independent force external to νοῦς: this would be totally repugnant, as I have said, to the cardinal doctrine of Platonism, that the only ἀρχὴ κινήσεως is ψυχή. For this reason we must not suppose that there is in matter as such any resisting power which thwarts the efforts of νοῦς: this is an absolute misconception. Matter, *qua* matter, being soulless, is entirely without any sort of power of its own: whatever power it has is of ψυχή. What then is ἀνάγκη or the πλανωμένη αἰτία? It signifies the forces of matter originated by νοῦς, the sum total of the physical laws which govern the material universe: that is to say, the laws which govern the existence of νοῦς in the form of plurality. Now these laws, once set in motion, must needs act constantly according to their nature; else would νοῦς be at variance with itself. Therefore all nature's forces must follow their proper impulse according to the conditions in which they are for the time being: if fire and a hayrick come in collision, it is ἀνάγκη that the rick be burnt, though fire was not designed to burn ricks. But this implies no originating power in matter; it means only that νοῦς, having once evolved itself in the pluralised form, the laws of its existence in that form are constant. Material nature is a machine wound up to go of itself; νοῦς is not for ever checking or correcting its action in detail—see *Laws* 903 B foll. But there is something more to be said. It is a necessary law for νοῦς to exist in the form of material nature: and within this sphere we see that things do not always work, at any rate immediately, ἐπὶ τὸ βέλτιστον. It was impossible, we must suppose, for νοῦς to assume the form of a multitude of physical forces, all in themselves and in their design beneficent, which should not, amid the infinite complexity of their interaction, inevitably under some conditions produce effects which are not beneficent. This necessity and this impossibility constitute ἀνάγκη. It is then in the final analysis the law by which νοῦς necessarily has a mode of existence to which imperfection attaches: and the very constancy with which the law acts is the cause of the friction which arises in its manifold and complex operation. But this is no law imposed upon νοῦς by any external cause, for there is none

we have been declaring the creations wrought through mind: we must now set by their side those things which come into being through necessity. For the generation of this universe was a mixed creation by a combination of necessity and reason. And whereas reason governed necessity, by persuading her to guide the greatest part of created things to the best end, on such conditions and principles, through necessity overcome by reasonable persuasion, this universe was fashioned in the beginning. If then we would really declare its creation in the manner whereby it has come to be, we must add also the nature of the Errant Cause, and its moving power. Thus then

such: it is in the very nature of νοῦς itself in its pluralised form. The problem of the πλανωμένη αἰτία is the same as the problem concerning the nature of evil, of which Plato has offered us no explicit solution.

6. ἡττωμένης ὑπὸ πειθοῦς ἔμφρονος] In these words is indicated the difference between the ἀνάγκη of Plato and the ἀνάγκη of Demokritos. For Plato, although the forces of nature are inevitable and inexorable in their action, yet these forces are themselves expressly designed by Intelligence for a good end. And though in detail evil may arise from their working, yet they are so ordained as to produce the best result that it was possible to attain. Necessity persuaded by intelligence means in fact that necessity is a mode of the operation of intelligence. The necessity of Demokritos, on the other hand, is an all-powerful unintelligent force working without design; and whether good or evil, as we term them, arises from its processes, this is entirely a matter of chance. Thus in Plato's scheme evil is deliberately limited to an irreducible minimum, while with that of Demokritos the whole question of good and evil has nothing to do.

8. τὸ τῆς πλανωμένης εἶδος αἰτίας] The name πλανωμένη αἰτία does not signify that Plato attributed any degree of uncertainty or caprice to the operation of ἀνάγκη. Every effect is the result of a cause; and just that effect and nothing else whatsoever must arise from just that cause. And were we omniscient, we could trace the connexion between cause and effect everywhere, and we could consequently predict everything that should happen. As it is, so obscure to us are the forces amid which we live, and so complex are the influences which work upon one another, that in innumerable instances we are unable to trace an effect back to its causes or to foresee the action of ἀνάγκη. Hence Plato calls ἀνάγκη the πλανωμένη αἰτία, because, though working strictly in obedience to a certain law, it is for the most part as inscrutable to us as if it acted from arbitrary caprice. We can detect the relation of cause and effect in results which are immediately due to the design of νοῦς, but frequently not in those which are indirectly due to it through the action of ἀνάγκη. It is extremely inaccurate in Stallbaum to say that the πλανωμένη αἰτία is 'materia corporum'.

ᾗ φέρειν πέφυκεν] Literally 'how it is its nature to set in motion'. The πλανωμένη αἰτία is the source of instability and uncertainty (relatively to us) in the order of things; whence Plato terms it the moving influence. What Stallbaum means or fails to mean by his rendering 'ea ratione, qua ipsius natura fert', it is difficult to conjecture.

ὧδε οὖν πάλιν ἀναχωρητέον, καὶ λαβοῦσιν αὐτῶν τούτων προσή- B
κουσαν ἑτέραν ἀρχὴν αὖθις αὖ, καθάπερ περὶ τῶν τότε, νῦν οὕτω
περὶ τούτων πάλιν ἀρκτέον ἀπ' ἀρχῆς. τὴν δὴ πρὸ τῆς οὐρανοῦ
γενέσεως πυρὸς ὕδατός τε καὶ ἀέρος καὶ γῆς φύσιν θεατέον αὐτὴν
5 καὶ τὰ πρὸ τούτου πάθη. νῦν γὰρ οὐδείς πω γένεσιν αὐτῶν μεμή-
νυκεν, ἀλλ' ὡς εἰδόσι, πῦρ ὅ τί ποτε ἔστι καὶ ἕκαστον αὐτῶν,
λέγομεν ἀρχὰς αὐτὰ τιθέμενοι, 'στοιχεῖα τοῦ παντός, προσῆκον
αὐτοῖς οὐδ' ἂν ὡς ἐν συλλαβῆς εἴδεσι μόνον εἰκότως ὑπὸ τοῦ καὶ C
βραχὺ φρονοῦντος ἀπεικασθῆναι. νῦν δὲ οὖν τό γε παρ' ἡμῶν
10 ὧδε ἐχέτω· τὴν μὲν περὶ ἁπάντων εἴτε ἀρχὴν εἴτε ἀρχὰς εἴτε ὅπῃ
δοκεῖ τούτων πέρι τὸ νῦν οὐ ῥητέον, δι' ἄλλο μὲν οὐδέν, διὰ δὲ τὸ
χαλεπὸν εἶναι κατὰ τὸν παρόντα τρόπον τῆς διεξόδου δηλῶσαι τὰ
δοκοῦντα. μήτ' οὖν ὑμεῖς οἴεσθε δεῖν ἐμὲ λέγειν, οὔτ' αὐτὸς αὖ
πείθειν ἐμαυτὸν εἴην ἂν δυνατός, ὡς ὀρθῶς ἐγχειροῖμ' ἂν τοσοῦτον
15 ἐπιβαλλόμενος ἔργον· τὸ δὲ κατ' ἀρχὰς ῥηθὲν διαφυλάττων, τὴν D
τῶν εἰκότων λόγων δύναμιν, πειράσομαι μηδενὸς ἧττον εἰκότα,
μᾶλλον δέ, καὶ ἔμπροσθεν ἀπ' ἀρχῆς περὶ ἑκάστων καὶ ξυμπάντων
λέγειν. θεὸν δὴ καὶ νῦν ἐπ' ἀρχῇ τῶν λεγομένων σωτῆρα ἐξ

2 ἑτέραν ἀρχήν: ἀρχὴν ἑτέραν S.
8 οὐδ' ἂν ὡς coniecit H. οὐδαμῶς A. οὐδ' ὡς SZ.

2. **καθάπερ περὶ τῶν τότε**] i.e. as we began at the beginning in expounding τὰ διὰ νοῦ δεδημιουργημένα, so we must begin at the beginning again in our exposition of τὰ δι' ἀνάγκης γιγνόμενα.

3. **πρὸ τῆς οὐρανοῦ γενέσεως**] The question next arises, what is meant by the nature of fire, &c before the generation of the universe, and the conditions anterior to this? Plato evidently means that we have to analyse these so-called elements into their primary constituents. Earlier thinkers had treated them as if they were simple primary substances: Plato, however, justly maintains that they are complex. Now as these substances exist in the κόσμος, they are everywhere more or less complete and in their finished forms; therefore in analysing them into their first beginnings, we are dealing with rudimentary forms which nowhere exist in the κόσμος, but which are analytically prior to those forms which do exist in the κόσμος. But the priority is in analysis only; there never was a time in which the elements existed in these forms. Indeed when we come to see the nature of Plato's στοιχεῖα, it will be apparent that they never could have an independent existence. πρὸ τούτου = πρὸ τοῦ γενέσθαι τὸν οὐρανόν—the state of fire, air, &c prior (in analysis) to their complete form.

8. **ἐν συλλαβῆς εἴδεσι**] This is an allusion to the common meaning of στοιχεῖα = letters of the alphabet. So far from belonging to this rank, fire and the rest are more composite even than syllables. For, as we shall see, Plato's ultimate στοιχεῖον is a particular kind of triangle, out of which is formed another triangle, and out of that again a regular solid figure, which is the corpuscule of fire.

10. **εἴτε ἀρχὴν εἴτε ἀρχάς**] Plato says he will not, like the early Ionians, attempt to find some principle or prin-

D] ΤΙΜΑΙΟΣ. 169

let us return upon our steps, and when we have found a second fitting cause for the things aforesaid, let us once more, proceeding in the present case as we did in the former, begin over again from the beginning. Now we must examine what came before the creation of the heavens, the very origin of fire and water and air and earth, and the conditions that were before them. For now no one has declared the manner of their generation; but we speak as if men knew what is fire and each of the others, and we treat them as beginnings, as elements of the whole; whereas by one who has ever so little intelligence they could not plausibly be represented as belonging even to the class of syllables. Now however let our say thus be said. The first principle or principles or whatever we may hold it to be which underlies all things we must not declare at present, for no other reason but that it is difficult according to the present method of our exposition to make clear our opinion. You must not then deem that I ought to discourse of this, nor could I persuade myself that I should be right in essaying so mighty a task. But holding fast the principle we laid down at the outset, the value of a probable account, I will strive to give an explanation that is no less probable than another, but more so; returning back to describe from the beginning each and all things. So now again at the outset of our quest let us call upon God to pilot us safe through a strange and un-

ciples to serve as an ἀρχή for matter, solely for the reason that in a physical inquiry (κατὰ τὸν παρόντα τρόπον τῆς διεξόδου) it is hardly possible to arrive at such an ἀρχή: a real ἀρχή can only be attained by dialectic. The Ionian ἀρχαί were no ἀρχαί at all. And so we may analyse matter into the ultimate geometrical forms, which are the law of its composition, but these are not properly speaking ἀρχαί. In the following chapter Plato, treating the subject metaphysically, does at least propound an ἀρχή for matter by far more recondite than any which had yet been conceived.

12. τῆς διεξόδου] Cf. *Parmenides* 136 E ἄνευ ταύτης τῆς διὰ πάντων διεξόδου τε καὶ πλάνης ἀδύνατον ἐντυχόντα τῷ ἀληθεῖ νοῦν ἔχειν.

17. καὶ ἔμπροσθεν ἀπ' ἀρχῆς] Stallbaum, who joins μᾶλλον δὲ with what follows, proposes to read κατὰ τὰ ἔμπροσθεν. But no change is necessary. ἔμπροσθεν means 'where we were before', viz. at the starting-point of the inquiry. I think Martin is justified in his rendering 'revenant sur mes pas jusqu'au commencement'. Lindau suggests μᾶλλον δ' ἢ κατ' ἔμπροσθεν, which is not Greek, as I think.

18. ἐξ ἀτόπου καὶ ἀήθους διηγήσεως] The metaphor is evidently taken from mariners embarking on a voyage of discovery in some new and unexplored ocean. Plato prays to be delivered from the perils of the voyage and brought safe

ἀτόπου καὶ ἀήθους διηγήσεως πρὸς τὸ τῶν εἰκότων δόγμα διασώζειν ἡμᾶς ἐπικαλεσάμενοι πάλιν ἀρχώμεθα λέγειν.

XVIII. Ἡ δ' οὖν αὖθις ἀρχὴ περὶ τοῦ παντὸς ἔστω μειζόνως τῆς πρόσθεν διῃρημένη. τότε μὲν γὰρ δύο εἴδη διειλόμεθα, νῦν δὲ τρίτον ἄλλο γένος ἡμῖν δηλωτέον. τὰ μὲν γὰρ δύο ἱκανὰ ἦν ἐπὶ τοῖς ἔμπροσθεν λεχθεῖσιν, ἓν μὲν ὡς παραδείγματος εἶδος ὑποτεθέν, νοητὸν καὶ ἀεὶ κατὰ ταὐτὰ ὄν, μίμημα δὲ παραδείγματος δεύτερον, γένεσιν ἔχον καὶ ὁρατόν· τρίτον δὲ τότε μὲν οὐ διειλόμεθα, νομίσαντες τὰ δύο ἕξειν ἱκανῶς, νῦν δὲ ὁ λόγος ἔοικεν εἰσαναγκάζειν χαλεπὸν καὶ ἀμυδρὸν εἶδος ἐπιχειρεῖν λόγοις ἐμφανίσαι. τίν' οὖν ἔχον δύναμιν κατὰ φύσιν αὐτὸ ὑποληπτέον; τοιάνδε μάλιστα, πάσης εἶναι γενέσεως ὑποδοχὴν αὐτήν, οἷον

1 ἀήθους: ἀληθοῦς A.

to the haven of probability. Martin is certainly mistaken in translating 'pour qu'elle nous préserve de discours incohérents et bizarres'. Plato shows himself fully alive to the difficulty of the subject he is about to treat and the entire novelty of his speculations. A glimpse of his theory of matter has been afforded in the *Philebus*, but here he carries his analysis far deeper. Compare 53 B, where he calls his very peculiar corpuscular theory ἀήθης λόγος.

48 E—52 D, c. xviii. We must extend the classification of all things which we formerly made. To the ideal model and the sensible copy which we then assumed must be added the substrate in which generation takes place. For consider: the four elements, as men call them, fire, air, water, earth, are continually changing places and passing one into another, so that we can never with any security say, this is fire, or this is water. Indeed we should not apply the word *this* to them at all, nor any other expression which signifies permanency: the most we can do is to say they are 'such-like'. To the substrate alone is it safe to apply the term 'this'. For it alone never changes its nature; but is as it were a matrix receiving all the forms that enter into it, which forms are the sensible semblances of the eternal ideas. So then we must distinguish these three, the eternal type, the generated copy, and the substrate wherein it is generated. This substrate must be without form or quality, else it would not faithfully express the images that enter into it, but would intrude its own attributes. It is not then fire nor any other of the elements, but a viewless and formless nature, which takes on it now the form of fire, anon the form of water, and all perceptible things. But since we talk of images entering in, we must ask, is there a type, an idea of fire and the rest whereof we behold the images? or are the visible images themselves the most real existence which is? We cannot dwell on this question at length: but we may briefly answer it thus. If knowledge differs from true opinion, then the ideas exist beyond the sensible images; if not, then sensibles alone are realities. Now it is a fact that knowledge differs from true opinion; for one is the result of teaching, the other of persuasion; one is the possession of all men, the other of the gods alone and but a few among mankind. Therefore the ideas exist eternally, neither passing forth of their own nature nor receiving aught therein, apprehensible by thought alone: next there are the

49 A]   ΤΙΜΑΙΟΣ.   171

familiar discourse to the haven of probability; and thus let us begin once more.

XVIII. Our new exposition of the universe then must be founded on a fuller classification than the former. Then we distinguished two forms, but now a third kind must be disclosed. The two were indeed enough for our former discussion, when we laid down one form as the pattern, intelligible and changeless, the second as a copy of the pattern, which comes into being and is visible. A third we did not then distinguish, deeming that the two would suffice: but now, it seems, by constraint of our discourse we must try to express and make manifest a form obscure and dim. What power then must we conceive that nature has given it? something like this. It is the receptacle, and as it were the nurse, of all becoming. This

images called after their names, sensible and perishable and ever in transition: thirdly the receptacle of all becoming, which is space, imperishable and imperceptible, apprehended by a kind of bastard reasoning. This third is the cause why, like men in a dream, we declare that everything which exists must be in some place, and what is nowhere in heaven or earth is nothing. And this dream we carry into the region of waking verity, even the ideas; we do not remember that, since an image is not its own type, it must be imaged in something else, or else be not at all: for true reason declares that, while the type is one, and the image another, they must be apart; for they cannot exist one in the other and so be one and two at once.

3. μειζόνως] i.e. the classification must be more comprehensive: the former left no room for one of the most important principles in nature.

4. τότε μὲν γάρ] The reference is to 28 A, where Timaeus divides the universe into ὄν and γιγνόμενον.

5. τὰ μὲν γὰρ δύο ἱκανὰ ἦν] This remark is most characteristic of Plato, who always confines himself to the limits of the subject in hand. He is like a good general, who does not call upon his re-

serves till they are wanted. So in the *Philebus* he carries his analysis of ἄπειρον no further than to describe it as indefinitely qualified, because that served all the purpose of that dialogue. And in the same way at the earlier stage of Timaeus's exposition he distinguishes only such principles of the universe as then concern the argument.

7. μίμημα] It may be as well to draw attention to the fact that throughout all the dialogue the relation of particular to idea is one of μίμησις: the old μέθεξις has disappeared never to return.

10. χαλεπὸν καὶ ἀμυδρὸν εἶδος] Plato repeatedly in the most emphatic language expresses his sense of the difficulty and obscurity attaching to this question concerning the substrate of material existence. The difficulty is recognised also in the *Philebus*, though in less forcible terms, cf. 24 A χαλεπὸν μὲν γὰρ καὶ ἀμφισβητήσιμον ὃ κελεύω σε σκοπεῖν. It must be remembered too that Plato's conception was an absolute novelty in philosophy. Aristotle has a curiously perverse reference to the theory of the *Timaeus* in *de gen. et corr.* II i 329^a 13 foll.

12. ὑποδοχήν] The substrate is the 'receptacle' of all things that become, inasmuch as it provides them with a place

172 ΠΛΑΤΩΝΟΣ [49 A—

τιθήνην. εἴρηται μὲν οὖν ἀληθές, δεῖ δὲ ἐναργέστερον εἰπεῖν
περὶ αὐτοῦ· χαλεπὸν δέ, ἄλλως τε καὶ διότι προαπορηθῆναι περὶ B
πυρὸς καὶ τῶν μετὰ πυρὸς ἀναγκαῖον τούτου χάριν· τούτων γὰρ
εἰπεῖν ἕκαστον, ὁποῖον ὄντως ὕδωρ χρὴ λέγειν μᾶλλον ἢ πῦρ καὶ
5 ὁποῖον ὁτιοῦν μᾶλλον ἢ καὶ ἅπαντα καθ' ἕκαστόν τε, οὕτως ὥστε
τινὶ πιστῷ καὶ βεβαίῳ χρήσασθαι λόγῳ, χαλεπόν. πῶς οὖν δὴ
τοῦτ' αὐτὸ καὶ πῇ καὶ τί περὶ αὐτῶν εἰκότως διαπορηθέντες ἂν
λέγοιμεν; πρῶτον μέν, ὃ δὴ νῦν ὕδωρ ὠνομάκαμεν, πηγνύμενον, ὡς C
δοκοῦμεν, λίθους καὶ γῆν γιγνόμενον ὁρῶμεν, τηκόμενον δὲ καὶ
10 διακρινόμενον αὖ ταὐτὸν τοῦτο πνεῦμα καὶ ἀέρα, ξυγκαυθέντα δὲ
ἀέρα πῦρ, ἀνάπαλιν δὲ πῦρ συγκριθὲν καὶ κατασβεσθὲν εἰς ἰδέαν
τε ἀπιὸν αὖθις ἀέρος, καὶ πάλιν ἀέρα ξυνιόντα καὶ πυκνούμενον
νέφος καὶ ὁμίχλην, ἐκ δὲ τούτων ἔτι μᾶλλον ξυμπιλουμένων ῥέον
ὕδωρ, ἐξ ὕδατος δὲ γῆν καὶ λίθους αὖθις, κύκλον τε οὕτω διαδι-
15 δόντα εἰς ἄλληλα, ὡς φαίνεται, τὴν γένεσιν. οὕτω δὴ τούτων D
οὐδέποτε τῶν αὐτῶν ἑκάστων φανταζομένων, ποῖον αὐτῶν ὡς ὂν
ὁτιοῦν τοῦτο καὶ οὐκ ἄλλο παγίως διισχυριζόμενος οὐκ αἰσχυ-
νεῖταί τις ἑαυτόν; οὐκ ἔστιν, ἀλλ' ἀσφαλέστατα μακρῷ περὶ
τούτων τιθεμένους ὧδε λέγειν· ἀεὶ ὃ καθορῶμεν ἄλλοτε ἄλλῃ
20 γιγνόμενον, ὡς πῦρ, μὴ τοῦτο ἀλλὰ τὸ τοιοῦτον ἑκάστοτε προσ-

1 ἀληθές: τἀληθές SZ.   6 πῶς οὖν δή: πῶς οὖν δή που A.

to become in: it is their 'nurse', because it fosters them, so to speak, and is the means of their existence; without it they could not exist in any way. Stallbaum's account of it as a vessel containing sensible things is most erroneous; indeed his treatment of the whole subject is as confused as it can well be. It will be convenient to defer a fuller discussion of Plato's ὑποδοχὴ until this conception receives its final development at the end of the chapter.

2. **προαπορηθῆναι περὶ πυρός**] This necessity arises because the conception of the ὑποδοχὴ as an unchanging substrate involves the conception of fire and the rest as merely transitory conditions of this substrate: therefore we must put the question, what is the real nature of this appearance which we call fire? And this in its turn raises the question of the existence of the ideas. τῶν μετὰ πυρὸς of course = air, water, earth.

5. **ἅπαντα καθ' ἕκαστόν τε**] i.e. to call it all or (some one) severally. The slight change of construction in καθ' ἕκαστον is not at all harsh, and certainly Stallbaum's plan of joining the words with the following is not an improvement. Seeing that the four elements are perpetually interchanging there can be no propriety in giving any fixed name to any one of them: while we apply the term appropriate to one form, the substance may have passed into another.

7. **εἰκότως** should be joined with **διαπορηθέντες**. 'raising what reasonable question'.

9. **λίθους καὶ γῆν**] Plato here speaks as if all four elements were interchangeable: this statement is corrected in 54 C, where we find that earth, as having a different base, will not pass into the other elements, nor they into it: the other

ΤΙΜΑΙΟΣ.

saying is true, but we must put it in clearer language: and this is hard; especially as for the sake of it we must needs inquire into fire and the substances that rank with fire. For it is hard to say which of all these we ought to call water any more than fire, or indeed which we ought to call by any given name, rather than all and each severally, in such a way as to employ any truthful and trustworthy mode of speech. How then are we to deal with this point, and what is the question that we should properly raise concerning it? In the first place, what we now have named water, by condensation, as we suppose, we see turning to stones and earth; and by rarefying and expanding this same element becomes wind and air; and air when inflamed becomes fire: and conversely fire contracted and quenched returns again to the form of air; also air concentrating and condensing becomes cloud and mist; and from these yet further compressed comes flowing water; and from water earth and stones once more: and so, it appears, they hand on one to another the cycle of generation. Thus then since these several bodies never assume one constant form, which of them can we positively affirm to be really *this* and not another without being shamed in our own eyes? It cannot be: it is far the safest course when we make a statement concerning them to speak as follows. What we see in process of perpetual transmutation, as for instance fire, we must not call *this*, but *such-like* is the

three however are interchangeable. Note however that the present statement is guarded with the qualification ὡς δοκοῦμεν. Of course this limitation of the interchangeability does not affect Plato's argument, which is probably the reason why it is not mentioned here.

11. ἀνάπαλιν δέ] This is just the ὁδὸς ἄνω κάτω μία of Herakleitos. Stallbaum wishes to omit τε after ἰδέαν and after κύκλον, which he would alter to κύκλῳ. There is really no occasion for any of these changes. The main participles in the sentence γιγνόμενον, συγκριθέν, κατασβεσθέν, ἀπιόν, διαδιδόντα, are governed by ὁρῶμεν, while the rest are subordinate to γιγνόμενον, which has to be supplied again with the clauses καὶ πάλιν...λίθους

αὖθις. κύκλον is perfectly right, being a predicate to γένεσιν: 'handing on their generation as a circle': the τε is also right, coupling διαδιδόντα and γιγνόμενον. There is more to be said for omitting τε after ἰδέαν; in which case συγκριθέν and κατασβεσθέν would be subordinate to ἀπιόν: but as it is in all the mss. I have not thought fit to expunge it.

20. μὴ τοῦτο ἀλλὰ τὸ τοιοῦτον] That is to say, we must not speak of it as a substance, but as a quality: in Aristotelian phrase, it is not ὑποκείμενον, but καθ' ὑποκειμένου. τοῦτο denotes what a thing is, τοιοῦτον what we predicate of it. Fire is merely an appearance which the ὑποδοχή assumes for the time being: we must not say then 'this portion of space

174   ΠΛΑΤΩΝΟΣ   [49 D—

ἀγορεύειν πῦρ, μηδὲ ὕδωρ τοῦτο ἀλλὰ τὸ τοιοῦτον ἀεί, μηδὲ ἄλλο
ποτὲ μηδὲν ὥς τινα ἔχον βεβαιότητα, ὅσα δεικνύντες τῷ ῥήματι E
τῷ τόδε καὶ τούτῳ προσχρώμενοι δηλοῦν ἡγούμεθά τι· φεύγει γὰρ
οὐχ ὑπομένον τὴν τοῦ τόδε καὶ τοῦτο καὶ τὴν τῷδε καὶ πᾶσαν ὅση
5 μόνιμα ὡς ὄντα αὐτὰ ἐνδείκνυται φάσις. ἀλλὰ ταῦτα μὲν ἕκαστα
μὴ λέγειν, τὸ δὲ τοιοῦτον ἀεὶ περιφερόμενον ὁμοίως ἑκάστου πέρι
καὶ ξυμπάντων οὕτω καλεῖν· καὶ δὴ καὶ πῦρ τὸ διὰ παντὸς τοιοῦτον καὶ ἅπαν ὁσονπερ ἂν ἔχῃ γένεσιν. ἐν ᾧ δὲ ἐγγιγνόμενα ἀεὶ
ἕκαστα αὐτῶν φαντάζεται καὶ πάλιν ἐκεῖθεν ἀπόλλυται, μόνον
10 ἐκεῖνο αὖ προσαγορεύειν τῷ τε τοῦτο καὶ τῷ τόδε προσχρωμένους 50 A
ὀνόματι, τὸ δὲ ὁποιονοῦν τι, θερμὸν ἢ λευκὸν ἢ καὶ ὁτιοῦν τῶν
ἐναντίων, καὶ πάνθ' ὅσα ἐκ τούτων, μηδὲν ἐκεῖνο αὖ τούτων καλεῖν. ἔτι δὲ σαφέστερον αὐτοῦ πέρι προθυμητέον αὖθις εἰπεῖν.
εἰ γὰρ πάντα τις σχήματα πλάσας ἐκ χρυσοῦ μηδὲν μεταπλάττων
15 παύοιτο ἕκαστα εἰς ἅπαντα, δεικνύντος δή τινος αὐτῶν ἓν καὶ
ἐρομένου τί ποτ' ἔστι, μακρῷ πρὸς ἀλήθειαν ἀσφαλέστατον εἰπεῖν B
ὅτι χρυσός, τὸ δὲ τρίγωνον ὅσα τε ἄλλα σχήματα ἐνεγίγνετο,

4 τοῦ τόδε καί : τοῦ τόδε καὶ τήν S.   τοῦτο: τούτου AS.
6 ὁμοίως scripsi suadente S. ceteri ὅμοιον.   16 ἐρομένου : προσερομένου S.

is fire', but 'this portion of space has the property of fire for its present condition'. For the same portion of space may presently assume the appearance of air and of water; whence we see that the only permanent thing is the space; fire, air, water are merely its transitory attributes derived from the ὁμοιώματα impressed upon it.

3. **τῷ τόδε καὶ τοῦτο**] Compare *Theaetetus* 157 B τὸ δ' οὐ δεῖ, ὥς ὁ τῶν σοφῶν λόγος, οὔτε τι ξυγχωρεῖν οὔτε του οὔτ' ἐμοῦ οὔτε τόδε οὔτ' ἐκεῖνο οὔτ' ἄλλο οὐδὲν ὄνομα, ὅ τι ἂν ἱστῇ. Also 183 A δεῖ δὲ οὐδὲ τοῦτο τὸ οὕτω λέγειν· οὐδὲ γὰρ ἂν ἔτι κινοῖτο τὸ οὕτω· οὐδ' αὖ μὴ οὕτω· οὐδὲ γὰρ τοῦτο κίνησις· ἀλλά τιν' ἄλλην φωνὴν θετέον τοῖς τὸν λόγον τοῦτον λέγουσιν, ὡς νῦν γε πρὸς τὴν αὑτῶν ὑπόθεσιν οὐκ ἔχουσι ῥήματα, εἰ μὴ ἄρα τὸ οὐδ' ὅπως. Thus we see that what is in the *Theaetetus* described as the οἰκειοτάτη διάλεκτος of the Herakleiteans is here expressly adopted by Plato as his own, when he speaks of material phenomena.

6. **μὴ λέγειν**] The infinitives still depend upon ἀσφαλέστατα in D.

περιφερόμενον ὁμοίως] On the suggestion of Stallbaum I have adopted ὁμοίως for ὅμοιον. The meaning is that the term τοιοῦτον keeping pace with the elements in their transformations (περιφερόμενον) can always be applied to any of them in the same sense (ὁμοίως). That is to say τοιοῦτον is a word which does not denote a permanent substance but a variable attribute: therefore we can apply it to fire &c without fear of treating such qualities as substantial fixities. If ὅμοιον be retained, it must be regarded as a predicate, and the sense will still be the same: but I think the construction is too awkward to have come from Plato. For περιφερόμενον compare *Theaetetus* 202 A ταῦτα μὲν γὰρ περιτρέχοντα πᾶσι προσφέρεσθαι: where ταῦτα = αὐτό, ἐκεῖνο, ἕκαστον and the like.

7. **τὸ διὰ παντός**] i.e. fire is the name

appellation we must confer on fire; nor must we call water *this*, but always *such;* nor must we apply to anything, as if it had any stability, such predicates as we express by the use of the terms *this* and *that* and suppose that we signify something thereby. For it flees and will not abide such terms as *this* and *that* and *relative to this*, and every phrase which represents it as stable. The word *this* we must not use of any of them; but *such*, applying in the same sense to all their mutations, we must predicate of each and all: fire we must call that which universally has that appearance; and so must we name all things such as come into being. That wherein they come to be severally and show themselves, and from whence again they perish, in naming that alone must we use the words *that* and *this;* but whatever has any quality, such as white or hot or any of two opposite attributes, and all combinations of these, we must denote by no such term.

But we must try to speak yet more clearly on this matter. Suppose a man having moulded all kinds of figures out of gold should unceasingly remould them, interchanging them all with one another, it were much the safest thing in view of truth to say that it is gold; but as to the triangles or any

we give to such and such a combination of attributes wheresoever in nature it may appear.

9. μόνον ἐκεῖνο] To the ὑποδοχή, on the other hand, we can and must apply the word τοῦτο, because it is ever unchanging. The manifold forms it assumes are merely impressed on it from without; underlying them all its own nature is the same.

11. ὁτιοῦν τῶν ἐναντίων] Not the opposites to hot and white, but any of the ἐναντιότητες which are the attributes predicable of matter. ὅσα ἐκ τούτων signifies any combination of simple qualities.

14. πλάσας ἐκ χρυσοῦ] Aristotle gives a strange turn to this, *de gen. et corr.* II i 329ᵃ 17. Referring to the illustration of the golden figures he says, καίτοι καὶ τοῦτο οὐ καλῶς λέγεται τοῦτον τὸν τρόπον λεγόμενον, ἀλλ' ὧν μὲν ἀλλοίωσις, ἔστιν οὕτως, ὧν δὲ γένεσις καὶ

φθορά, ἀδύνατον ἐκεῖνο προσαγορεύεσθαι ἐξ οὗ γέγονεν. καίτοι γέ φησι μακρῷ ἀληθέστατον εἶναι χρυσὸν λέγειν ἕκαστον εἶναι. How this criticism applies I fail to see. That which suffers γένεσις καὶ φθορά is the shapes, whether in the ὑποδοχή or in the gold. These shapes have not their γένεσις from the ὑποδοχή nor from the gold: Plato accurately describes the ὑποδοχή not as τὸ ἐξ οὗ, which it is not, but as τὸ ἐν ᾧ γίγνεται, which it is. When Plato bids us say 'this is gold', not 'this is a cube', he does not mean that the cubic shape is gold, or that a cubic shape is generated out of gold; but that in calling it gold we designate the substance, whereas if we call it a cube, we are designating an attribute which is accidental and transitory. In the golden cube the gold is (or rather serves to illustrate) τοῦτο, the substance, the cubic form is τοιοῦτον, the quality.

μηδέποτε λέγειν ταῦτα ὡς ὄντα, ἅ γε μεταξὺ τιθεμένου μεταπίπτει, ἀλλ᾽ ἐὰν ἄρα καὶ τὸ τοιοῦτον μετ᾽ ἀσφαλείας ἐθέλῃ δέχεσθαί τινος, ἀγαπᾶν. ὁ αὐτὸς δὴ λόγος καὶ περὶ τῆς τὰ πάντα δεχομένης σώματα φύσεως· ταὐτὸν αὐτὴν ἀεὶ προσρητέον· ἐκ γὰρ
5 τῆς ἑαυτῆς τὸ παράπαν οὐκ ἐξίσταται δυνάμεως. δέχεταί τε γὰρ ἀεὶ τὰ πάντα, καὶ μορφὴν οὐδεμίαν ποτὲ οὐδενὶ τῶν εἰσιόντων C ὁμοίαν εἴληφεν οὐδαμῇ οὐδαμῶς· ἐκμαγεῖον γὰρ φύσει παντὶ κεῖται, κινούμενόν τε καὶ διασχηματιζόμενον ὑπὸ τῶν εἰσιόντων, φαίνεται δὲ δι᾽ ἐκεῖνα ἄλλοτε ἀλλοῖον· τὰ δὲ εἰσιόντα καὶ ἐξιόντα
10 τῶν ὄντων ἀεὶ μιμήματα, τυπωθέντα ἀπ᾽ αὐτῶν τρόπον τινὰ δύσφραστον καὶ θαυμαστόν, ὃν εἰσαῦθις μέτιμεν. ἐν δ᾽ οὖν τῷ παρόντι χρὴ γένη διανοηθῆναι τριττά, τὸ μὲν γιγνόμενον, τὸ δ᾽ ἐν ᾧ γίγνεται, τὸ δ᾽ ὅθεν ἀφομοιούμενον φύεται τὸ γιγνόμενον· καὶ D δὴ καὶ προσεικάσαι πρέπει τὸ μὲν δεχόμενον μητρί, τὸ δ᾽ ὅθεν
15 πατρί, τὴν δὲ μεταξὺ τούτων φύσιν ἐκγόνῳ, νοῆσαί τε, ὡς οὐκ ἂν ἄλλως, ἐκτυπώματος ἔσεσθαι μέλλοντος ἰδεῖν ποικίλου πάσας ποικιλίας, τοῦτ᾽ αὐτό, ἐν ᾧ ἐκτυπούμενον ἐνίσταται, γένοιτ᾽ ἂν παρεσκευασμένον εὖ, πλὴν ἄμορφον ὂν ἐκείνων ἁπασῶν τῶν ἰδεῶν,

10 ὄντα post ἀεί dedit A.

2. ἐὰν ἄρα καὶ τὸ τοιοῦτον] Plato warns us that we have gone to the uttermost verge of security in venturing to describe phenomena even in terms of quality: the advanced Herakleitean point of view is as conspicuous here as in the passages quoted above from the *Theaetetus*.

4. ταὐτὸν αὐτὴν ἀεὶ προσρητέον] We are not here to take ταὐτὸν in the technical sense in which it is used in 35 A. For as the ὑποδοχή is the home of γιγνόμενα, as it is the region of thought as pluralised in material objects, it must belong to the domain of θάτερον: and thus ταὐτὸν will simply denote the changelessness of the substrate contrasted with the mutability of the phenomena. Nevertheless, as we saw that there is a sense in which time may be spoken of as eternal (see 37 D), so there is a sense in which the principle of ταὐτὸν may be said to inhere in θάτερον. The phenomena which belong to the sphere of pluralised thought are transient, but this mode or law of their appearance under the form of space is changeless. Considered as the law or principle of pluralised existence the ὑποδοχή may be termed eternal.

ἐκ γὰρ τῆς ἑαυτῆς] Thus we have two immutable fixities, the ideas and the ὑποδοχή, between which is the fluctuating mass of sensible appearances.

7. ἐκμαγεῖον] That is to say, as it were a plastic material capable of being moulded into any form, like a mass of soft wax or the molten gold in the simile above. Plato seeks by frequently varying his metaphor to bring home to the understanding his novel and unfamiliar conception of the substrate.

9. τὰ δὲ εἰσιόντα καὶ ἐξιόντα] These forms which pass in and out of the substrate are of course not the ideas, which go not forth into aught else: here comes in the difference between the Platonism of the *Timaeus* and that of the *Republic* and

other shapes that were impressed on it, never to speak of them as existing, seeing that they change even as we are in the act of defining them; but if it will admit the term *such* with any tolerable security, we must be content. The same language must be applied to the nature which receives into it all material things: we must call it always the same; for it never departs from its own function at all. It ever receives all things into it and has nowhere any form in any wise like to aught of the shapes that enter into it. For it is as the substance wherein all things are naturally moulded, being stirred and informed by the entering shapes; and owing to them it appears different from time to time. But the shapes which pass in and out are likenesses of the eternal existences, being copied from them in a fashion wondrous and hard to declare, which we will follow up later on. For the present however we must conceive three kinds: first that which comes to be, secondly that wherein it comes to be, third that from which the becoming is copied when it is created. And we may liken the recipient to a mother, the model to a father, and that which is between them to a child; and we must remember that if a moulded copy is to present to view all varieties of form, the matter in which it is moulded cannot be rightly prepared unless it be entirely bereft of all those

Phaedo: they are, like the πέρας ἔχοντα of the *Philebus*, the form, as distinguished from the substance of material objects, apart from which they have no independent existence; they are in fact (apart from their relation to the ideas) practically indistinguishable from Aristotle's εἶδος as opposed to ὕλη. These are the visible semblances of the invisible verities of the ideal world, whereupon they are modelled in a mysterious manner hard to explain: for it is not easy to understand how the immaterial is expressed in terms of matter, or the invisible represented by a visible symbol. The εἰσιόντα must then be distinguished (logically, for they are never actually separable) from the material objects which they inform; these objects are εἰσιόντα + ἐκμαγεῖον.

11. ὃν εἰσαῦθις μέτιμεν] This refers probably to the conclusion of the chapter, 52 C.

15. ἐκγόνῳ] The ἔκγονα are the material phenomena formed by the impress of the εἰσιόντα upon the ἐκμαγεῖον.

16. ἰδεῖν ποικίλου] ἰδεῖν follows ποικίλου, to which πάσας ποικιλίας is a cognate accusative. Plato is rather fond of this construction with ἰδεῖν, cf. *Phaedo* 84 C, *Republic* 615 E, *Phaedrus* 250 D.

18. ἄμορφον ὄν] Aristotle has derived from hence his description of the thinking faculty, *de anima* III iv 429ᵃ 15 ἀπαθὲς ἄρα δεῖ εἶναι, δεκτικὸν δὲ τοῦ εἴδους καὶ δυνάμει τοιοῦτον, ἀλλὰ μὴ τοῦτο.... ἀνάγκη ἄρα, ἐπεὶ πάντα νοεῖ, ἀμιγῆ εἶναι, ὥσπερ φησὶν Ἀναξαγόρας, ἵνα κρατῇ, τοῦτο δ' ἐστὶν ἵνα γνωρίζῃ—παρεμφαινόμενον γὰρ

ὅσας μέλλοι δέχεσθαί ποθεν. ὅμοιον γὰρ ὂν τῶν ἐπεισιόντων τινὶ E
τὰ τῆς ἐναντίας τά τε τῆς τὸ παράπαν ἄλλης φύσεως, ὁπότ᾽ ἔλθοι,
δεχόμενον κακῶς ἂν ἀφομοιοῖ, τὴν αὐτοῦ παρεμφαῖνον ὄψιν. διὸ
καὶ πάντων ἐκτὸς εἰδῶν εἶναι χρεὼν τὸ τὰ πάντα ἐκδεξόμενον ἐν
5 αὑτῷ γένη, καθάπερ περὶ τὰ ἀλείμματα, ὁπόσα εὐώδη, τέχνῃ
μηχανῶνται πρῶτον τοῦτ᾽ αὐτὸ ὑπάρχον, ποιοῦσιν ὅ τι μάλιστα
ἀνώδη τὰ δεξόμενα ὑγρὰ τὰς ὀσμάς· ὅσοι τε ἔν τισι τῶν μαλακῶν
σχήματα ἀπομάττειν ἐπιχειροῦσι, τὸ παράπαν σχῆμα οὐδὲν ἔν-
δηλον ὑπάρχειν ἐῶσι, προομαλύναντες δὲ ὅ τι λειότατον ἀπερ-
10 γάζονται. ταὐτὸν οὖν καὶ τῷ τὰ τῶν πάντων ἀεί τε ὄντων κατὰ 51 A
πᾶν ἑαυτοῦ πολλάκις ἀφομοιώματα καλῶς μέλλοντι δέχεσθαι
πάντων ἐκτὸς αὐτῷ προσήκει πεφυκέναι τῶν εἰδῶν. διὸ δὴ τὴν
τοῦ γεγονότος ὁρατοῦ καὶ πάντως αἰσθητοῦ μητέρα καὶ ὑποδοχὴν
μήτε γῆν μήτε ἀέρα μήτε πῦρ μήτε ὕδωρ λέγωμεν, μήτε ὅσα ἐκ
15 τούτων μήτε ἐξ ὧν ταῦτα γέγονεν· ἀλλ᾽ ἀνόρατον εἶδός τι καὶ
ἄμορφον, πανδεχές, μεταλαμβάνον δὲ ἀπορώτατά πῃ τοῦ νοητοῦ
καὶ δυσαλωτότατον αὐτὸ λέγοντες οὐ ψευσόμεθα· καθ᾽ ὅσον δ᾽ ἐκ B
τῶν προειρημένων δυνατὸν ἐφικνεῖσθαι τῆς φύσεως αὐτοῦ, τῇδ᾽ ἂν

7 ἀνώδη: εὐώδη A. ἀώδη ΗΖ.

κωλύει τὸ ἀλλότριον καὶ ἀντιφράττει. It will be observed that the passage of Aristotle is full of verbal echoes of the *Timaeus*: and his ἀπαθὲς applied to the mind is exactly equivalent to Plato's ἄμορφον applied to the ὑποδοχή.

18. τῶν ἰδεῶν] Not the ideas, which do not enter into the ὑποδοχή, but the shapes which symbolise them—the εἰσιόντα καὶ ἐξιόντα.

3. τὴν αὐτοῦ παρεμφαῖνον ὄψιν] If the ὑποδοχή had any quality of its own, this quality would mingle with that impressed upon it by any of the εἰσιόντα and mar the faithfulness of the μίμημα. The only condition which the ὑποδοχή imposes upon our sensuous perceptions is that they shall exist in what we term space: we can perceive nothing that is not in space. Sensuous perceptions, as we have said, are symbols of the ideas: now it is quite free to the senses to symbolise an idea by the perception of round or square or any other shape, without any interference from the ὑποδοχή. The latter παρεμφαίνει τὴν αὐτῆς ὄψιν just in so far as round square and the like are and must be shapes that have extension.

6. μηχανῶνται...ποιοῦσιν] These two words are in a kind of apposition. Compare Euripides *Heraclidae* 181 ἄναξ, ὑπάρχει μὲν τόδ᾽ ἐν τῇ σῇ χθονί, | εἰπεῖν ἀκοῦσαί τ᾽ ἐν μέρει πάρεστί μοι. This same simile of the unguent is used by Lucretius II 848 to illustrate the necessary absence of secondary qualities from his atoms.

10. τῶν πάντων ἀεί τε ὄντων] Stallbaum would omit the τε, and νοητῶν has been proposed instead of πάντων. But πάντων is indispensable: it is because the ἐκμαγεῖον has to receive all forms that it can have no form of its own. Nor is the omission of τε satisfactory. Plato would probably have written πάντων τῶν ἀεὶ ὄντων. I think the text may be defended as it stands, ἀεί τε ὄντων being added to explain what is meant by τῶν πάντων—

forms which it is about to receive from without. For were it like any one of the entering shapes, whenever that of an opposite or entirely different nature came upon it, it would in receiving it give the impression badly, intruding its own form. Wherefore that which shall receive all forms within itself must be utterly without share in any of the forms; just as in the making of sweet unguents, men purposely contrive, as the beginning of the work, to make the fluids that are to receive the perfumes perfectly scentless: and those who set about moulding figures in any soft substance do not suffer any shape to show itself therein at the beginning, but they first knead it smooth and make it as uniform as they can. In the same way it behoves that which is fitly to receive many times over its whole extent likenesses of all things, that is of all eternal existences, to be itself naturally without part or lot in any of the forms. Therefore the mother and recipient of creation which is visible and by any sense perceptible we must call neither earth nor air nor fire nor water, nor the combinations of these nor the elements of which they are formed: but we shall not err in affirming it to be a viewless nature and formless, all-receiving, in some manner most bewildering and hard to comprehend partaking of the intelligible. But so far as from what has been said we may arrive at its nature, this would be the most just account

all things, that is, all eternal existences. Perhaps however we should read δεί ποτε ὄντων.

12. αὐτῷ προσήκει] Stallbaum erroneously considers αὐτῷ to be redundant: it is emphatic—'must itself be destitute of all forms'.

14. μήτε γῆν] It is indeed hard to conceive how Aristotle would attempt to justify his assertion in *de gen. et corr.* II i 329ᵃ 13 ὡς δ' ἐν τῷ Τιμαίῳ γέγραπται οὐδένα ἔχει διορισμόν· οὐ γὰρ εἴρηκε σαφῶς τὸ πανδεχές, εἰ χωρίζεται τῶν στοιχείων. If Plato has not most explicitly characterised the relation between the πανδεχές and the στοιχεῖα, then there is no such thing as precision in language. But the truth is, as not rarely happens when Aristotle is at cross purposes with Plato, that Aristotle is treating from a physical point of view a subject which Plato deals with metaphysically.

16. μεταλαμβάνον δὲ ἀπορώτατά πῃ τοῦ νοητοῦ] Plato's meaning is more fully expressed in 52 B. The puzzle arises from the fact that this ὑποδοχή, though it does not form part of real existence, is yet grasped by the reason and not by the senses. In the metaphysical scheme represented by the *Phaedo* we should find that constituting the test of reality, the object of reason being a real existence, the object of sense an unreality. But now we have found an anomalous principle which defies this test. It is not surprising then that Plato describes it as δυσαλωτότατον.

180   ΠΛΑΤΩΝΟΣ   [51 B—

τις ὀρθότατα λέγοι, πῦρ μὲν ἑκάστοτε αὐτοῦ τὸ πεπυρωμένον μέρος
φαίνεσθαι, τὸ δὲ ὑγρανθὲν ὕδωρ, γῆν δὲ καὶ ἀέρα, καθ' ὅσον ἂν
μιμήματα τούτων δέχηται. λόγῳ δὲ δὴ μᾶλλον τὸ τοιόνδε διοριζομένους περὶ αὐτῶν διασκεπτέον· ἆρ' ἔστι τι πῦρ αὐτὸ ἐφ'
5 ἑαυτοῦ καὶ πάντα, περὶ ὧν ἀεὶ λέγομεν οὕτως αὐτὰ καθ' αὑτὰ C
ὄντα ἕκαστα, ἢ ταῦτα, ἅπερ καὶ βλέπομεν ὅσα τε ἄλλα διὰ τοῦ
σώματος αἰσθανόμεθα, μόνα ἐστὶ τοιαύτην ἔχοντα ἀλήθειαν, ἄλλα
δὲ οὐκ ἔστι παρὰ ταῦτα οὐδαμῇ οὐδαμῶς, ἀλλὰ μάτην ἑκάστοτε
εἶναί τί φαμεν εἶδος ἑκάστου νοητόν, τὸ δὲ οὐδὲν ἄρ' ἦν πλὴν
10 λόγος; οὔτε οὖν δὴ τὸ παρὸν ἄκριτον καὶ ἀδίκαστον ἀφέντα ἄξιον
φάναι διισχυριζόμενον ἔχειν οὕτως, οὔτ' ἐπὶ λόγου μήκει πάρερ-
γον ἄλλο μῆκος ἐπεμβλητέον· εἰ δέ τις ὅρος ὁρισθεὶς μέγας διὰ D
βραχέων φανείη, τοῦτο μάλιστ' ἐγκαιριώτατον γένοιτ' ἄν. ὧδε
οὖν τήν γ' ἐμὴν αὐτὸς τίθεμαι ψῆφον· εἰ μὲν νοῦς καὶ δόξα ἀληθής
15 ἐστον δύο γένη, παντάπασιν εἶναι καθ' αὑτὰ ταῦτα, ἀναίσθητα
ὑφ' ἡμῶν εἴδη, νοούμενα μόνον· εἰ δ', ὥς τισι φαίνεται, δόξα
ἀληθὴς νοῦ διαφέρει τὸ μηδέν, πάνθ' ὁπόσ' αὖ διὰ τοῦ σώματος
αἰσθανόμεθα, θετέον βεβαιότατα. δύο δὴ λεκτέον ἐκείνω, διότι E
χωρὶς γεγόνατον ἀνομοίως τε ἔχετον. τὸ μὲν γὰρ αὐτῶν διὰ
20 διδαχῆς, τὸ δ' ὑπὸ πειθοῦς ἡμῖν ἐγγίγνεται· καὶ τὸ μὲν ἀεὶ μετ'
ἀληθοῦς λόγου, τὸ δὲ ἄλογον· καὶ τὸ μὲν ἀκίνητον πειθοῖ, τὸ δὲ

2 γῆν δέ: γῆν τε Λ.   3 δέχηται: δέχεται H typographi culpa.
διοριζομένους: διοριζομένοις S.

3. **μιμήματα τούτων**] i.e. τοῦ ὅ ἐστιν ἀὴρ and τοῦ ὅ ἐστι γῆ.

4. **ἆρ' ἔστι τι πῦρ**] When we say the ὑποδοχὴ receives the μίμημα of fire, we are assuming the existence of an essential idea of fire: it is now time to justify this assumption. The list of ideas in the *Timaeus* includes, in addition to ideas of living creatures, only the ideas of fire air water and earth: see Introduction § 33. Presently in the words εἶδος ἑκάστου νοητὸν we are to understand by ἑκάστου only every class naturally determined, τῶν ὁπόσα φύσει.

9. **τὸ δὲ οὐδὲν ἄρ' ἦν πλὴν λόγος**] By λόγος Plato means a mental concept, or universal: the question is in fact between Sokraticism and Platonism; that is to say, between conceptualism and idealism.

11. **διισχυριζόμενον ἔχειν οὕτως**] It is not often that Plato addresses himself to prove the existence of the ideas; the mere fact that it is impossible to find any stable reality or basis of knowledge in the material world is sufficient warrant for affirming the existence of the immaterial. Here the existence of ideas stands or falls with the distinction between knowledge and true opinion. Compare the discussion in *Republic* 476 E—480 A, also *Meno* 97 A foll. In the *Phaedo* a different line is taken, the existence of the ideas being deduced from ἀνάμνησις.

18. **θετέον βεβαιότατα**] i.e. we must accept them for the truest realities that exist, however fleeting and mutable they may be. For if there are no ideas, particulars are more real than the λόγοι,

of it. That part of it which is enkindled from time to time appears as fire, and that which is made liquid as water, and as earth and air such part of it as receives the likenesses of these.

But in our inquiry concerning these we must deliver a stricter statement. Is there an absolute idea of fire, and do all those absolute ideas exist to which in every case we always ascribe absolute being? Or do those things which we actually see or perceive with any other bodily sense alone possess such reality? and is it true that there are no manner of real existences beyond these at all, but we talk idly when we speak of an intelligible idea as actually existent, whereas it was nothing but a conception? Now it does not become us either to dismiss the present question unjudged and undecided, simply asserting that the ideas exist, nor yet must we add to our already long discourse another as long which is subordinate. But if we could see our way to a great definition couched in brief words, that would be most seasonable for our present purpose. Thus then do I give my own verdict: if reason and true opinion are of two different kinds, then the ideas do surely exist, forms not perceptible by our senses, the objects of thought alone; but if, as some hold, true opinion differs nothing from reason, then all that we apprehend by our bodily organs we must affirm to be the most real existence. Now we must declare them to be two, because they are different in origin and unlike in nature. The one is engendered in us by instruction, the other by persuasion; the one is ever accompanied by right understanding, the other is without understanding; the one is not to be moved by per-

which are merely formed from observation of them: but if the ideas exist, then λόγοι are more real than particulars, because the former are the intellectual, the latter only the sensible images of the ideas: cf. *Phaedo* 99 E.

19. χωρὶς γεγόνατον ἀνομοίως τε ἔχετον] They are of diverse origin, because one springs from instruction and the other from persuasion; of diverse nature, because one is immovable by persuasion, the other yields to it. You may persuade a man that pinchbeck is gold, but you never can persuade him that two straight lines enclose a space. It will be observed that the difference between knowledge and opinion rests here upon the same reasoning as the final rejection of the claims of ἀληθὴς δόξα in *Theaetetus* 201 A—C, where Sokrates, after showing that a jury may be persuaded by a skilful advocate to hold a right opinion on a case the facts of which they do not know, concludes his argument thus: οὐκ ἄν, ὦ φίλε, εἴ γε ταὐτὸν ἦν δόξα τε ἀληθὴς καὶ ἐπιστήμη, ὀρθά ποτ' ἂν δικαστὴς ἄκρος ἐδόξαζεν ἄνευ ἐπιστήμης· νῦν δὲ ἔοικεν ἄλλο τι ἑκάτερον εἶναι.

182   ΠΛΑΤΩΝΟΣ   [51 E—

μεταπειστόν· καὶ τοῦ μὲν πάντα ἄνδρα μετέχειν φατέον, νοῦ δὲ
θεούς, ἀνθρώπων δὲ γένος βραχύ τι. τούτων δὲ οὕτως ἐχόντων
ὁμολογητέον ἓν μὲν εἶναι τὸ κατὰ ταὐτὰ εἶδος ἔχον, ἀγένητον καὶ 52 A
ἀνώλεθρον, οὔτε εἰς ἑαυτὸ εἰσδεχόμενον ἄλλο ἄλλοθεν οὔτε αὐτὸ εἰς
5 ἄλλο ποι ἰόν, ἀόρατον δὲ καὶ ἄλλως ἀναίσθητον, τοῦτο ὃ δὴ νόησις
εἴληχεν ἐπισκοπεῖν· τὸ δ' ὁμώνυμον ὅμοιόν τε ἐκείνῳ δεύτερον,
αἰσθητόν, γεννητόν, πεφορημένον ἀεί, γιγνόμενόν τε ἔν τινι τόπῳ
καὶ πάλιν ἐκεῖθεν ἀπολλύμενον, δόξῃ μετ' αἰσθήσεως περιληπτόν·
τρίτον δὲ αὖ γένος ὂν τὸ τῆς χώρας ἀεί, φθορὰν οὐ προσδεχόμενον,

3 ἀγένητον: ἀγέννητον HSZ. sed cf. *Phaedr.* 245 D.
7 πεφορημένον: πεφωνημένον A.

1. **πάντα ἄνδρα μετέχειν**] cf. *Theaetetus* 206 D.

4. **οὔτε αὐτὸ εἰς ἄλλο ποι ἰόν**] Here we have a perfectly unmistakable assertion of the solely transcendental existence of the ideas. The difficulties raised against the doctrine of immanent ideas in *Parmenides* 131 A are fatal and insurmountable. From that time forth παρουσία and μέθεξις (in connexion with αὐτὰ καθ' αὑτὰ εἴδη) disappear from Plato's vocabulary, and μίμησις takes their place. It may be added that the previous words οὔτε εἰς ἑαυτὸ εἰσδεχόμενον ἄλλο ἄλλοθεν would seem enough in themselves to dispose of Zeller's theory of particulars inherent in the ideas.

8. **δόξῃ μετ' αἰσθήσεως**] Cf. 28 A, where ἀλόγου is added.

9. **τὸ τῆς χώρας ἀεί**] Thus then we have materiality in its ultimate analysis reduced to space or extension. It may now be desirable to scrutinize Plato's conception a little more closely. First then as to the relation of χώρα to the absolute intelligence and to finite intelligences. Absolute νοῦς or ψυχὴ evolves itself into the form of a multitude of finite intelligences. For these it is a necessity of their nature that they should apprehend, *qua* finite, under certain unalterable forms, which we call time and space. Therefore whatever they perceive, they perceive somewhere. But this *somewhere* is relative to them and purely subjective (for we know that Plato's Herakleiteanism so far as concerns the region of sensibles was complete). All sensible perceptions then have no existence except in the consciousness of the percipient. But the law which binds particular ψυχαὶ to apprehend in this mode is immutable and eternal: hence space must be eternal; for ψυχὴ must exist not only in the mode of unity but in the mode of plurality, in the form of limited souls. There must then always be finite intelligences percipient of a material universe existing in space. So far then as we confine our view to the relation of the material universe to the finite percipients, we find Plato's position to be a form of subjective idealism. But as soon as we consider the relation of finite percipients and their perceptions to the absolute intelligence, we shall find that the subjective is merged in an absolute idealism. For these percipients and percepts with the law which binds them to perceive and be perceived in this mode, though regarded as individuals they are severally transient and subject to time and space, yet regarded as a whole constitute one element in the eternal and spaceless process of thought, the element of θάτερον. And thus are material phenomena said to be μιμήματα τῶν ὄντων: they are perceptions existing in the consciousness of finite intelligences, which perceptions are the mode in which finite intelligences, acting through the senses, apprehend the ideas

suasion, the other yields to persuasion; true opinion we must admit is shared by all men, but reason by the gods alone and a very small portion of mankind. This being so, we must agree that there is first the unchanging idea, unbegotten and imperishable, neither receiving aught into itself from without nor itself entering into aught else, invisible, nor in any wise perceptible— even that whereof the contemplation belongs to thought. Second is that which is named after it and is like to it, sensible, created, ever in motion, coming to be in a certain place and again from thence perishing, apprehensible by opinion with sensation. And the third kind is space everlasting, admitting not destruction, but

as existing in infinite intelligence. The phenomena are material symbols of ideal truths: and it is only by these symbols that a finite intelligence, so far as it acts through the senses, can apprehend such truths.

Plato's identification of the ὑποδοχὴ with χώρα arises from the absolute ἀπάθεια of the former. The manner of approaching it may perhaps be most readily seen in the following way. Let us take any material object, say a ball of bronze. Now every one of the qualities belonging to the bronze we know to be due to the μίμημα which informs the ὑποδοχή: therefore to reach the ὑποδοχὴ we must abstract, one after another, all the attributes which belong to the bronze. When these are stripped away, what have we remaining? simply a spherical space of absolute vacancy. The ὑποδοχή then, as regards the bronze ball, is that sphere of empty space. But still this void sphere is something; because it is defined by the limits of the air surrounding it: it is in fact a sphere of emptiness. But now suppose, instead of abstracting the qualities from the bronze alone, we abstract them from the whole universe and all its contents: then we have vacancy coextensive with the universe. But mark the difference. The empty sphere we could speak of as something, because it was the interval between the limits of the surrounding air. But our universal vacancy there is nothing to

limit, there is nothing to be contrasted with it to give it a *differentia*, it is vacancy undefined: that is to say, it is just nothing at all. Thus we see that space pure and simple is an abstract logical conception; extension without the extended is nothing, for space can no more exist independently of the things in it than time can exist without events to measure it. Thus in its most abstract significance χώρα is the eternal law or necessity constraining pluralised ψυχὴ to have its perceptions under the form we call space: since then ψυχὴ does, and therefore must, evolve itself under this form and not another, χώρα ultimately represents the law that ψυχὴ shall pluralise itself.

Between Plato's χώρα and Aristotle's ὕλη the only difference physically seems to me to lie in the superior distinctness and definiteness of Plato's conception: it was the intense vividness of Plato's insight that led him to the identification of the substrate with space. Aristotle, whose ὕλη is taken bodily from Plato, ought to have made the same identification: that he did not do so is due to the mistiness which pervades his whole thought as compared with Plato's.

A few words are demanded by Aristotle's reference to the Platonic theory in *physica* IV ii 209^b 11. Aristotle there affirms that Plato identifies the μεταληπτικὸν with χώρα, but that he gives one account of the μεταληπτικὸν in the *Ti-*

ἕδραν δὲ παρέχον ὅσα ἔχει γένεσιν πᾶσιν, αὐτὸ δὲ μετ' ἀναισθη- B
σίας ἁπτὸν λογισμῷ τινὶ νόθῳ, μόγις πιστόν· πρὸς ὃ δὴ καὶ
ὀνειροπολοῦμεν βλέποντες καί φαμεν ἀναγκαῖον εἶναί που τὸ ὂν
ἅπαν ἔν τινι τόπῳ καὶ κατέχον χώραν τινά, τὸ δὲ μήτ' ἐν γῇ
5 μήτε που κατ' οὐρανὸν οὐδὲν εἶναι. ταῦτα δὴ πάντα καὶ τούτων
ἄλλα ἀδελφὰ καὶ περὶ τὴν ἄυπνον καὶ ἀληθῶς φύσιν ὑπάρχουσαν
ὑπὸ ταύτης τῆς ὀνειρώξεως οὐ δυνατοὶ γιγνόμεθα ἐγερθέντες διο- C
ριζόμενοι τἀληθὲς λέγειν, ὡς εἰκόνι μέν, ἐπείπερ οὐδ' αὐτὸ τοῦτο,
ἐφ' ᾧ γέγονεν, ἑαυτῆς ἐστίν, ἑτέρου δέ τινος ἀεὶ φέρεται φάντασμα,
10 διὰ ταῦτα ἐν ἑτέρῳ προσήκει τινὶ γίγνεσθαι, οὐσίας ἁμῶς γέ πως
ἀντεχομένην, ἢ μηδὲν τὸ παράπαν αὐτὴν εἶναι, τῷ δὲ ὄντως ὄντι
βοηθὸς ὁ δι' ἀκριβείας ἀληθὴς λόγος, ὡς ἕως ἄν τι τὸ μὲν ἄλλο ᾖ,
τὸ δὲ ἄλλο, οὐδέτερον ἐν οὐδετέρῳ ποτὲ γενόμενον ἓν ἅμα ταὐτὸν
καὶ δύο γενήσεσθον. D

13 γενόμενον: γεγενημένον HSZ.

maeus, another ἐν τοῖς λεγομένοις ἀγράφοις δόγμασιν. What the account in the ἄγραφα δόγματα was, Aristotle does not tell us; presently however he says, 209ᵇ 34, Πλάτωνι μέντοι λεκτέον, εἰ δεῖ παρεκβάντας εἰπεῖν, διὰ τί οὐκ ἐν τόπῳ τὰ εἴδη καὶ οἱ ἀριθμοί, εἴπερ τὸ μεθεκτικὸν ὁ τόπος, εἴτε τοῦ μεγάλου καὶ τοῦ μικροῦ ὄντος τοῦ μεθεκτικοῦ εἴτε τῆς ὕλης, ὥσπερ ἐν τῷ Τιμαίῳ γέγραφεν. Now as to this ἀπορία, it may be observed that it does not affect Plato at all: by the time his theory of χώρα was worked out, the doctrine of μέθεξις was abandoned: Aristotle has in fact no right to apply to the ὑποδοχὴ the terms μεθεκτικόν, μεταληπτικόν, in relation to the ideas. Next it will be evident to any one who reads the whole discussion in the *physica* that the object of Aristotle's inquiry is a purely physical one, what is τόπος? meaning by τόπος the place in which any object is situate, which he ultimately defines to be τὸ πέρας τοῦ περιέχοντος σώματος. This has evidently nothing in the world to do with the metaphysical question of the *Timaeus*: yet Aristotle makes as though it were the same. Zeller is perfectly just in his criticism (platonische Studien p. 212): 'während also Platon im Timäus die Frage aufwirft: was ist die Materie? und darauf antwortet: der Raum; so fragt Aristoteles: was ist der Raum? und lässt Platon darauf antworten: die Materie'.

1. **μετ' ἀναισθησίας ἁπτὸν λογισμῷ τινὶ νόθῳ**] None of our senses can intimate to us the existence or nature of space; it is attained only by an effort of logical analysis, λογισμῷ. Yet space is no real existence; therefore it cannot be the object of reason properly so called, which deals with ideal truth. Plato says then it is reached by a kind of bastard reasoning, which is indeed a purely mental process, unaided by the senses, yet distinct from the true activity of the soul when she is engaged on her proper objects of cognition. It is, as I have said, the anomaly of these conditions from which the obscurity of the subject arises. The compiler of the *Timaeus Locrus* (94 B) seeks to explain νόθῳ by the words τῷ μήπω κατ' εὐθυωρίαν νοῆσθαι ἀλλὰ κατ' ἀναλογίαν.

2. **μόγις πιστόν**] πίστις is the word used in the sixth book of the *Republic* to denote the mental πάθημα which deals with sensible objects. Space then is μόγις πιστόν, because, although it is the mode

ΤΙΜΑΙΟΣ.

affording place for all things that come into being, itself apprehensible without sensation by a sort of bastard reasoning, hardly matter of belief. It is with this in view that dreaming we say that all which exists must be in some place and filling some space, and that what is neither on earth nor in heaven anywhere is nought. All these and many kindred fancies have we even concerning that unsleeping essence and truly existing, for that by reason of this dreaming state we become impotent to arouse ourselves and affirm the truth; namely, that to an image it belongs, seeing that it is not the very model of itself, on which itself has been created, but is ever the fleeting semblance of another, in another to come into being, clinging to existence as best it may, on pain of being nothing at all; but to the really existent essence reason in all exactness true comes as an ally, declaring that so long as one thing is one and another thing is other, neither of them shall come to be in the other, so that the same becomes at once one and two.

in which sensible things are perceived, it is not itself an object of sensation: it is an ambiguous and doubtful form, hard to grasp and hard to trust.

πρὸς δ δή] It is this that causes our vague and dreamy state of mind regarding existence. Because everything of which our senses affirm the existence exists in space, we rashly assume that all things which exist exist in space, and that what is not somewhere is nothing. For we are held fast in the thraldom of our own subjective perceptions, and suppose, as dreamers do, that the visions within our own consciousness are external realities. It must be remembered that Plato was the very first who had any real conception of immaterial existence.

6. τὴν ἄυπνον] i.e. the region of objective truth, which we apprehend with our waking faculties, that is to say, by pure reason unhampered by sensation. We do not conceive of the ideal world as it really is, independent of all conditions of time and space.

8. ἐπείπερ οὐδ' αὐτὸ τοῦτο] I believe the true construction of these words has escaped all the editors and translators, who are consequently in sore straits what to make of ἑαυτῆς. The construction seems to me to be a very simple and very Platonic σχῆμα πρὸς τὸ σημαινόμενον. What is meant by αὐτὸ τοῦτο ἐφ' ᾧ γέγονεν? of course the παράδειγμα, and the whole phrase governs ἑαυτῆς just as if παράδειγμα had been written: 'since it is not the original-upon-which-it-is-modelled of itself'.

10. ἐν ἑτέρῳ τινί] Since the image is not identical with the type, it must be manifested in some mode external to the type, that it may be numerically different. This external mode is what we term space. Space then is that which differentiates the image from the idea and thereby enables the former to exist, οὐσίας ἀμωσγέπως ἀντεχομένη. It is a dubious kind of existence that is in space: but, such as it is, it is owing to space: for did not space exist, nothing would remain but the idea: and since the image cannot be in that, it could not be at all.

13. οὐδέτερον ἐν οὐδετέρῳ] Here again we have a distinct repudiation of

XIX. Οὗτος μὲν οὖν δὴ παρὰ τῆς ἐμῆς ψήφου λογισθεὶς ἐν κεφαλαίῳ δεδόσθω λόγος, ὅν τε καὶ χώραν καὶ γένεσιν εἶναι, τρία τριχῇ, καὶ πρὶν οὐρανὸν γενέσθαι· τὴν δὲ δὴ γενέσεως τιθήνην ὑγραινομένην καὶ πυρουμένην καὶ τὰς γῆς τε καὶ ἀέρος μορφὰς
5 δεχομένην, καὶ ὅσα ἄλλα τούτοις πάθη ξυνέπεται πάσχουσαν, παντοδαπὴν μὲν ἰδεῖν φαίνεσθαι, διὰ δὲ τὸ μήθ᾽ ὁμοίων δυνάμεων Ε μήτε ἰσορρόπων ἐμπίπλασθαι κατ᾽ οὐδὲν αὐτῆς ἰσορροπεῖν, ἀλλ᾽ ἀνωμάλως πάντῃ ταλαντουμένην σείεσθαι μὲν ὑπ᾽ ἐκείνων αὐτήν, κινουμένην δ᾽ αὖ πάλιν ἐκεῖνα σείειν· τὰ δὲ κινούμενα ἄλλα ἄλλοσε
10 ἀεὶ φέρεσθαι διακρινόμενα, ὥσπερ τὰ ὑπὸ τῶν πλοκάνων τε καὶ ὀργάνων τῶν περὶ τὴν τοῦ σίτου κάθαρσιν σειόμενα καὶ ἀναλικμώμενα τὰ μὲν πυκνὰ καὶ βαρέα ἄλλῃ, τὰ δὲ μανὰ καὶ κοῦφα εἰς 53 Α

3 τὴν δὲ δή: δή omittunt ASZ.   5 ἄλλα τούτοις: τούτοις ἄλλα S.   7 ἐμπίπλασθαι: ἐμπίπλασθαι A.   11 ἀναλικμώμενα: ἀναλικνώμενα pr. AS. ἀνικμώμενα H.

the old doctrine of παρουσία. That doctrine affirmed that the idea existed (1) in its own independent nature, (2) inherent in the particulars. The latter mode is now declared to be impossible for the plain reason that things cannot be two and one at the same time, nor can the same thing be at once original and copy. If the copy were inherent in the original, or the original in the copy, the difference between them would be lost; and we should once more be reduced to a bare denial of the existence of the material world. It will be observed that the rejection of μέθεξις is here based upon a different ground from that taken up in the *Parmenides*, although the criticism in that dialogue remains perfectly valid. We see then the truth of Aristotle's statement in *metaph.* I vi that Plato was led, in opposition to the Pythagoreans, to place the ideas παρὰ τὰ αἰσθητά through his logical speculations, διὰ τὴν ἐν τοῖς λόγοις σκέψιν.

52 D—53 C, *c.* xix. All the universe then is divided into Being Space and Becoming, these three. And space, receiving the forms that enter in, and being thereby filled with unbalanced forces, is nowhere in equipoise but ever swaying to and fro over its whole expanse. And thus too it sways in turn the things that arise in it and sifts them, so that the lighter bodies fly off to one region, and the heavier settle in another. Thus, even in the rudimentary state, wherein without the working of intelligence they would have been, the different bodies tend to occupy different regions in space; and yet more, when all is ordered by intelligence for the best, as we affirm to be the truth. And now we must set forth the order and generation of them.

1. λογισθείς...λόγος] Compare 34 A λογισμὸς θεοῦ περὶ τὸν ποτὲ ἐσόμενον θεὸν λογισθείς.

2. τρία τριχῇ] This seems to mean no more than 'three things with three distinct natures': cf. 89 E τρία τριχῇ ψυχῆς ἐν ἡμῖν εἴδη κατῴκισται. Of course this triad is not in any way to be confounded with the former triad of ταὐτὸν θάτερον and οὐσία.

3. καὶ πρὶν οὐρανὸν γενέσθαι] This, it need hardly be said, is again to be taken logically: these three are prior in analysis.

6. μήθ᾽ ὁμοίων δυνάμεων] The manifold bodies which are generated in space have most diverse and unequal forces, and inequality is the parent of motion, as

XIX. Such then is the statement for which I give my sentence, as we have briefly reasoned it out: that there are Being and Space and Becoming, three in number with threefold nature, even before the heavens were created. And the nurse of becoming, being made liquid and fiery and putting on the forms of earth and air, and undergoing all the conditions that attend thereupon, displays to view all manner of semblances; and because she is filled with powers that are not similar nor equivalent, she is at no part of her in even balance, but being swayed in all directions unevenly, she is herself shaken by the entering forms, and by her motion shakes them again in turn: and they, being thus stirred, are carried in different directions and separated, just as by sieves and instruments for winnowing corn the grain is shaken and sifted, and the dense and heavy parts go one way, and the rare and light are carried to a different

we are informed in 58 A. Thus a vibratory motion is set up throughout the whole extent of the ὑποδοχή and communicated to the objects contained in it, which are thereby sifted as by a winnowing machine. This vibration of the ὑποδοχή and the πλῆσις hereafter to be mentioned are the two most important physical forces in Plato's scheme; nearly all the processes of nature being due to them in one way or another.

9. κινουμένην δ' αὖ πάλιν ἐκεῖνα σείειν] What Plato means by this action and reaction existing between the ὑποδοχή and its contents may thus be explained. If we abstract every sort of determination from sensuous perception, the residuum is space pure and simple. Now this, being without content, can of course have no motion. But once it is determined by the εἰσιόντα καὶ ἐξιόντα, motion becomes possible; so that it is from these that the ὑποδοχή receives motive power. On the other hand the motion thus initiated has to obey the law of existence in space: i.e. (1) it is a φορά, or motion in respect of place, (2) it sifts the divers objects into different regions. Motion then begins with the εἰσιόντα καὶ ἐξιόντα, but once begun it is controlled by the law of the ὑποδοχή. In starting motion with the εἰσιόντα καὶ ἐξιόντα Plato distinctly intimates that there is no independent force in matter: therefore the πλανωμένη αἰτία cannot be regarded as an independent principle of causation.

10. πλόκανον] This was a kind of wicker sieve used for winnowing. Plato may have got the hint for his sifting motion from Demokritos: compare a fragment given by Sextus Empiricus *adv. math.* VII §§ 117, 118 καὶ γὰρ ζῷα ὁμογενέσι ζῴοισι ξυναγελάζεται, ὡς περιστεραὶ περιστερῇσι καὶ γέρανοι γεράνοισι, καὶ ἐπὶ τῶν ἄλλων ἀλόγων. ὡσαύτως δὲ καὶ περὶ τῶν ἀψύχων, καθάπερ ὁρῆν πάρεστι ἐπί τε τῶν κοσκινευομένων σπερμάτων καὶ ἐπὶ τῶν παρὰ τῇσι κυματωγῇσι ψηφίδων· ὅκου μὲν γὰρ παρὰ τὸν τοῦ κοσκίνου δῖνον διακριτικῶς φακοὶ μετὰ φακῶν τάσσονται καὶ κριθαὶ μετὰ κριθέων καὶ πυροὶ μετὰ πυρῶν· ὅκου δὲ κατὰ τὴν τοῦ κύματος κίνησιν αἱ μὲν ἐπιμηκέες ψηφῖδες εἰς τὸν αὐτὸν τόπον τῇσι ἐπιμηκέσι ὠθέονται, αἱ δὲ περιφερέες τῇσι περιφερέσι· ὡς ἂν ξυναγωγόν τι ἐχούσης τῶν πρηγμάτων τῆς ἐν τούτοισι ὁμοιότητος. Cf. Diogenes Laertius IX §§ 31, 32. As Mr Heath observes (*Journal of Philology*

ἑτέραν ἵζει φερόμενα ἕδραν· τότε οὕτω τὰ τέτταρα γένη σειόμενα ὑπὸ τῆς δεξαμένης, κινουμένης αὐτῆς οἷον ὀργάνου σεισμὸν παρέχοντος, τὰ μὲν ἀνομοιότατα πλεῖστον αὐτὰ ἀφ' αὑτῶν ὁρίζειν, τὰ δ' ὁμοιότατα μάλιστα εἰς ταὐτὸν ξυνωθεῖν· διὸ δὴ καὶ χώραν
5 ταῦτα ἄλλα ἄλλην ἴσχειν, πρὶν καὶ τὸ πᾶν ἐξ αὐτῶν διακοσμηθὲν γενέσθαι. καὶ τὸ μὲν δὴ πρὸ τούτου πάντα ταῦτ' ἔχειν ἀλόγως καὶ ἀμέτρως· ὅτε δ' ἐπεχειρεῖτο κοσμεῖσθαι τὸ πᾶν, πῦρ πρῶτον B καὶ ὕδωρ καὶ γῆν καὶ ἀέρα, ἴχνη μὲν ἔχοντα αὑτῶν ἄττα, παντάπασί γε μὴν διακείμενα ὥσπερ εἰκὸς ἔχειν ἅπαν, ὅταν ἀπῇ τινὸς
10 θεός, οὕτω δὴ τότε πεφυκότα ταῦτα πρῶτον διεσχηματίσατο εἴδεσί τε καὶ ἀριθμοῖς. τὸ δὲ ᾗ δυνατὸν ὡς κάλλιστα ἄριστά τε ἐξ οὐχ οὕτως ἐχόντων τὸν θεὸν αὐτὰ ξυνιστάναι, παρὰ πάντα ἡμῖν ὡς ἀεὶ τοῦτο λεγόμενον ὑπαρχέτω· νῦν δ' οὖν τὴν διάταξιν αὐτῶν ἐπιχειρητέον ἑκάστων καὶ γένεσιν ἀήθει λόγῳ πρὸς ὑμᾶς δηλοῦν, ἀλλὰ C
15 γὰρ ἐπεὶ μετέχετε τῶν κατὰ παίδευσιν ὁδῶν, δι' ὧν ἐνδείκνυσθαι τὰ λεγόμενα ἀνάγκη, ξυνέψεσθε.

2 δεξαμένης: δεξαμενῆς ASZ.   8 ὕδωρ καὶ γῆν καὶ ἀέρα: γῆν καὶ ἀέρα καὶ ὕδωρ S.
αὑτῶν ἄττα: αὑτῶν αὐτά A.   14 ἀήθει: ἀληθεῖ corr. A.

VIII p. 162), 'it is remarkable that Plato sees the dynamical reason of the thing; while Democritus draws the fanciful and false inference that "like seeks its like".'

2. **ὑπὸ τῆς δεξαμένης**] Stallbaum is unquestionably wrong in reading δεξαμενῆς, which means a cistern and nothing else: cf. *Critias* 117 B.

5. **πρὶν καὶ τὸ πᾶν**] Plato's meaning I take to be as follows. From the pluralisation of Being as such (the nature of Being remaining undefined) we get only the necessity of material perceptions: and all that is thereby necessarily involved is the existence of matter in some chaotic or rudimentary form. But when Being is defined to be Intelligence, the pluralisation of it must involve the ordering of matter according to some intelligent design. This metaphysical meaning Plato clothes in a mythical form borrowed from Anaxagoras. In this chapter he gives us a completion of Anaxagoras and a polemic against Demokritos. Anaxagoras, though he postulated νοῦς as a motive cause, failed to represent the universe as the orderly evolution of intelligence everywhere working ἐπὶ τὸ βέλτιστον: he confined himself to giving an account of the physical agencies through which he supposed νοῦς to work. Plato, in explaining these physical agencies, is careful to insist that they are merely subsidiary to the final cause: the real explanation of each thing is to be found in its motive. Demokritos held that the present order of the universe was the effect of a blind force working without intelligence, which by fortuitous collisions and combinations formed a symmetrical system. This view Plato controverts, urging that such fortuitous conjunctions could not amount to more than a rudimentary and chaotic condition of material existence: form, arrangement, symmetry imply intelligence in the motive power. Properly interpreted then, matter as it is πρὶν γενέσθαι τὸν οὐρανὸν is matter evolved on the Demokritean plan as contrasted with the Platonic. Plato does not mean that there was a time when matter existed in this form.

place and settle there. Even so when the four kinds are shaken by the recipient, which by the motion she has received acts as an instrument for shaking, she separates the most dissimilar elements furthest apart from one another, and the most similar she draws chiefly together; for which cause these elements had different regions even before the universe was ordered out of them and created. Before that came to pass all these things were without method or measure; but when an essay was being made to order the universe, first fire and water and earth and air, which had certain vestiges of their own nature, yet were altogether in such a condition as we should expect for everything when God is not in it, being by nature in the state we have said, were then first by the creator fashioned forth with forms and numbers. And that God formed them to be most fair and perfect, not having been so heretofore, must above all things be the foundation whereon our account is for ever based. But now the disposition of each and their generation is what I must strive to make known to you in speech unwonted: but seeing ye are no strangers to the paths of learning, through which my sayings must be revealed to you, ye will follow me.

8. αὐτῶν ἄττα] This is an obviously certain correction of the senseless αὐτῶν αὐτὰ of the mss. Fire and the rest, before the universe was framed,—that is in a universe framed on the Demokritean theory—had some incipient indications of their present nature, but only in an inchoate condition.

9. ὅταν ἀπῇ τινὸς θεός] i.e. in a world which is not the evolution of θεός, but the result of mere chance and coincidence.

10. εἴδεσί τε καὶ ἀριθμοῖς] 'with forms and measures'; i.e. with bodies definitely qualified and quantified. ἀριθμοί has not the meaning it so frequently bears in Aristotle, 'the ideal numbers'; for this never occurs in the Platonic writings.

14. ἀήθει λόγῳ] Plato's expression is fully justified. When we come to examine his atomic theory (if so it may be called), we shall find it exceedingly peculiar and totally unlike any other that has ever been propounded.

15. τῶν κατὰ παίδευσιν ὁδῶν] Probably with especial reference to geometry, without some knowledge of which Plato's theory could not be comprehended. ὁδῶν is here practically equivalent to μεθόδων, a sense in which it is not unfrequently found; cf. *Phaedrus* 263 B οὐκοῦν τὸν μέλλοντα τέχνην ῥητορικὴν μετιέναι πρῶτον μὲν δεῖ ταῦτα ὁδῷ διῃρῆσθαι: and *Cratylus* 425 B ἄλλως δὲ συνείρειν μὴ φαῦλον ᾖ καὶ οὐ καθ' ὁδόν.

53 C—55 C, *c*. xx. This is the generation of fire air water and earth. All these are solid bodies, and solid bodies are bounded by plane surfaces. Every rectilinear plane surface can be divided into triangles: the triangle then is the primary plane figure. The triangles which we affirm to be the fundamental form of all matter are two in number, the rectangular isosceles, and the rectangular scalene which is obtained by bisecting an equi-

190   ΠΛΑΤΩΝΟΣ   [53 C—

XX. Πρῶτον μὲν δὴ πῦρ καὶ γῆ καὶ ὕδωρ καὶ ἀὴρ ὅτι σώματά ἐστι, δῆλόν που καὶ παντί· τὸ δὲ τοῦ σώματος εἶδος πᾶν καὶ βάθος ἔχει· τὸ δὲ βάθος αὖ πᾶσα ἀνάγκη τὴν ἐπίπεδον περιειληφέναι φύσιν· ἡ δὲ ὀρθὴ τῆς ἐπιπέδου βάσεως ἐκ τριγώνων 5 συνέστηκε. τὰ δὲ τρίγωνα πάντα ἐκ δυοῖν ἄρχεται τριγώνοιν, D μίαν μὲν ὀρθὴν ἔχοντος ἑκατέρου γωνίαν, τὰς δὲ ὀξείας· ὧν τὸ μὲν ἕτερον ἑκατέρωθεν ἔχει μέρος γωνίας ὀρθῆς πλευραῖς ἴσαις διῃρημένης, τὸ δ' ἕτερον ἀνίσοις ἄνισα μέρη νενεμημένης. ταύτην δὴ πυρὸς ἀρχὴν καὶ τῶν ἄλλων σωμάτων ὑποτιθέμεθα κατὰ τὸν 10 μετ' ἀνάγκης εἰκότα λόγον πορευόμενοι· τὰς δ' ἔτι τούτων ἀρχὰς ἄνωθεν θεὸς οἶδε καὶ ἀνδρῶν ὃς ἂν ἐκείνῳ φίλος ᾖ. δεῖ δὴ λέγειν, ποῖα κάλλιστα σώματα γένοιτ' ἂν τέτταρα, ἀνόμοια μὲν ἑαυτοῖς, E δυνατὰ δὲ ἐξ ἀλλήλων αὐτῶν ἄττα διαλυόμενα γίγνεσθαι. τούτου γὰρ τυχόντες ἔχομεν τὴν ἀλήθειαν γενέσεως πέρι γῆς τε 15 καὶ πυρὸς τῶν τε ἀνὰ λόγον ἐν μέσῳ· τόδε γὰρ οὐδενὶ συγχωρησόμεθα, καλλίω τούτων ὁρώμενα σώματα εἶναί που καθ' ἓν γένος ἕκαστον ὄν. τοῦτ' οὖν προθυμητέον, τὰ διαφέροντα κάλλει σωμάτων τέτταρα γένη συναρμόσασθαι καὶ φάναι τὴν τούτων ἡμᾶς φύσιν ἱκανῶς εἰληφέναι. τοῖν δὴ δυοῖν τριγώνοιν τὸ μὲν 54 A

5 δυοῖν : δυεῖν S.   6 τὰς δέ : τὰς δὲ δύο S.   15 τόδε : τότε SZ.

lateral triangle. From the latter the three elements fire air and water, are framed: from the former earth alone. It follows then that while fire air and water can interchange and pass one into another, earth cannot pass into any of them nor they into it, because its base is different. But since the other three are formed on the same triangle, they can interchange, when a figure formed of many triangles breaks up into several formed of fewer, or *vice versa*. The way in which the figures are formed is as follows. Six of the primary scalenes placed together constitute an equilateral triangle; and four equilaterals form the sides of a regular solid, the tetrahedron or pyramid, which is the constituent particle of fire: eight such equilaterals are the sides of the octahedron, which is the particle of air; twenty equilaterals are the sides of the icosahedron, being the particle of water. These are all the forms constructed on the rectangular scalene. From the rectangular isosceles, by placing four together, is formed a square; and six squares are the sides of a fourth regular solid called the cube, which is the particle proper to earth. A fifth regular solid still exists, namely the dodecahedron, which does not form the element of any substance; but God used it as a pattern for dividing the zodiac into its twelve signs.

3. **τὴν ἐπίπεδον**] Every solid is bounded by plane surfaces. Aristotle, in criticising the Platonic theory (see *de caelo* III i 298[b] 33; *de gen. et corr.* I ii 315[b] 30), objects (1) that you cannot make solid matter out of planes, (2) that there are no such things as indivisible magnitudes. To the first objection it is sufficient to reply that Plato, who was presumably as well aware as every one else of the impossibility of forming solids by an aggregation of mathematical planes, does not attempt to do anything of the

XX. In the first place, that fire and earth and water and air are material bodies is evident to all. Every form of body has depth: and depth must be bounded by plane surfaces. Now every rectilinear plane is composed of triangles. And all triangles are derived from two triangles, each having one right angle and the others acute: and one triangle has on each side a moiety of a right angle marked off by equal sides, the other has it divided into unequal parts by unequal sides. These we conceive to be the basis of fire and the other bodies, following up the probable account which is concerned with necessity: but the principles yet more remote than these are known but to God and to whatsoever man is a friend of God. Now we must declare what are the four fairest bodies that could be created, unlike one another, but capable, some of them, of being generated out of each other by their dissolution: for if we succeed in this, we have come at the truth concerning earth and fire and the intermediate proportionals. For we will concede to no one that there exist any visible bodies fairer than these, each after its own kind. We must do our diligence then to put together these four kinds of bodies most excellent in beauty, and so we shall say that we have a full comprehension of their nature.

Now of the two triangles the isosceles has but one kind,

sort: to the second, that Plato's solids are not indivisible, but are the minutest forms of organised matter which exist. When they are broken up, they are either reformed into another figure, or the matter of which they are composed goes on existing in a formless condition. There is however a real difficulty not noticed by Aristotle, which will be discussed on 56 D.

4. ἐκ τριγώνων συνέστηκε] Because every rectilinear plane of whatever shape can be divided up into triangles, three straight lines being the fewest that can enclose a space.

5. ἐκ δυοῖν ἄρχεται τριγώνοιν] All triangles are reducible to two, the rectangular isosceles and the rectangular scalene, because any triangle can be di-vided into one or other of these by simply drawing a perpendicular from one of the angles to the opposite side. Of the rectangular isosceles there is of course but one kind; of the rectangular scalene an endless variety. Out of these Plato chooses as best that which is obtained by bisecting an equilateral triangle; the reason for this choice becomes presently obvious.

10. τὰς δ' ἔτι τούτων ἀρχάς] Plato will not affirm that there is any physical ἀρχή which is absolutely ultimate.

13. αὐτῶν ἄττα] This anticipates the correction given in 54 B of the statement in 49 C.

15. τῶν τε ἀνὰ λόγον] i.e. the mean proportionals, air and water, between fire and earth; see 32 A.

192   ΠΛΑΤΩΝΟΣ   [54 A—

ἰσοσκελὲς μίαν εἴληχε φύσιν, τὸ δὲ πρόμηκες ἀπεράντους· προ-
αιρετέον οὖν αὖ τῶν ἀπείρων τὸ κάλλιστον, εἰ μέλλομεν ἄρξεσθαι
κατὰ τρόπον. ἂν οὖν τις ἔχῃ κάλλιον ἐκλεξάμενος εἰπεῖν εἰς τὴν
τούτων ξύστασιν, ἐκεῖνος οὐκ ἐχθρὸς ὢν ἀλλὰ φίλος κρατεῖ· τιθέ-
5 μεθα δ' οὖν τῶν πολλῶν τριγώνων κάλλιστον ἕν, ὑπερβάντες
τἆλλα, ἐξ οὗ τὸ ἰσόπλευρον τρίγωνον ἐκ τρίτου συνέστηκε. διότι B
δέ, λόγος πλείων· ἀλλὰ τῷ τοῦτο ἐξελέγξαντι καὶ ἀνευρόντι μὴ
οὕτως ἔχον κεῖται φίλια τὰ ἆθλα. προῃρήσθω δὴ δύο τρίγωνα,
ἐξ ὧν τό τε τοῦ πυρὸς καὶ τὰ τῶν ἄλλων σώματα μεμηχάνηται,
10 τὸ μὲν ἰσοσκελές, τὸ δὲ τριπλῆν κατὰ δύναμιν ἔχον τῆς ἐλάττονος
τὴν μείζω πλευρὰν ἀεί. τὸ δὴ πρόσθεν ἀσαφῶς ῥηθὲν νῦν μᾶλλον
διοριστέον. τὰ γὰρ τέτταρα γένη δι' ἀλλήλων εἰς ἄλληλα ἐφαίνετο πάντα γένεσιν ἔχειν, οὐκ ὀρθῶς φανταζόμενα· γίγνεται μὲν C
γὰρ ἐκ τῶν τριγώνων ὧν προῃρήμεθα γένη τέτταρα, τρία μὲν ἐξ
15 ἑνὸς τοῦ τὰς πλευρὰς ἀνίσους ἔχοντος, τὸ δὲ τέταρτον ἓν μόνον
ἐκ τοῦ ἰσοσκελοῦς τριγώνου ξυναρμοσθέν. οὔκουν δυνατὰ πάντα
εἰς ἄλληλα διαλυόμενα ἐκ πολλῶν σμικρῶν ὀλίγα μεγάλα καὶ
τοὐναντίον γίγνεσθαι, τὰ δὲ τρία οἷόν τε· ἐκ γὰρ ἑνὸς ἅπαντα

2 μέλλομεν: μέλλοιμεν A.
μή: δή A.   δὴ μή SZ.

7 λόγος: ὁ λόγος SZ.   δὲ ὁ erasit Λ.
8 φίλια: φιλία AHSZ.

1. τὸ δὲ πρόμηκες] i.e. the scalene. πρόμηκες denotes that one side exceeds the other in length: the word is applied to almost any shape which is longer than it is broad; in *Theaetetus* 148 A to a rectangle which is not a square; there and in *Republic* 546 C to a number expressing such a rectangle; to a long vault, *Laws* 947 D; to the elongated heads of beasts, *Timaeus* 91 E: στρογγύλα καὶ προμήκη = cylindrical, said of the spine, *Timaeus* 73 D.

6. ἐκ τρίτου συνέστηκε] i.e. the two triangles combined form a third, which is equilateral.

The extreme ἀήθεια of Plato's theory will be at once seen by a brief comparison with those of his predecessors. Empedokles limited the primal elements to four and conceived them as indefinitely divisible; and he treats as primary those which Plato says are οὐδ' ἐν συλλαβῆς εἴδεσιν. Anaxagoras reduces matter to qualitatively determinate corpuscles, infinitely numerous, infinitely various, and infinitely divisible. The atoms of Demokritos are infinite in number, indefinitely varying in size shape and weight, in other respects perfectly similar, and indivisible. Plato differs (1) in the derivation of his particles from his two primal triangles; (2) in limiting their varieties to four; (3) in assigning to these four certain specified geometrical forms; (4) in the peculiar conditions he imposes upon their divisibility; (5) in allowing two or more of the smaller particles to coalesce into one larger—this is directly contrary to the view of Demokritos; (6) in allowing within limits a diversity of size in the primal triangles, Plato seeks to explain differences of qualities which Demokritos ascribes to

c] ΤΙΜΑΙΟΣ. 193

but the scalene an endless number. Out of this infinite multitude then we must choose the fairest, if we are to begin upon our own principles. If then any man can tell of a fairer kind that he has selected for the composition of these bodies, it is no enemy but a friend who vanquishes us: however of all these triangles we declare one to be the fairest, passing over the rest; that namely of which two conjoined form an equilateral triangle. The reason it were too long to tell: but if any man convict us in this and find that it is not so, the palm is ready for him with our right good will. Let then two triangles be chosen whereof the substance of fire and of the other elements has been wrought; the one isosceles, the other always having the square on the greater side three times the square on the lesser. And now we must more strictly define something which we expressed not quite clearly enough before. For it appeared as though all the four classes had generation through each other and into each other, but this appearance was delusive. For out of the triangles we have chosen arise four kinds, three from one of them, that which has unequal sides, and the fourth one alone composed of the isosceles triangle. It is not then possible for all of them by dissolution to pass one into another, a few large bodies being formed of many small, and the converse: but for three of them it is possible.

varieties in the size and shape of the atoms; (7) whereas Demokritos insisted upon the necessity of void, Plato eliminates it so far as possible and makes no mechanical use of it; (8) though Plato agrees with Demokritos as to the sifting of like bodies into their proper region, he differs from him *toto caelo* on the subject of gravitation. There is moreover a still more fundamental peculiarity in the Platonic theory, which will be discussed later: see 56 D.

10. τριπλῆν κατὰ δύναμιν] i.e. having the square on the longer side three times the square on the shorter.

Let ABC be an equilateral triangle bisected by the perpendicular AD. Then the square on the hypotenuse $AC=(AD)^2+(DC)^2$. But $AC=2DC$, therefore $(AC)^2=4(DC)^2$; therefore

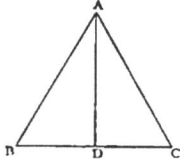

$(AD)^2=3(DC)^2$, or $AD:DC::\sqrt{3}:1$. cf. *Timaeus Locrus* 98 A.

11. τὸ δὴ πρόσθεν] Referring to the statement in 49 C that all the elements are interchangeable. Aristotle makes all four interchangeable: see for instance *meteorologica* I iii 339ᵃ 37 φαμὲν δὲ πῦρ καὶ ἀέρα καὶ ὕδωρ καὶ γῆν γίνεσθαι ἐξ ἀλλήλων, καὶ ἕκαστον ἐν ἑκάστῳ ὑπάρχειν τούτων δυνάμει.

P. T. 13

194   ΠΛΑΤΩΝΟΣ   [54 C—

πεφυκότα λυθέντων τε τῶν μειζόνων πολλὰ σμικρὰ ἐκ τῶν αὐτῶν ξυστήσεται, δεχόμενα τὰ προσήκοντα ἑαυτοῖς σχήματα, καὶ D
σμικρὰ ὅταν αὖ πολλὰ κατὰ τὰ τρίγωνα διασπαρῇ, γενόμενος εἷς
ἀριθμὸς ἑνὸς ὄγκου μέγα ἀποτελέσειεν ἂν ἄλλο εἶδος ἕν. ταῦτα
5 μὲν οὖν λελέχθω περὶ τῆς εἰς ἄλληλα γενέσεως· οἷον δὲ ἕκαστον
αὐτῶν γέγονεν εἶδος καὶ ἐξ ὅσων συμπεσόντων ἀριθμῶν, λέγειν
ἂν ἑπόμενον εἴη. ἄρξει δὴ τό τε πρῶτον εἶδος καὶ σμικρότατον
ξυνιστάμενον, στοιχεῖον δ' αὐτοῦ τὸ τὴν ὑποτείνουσαν τῆς ἐλάττονος πλευρᾶς διπλασίαν ἔχον μήκει· ξύνδυο δὲ τοιούτων κατὰ
10 διάμετρον ξυντιθεμένων καὶ τρὶς τούτου γενομένου, τὰς διαμέτρους E
καὶ τὰς βραχείας πλευρὰς εἰς ταὐτὸν ὡς κέντρον ἐρεισάντων, ἓν
ἰσόπλευρον τρίγωνον ἐξ ἓξ τὸν ἀριθμὸν ὄντων γέγονε· τρίγωνα δὲ
ἰσόπλευρα ξυνιστάμενα τέτταρα κατὰ σύντρεις ἐπιπέδους γωνίας
μίαν στερεὰν γωνίαν ποιεῖ, τῆς ἀμβλυτάτης τῶν ἐπιπέδων γωνιῶν 55 A
15 ἐφεξῆς γεγονυῖαν· τοιούτων δὲ ἀποτελεσθεισῶν τεττάρων πρῶτον
εἶδος στερεόν, ὅλον περιφεροῦς διανεμητικὸν εἰς ἴσα μέρη καὶ

3 σμικρά: οὐ σμικρά A.   κατὰ τὰ τρίγωνα: τὰ omittit A.   6 ὅσων: ὧν S.

8. **τὴν ὑποτείνουσαν**] The same triangle given above, having its sides in the proportion 1, √3, 2.

9. **ξύνδυο δέ**] Take two equal rectangular scalenes *AOF*, *AOE*, of the form aforesaid, and place them so that their hypotenuses coincide. Thus we

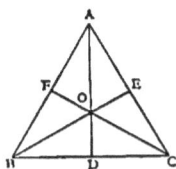

have a trapezium *AFOE*. In the same way form two other equal and similar trapeziums *BFOD*, *CEOD*, and place them so that in each of them the two sides which are the shortest sides of the triangles coincide severally with a similar side in each of the two others, *FO*, *EO*, *DO*. The juxtaposition of these three trapeziums gives us an equilateral triangle *ABC* formed of six rectangular scalenes similar in all respects to the triangle obtained by bisecting *ABC*. For let *ABC* be an equilateral triangle, and draw the three perpendiculars *AD*, *BE*, *CF*, each bisecting it. Then it is easy to prove that the three perpendiculars intersect in the point *O*: and since in the triangle *AOF* the angle *AFO* is a right angle and the angle *FAO* is ⅓ of a right angle, therefore the angle *AOF* must be ⅔ of a right angle; and the triangle *AOF* is consequently similar to *ADB*, as also are the other five. Accordingly the juxtaposition of six rectangular scalenes of the form and in the manner described will make up a single equilateral triangle.

κατὰ διάμετρον] That is, placed so that the hypotenuse of one coincides with that of the other: the common hypotenuse *AO* of the two triangles *AOF*, *AOE* becomes the diagonal of the trapezium *AFOE*.

11. **εἰς ταὐτὸν ὡς κέντρον**] i.e. at the point *O*.

12. **ἐξ ἓξ τὸν ἀριθμόν**] It is notable that Plato uses six of the primary scalenes to compose his equilateral triangle, when

For since they all arise from one basis, when the larger bodies are broken up, a number of small ones will be formed from the same elements, putting on the shapes proper to them; and again when a number of small bodies are resolved into their triangles, they will become one in number and constitute a single large body of a different form. So much for their generation into one another: the next thing will be to say what is the form in which each has been created, and by the combination of what numbers. We will begin with the form which is simplest and smallest in its construction. Its element is the triangle which has the hypotenuse double of the shorter side in length. If a pair of these are put together so that their hypotenuses coincide, and this is done three times, in such a way that the hypotenuses and the shorter sides meet in one point as a centre, thus one equilateral triangle has been formed out of the other six triangles: and if four equilateral triangles are combined, so that three plane angles meet in a point, they make at each point one solid angle, that which comes immediately next to the most obtuse of plane angles; and when four such angles are produced there is formed the first solid figure, dividing its whole surface into four equal and similar

he could have done it equally well with two. Similarly he uses four rectangular isosceles to compose the square, whereas he could have formed it of two. The reason is probably this: the sides of the primary triangles mark the lines along which the equilaterals are broken up in case of dissolution. Now had Plato formed his equilaterals of two scalenes only, it would have been left in doubt whether the triangle ABC would be broken up along the line AD, or along BE, or CF. But if they are composed of six, the lines along which dissolution takes place is positively determined; since there is only one way in which six can be joined so as to form one equilateral. The same remark applies to the composition of the square. Also by taking one-sixth of the equilateral, instead of one-half, we get the smallest element possible for our primal base.

12. τρίγωνα δὲ ἰσόπλευρα] Next we take four equilateral triangles thus constructed each of six elementary scalenes, and place them so as to make a regular tetrahedron or pyramid; each of whose solid angles is bounded by three planes meeting in a point. The pyramid is the simplest of the regular solids, having four equilateral triangles for its sides, and therefore containing 24 of the primal scalenes. This is the corpuscule composing fire.

14. τῆς ἀμβλυτάτης] The most obtuse plane angle (expressed in integral numbers) is 179 degrees, one degree short of two right angles, or a straight line. The solid angle of a pyramid is, as we have seen, bounded by three equilateral triangles. The angle of an equilateral triangle is two-thirds of a right angle, that is, 60 degrees. Therefore the angle of the pyramid contains 180 degrees, or

ὅμοια, ξυνίσταται. δεύτερον δὲ ἐκ μὲν τῶν αὐτῶν τριγώνων, κατὰ δὲ ἰσόπλευρα τρίγωνα ὀκτὼ ξυστάντων, μίαν ἀπεργασαμένων στερεὰν γωνίαν ἐκ τεττάρων ἐπιπέδων· καὶ γενομένων ἐξ τοιούτων τὸ δεύτερον αὖ σῶμα οὕτως ἔσχε τέλος. τὸ δὲ τρίτον ἐκ δὶς
5 ἑξήκοντα τῶν στοιχείων ξυμπαγέντων, στερεῶν δὲ γωνιῶν δώδεκα, ὑπὸ πέντε ἐπιπέδων τριγώνων ἰσοπλεύρων περιεχομένης ἑκάστης, εἴκοσι βάσεις ἔχον ἰσοπλεύρους τριγώνους γέγονε. καὶ τὸ μὲν ἕτερον ἀπήλλακτο τῶν στοιχείων ταῦτα γεννῆσαν· τὸ δὲ ἰσοσκελὲς τρίγωνον ἐγέννα τὴν τοῦ τετάρτου φύσιν, κατὰ τέτταρα
10 ξυνιστάμενον, εἰς τὸ κέντρον τὰς ὀρθὰς γωνίας ξυνάγον, ἓν ἰσόπλευρον τετράγωνον ἀπεργασάμενον· ἓξ δὲ τοιαῦτα ξυμπαγέντα γωνίας ὀκτὼ στερεὰς ἀπετέλεσε, κατὰ τρεῖς ἐπιπέδους ὀρθὰς ξυναρμοσθείσης ἑκάστης· τὸ δὲ σχῆμα τοῦ ξυστάντος σώματος γέγονε κυβικόν, ἐξ ἐπιπέδους τετραγώνους ἰσοπλεύρους βάσεις ἔχον. ἔτι
15 δὲ οὔσης ξυστάσεως μιᾶς πέμπτης, ἐπὶ τὸ πᾶν ὁ θεὸς αὐτῇ κατεχρήσατο ἐκεῖνο διαζωγραφῶν.

8 ταῦτα γεννῆσαν : γεννῆσαν ταῦτα S.

one degree more than the obtusest possible of plane angles.

2. **ἰσόπλευρα τρίγωνα ὀκτώ**] The next figure is the octahedron, the second regular solid, having eight equilateral triangular sides, and six angles, each of them bounded by four planes: this then contains 48 of the primal scalenes. This is the constituent corpuscule of air.

4. **τὸ δὲ τρίτον**] The third regular solid is the icosahedron, which has twenty sides, of the same shape as the former, and twelve angles, each bounded by five of the equilateral planes; this consequently contains no less than 120 primal scalenes. This forms the element of water. And now the rectangular scalene, out of which the equilateral is formed, has finished its work: since these three are the only regular solids whose sides are equilateral triangles.

9. **κατὰ τέτταρα ξυνισταμένον**] The corpuscule of which earth is formed is based upon the other element, the rectangular isosceles: four of which, joined in the manner shewn in the accompanying figure, make a square. Six of these squares set together form the fourth regu-

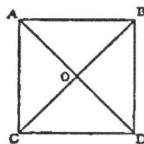

lar solid, which is the cube, having eight solid angles each bounded by three planes: the cube then contains 24 of the elementary isosceles. The reason why Plato forms his square of four instead of two triangles has been already suggested: it is obvious however that he might have constructed it of any number he chose: for by bisecting the triangle AOB we should obtain two precisely similar triangles, which again might be bisected into precisely similar triangles *usque ad infinitum*. Plato however had to stop short somewhere in the number of triangles which he assigned to the square; and naturally enough he stopped short at the smallest number which gave him

ΤΙΜΑΙΟΣ.

parts. The second is formed of the same triangles in sets of eight equilateral triangles, bounding every single solid angle by four planes; and with the formation of six such solid angles the second figure is also complete. The third is composed of 120 of the elementary triangles united, and of twelve solid angles, each contained by five plane equilateral triangles; and it has twenty equilateral surfaces. And the first element, when it had generated these figures, had done its part: the isosceles triangle generated the fourth, combined in sets of four, with the right angles meeting at the centre, thus forming a single square. Six of these squares joined together formed eight solid angles, each produced by three plane right angles: and the shape of the body thus formed was cubical, having six square planes for its surfaces. And whereas a fifth figure yet alone remained, God used it for the universe in embellishing it with signs.

determinate lines of cleavage.

14. ἔτι δὲ οὔσης ξυστάσεως μιᾶς πέμπτης] There is in existence yet a fifth regular solid, the dodecahedron. This has twelve sides, each of which is an equilateral pentagon; it has twenty solid angles each contained by three planes. This is of course not based upon either of the elementary triangles; nor is it the corpuscule of any material substance. God, says Plato, used it for a pattern in diversifying the universe with signs: that is it served as a model for the twelvefold division of the zodiac. The writer of the *Timaeus Locrus* (see 98 E τὸ δὲ δωδεκάεδρον εἰκόνα τοῦ παντὸς ἐστάσατο, ἔγγιστα σφαίρας ἐόν) is quite in error in supposing that the shape of the dodecahedron has anything to do with that of the universe: the spherical shape of the latter is the material symbol of the αὐτὸ ζῷον. Plato was bound to find some significance for the only remaining regular solid; and he found it as suggesting the twelve signs of the heavens. Compare *Phaedo* 110 B πρῶτον μὲν εἶναι τοιαύτη ἡ γῆ αὐτὴ ἰδεῖν, εἴ τις ἄνωθεν θεῷτο, ὥσπερ αἱ δωδεκάσκυτοι σφαῖραι, where obviously the 'twelve-patched

ball' represents the duodenary division. There is a curious blunder in Plutarch *quaestiones platonicae* V i: συνήρμοσται δὲ καὶ συμπέπηγεν ἐκ δώδεκα πενταγώνων ἰσογωνίων καὶ ἰσοπλεύρων, ὧν ἕκαστον ἐκ τριάκοντα τῶν πρώτων σκαληνῶν τριγώνων συνέστηκε· διὸ καὶ δοκεῖ τὸν ζῳδιακὸν ἅμα καὶ τὸν ἐνιαυτὸν ἀπομιμεῖσθαι ταῖς διανομαῖς τῶν μοιρῶν ἰσαρίθμοις οὖσιν. Alkinoos has a similar statement: this would involve the consequence that every side of the dodecahedron can be divided into five equilateral triangles, each consisting of six primal scalenes; an opinion which Stallbaum welcomes with joy, saying that it 'mirifice convenit' with the 360 degrees into which the circle is divided. It is perhaps strange that neither Stallbaum Plutarch nor Alkinoos took the trouble even to draw a regular pentagon in order to verify this theory, which is of course geometrically absurd: Martin goes so far as to give, not without sarcasm, a mathematical demonstration of its impossibility.

55 C—56 C, c. xxi. Now if the question be put, are there more cosmical systems than one? the reply that there are an indefinite number would be a very in-

XXI. Ἃ δή τις εἰ πάντα λογιζόμενος ἐμμελῶς ἀποροῖ, πότερον ἀπείρους χρὴ κόσμους εἶναι λέγειν ἢ πέρας ἔχοντας, τὸ μὲν
ἀπείρους ἡγήσαιτ᾽ ἂν ὄντως ἀπείρου τινὸς εἶναι δόγμα ὧν ἔμπει- D
ρον χρεὼν εἶναι· πότερον δὲ ἕνα ἢ πέντε αὐτοὺς ἀληθείᾳ πεφυ-
5 κότας λέγειν προσήκει, μᾶλλον ἂν ταύτῃ στὰς εἰκότως διαπορήσαι.
τὸ μὲν οὖν δὴ παρ᾽ ἡμῶν ἕνα αὐτὸν κατὰ τὸν εἰκότα λόγον πεφυκότα μηνύει, ἄλλος δὲ εἰς ἄλλα πῃ βλέψας ἕτερα δοξάσει. καὶ
τοῦτον μὲν μεθετέον, τὰ δὲ γεγονότα νῦν τῷ λόγῳ γένη διανεί-
μωμεν εἰς πῦρ καὶ γῆν καὶ ὕδωρ καὶ ἀέρα. γῇ μὲν δὴ τὸ κυβικὸν
10 εἶδος δῶμεν· ἀκινητοτάτη γὰρ τῶν τεττάρων γενῶν γῆ καὶ τῶν E
σωμάτων πλαστικωτάτη, μάλιστα δὲ ἀνάγκη γεγονέναι τοιοῦτον
τὸ τὰς βάσεις ἀσφαλεστάτας ἔχον· βάσις δὲ ἥ τε τῶν κατ᾽ ἀρχὰς
τριγώνων ὑποτεθέντων ἀσφαλεστέρα κατὰ φύσιν, ἡ τῶν ἴσων

5 ποτέ post λέγειν dat A, quod inclusum retinet H. cum SZ eieci. στάς :
πᾶς S.   7 θεὸς post μηνύει addit A. uncis inclusum servat H.   8 τοῦτον :
τούτων SZ.

definite answer: but to affirm that there are five might be more reasonable. We however in conformity with our principles assert that there is but one. We must now assign our elementary solids to the natural substances which they severally compose. Earth is the most unyielding of the four; therefore to it we assign the cube as its constituent; for this is the most stable solid, being formed of the rectangular isosceles. To water, which next to earth is the most sluggish, we give the icosahedron; and to fire, which is of all the most mobile, the pyramid; while for air there remains the intermediate form of the octahedron. Now all these corpuscles are separately so small as to be invisible; it is only when they are collected in large numbers that they can be seen by us: but God assigned them to the four substances with due regard to proportion in respect of multitude and motion and all other powers.

3. ἀπείρους...ἀπείρου] For the play on the word compare *Philebus* 17 E τὸ δὲ ἄπειρόν σε ἑκάστων καὶ ἐν ἑκάστοις πλῆθος ἄπειρον ἑκάστοτε ποιεῖ τοῦ φρονεῖν καὶ οὐκ ἐλλόγιμον οὐδ᾽ ἐνάριθμον, ἅτ᾽ οὐκ εἰς ἀριθμὸν οὐδένα ἐν οὐδενὶ πώποτε ἀπιδόντα. Plato is at issue with Demokritos, who consistently with his whole physical theory maintained that the number of κόσμοι was infinite: Plato is equally consistent in affirming that there is only one. The oddest fancy in this way is one ascribed by Plutarch *de defectu oraculorum* § 22 to Petron of Himera, who declared there were 183 κόσμοι, disposed in the form of an equilateral triangle. The eternal fitness of this arrangement is not explained by Plutarch.

4. πότερον δὲ ἕνα ἢ πέντε] Plato regards as a comparatively reasonable supposition the view that there may be five κόσμοι, because there exist in nature five regular rectilinear solids. Compare Plutarch *de ei apud Delphos* § 11 πολλὰ δ᾽ ἄλλα τοιαῦτα, ἔφην ἐγώ, παρελθών, τὸν Πλάτωνα προσάξομαι λέγοντα κόσμον ἕνα, ὡς εἴπερ εἰσὶ παρὰ τοῦτον ἕτεροι καὶ μὴ μόνος οὗτος εἷς, πέντε τοὺς πάντας ὄντας καὶ μὴ πλείονας. οὐ μὴν ἀλλὰ κἂν εἷς οὗτος ᾖ μονογενής, ὡς οἴεται καὶ Ἀριστοτέλης, τρόπον τινὰ καὶ τοῦτον ἐκ πέντε συγκειμένων κόσμων καὶ συνηρμοσμένων

ΤΙΜΑΙΟΣ.   199

XXI. Now if any man, reflecting upon all these things, should fairly ask himself whether the number of cosmic systems is indefinite or definite, he would deem that to believe them indefinite was the opinion of one who thought very indefinitely on a matter where he ought to be most definitely informed: but whether we ought to say that there is but one, or that there are really in nature five, he might, if he stopped short there, with more justice feel doubtful. Our verdict declares that according to the probable theory it is by nature one; another however, looking to some other guide, may have a different view. But no more of him; let us assign the figures that have come into being in our theory to fire and earth and water and air. To earth let us give the cubical form; for earth is least mobile of the four and most plastic of bodies: and that substance must possess this nature in the highest degree which has its bases most stable. Now of the triangles which we assumed as our starting-point that with equal sides is more stable than that

εἶναι· ὧν ὁ μέν ἐστι γῆς, ὁ δ' ὕδατος, τρίτος δὲ πυρὸς, καὶ τέταρτος ἀέρος· τὸν δὲ πέμπτον, οὐρανόν, οἱ δὲ φῶς, οἱ δ' αἰθέρα καλοῦσιν, οἱ δ' αὐτὸ τοῦτο, πέμπτην οὐσίαν, ᾗ τὸ κύκλῳ περιφέρεσθαι μόνῃ τῶν σωμάτων κατὰ φύσιν ἐστίν, οὐκ ἐξ ἀνάγκης οὐδ' ἄλλως συμβεβηκός. The latter part of this extract does not accurately represent Plato's opinion, since the dodecahedron was not a constituent of any substance existing in nature, but simply the model for the distribution of the zodiac into twelve signs.

5. ταύτῃ στάς] This is evidently the true reading. If the inquirer were to stop short at the number five and declare that so many κόσμοι existed, he would be more reasonable, says Plato, than he who should go on to a larger or indefinite number. Stallbaum's πᾶς, which has but slight support, is quite inappropriate: Plato could not say that it was reasonable for every one to doubt whether there are five κόσμοι or one; it would not be reasonable in his own case, as we see in 31 D.

6. ἕνα αὐτὸν κατὰ τὸν εἰκότα λόγον]

It will be noted that here, where he is dealing with physics and the region of opinion, Plato only pronounces the unity of the universe to be probable and consonant to his theory of nature. But at 31 B it is authoritatively declared to be one on the infallible principle of metaphysical necessity. After μηνύει, θεὸς cannot possibly be genuine.

7. ἄλλος δὲ εἰς ἄλλα πῃ] Obviously aimed at Demokritos: a philosopher who has no place for νοῦς in his system may very well maintain an infinity of κόσμοι.

8. τοῦτον] i.e. Demokritos, who is dismissed with something more like contempt than Plato is wont to show for other thinkers.

τὰ δὲ γεγονότα νῦν τῷ λόγῳ] Compare 27 A ἀνθρώπους τῷ λόγῳ γεγονότας.

11. πλαστικωτάτη] The other three are too subtle to be plastic. Aristotle's objections to the present theory will be found in de caelo III viii 306^b 3: they are not for the most part very forcible. The most pertinent is that of Plato's geometrical figures only the pyramid and the cube can fill up space continuously: the

200   ΠΛΑΤΩΝΟΣ   [55 E—

πλευρῶν, τῆς τῶν ἀνίσων, τό τε ἐξ ἑκατέρου ξυντεθὲν ἐπίπεδον
ἰσόπλευρον ἰσοπλεύρου τετράγωνον τριγώνου κατά τε μέρη καὶ
καθ' ὅλον στασιμωτέρως ἐξ ἀνάγκης βέβηκε. διὸ γῇ μὲν τοῦτο
ἀπονέμοντες τὸν εἰκότα λόγον διασῴζομεν, ὕδατι δ' αὖ τῶν λοιπῶν 56 A
5 τὸ δυσκινητότατον εἶδος, τὸ δ' εὐκινητότατον πυρί, τὸ δὲ μέσον
ἀέρι· καὶ τὸ μὲν σμικρότατον σῶμα πυρί, τὸ δ' αὖ μέγιστον ὕδατι,
τὸ δὲ μέσον ἀέρι· καὶ τὸ μὲν ὀξύτατον αὖ πυρί, τὸ δὲ δεύτερον
ἀέρι, τὸ δὲ τρίτον ὕδατι. ταῦτ' οὖν δὴ πάντα, τὸ μὲν ἔχον ὀλιγίστας βάσεις εὐκινητότατον ἀνάγκη πεφυκέναι, τμητικώτατόν τε
10 καὶ ὀξύτατον ὂν πάντῃ πάντων, ἔτι τε ἐλαφρότατον, ἐξ ὀλιγίστων B
ξυνεστὸς τῶν αὐτῶν μερῶν· τὸ δὲ δεύτερον δευτέρως τὰ αὐτὰ ταῦτ'
ἔχειν, τρίτως δὲ τὸ τρίτον. ἔστω δὴ κατὰ τὸν ὀρθὸν λόγον καὶ
κατὰ τὸν εἰκότα τὸ μὲν τῆς πυραμίδος στερεὸν γεγονὸς εἶδος
πυρὸς στοιχεῖον καὶ σπέρμα· τὸ δὲ δεύτερον κατὰ γένεσιν εἴπω-
15 μεν ἀέρος, τὸ δὲ τρίτον ὕδατος. πάντα οὖν δὴ ταῦτα δεῖ διανοεῖσθαι σμικρὰ οὕτως, ὡς καθ' ἓν ἕκαστον μὲν τοῦ γένους ἑκάστου
διὰ σμικρότητα οὐδὲν ὁρώμενον ὑφ' ἡμῶν, ξυναθροισθέντων δὲ C
πολλῶν τοὺς ὄγκους αὐτῶν ὁρᾶσθαι· καὶ δὴ καὶ τὸ τῶν ἀναλογιῶν περί τε τὰ πλήθη καὶ τὰς κινήσεις καὶ τὰς ἄλλας δυνάμεις
20 πανταχῇ τὸν θεόν, ὅπῃπερ ἡ τῆς ἀνάγκης ἑκοῦσα πεισθεῖσά τε

8 ὀλιγίστας : ὀλίγας τὰς A. ὀλιγοστὰς S.   10 καὶ ante ὀξύτατον omittit S.
14 εἴπωμεν : εἴπομεν A.

bearing of this will be discussed a little later; see note on 58 A.

2. κατά τε μέρη καὶ καθ' ὅλον] i.e. as the rectangular isosceles is more stable, owing to the equality of its sides, than the rectangular scalene, so the solid based on the former is more steady than that based on the latter.

6. τὸ μὲν σμικρότατον] No comparison in point of size is made with the corpuscules of earth, because the latter has a different base: but in the case of the other three the size of the figure varies according to the number of the radical triangles contained in it.

8. ὀλιγίστας βάσεις] Stallbaum seems perverse in reading ὀλιγοστάς. For even if ὀλιγοστὰς could mean 'very small' (which is quite dubious: see Campbell on Sophokles *Antigone* 625), this is not the right meaning; the sense requires 'very few': for the mobile and penetrating nature of fire is due to the small number of its sides and the consequent acuteness of its angles. Plato evidently considers that the sharp points of the pyramid most readily cleave their way through other bodies; and so Aristotle understood him to mean, *de caelo* III viii 307ª 2. It is curious to observe how the meaning of πολλοστὸς and of ὀλιγοστὸς sometimes seems to be inverted: compare the passage of the *Antigone* aforesaid, πράσσει δ' ὀλιγοστὸν χρόνον ἐκτὸς ἄτας (v. l. ὀλίγιστον) with Demosthenes κατὰ Τιμοκράτους § 196 τὸ τὰ τούτων πολλοστῷ χρόνῳ μόλις καὶ ἄκοντας...κατατιθέναι. In the first case the meaning will be 'he is free from woe for a time which is one of a few (sc. of a few times when he is free)'; i.e. he is

56 c] ΤΙΜΑΙΟΣ. 201

with unequal ; and of the surfaces composed of the two triangles the equilateral quadrangle necessarily is more stable than the equilateral triangle, both in its parts and as a whole. Therefore in assigning this to earth we preserve the probability of our account; and also in giving to water the least mobile and to fire the most mobile of those which remain ; while to air we give that which is intermediate. Again we shall assign the smallest figure to fire, and the largest to water and the intermediate to air: and the keenest to fire, the next to air, and the third to water. Now among all these that which has the fewest bases must naturally in all respects be the most cutting and keen of all, and also the most nimble, seeing it is composed of the smallest number of similar parts; and the second must have these same qualities in the second degree, and the third in the third degree. Let it be determined then, according to the right account and the probable, that the solid body which has taken the form of the pyramid is the element and seed of fire ; and the second in order of generation let us say to be that of air, and the third that of water. Now all these bodies we must conceive as being so small that each single body in the several kinds cannot for its smallness be seen by us at all; but when many are heaped together, their united mass is seen: and we must suppose that the due proportion in respect of their multitude and motions and all their other powers, when God had completed them with all perfection, in so far as the nature of neces-

seldom free; the second 'they paid at a moment which is one of many moments (sc. in which they had not paid)', i.e. after a long interval. But neither of these constructions countenances ὀλιγοστὰς here. In assigning the pyramidal form to fire Plato differs from Demokritos, who attributed the mobility of fire to the roundness of its atoms: cf. Aristotle *de caelo* 307ᵃ 16.

10. **ἐλαφρότατον**] Not light, but nimble, mobile.

13. **στερεὸν γεγονός**] For the bearing of this see note on 56 D. κατὰ γένεσιν, i.e. in order of generation, having the next fewest sides.

16. **σμικρὰ οὕτως**] Here Plato is in agreement with Demokritos, in making his atoms so small as to be individually invisible, and only perceptible in masses.

18. **τὸ τῶν ἀναλογιῶν**] That is to say, observing the proportional relations propounded in 32 A, B.

20. **πεισθεῖσα**] cf. 48 A. ξυνηρμόσθαι is sometimes regarded as an anacoluthon; but there can be hardly a doubt that it is a middle. The middle of this word is used twice elsewhere by Plato, each time in the aorist: see above 53 E σωμάτων τέτταρα γένη συναρμόσασθαι, and *Politicus* 309 C θείῳ ξυναρμοσαμένη δεσμῷ.

56 C—57 D, *c.* xxii. When earth then is resolved by fire, it drifts about until it can reunite with earthy elements, and so

φύσις ὑπεῖκεν, ταύτῃ πάντῃ δι' ἀκριβείας ἀποτελεσθεισῶν ὑπ'
αὐτοῦ, ξυνηρμόσθαι ταῦτ' ἀνὰ λόγον.

XXII. Ἐκ δὴ πάντων ὧν περὶ τὰ γένη προειρήκαμεν ὧδ' ἂν
κατὰ τὸ εἰκὸς μάλιστ' ἂν ἔχοι. γῆ μὲν ξυντυγχάνουσα πυρὶ δια- D
5 λυθεῖσά τε ὑπὸ τῆς ὀξύτητος αὐτοῦ φέροιτ' ἄν, εἴτ' ἐν αὐτῷ πυρὶ
λυθεῖσα εἴτ' ἐν ἀέρος εἴτ' ἐν ὕδατος ὄγκῳ τύχοι, μέχριπερ ἂν αὐτῆς
πῃ ξυντυχόντα τὰ μέρη, πάλιν ξυναρμοσθέντα αὐτὰ αὑτοῖς, γῆ
γένοιτο· οὐ γὰρ εἰς ἄλλο γε εἶδος ἔλθοι ποτ' ἄν. ὕδωρ δὲ ὑπὸ
πυρὸς μερισθέν, εἴτε καὶ ὑπ' ἀέρος, ἐγχωρεῖ γίγνεσθαι ξυστάντα

1 ὑπεῖκεν: ὑπεῖκε HSZ.   4 ἂν post μάλιστ' omittit S.
6 μέχριπερ: ὃ μέχριπερ A.

resume the form of earth; for, owing to the dissimilarity of base, it cannot be changed to any of the other three. But when water is resolved by fire or air, it can be reformed in the shape of fire and air. So when air is resolved, one of its particles make two of fire, or two particles and a half form one of water. Of fire also two particles may coalesce into one of air. And, in general, when a smaller mass of any of the three is overcome by a larger mass of any other and resolved, its resolution ceases the moment it assumes the form of the victorious element, but not until then. So the vanquished element must either escape away and seek its own region in space, or else accept the form of the other. It follows then that, owing to this incessant conflict between the elements, perpetual changes of form are taking place, and perpetual changes of position in space.

All this has been said in view of the primary and typical kinds in the four forms, fire, air, water, earth: but a variety of kinds are found within the limits of each form. These are due to a variation of size in the primal triangles, of which there are so many sizes as there are kinds in each form. Such kinds by manifold intermixture produce an endless number of varieties in phenomena, which it is our business to investigate.

5. **φέροιτ' ἄν**] Earth has not the alternative, which is open to the other three, of coalescing with the dominant element: it must therefore drift about in a chaotic condition, until it can escape into its own place and so regain its proper form.

6. **εἴτ' ἐν ἀέρος**] The form of this sentence suggests that the dissolution takes place by the agency of fire within a mass of air or of water. But clearly the same result follows whether the agent be fire air or water.

9. **ξυστάντα**] Ast and Stallbaum would read ξυστάν. But ξυστάντα agrees, by an easy attraction, with ἐν μὲν δύο δὲ following. It might be considered however that, since the single particle of water is resolved into two of air and one of fire, διαλυθέντα would be more correct than ξυστάντα. Plato's word however is perfectly accurate, if his theory be rightly understood. And this leads to a discussion of the chief peculiarity and difficulty of that theory.

First then Aristotle *de caelo* III i 299ᵃ 1 brings against it the fundamental objection that it is impossible to form solid matter out of mathematical planes. Now it is entirely preposterous to suppose that the most accomplished mathematician of his time was not fully alive to a truth which, as Aristotle himself admits, ἐπιπολῆς ἐστιν ἰδεῖν. The theory of an oversight in this respect must therefore be

D] ΤΙΜΑΙΟΣ. 203

sity, consenting and yielding to persuasion, suffered, were everywhere by him ordained in fitting measure.

XXII. From all that we have already said in the matter of these four kinds, the facts would seem to be as follows. When earth meets with fire and is dissolved by the keenness of it, it would drift about, whether it were dissolved in fire itself, or in some mass of air or water, until the parts of it meeting and again being united became earth once more; for it never could pass into any other kind. But when water is divided by fire or by air, it may be formed again and become one particle of fire and

dismissed out of hand. Howbeit, if we regard these geometrical figures as solid bodies which interchange their forms, they will not produce the combinations required. For instance, the apposition of two pyramids will not produce an octahedron, as it ought according to Plato, but an irregular six-sided figure: and by dividing the octahedron we obtain not a regular tetrahedron, but a five-sided figure having four equilateral triangles meeting in the apex, and a square for the base. Similarly the icosahedron refuses to play its prescribed part. Again it is incredible that Plato was unaware or oblivious of these elementary facts.

Martin has a theory so neat and ingenious that, although I do not see my way to accepting it, yet it ought not to be left unnoticed. His view is that Plato's ἐπίπεδα are not mathematical planes at all, but thin laminae of matter, 'feuilles minces taillées suivant les figures rectilignes qu'il a décrites.' Thus our four geometrical figures are not solid bodies, but merely envelopes or shells, void within. In this way no doubt Plato's transformations would be perfectly practicable. Supposing that an octahedron were shattered κατὰ τὰ τρίγωνα, then its eight triangular sides would be recomposed in the form of two pyramids; and all the other transmutations would be equally feasible. This explanation, despite its ingenuity, is nevertheless not to my mind satisfactory. For Plato eli-

minates void as far as possible from his material system; and though we shall presently see that it cannot be entirely banished, it is reduced to an absolute minimum. It is hardly credible then that he should have admitted an admixture of void into the very foundation of his structure of matter. Again, if he had intended to propound so very novel and extraordinary a theory as the construction of matter out of hollow particles, surely he must have stated it with a little more definiteness. Moreover on this hypothesis Plato sadly misuses technical terms: he denominates planes what are really solid bodies, though very thin; and he terms solid what is really but a hollow shell: for the phrase in 56 B is quite definite as to this point, τὸ μὲν τῆς πυραμίδος στερεὸν γεγονὸς εἶδος. Finally how could hollow particles escape being crushed by the tremendous constricting force described in 58 A? In the face of all these objections, the force of which is in part admitted by Martin himself, it seems difficult to accept this explanation.

The following is the solution which I should propound as less open to exception. We must bear in mind that matter in its ultimate analysis is just space. We must not look upon the geometrical solids as so much stuff which is put up into parcels, now of one shape, now of another; but as the expression of the geometrical law which rules the con-

ἐν μὲν πυρὸς σῶμα, δύο δὲ ἀέρος· τὰ δὲ ἀέρος τμήματα ἐξ ἑνὸς
μέρους διαλυθέντος δύ᾽ ἂν γενοίσθην σώματα πυρός. καὶ πάλιν, E
ὅταν ἀέρι πῦρ ὕδασίν τε ἤ τινι γῇ περιλαμβανόμενον ἐν πολλοῖς
ὀλίγον, κινούμενον ἐν φερομένοις, μαχόμενον καὶ νικηθὲν κατα-
5 θραυσθῇ, δύο πυρὸς σώματα εἰς ἓν ξυνίστασθον εἶδος ἀέρος· καὶ
κρατηθέντος ἀέρος κερματισθέντος τε ἐκ δυοῖν ὅλοιν καὶ ἡμίσεος
ὕδατος εἶδος ἓν ὅλον ἔσται ξυμπαγές. ὧδε γὰρ δὴ λογισώμεθα
αὐτὰ πάλιν, ὡς ὅταν ἐν πυρὶ λαμβανόμενον τῶν ἄλλων ὑπ᾽ αὐτοῦ 57 A
τι γένος τῇ τῶν γωνιῶν καὶ κατὰ τὰς πλευρὰς ὀξύτητι τέμνηται,
10 ξυστὰν μὲν εἰς τὴν ἐκείνου φύσιν πέπαυται τεμνόμενον· τὸ γὰρ
ὅμοιον καὶ ταὐτὸν αὑτῷ γένος ἕκαστον οὔτε τινὰ μεταβολὴν ἐμ-
ποιῆσαι δυνατὸν οὔτε τι παθεῖν ὑπὸ τοῦ κατὰ ταὐτὰ ὁμοίως τε
ἔχοντος· ἕως δ᾽ ἂν εἰς ἄλλο τι γιγνόμενον ἧττον ὂν κρείττονι
μάχηται, λυόμενον οὐ παύεται. τά τε αὖ σμικρότερα ὅταν ἐν

3 ὕδασίν: ὕδασί HSZ.   6 δυοῖν: δυεῖν S.

stitution of matter: they are definite forms under which space by the law of nature appears in various circumstances. The planes are real planes; but they do not compose the solid; they merely express the law of its formation. Given certain conditions, the geometrical law obtains that matter shall receive form as pyramids: alter the conditions, e.g. increase the pressure, and the pyramids disappear, their place being taken by octahedrons; and so forth. It is not then that two of the former particles have combined to make one of the latter, but that the matter in its new condition assumes a shape in which the radical form, the rectangular scalene, appears twice as many times as in the former. Increase the pressure again, and the triangle will appear five times as often as in the first. And if the triangles are equal, the second and third contain twice and five times as much stuff as the first. In short, when matter which has been existing in the pyramidal form is prevented from doing so any longer, it must not assume any random figure, but one which is constructed on either twice or five times as many primal triangles as the pyramid. The ἐπίπεδα then are, I believe, neither to be regarded with Aristotle as planes out of which we are expected to construct solids, nor with Martin as thin solids; but as the law of the structure of matter. Thus, instead of having two or more corpuscles combined into one, or one resolved into several, we have the whole mass fused, as it were, and remoulded. This interchange however can only take place where the law of formation is one and the same. Earth, obeying a different formative law, cannot go beyond one sole form. For matter which has once been impressed with either of the primal figures can never pass into the other figure: in the rudimentary condition to which it is reduced by the fracture of its particles, the force which forms it as a pyramid or a cube is in abeyance, but not the law which impressed it with the rectangular scalene or the rectangular isosceles.

On this showing then the correctness of ξυστάντα is clear: though I admit it is equally justified by Martin's hypothesis, could the objections which I have urged against the latter be overcome.

1. **ἐν μὲν πυρός**] The sides of the

two of air: and the divisions of air may become for every particle broken up two particles of fire. And again when fire is caught in air or in waters or in earth, a little in a great bulk, moving amid a rushing body, and contending with it is vanquished and broken up, two particles of fire combine into one figure of air: and when air is vanquished and broken small, from two whole and one half particle one whole figure of water will be composed. Let us also reckon it once again thus: when any of the other kinds is intercepted in fire and is divided by it through the sharpness of its angles and its sides, if it forms into the shape of fire, it at once ceases from being divided: for a kind which is uniform and identical, of whatever sort it be, can neither be the cause of any change nor can it suffer any from that which is identical and uniform with itself; but so long as passing into another kind a lesser bulk contends with the greater, it ceases never from being broken. And when the

icosahedron, being 20 in number, are equal to the sum of the sides of two octahedrons and one pyramid.

2. **καὶ πάλιν**] Having given instances of smaller corpuscules arising from the resolution of larger, Plato now passes to the formation of larger particles from the resolution of smaller.

4. **καταθραυσθῇ**] This is the converse of ξυστάντα above: the pyramids, being the smallest particle, could not literally be 'broken up' into the larger bodies. The same applies to κατακερματισθέντος ἀέρος below.

7. **ὧδε γὰρ δὴ λογισώμεθα**] Having set forth the rules governing the transition of one kind of particle into another, Plato proceeds to point out that, when one element is overpowered by another, the only mode in which it can recover any form, in default of escape to its own region, is to assimilate itself to the victorious body.

9. **κατὰ τὰς πλευράς**] i.e. cleft by the sharp edges of the sides.

10. **τὸ γὰρ ὅμοιον**] This view was universally held, with the sole exception of Demokritos: cf. Aristotle *de gen. et corr.* I vii 323^b 3 οἱ μὲν γὰρ πλεῖστοι τοῦτό γε ὁμονοητικῶς λέγουσιν, ὡς τὸ μὲν ὅμοιον ὑπὸ τοῦ ὁμοίου πᾶν ἀπαθές ἐστι διὰ τὸ μηδὲν μᾶλλον ποιητικὸν ἢ παθητικὸν εἶναι θάτερον θατέρου (πάντα γὰρ ὁμοίως ὑπάρχειν ταὐτὰ τοῖς ὁμοίοις), τὰ δ' ἀνόμοια καὶ τὰ διάφορα ποιεῖν καὶ πάσχειν εἰς ἄλληλα πέφυκεν. ...Δημόκριτος δὲ παρὰ τοὺς ἄλλους ἰδίως ἔλεξε μόνος· φησὶ γὰρ τὸ αὐτὸ καὶ ὅμοιον εἶναι τό τε ποιοῦν καὶ τὸ πάσχον· οὐ γὰρ ἐγχωρεῖν τὰ ἕτερα καὶ διαφέροντα πάσχειν ὑπ' ἀλλήλων, ἀλλὰ κἂν ἕτερα ὄντα ποιῇ τι εἰς ἄλληλα, οὐκ ᾗ ἕτερα, ἀλλ' ᾗ ταὐτόν τι ὑπάρχει, ταύτῃ τοῦτο συμβαίνειν αὐτοῖς. Theophrastos however considers that the view of Demokritos is uncertain: see *de sensu* § 49. This doctrine of μηδὲν παθεῖν τὸ ὅμοιον ὑπὸ τοῦ ὁμοίου only refers to physical change, and does not affect the principle 'like is known by like'.

14. **τά τε αὖ σμικρότερα**] There seems at first sight a good deal of iteration in this chapter; but there is no real tautology. Plato (1) explains how (α) the larger figures are dissolved by the smaller, (β) how the smaller are dissolved by the larger; (2) he declares that (α) a small mass of the larger figures, intercepted by

τοῖς μείζοσι πολλοῖς περιλαμβανόμενα ὀλίγα διαθραυόμενα κα- B
τασβεννύηται, ξυνίστασθαι μὲν ἐθέλοντα εἰς τὴν τοῦ κρατοῦντος
ἰδέαν πέπαυται κατασβεννύμενα γίγνεταί τε ἐκ πυρὸς ἀήρ, ἐξ
ἀέρος ὕδωρ· ἐὰν δ' εἰς αὐτὰ ἴῃ καὶ τῶν ἄλλων τι ξυνιὸν γενῶν
5 μάχηται, λυόμενα οὐ παύεται, πρὶν ἢ παντάπασιν ὠθούμενα καὶ
διαλυθέντα ἐκφύγῃ πρὸς τὸ ξυγγενές, ἢ νικηθέντα, ἓν ἐκ πολλῶν
ὅμοιον τῷ κρατήσαντι γενόμενον, αὐτοῦ ξύνοικον μείνῃ. καὶ δὴ
καὶ κατὰ ταῦτα τὰ παθήματα διαμείβεται τὰς χώρας ἅπαντα· C
διέστηκε μὲν γὰρ τοῦ γένους ἑκάστου τὰ πλήθη κατὰ τόπον ἴδιον
10 διὰ τὴν τῆς δεχομένης κίνησιν, τὰ δὲ ἀνομοιούμενα ἑκάστοτε ἑαυτοῖς, ἄλλοις δὲ ὁμοιούμενα, φέρεται διὰ τὸν σεισμὸν πρὸς τὸν ἐκείνων οἷς ἂν ὁμοιωθῇ τόπον.

Ὅσα μὲν οὖν ἄκρατα καὶ πρῶτα σώματα, διὰ τοιούτων αἰτιῶν
γέγονε· τοῦ δ' ἐν τοῖς εἴδεσιν αὐτῶν ἕτερα ἐμπεφυκέναι γένη τὴν
15 ἑκατέρου τῶν στοιχείων αἰτιατέον σύστασιν, μὴ μόνον ἓν ἑκατέραν
μέγεθος ἔχον τὸ τρίγωνον φυτεῦσαι κατ' ἀρχάς, ἀλλ' ἐλάττω τε D
καὶ μείζω, τὸν ἀριθμὸν δὲ ἔχοντα τοσοῦτον, ὅσαπερ ἂν ᾖ τἀν τοῖς
εἴδεσι γένη. διὸ δὴ συμμιγνύμενα αὐτά τε πρὸς αὑτὰ καὶ πρὸς

14 τοῦ: τὸ A.   ἐν: ἓν A.

a large mass of the smaller, (β) a small mass of the smaller, intercepted by a large mass of the larger, can recover a definite form by becoming assimilated to the victorious element.

4. **ἐὰν δ' εἰς αὐτὰ ἴῃ**] The case put here seems to differ from the foregoing in this. Hitherto we have supposed a small mass of one kind intercepted by a large mass of the other: now we take the case of a prolonged struggle between pretty equal forces, when the process of dissolution continues without intermission, until one side is vanquished and either escapes away or is assimilated.

6. **ἓν ἐκ πολλῶν**] This ensues of course only if the victorious side is the kind formed of the larger figures.

8. **διαμείβεται τὰς χώρας**] Any kind by changing its figure changes the region of its affinity, as will be explained in the following chapter.

9. **τὰ πλήθη**] i.e. the main bulk of the substance. Detached portions of every kind may from various causes be found scattered everywhere through space, but the great mass of each is in its own region: cf. 63 B οὗ καὶ πλεῖστον ἂν ἠθροισμένον εἴη πρὸς ὃ φέρεται.

10. **τὴν τῆς δεχομένης κίνησιν**] The vibration of the ὑποδοχή described at 52 E.

13. **ὅσα μὲν οὖν ἄκρατα καὶ πρῶτα σώματα**] i.e. the primary and typical forms of the four so-called elements. Hitherto we have been dealing merely with the broad distinctions between fire, air, water, and earth. We shall hereafter find it necessary to treat of a number of different varieties. These diversities are accounted for by a diversity in the magnitude of the primary triangles.

17. **ὅσαπερ ἂν ᾖ τἀν τοῖς εἴδεσι γένη**] The εἶδος of course signifies some one of the four, as distinguished from the other three; say fire. There are a certain number of sizes in the radical triangles, and consequently an equal number of

smaller figures few in number are caught in a multitude of larger figures and are being broken in pieces and quenched, if they consent to combine into the form of the stronger they then and there cease from being quenched; and from fire arises air, from air water. But if they assail the others, and another sort meet and contend with them, they cease not from being shattered until, being entirely repelled and dissolved, they find refuge with some of their own kind, or being overcome, form from many of their own figures one similar to the victorious element, and there remain and abide with it. Moreover on account of these conditions they all are changing their places; for the bulk of every kind are sorted into separate regions of their own through the motion of the recipient: and those which are altered from their own nature and made like some other are carried by reason of this movement to the region proper to the element to which they are assimilated.

All unmixed and primary bodies have thus come into being through the causes we have described: but for the fact that within the several classes different kinds exist we must assign as its cause the structure of the elementary triangles; it does not originally produce in each kind of triangle one and the same size only, but some greater and some less; and there are just so many sizes as there are kinds in the classes: and when these

sizes in the pyramid. Now every substance which is composed entirely from pyramids of some one size constitutes a γένος of fire; there are therefore just so many γένη of fire as there are sizes of pyramids. But there are also substances which are composed of pyramids of different sizes: such substances will not be typical of any γένος, but will approximate to some γένος according as any special size of pyramid preponderates in its fabric. Accordingly we have in nature an indefinite number of substances belonging to each εἶδος, graduating from one γένος to another. The investigation of these begins in chapter xxiv. It is obvious that the variation in the size of the triangles must be confined within definite limits, for the largest pyramid is always smaller than the smallest octahedron, and the largest octahedron than the smallest icosahedron—for instance we find in 66 D that the φλέβες of the nostrils are too wide for the densest form of air and too narrow for the subtlest form of water.

57 D—58 C, c. xxiii. Our discourse now requires that we should set forth the causes of rest and motion. Motion implies the mover and the moved, without which two it cannot be. These two must be dissimilar; therefore dissimilarity is an essential condition of motion. And the cause of dissimilarity is inequality. Now the reason why all things are not sifted once for all into their proper regions and so become at rest is as follows. The whole globe of the universe is subject to a mighty constricting centripetal force,

ἄλληλα τὴν ποικιλίαν ἐστὶν ἄπειρα· ἧς δὴ δεῖ θεωροὺς γίγνεσθαι
τοὺς μέλλοντας περὶ φύσεως εἰκότι λόγῳ χρήσεσθαι.

XXIII. Κινήσεως οὖν στάσεώς τε πέρι, τίνα τρόπον καὶ μεθ᾽
ὧντινων γίγνεσθον, εἰ μή τις διομολογήσεται, πόλλ᾽ ἂν εἴη ἐμπο-
5 δὼν τῷ κατόπισθεν λογισμῷ. τὰ μὲν οὖν ἤδη περὶ αὐτῶν εἴρηται, E
πρὸς δ᾽ ἐκείνοις ἔτι τάδε, ἐν μὲν ὁμαλότητι μηδέποτε ἐθέλειν κίνησιν ἐνεῖναι. τὸ γὰρ κινησόμενον ἄνευ τοῦ κινήσοντος ἢ τὸ
κινῆσον ἄνευ τοῦ κινησομένου χαλεπόν, μᾶλλον δὲ ἀδύνατον εἶναι·
κίνησις δὲ οὐκ ἔστι τούτων ἀπόντων· ταῦτα δὲ ὁμαλὰ εἶναί ποτε
10 ἀδύνατον. οὕτω δὴ στάσιν μὲν ἐν ὁμαλότητι, κίνησιν δὲ εἰς ἀνωμαλότητα ἀεὶ τιθῶμεν· αἰτία δὲ ἀνισότης αὖ τῆς ἀνωμάλου φύ- 58 A
σεως. ἀνισότητος δὲ γένεσιν μὲν διεληλύθαμεν· πῶς δέ ποτε οὐ
κατὰ γένη διαχωρισθέντα ἕκαστα πέπαυται τῆς δι᾽ ἀλλήλων
κινήσεως καὶ φορᾶς, οὐκ εἴπομεν. ὧδε οὖν πάλιν ἐροῦμεν. ἡ
15 τοῦ παντὸς περίοδος, ἐπειδὴ συμπεριέλαβε τὰ γένη, κυκλοτερὴς
οὖσα καὶ πρὸς αὑτὴν πεφυκυῖα βούλεσθαι ξυνιέναι, σφίγγει πάντα

11 ἡ ante ἀνισότης dederunt SZ.

which crushes its whole mass together and will not suffer any vacant space within it. This forces the subtler elements into the interstices of the coarser; and so by the admixture of larger and smaller forms, dilation and compression is everywhere at work; thereupon ensues the transmutation of one element into another, and by consequence a change of its proper region to which it tends. Thus a perpetual shifting of forms ensures a perpetual shifting of place.

3. **κινήσεως οὖν**] Concerning motion Plato sets forth in this chapter (1) whence it originates, (2) why it never ceases.

6. **ἐν μὲν ὁμαλότητι**] We saw above at 57 A that like could not affect like nor be affected by it: it follows then that in a perfectly uniform mass motion cannot arise, since motion is the effect of a moving cause upon the object moved. The κινοῦν then and κινούμενον must be ἀνώμαλα, heterogeneous.

7. **τὸ γὰρ κινησόμενον**] cf. Aristotle *physica* III i 200[b] 31 τὸ γὰρ κινητικὸν κινητικὸν τοῦ κινητοῦ καὶ τὸ κινητὸν κινητὸν ὑπὸ τοῦ κινητικοῦ: and below 202[a] 13 ἐστὶν ἡ κίνησις ἐν τῷ κινητῷ· ἐντελέχεια γάρ ἐστι τούτου· καὶ ὑπὸ τοῦ κινητικοῦ.

9. **ταῦτα**] sc. τὸ κινῆσον καὶ τὸ κινησόμενον.

10. **ἐν ὁμαλότητι...εἰς ἀνωμαλότητα**] Rest exists in uniformity, motion is attributed to dissimilarity: thus we may express the change of preposition.

12. **ἀνισότητος δὲ γένεσιν**] How inequality originates we have seen in the account of the structure of matter. It arises (1) from the dissimilarity of the two primal triangles, (2) from the different geometrical figures which are based upon one of the triangles, (3) from inequality in size of the triangles themselves.

πῶς δέ ποτε οὐ κατὰ γένη] This sentence is misunderstood by Lindau and Stallbaum. Plato means to explain how it is that the four εἴδη have not settled each in its proper sphere, and thus avoided interfering with each other and so producing irregularity and consequently motion. For the vibration of the ὑπο-

are mixed up with themselves or with one another, an endless diversity arises, which must be examined by those who would put forward a probable theory concerning nature.

XXIII. Now concerning rest and motion, how they arise and under what conditions, we must come to an agreement, else many difficulties will stand in the way of our argument that is to follow. This has been already in part set forth, but we have yet to add that in uniformity no movement will ever exist. For that what is to be moved should exist without that which is to move it, or what is to move without that which is to be moved, is difficult or rather impossible: but without these there can be no motion, and for these to be uniform is not possible. So then let us always assign rest to uniformity and motion to its opposite. Now the opposite of uniformity is caused by inequality; and of inequality we have discussed the origin. But how it comes to pass that all bodies are not sorted off into their several kinds and cease from passing through one another and changing their place, this we have not explained. Let us put it again in this way. The revolution of the whole, when it had embraced the four kinds, being circular, with a natural tendency to return upon itself, compresses everything and suffers

δοχὴ tends to keep them all assorted and apart from each other; and this would actually be the condition of things, were it not for the πίλησις presently to be mentioned. Stallbaum supposes that the elements are κατὰ γένη διαχωρισθέντα: but Plato's reasoning turns precisely on the point that they are not: never completely, that is; for the bulk of each is to be found in its own home.

16. πρὸς αὑτὴν πεφυκυῖα] The notion is that the whole universe globes itself about its centre with a mighty inward pressure, εἱλεῖται περὶ τὸν διὰ παντὸς τεταμένον πόλον, so that everything within it is packed as tightly as possible. The force may be compared to that exerted in winding a hank of string into a round ball. This is the second of Plato's two great dynamic powers: we shall afterwards see what varied and extensive use

he makes of it.

σφίγγει πάντα] Compare Empedokles 185 (Karsten) Τιτὰν ἠδ' αἰθὴρ σφίγγων περὶ κύκλον ἅπαντα. This vast circular constriction squeezes all matter together with so overpowering force, that no vacancy is allowed to remain anywhere; but wherever there is room for a smaller particle to penetrate the interstices between the larger, it is at once forced in. So that not only are heterogeneous elements forced into combination, but the subtler and acuter figures divide the larger κατὰ τὰ τρίγωνα and so change their structure: while they in turn are themselves compressed by the larger until they assume the form of the latter. Consequently we have side by side perpetually the ὁδὸς κάτω, fire through air to water, and the ὁδὸς ἄνω, water through air to fire.

210   ΠΛΑΤΩΝΟΣ   [58 A—

καὶ κενὴν χώραν οὐδεμίαν ἐᾷ λείπεσθαι. διὸ δὴ πῦρ μὲν εἰς
ἅπαντα διελήλυθε μάλιστα, ἀὴρ δὲ δεύτερον, ὡς λεπτότητι δεύτερον B
ἔφυ, καὶ τἆλλα ταύτῃ· τὰ γὰρ ἐκ μεγίστων μερῶν γεγονότα μεγίστην κενότητα ἐν τῇ ξυστάσει παραλέλοιπε, τὰ δὲ σμικρότατα
5 ἐλαχίστην. ἡ δὴ τῆς πιλήσεως ξύνοδος τὰ σμικρὰ εἰς τὰ τῶν
μεγάλων διάκενα ξυνωθεῖ. σμικρῶν οὖν παρὰ μεγάλα τιθεμένων
καὶ τῶν ἐλαττόνων τὰ μείζονα διακρινόντων, τῶν δὲ μειζόνων ἐκεῖνα
συγκρινόντων, πάντ' ἄνω κάτω μεταφέρεται πρὸς τοὺς ἑαυτῶν
τόπους· μεταβάλλον γὰρ τὸ μέγεθος ἕκαστον καὶ τὴν τόπων μετα- C
10 βάλλει στάσιν. οὕτω δὴ διὰ ταῦτά τε ἡ τῆς ἀνωμαλότητος δια-
σῳζομένη γένεσις ἀεὶ τὴν ἀεὶ κίνησιν τούτων οὖσαν ἐσομένην τε
ἐνδελεχῶς παρέχεται.

XXIV. Μετὰ δὴ ταῦτα δεῖ νοεῖν, ὅτι πυρός τε γένη πολλὰ
γέγονεν, οἷον φλὸξ τό τε ἀπὸ τῆς φλογὸς ἀπιόν, ὃ καίει μὲν οὔ,
15 φῶς δὲ τοῖς ὄμμασι παρέχει, τό τε φλογὸς ἀποσβεσθείσης ἐν τοῖς
διαπύροις καταλειπόμενον αὐτοῦ. κατὰ ταὐτὰ δὲ ἀέρος τὸ μὲν D

2 δεύτερον...δεύτερον: δευτέρως...δεύτερος S.   9 μεταβάλλον: μεταβαλὸν A pr. m.
14 ἀπιόν: ἀπτὸν A.   καίει: κάει ASZ.

3. **μεγίστην κενότητα**] This expression shows plainly enough that Plato was well aware of the fact which Aristotle urges as a flaw in his theory, namely that it is impossible for all his figures to fill up space with entire continuity. In the structure of air and of water there must be minute interstices of void; there must also be a certain amount of void for the reason that, the universe being a sphere, it is impossible for rectilinear figures exactly to fill it up. But, it is to be observed, Plato's theory does not demand that void shall be absolutely excluded from his system, but only that there shall be no vacant space large enough to contain the smallest existing corpuscule of matter. The larger corpuscules have larger interstices between them than the smaller. So long however as these interstices are not large enough to afford entrance to the smallest particle of any element, the effect is the same as of a solid mass without any cavities; but when once they are large enough to contain any particle, πίλησις instantly forces one into the vacancy. This is all Plato means by κενὴν χώραν οὐδεμίαν ἐᾷ λείπεσθαι: he denies void as a mechanical principle, but not its existence altogether in the nature of things.

Besides the atomists, the existence of void was affirmed by the Pythagoreans; see above, 33 C, and Aristotle *physica* IV vi 213[b] 22: it was denied by the Eleatics, by Empedokles, by Anaxagoras, and by Aristotle: see *physica* IV vii.

5. **ἡ τῆς πιλήσεως ξύνοδος**] cf. *Phaedo* 97 A ἡ ξύνοδος τοῦ πλησίον ἀλλήλων τεθῆναι.

9. **μεταβάλλον γὰρ τὸ μέγεθος**] For example, particles of fire, by being transformed into particles of water, not only changed their magnitude, but also the region of space to which they belonged. Hence any fire in the home of fire which became water would instantly struggle to reach the home of water; and similarly with air and water; so that a perpetual flux and reflux is kept up between one region and another. In this manner the production of heterogeneity (ἀνωμαλότη-

no vacant space to be left. Therefore fire penetrates most of all through all things, and in the second degree air, since it is second in fineness, and the rest in proportion. For the substances which are formed of the largest parts have the most void left in their structure, and those made of the smallest have the least. Now the constriction of this contracting force thrusts the small particles into the interspaces between the larger: so that when small are set side by side with great, and the lesser particles divide the greater, while the greater compress the smaller, all things keep rushing backwards and forwards to their own region; since in changing its bulk each changes its proper position in space. Thus owing to these causes a perpetual disturbance of uniformity is always kept up and so preserves the perpetual motion of matter now and henceforth without cessation.

XXIV. Next we must remember that of fire there are many kinds: for instance flame and that effluence from flame, which burns not but gives light to the eyes, and that which remains in the embers when the flame is out. And so with air: the purest

τος γένεσις) is maintained, and the perpetuation of motion secured. Compare Aristotle *de gen. et corr.* II x 337ª 7 ἅμα δὲ δῆλον ἐκ τούτων ὅ τινες ἀποροῦσιν, διὰ τί ἑκάστου τῶν σωμάτων εἰς τὴν οἰκείαν φερομένου χώραν ἐν τῷ ἀπείρῳ χρόνῳ οὐ διεστᾶσι τὰ σώματα. αἴτιον γὰρ τούτου ἐστὶν ἡ εἰς ἄλληλα μετάβασις· εἰ γὰρ ἕκαστον ἔμενεν ἐν τῇ αὑτοῦ χώρᾳ καὶ μὴ μετέβαλλεν ὑπὸ τοῦ πλησίον, ἤδη ἂν διεστήκεσαν. μεταβάλλει μὲν οὖν διὰ τὴν φορὰν διπλῆν οὖσαν· διὰ δὲ τὸ μεταβάλλειν οὐκ ἐνδέχεται μένειν οὐδὲν αὐτῶν ἐν οὐδεμιᾷ χώρᾳ τεταγμένῃ.

58 C—60 B, *c.* xxiv. Of fire there are three kinds, the flame, the light radiated from it, and the glow remaining after the flame is extinct. Of air there are many kinds, the purest being aether, the grossest mist and cloud. Water falls into two main classes, liquid and fusible: the first is ever unstable and flowing; the second is hard and compact, but can be fused and liquefied by the action of fire aided by air. Of fusible water that which is formed of the finest and most even par- ticles is gold, an offshoot of which is adamant. A metal resembling gold, but harder owing to an admixture of earth, is bronze. And so we might describe all the rest, following our theory of probability, which serves us as a harmless and rational diversion in the intervals of more serious speculations. To proceed: when water is mingled with fire and flows freely, we call it liquid: but when fire abandons it and the surrounding air compresses and solidifies it, according to the degree of solidification we call it on the earth ice or hoar-frost, in the air hail or snow. The forms of water which circulate in the structure of plants we call in general sap: four only have peculiar names, wine, oil, honey, and verjuice.

14. τό τε ἀπὸ τῆς φλογὸς ἀπιόν] The reading ἀπιόν is unquestionably right although confirmed by only one ms. and by Galen. Plato then regards light as an effluence, issuing from the flame; the third species of fire being the red glow left in the embers when the flame has burnt down.

16. αὐτοῦ] sc. πυρός.

212   ΠΛΑΤΩΝΟΣ   [58 D—

εὐαγέστατον ἐπίκλην αἰθὴρ καλούμενος, ἡ δὲ θολερώτατος ὀμίχλη
τε καὶ σκότος, ἕτερά τε ἀνώνυμα εἴδη γεγονότα διὰ τὴν τῶν τριγώνων ἀνισότητα. τὰ δὲ ὕδατος διχῇ μὲν πρῶτον, τὸ μὲν ὑγρόν,
τὸ δὲ χυτὸν γένος αὐτοῦ. τὸ μὲν οὖν ὑγρὸν διὰ τὸ μετέχον εἶναι
5 τῶν γενῶν τῶν ὕδατος, ὅσα σμικρά, ἀνίσων ὄντων, κινητὸν αὐτό
τε καθ᾽ αὑτὸ καὶ ὑπ᾽ ἄλλου διὰ τὴν ἀνωμαλότητα καὶ τὴν τοῦ
σχήματος ἰδέαν γέγονε· τὸ δὲ ἐκ μεγάλων καὶ ὁμαλῶν στασιμώ- E
τερον μὲν ἐκείνου καὶ βαρὺ πεπηγὸς ὑπὸ ὁμαλότητός ἐστιν, ὑπὸ δὲ
πυρὸς εἰσιόντος καὶ διαλύοντος αὐτὸ τὴν ὁμαλότητα [ἀποβάλλει,
10 ταύτην δὲ] ἀπολέσαν μετίσχει μᾶλλον κινήσεως, γενόμενον δὲ
εὐκίνητον, ὑπὸ τοῦ πλησίον ἀέρος ὠθούμενον καὶ κατατεινόμενον
ἐπὶ γῆν, τήκεσθαι μὲν τὴν τῶν ὄγκων καθαίρεσιν, ῥοὴν δὲ τὴν
κατάτασιν ἐπὶ γῆν ἐπωνυμίαν ἑκατέρου τοῦ πάθους ἔλαβε. πάλιν
δὲ ἐκπίπτοντος αὐτόθεν τοῦ πυρός, ἅτε οὐκ εἰς κενὸν ἐξιόντος, 59 A
15 ὠθούμενος ὁ πλησίον ἀὴρ εὐκίνητον ὄντα ἔτι τὸν ὑγρὸν ὄγκον εἰς
τὰς τοῦ πυρὸς ἕδρας ξυνωθῶν αὐτὸν αὑτῷ ξυμμίγνυσιν· ὁ δὲ ξυνω-
θούμενος ἀπολαμβάνων τε τὴν ὁμαλότητα πάλιν, ἅτε τοῦ τῆς
ἀνωμαλότητος δημιουργοῦ πυρὸς ἀπιόντος, εἰς ταὐτὸν αὑτῷ καθί-
σταται· καὶ τὴν μὲν τοῦ πυρὸς ἀπαλλαγὴν ψῦξιν, τὴν δὲ ξύνοδον
20 ἀπελθόντος ἐκείνου πεπηγὸς εἶναι γένος προσερρήθη. τούτων δὴ
πάντων, ὅσα χυτὰ προσείπομεν ὕδατα, τὸ μὲν ἐκ λεπτοτάτων καὶ B

5 κινητόν: κινητικὸν AH.   9 ἀποβάλλει, ταύτην δὲ habet corr. A. omittunt SZ.
13 κατάτασιν: κατάστασιν A.   19 τὴν μὲν : τὸν μὲν H per typographi incuriam.
20 ἀπελθόντος ἐκείνου : ἐκείνου ἀπελθόντος S.   21 λοιπὸν post τὸ μὲν habet A.

1. **αἰθὴρ καλούμενος**] Hence it is evident that Plato did not regard aether as a distinct element: cf. *Phaedo* 111 A, where αἰθήρ is simply the pure air of which our atmosphere is the sediment.

ὀμίχλη καὶ σκότος] This is the ἀὴρ βίᾳ ξυστὸς of 61 C.

3. **τὸ μὲν ὑγρόν, τὸ δὲ χυτόν**] The ὑγρόν includes all fluids which are ordinarily so regarded by us: that is to say, all substances which at the normal temperature are liquid and flowing: χυτὸν comprises metals, which are normally solid but are liquefied by the application of strong heat. To rank metals as forms of water seems no doubt a strange classification: it is however adopted by Theophrastos also: see *de lapidibus* § 1 τῶν ἐν

τῇ γῇ συνισταμένων τὰ μέν ἐστιν ὕδατος, τὰ δὲ γῆς. ὕδατος μὲν τὰ μεταλλευόμενα καθάπερ ἄργυρος καὶ χρυσὸς καὶ τἆλλα.

5. **τῶν γενῶν τῶν ὕδατος**] This seems a very strange phrase to denote the corpuscules which constitute water : ought we perhaps to read τῶν μερῶν ?

9. **τὴν ὁμαλότητα ἀπολέσαν**] Martin quite mistakes the meaning of this. He supposes that fire has the power of dilating the elementary triangles and so introducing a difference of size in the corpuscules of water. This can in no wise be admitted by the theory. Plato's meaning is that the particles of fire by interposing themselves between those of water, to which they are of course greatly inferior in size, destroy the homogeneous-

is that which is called by the name of aether, and the most turbid is mist and gloom; and there are other kinds which have no name, arising from the inequality of the triangles. Of water there are two primary divisions, the liquid and the fusible kind. The liquid sort owes its nature to possessing the smaller kinds of watery atoms, unequal in size; and so it can readily either move of itself or be moved by something else, owing to its lack of uniformity and the peculiar shape of its atoms. But that which consists of larger and uniform particles is more stable than the former and heavy, being stiffened by its uniformity: but when fire enters into it and breaks it up, it loses its uniformity and gains more power of motion: and as soon as it has become mobile, it is thrust by the surrounding air and spread out upon the earth: and it has received names descriptive of either process, *melting* of the dissolution of the mass, *flowing* of the extension on the ground. But when the fire goes forth from it again, seeing that it does not issue into empty space, the neighbouring air receives a thrust, and while the liquid mass is still mobile, it forces it to fill up the vacant places of the fire and unites it with itself. And being thus compressed and recovering its uniformity, seeing that fire the creator of inequality is quitting it, it settles into its normal state. And the departure of fire we call *cooling*, and the contraction that ensues on its withdrawal we class as *solidification*. Of all the substances which we have ranked as fusible kinds of water, that which is densest

ness of the whole mass. At the same time, by the interposition of the fiery particles its bulk is expanded, so that it comes into forcible collision with the surrounding air, which gives it the impulse that sheds it (κατατείνει) on the ground. It now is subject to the same conditions as ὑγρὸν ὕδωρ, which flows owing to the inequality of its own particles. Thus the fusion and flowing of molten metal is due to two causes: (1) the intrusion of particles of fire and consequent dislocation of the particles of water, rendering the mass ἀνώμαλον and therefore εὐκίνητον—this we call melting; (2) the yielding of the now heterogeneous substance to the pressure of the air, which we call flowing.

13. πάλιν δ' ἐκπίπτοντος] Solidification is explained thus. The particles of fire, on quitting their place amid those of water, thrust against the immediately surrounding particles of air, since of course there is no vacant space to receive them. Now the metal, though the fire has left it, is still mobile and yielding, because its particles are dislocated. The air then, on the impulse of the outgoing fire, thrusts against the metal and compresses it, forcing its particles to fill up the vacancies left by the fire. Thereby the particles are restored to their old places and the metal regains its equilibrium and solidity.

ΠΛΑΤΩΝΟΣ [59 B—

ὁμαλωτάτων πυκνότατον γιγνόμενον, μονοειδὲς γένος, στίλβοντι καὶ
ξανθῷ χρώματι κοινωθέν, τιμαλφέστατον κτῆμα χρυσὸς ἠθημένος
διὰ πέτρας ἐπάγη· χρυσοῦ δὲ ὄζος, διὰ πυκνότητα σκληρότατον ὂν
καὶ μελανθέν, ἀδάμας ἐκλήθη. τὸ δ' ἐγγὺς μὲν χρυσοῦ τῶν μερῶν,
5 εἴδη δὲ πλέονα ἑνὸς ἔχον, πυκνότητι δ' ἔτι μὲν χρυσοῦ πυκνότερον
ὄν, καὶ γῆς μόριον ὀλίγον καὶ λεπτὸν μετασχόν, ὥστε σκληρότερον C
εἶναι, τῷ δὲ μεγάλα ἐντὸς αὐτοῦ διαλείμματα ἔχειν κουφότερον, τῶν
λαμπρῶν πηκτῶν τε ἓν γένος ὑδάτων χαλκὸς συσταθεὶς γέγονε·
τὸ δ' ἐκ γῆς αὐτῷ μιχθέν, ὅταν παλαιουμένῳ διαχωρίζησθον πάλιν
10 ἀπ' ἀλλήλων, ἐκφανὲς καθ' αὑτὸ γιγνόμενον ἰὸς λέγεται. τἆλλα
δὲ τῶν τοιούτων οὐδὲν ποικίλον ἔτι διαλογίσασθαι τὴν τῶν εἰκότων
μύθων μεταδιώκοντα ἰδέαν, ἣν ὅταν τις ἀναπαύσεως ἕνεκα τοὺς
περὶ τῶν ὄντων ἀεὶ κατατιθέμενος λόγους τοὺς γενέσεως πέρι δια- D
θεώμενος εἰκότας ἀμεταμέλητον ἡδονὴν κτᾶται, μέτριον ἂν ἐν τῷ
15 βίῳ παιδιὰν καὶ φρόνιμον ποιοῖτο. ταύτῃ δὴ καὶ τὰ νῦν ἐφέντες
τὸ μετὰ τοῦτο τῶν αὐτῶν πέρι τὰ ἑξῆς εἰκότα διιμεν τῇδε. τὸ πυρὶ

4 ἐκλήθη omittit Λ.   5 δ' ἔτι : δὲ τῇ Α.   omittunt SZ.   13 κατατιθέμενος : καταθέμενος SZ.   15 παιδιάν : παιδείαν Λ.   ἐφέντες : ἀφέντες AZ.

1. **στίλβοντι καὶ ξανθῷ**] 'infused with a glittering and yellow hue.' στίλβον, as Lindau says, is a χρόα coordinate with ξανθόν: its γένεσις is described in 68 A.

3. **χρυσοῦ δὲ ὄζος**] What this substance was it is very difficult to determine, further than that it is some hard dark metal always found, as Plato supposes, with gold and closely akin to it. It is mentioned again in *Politicus* 303 E μετὰ δὲ ταῦτα λείπεται ξυμμεμιγμένα τὰ ξυγγενῆ τοῦ χρυσοῦ τίμια καὶ πυρὶ μόνον ἀφαιρετά, χαλκὸς καὶ ἄργυρος, ἔστι δ' ὅτε καὶ ἀδάμας. In Hesiod *Scut. Her.* 137, 231, and *Theog.* 161 it signifies a hard metal, probably something like steel, of which armour and cutting instruments were made. This cannot be meant here, far less a mixture of copper and gold, as Stallbaum thinks. Pliny *nat. hist.* xxxvii 15 says maximum in rebus humanis, non solum inter gemmas, pretium habet adamas, diu non nisi regibus et iis admodum paucis cognitus; ita appellabatur auri nodus in metallis repertus perquam raro, comes auri, nec nisi in auro nasci videbatur. The six kinds he goes on to describe are evidently all crystals. It is clear that Plato's χρυσοῦ ὄζος was not a crystal: for the term ἀδάμας is not applied to any precious stone by writers before Theophrastos; moreover a crystal could not be a species of χυτὸν ὕδωρ, all such being forms of earth. Professor W. J. Lewis, who has been kind enough to make some inquiry into this matter on my behalf, formed the opinion, on such data as I was able to lay before him, that Plato's ἀδάμας was probably haematite.

5. **πυκνότητι δ' ἔτι μέν**] This is Baiter's conjecture, followed by Hermann. I have adopted it as possibly accounting for the τῇ μὲν of Λ.

7. **μεγάλα ἐντὸς διαλείμματα**] These would appear to be cavities in the substance of the metal filled with air, which cause bronze, notwithstanding its superior density, to be lighter than gold. Plato is of course mistaken in supposing that bronze is denser than gold. He attri-

D] ΤΙΜΑΙΟΣ. 215

and formed of the finest and most uniform particles, a unique kind, combining brightness with a yellow hue, is gold, a most precious treasure, which has filtered through rocks and there congealed: and the 'offspring of gold', which is extremely hard owing to its density and has turned black, is called adamant. Another has particles resembling those of gold, but more than one kind; in density it even surpasses gold and has a small admixture of fine earth, so that it is harder, but lighter, because it has large interstices within; this formation is one of the shining and solid kinds of water and is called bronze. The earth which is mingled with it, when the two through age begin to separate again, becomes visible by itself and is named rust. And it were no intricate task to explain all the other substances of this kind, following the outline of our probable account. For if we pursue this as a recreation, and while laying down the principles of eternal being find in plausible theories of becoming a pleasure that brings no remorse in its train, we may draw from it a sober and sensible amusement during our life. Now therefore setting out in this way let us go on to discuss the probabilities that lie next on the same subject.

buted the greater hardness of bronze partly to its superior density, partly to the admixture of earth: he was not aware that hardness does not depend upon density. As to the διαλείμματα, compare Theophrastos *de sensu* § 61, speaking of Demokritos, σκληρότερον μὲν εἶναι σίδηρον βαρύτερον δὲ μόλυβδον· τὸν μὲν γὰρ σίδηρον ἀνωμάλως συγκεῖσθαι καὶ τὸ κενὸν ἔχειν πολλαχῇ καὶ κατὰ μεγάλα πεπυκνῶσθαι καὶ κατὰ ἔνια *, ἁπλῶς δὲ πλέον ἔχειν κενόν· τὸν δὲ μόλυβδον ἔλαττον ἔχοντα κενὸν ὁμαλῶς συγκεῖσθαι κατὰ πᾶν ὁμοίως, διὸ βαρύτερον μὲν μαλακώτερον δὲ τοῦ σιδήρου. This is identical with Plato's view, except that Demokritos held the cavities to be absolutely void.

9. ὅταν παλαιουμένω διαχωρίζησθον] Plato considered that the rust on bronze, or verdigris, was the intermingled earth, which in course of time works its way to the surface.

12. ἣν ὅταν τις] Here we have in the plainest terms Plato's opinion of the value of physical science. In itself it is but a harmless recreation, a pleasure leaving behind it no regrets, with which a philosopher may reasonably solace himself, when wearied with his incessant struggle after the truth. This passage should be read in connexion with 68 E διὸ δὴ χρὴ δύ' αἰτίας εἴδη διορίζεσθαι κ.τ.λ., where we learn that the study of ἀναγκαῖον, that is to say, of the forces of material nature, is useful just so far as it bears upon the investigation of θεῖον, that is, of primary causes. Physical speculations then are profitable only in so far as they can be made subservient to metaphysical science; to suppose that they have any intrinsic merit is an egregious error: they can only be pursued for their own sake with a view to recreation. As regards the construction there is a slight anacoluthon; ἣν being presently superseded by τοὺς γενέσεως πέρι.

ΠΛΑΤΩΝΟΣ [59 D—

μεμιγμένον ὕδωρ, ὅσον λεπτὸν ὑγρόν τε διὰ τὴν κίνησιν καὶ τὴν
ὁδόν, ἣν κυλινδούμενον ἐπὶ γῆς ὑγρὸν λέγεται, μαλακόν τε αὖ τῷ
τὰς βάσεις ἧττον ἑδραίους οὔσας ἢ τὰς γῆς ὑπείκειν, τοῦτο ὅταν
πυρὸς ἀποχωρισθὲν ἀέρος τε μονωθῇ, γέγονε μὲν ὁμαλώτερον, E
5 ξυνέωσται δὲ ὑπὸ τῶν ἐξιόντων εἰς αὑτό, παγέν τε οὕτως τὸ μὲν
ὑπὲρ γῆς μάλιστα παθὸν ταῦτα χάλαζα, τὸ δ' ἐπὶ γῆς κρύσταλλος,
τὸ δὲ ἧττον ἡμιπαγές τε ὂν ἔτι, τὸ μὲν ὑπὲρ γῆς αὖ χιών, τὸ δ' ἐπὶ
γῆς ξυμπαγὲν ἐκ δρόσου γενόμενον πάχνη λέγεται. τὰ δὲ δὴ
πλεῖστα ὑδάτων εἴδη μεμιγμένα ἀλλήλοις, ξύμπαν μὲν τὸ γένος,
10 διὰ τῶν ἐκ γῆς φυτῶν ἠθημένα, χυμοὶ λεγόμενοι· διὰ δὲ τὰς μίξεις 60 A
ἀνομοιότητα ἕκαστοι σχόντες τὰ μὲν ἄλλα πολλὰ ἀνώνυμα γένη
παρέσχοντο, τέτταρα δέ, ὅσα ἔμπυρα εἴδη, διαφανῆ μάλιστα γενόμενα εἴληφεν ὀνόματα αὐτῶν, τὸ μὲν τῆς ψυχῆς μετὰ τοῦ σώματος
θερμαντικὸν οἶνος, τὸ δὲ λεῖον καὶ διακριτικὸν ὄψεως διὰ ταῦτά τε
15 ἰδεῖν λαμπρὸν καὶ στίλβον λιπαρόν τε φανταζόμενον ἐλαιηρὸν
εἶδος, πίττα καὶ κίκι καὶ ἔλαιον αὐτὸ ὅσα τ' ἄλλα τῆς αὐτῆς
δυνάμεως· ὅσον δὲ διαχυτικὸν μέχρι φύσεως τῶν περὶ τὸ στόμα B

2 αὖ τῷ: αὐτῷ A. 3 τοῦτο: τοῦτο δ' S.
9 τῶν ante ὑδάτων habet A. 16 κίκι: τήκει A pr. m.

1. **ὅσον λεπτὸν ὑγρόν τε**] Although Stallbaum asserts that this sentence is 'turpi labe contaminatus', I see no necessity for alteration: his own attempts are certainly far from fortunate. The repetition of ὑγρόν, which offends him so sorely, is, I think, due to the fact that we have, as Lindau saw, an etymology implied in the words ἣν...λέγεται 'the mode of rolling on the earth which has in fact gained it the name of ὑγρόν': as if ὑγρὸν = ὑπὲρ γῆς ῥέον. Thus understood, the objection to the second ὑγρόν vanishes. μαλακόν τε is then coordinate with λεπτὸν ὑγρόν τε, and τῷ...ὑπείκειν with διὰ τὴν κίνησιν.

4. **πυρὸς ἀποχωρισθέν**] Water then in its pure and unmixed form is in a state of congelation: the liquid condition being due to the intermixture of fire which disturbs the uniformity of the whole. What we ordinarily term water then is a compound of fire and water.

ἀέρος τε] It is rather hard to see what air has to do with the matter: no air entered into the composition of the ὑγρὸν ὕδωρ, which merely yielded to the impact of the air which pushed it from without. May not ἀέρος τε be an interpolation from the hand of some copyist who thought it necessary to separate water from both the kindred elements? The copyists have an unconquerable desire to drag in all the elements, whether they are wanted or not: see note on 61 B, where there is an indisputable interpolation.

5. **ὑπὸ τῶν ἐξιόντων**] That is to say, by the agency of the outgoing fire that thrusts the surrounding air, which in turn communicates the impulse to the water. Plato classifies the congealed forms of water according to the intensity of the compression and to the situation: when completely condensed it is on the earth ice, in the air hail; if partially condensed, it is on the earth hoar-frost, in the air snow.

Water mingled with fire, such as is rare and liquid (owing to its mobility and its way of rolling along the ground, which gets it the name of liquid), and is also soft, because its bases give way, being less stable than those of earth,—when relinquished by fire and deserted of air, becomes more uniform and is compressed by the outgoing elements; thus it is congealed, and when above the earth this process takes place in an extreme degree, the result is hail; if upon the earth, it is ice: but when the process has not gone so far but leaves it half-congealed, above the earth it is snow, and when congealed from dew upon the earth, it is called hoarfrost. Most forms of water, which are intermingled with one another, filtered through the plants of the earth, are called by the class-name of *saps*; but owing to their intermixture they are all of diverse natures and the great multitude of them are accordingly unnamed: four kinds however which are of a fiery nature, being more conspicuous, have obtained names: one that heats the soul and body together, namely wine; next a kind which is smooth and divides the visual current and therefore appears bright and shining to view and glistening, I mean the class of oils, resin and castor oil and olive oil itself and all others that have the same properties; thirdly that which expands the contracted

7. τὸ δὲ ἧττον] sc. παθὸν τοῦτο. Cf. Aristotle *meteorologica* I x 347ª 16 πάχνη μὲν ὅταν ἡ ἀτμὶς παγῇ πρὶν εἰς ὕδωρ συγκριθῆναι πάλιν.

8. τὰ δὲ δὴ πλεῖστα] A complex form of water, composed of many sorts combined, are the juices of plants of which the general appellation is sap. Of these Plato distinguishes four kinds, having peculiar properties and specific names.

12. ὅσα ἔμπυρα εἴδη] Plato infers the presence of fire from the brightness and transparency of these saps, not from any pungent or burning quality, which olive oil, for example, does not possess.

14. διακριτικὸν ὄψεως] That is to say, having a bright and glistening appearance, see 68 E, 69 A. We must understand Plato to mean διακριτικὸν ὄψεως μέχρι τῶν ὀμμάτων, for what is merely διακριτικὸν ὄψεως is white. ὄψις here = ὄψεως ῥεῦμα.

16. κίκι] This is castor oil, obtained from the Ricinus communis. See Herodotus II 94, where he says that the Egyptians use this oil for anointing themselves and for illuminating purposes: it is said to be still put to the latter use in India. The word κίκι is affirmed by Herodotus to be Egyptian. Cf. Pliny *nat. hist.* XV 7.

17. ὅσον δὲ διαχυτικὸν μέχρι φύσεως] The construction and meaning of these words seem to have escaped all the editors. τῶν περὶ τὸ στόμα ξυνόδων depends upon διαχυτικόν, not upon φύσεως, and the meaning is 'that which expands the contracted pores of the mouth to their natural condition'. In 64 D we learn that a pleasurable sensation is the perceptible transition from an abnormal to a normal state: τὸ δ' εἰς φύσιν ἀπιὸν

218   ΠΛΑΤΩΝΟΣ   [60 B—

ξυνόδων, ταύτῃ τῇ δυνάμει γλυκύτητα παρεχόμενον, μέλι τὸ κατὰ
πάντων μάλιστα πρόσρημα ἔσχε· τὸ δὲ τῆς σαρκὸς διαλυτικὸν τῷ
καίειν ἀφρῶδες γένος ἐκ πάντων ἀφορισθὲν τῶν χυμῶν ὀπὸς ἐπωνομάσθη.

5 XXV. Γῆς δὲ εἴδη, τὸ μὲν ἠθημένον διὰ ὕδατος τοιῷδε τρόπῳ
γίγνεται σῶμα λίθινον. τὸ ξυμμιγὲς ὕδωρ ὅταν ἐν τῇ ξυμμίξει
κοπῇ, μετέβαλεν εἰς ἀέρος ἰδέαν· γενόμενος δὲ ἀὴρ εἰς τὸν ἑαυτοῦ
τόπον ἀναθεῖ. κενὸν δ' οὐ περιεῖχεν αὐτὸν οὐδέν· τὸν οὖν πλησίον C
ἔωσεν ἀέρα· ὁ δὲ ἅτε ὢν βαρύς, ὠσθεὶς καὶ περιχυθεὶς τῷ τῆς γῆς
10 ὄγκῳ, σφόδρα ἔθλιψε ξυνέωσέ τε αὐτὸν εἰς τὰς ἕδρας, ὅθεν ἀνῄει ὁ
νέος ἀήρ· ξυνωσθεῖσα δὲ ὑπ' ἀέρος [ἀλύτως ὕδατι] γῇ ξυνίσταται
πέτρα, καλλίων μὲν ἡ τῶν ἴσων καὶ ὁμαλῶν διαφανὴς μερῶν,
αἰσχίων δὲ ἡ ἐναντία. τὸ δὲ ὑπὸ πυρὸς τάχους τὸ νοτερὸν πᾶν

3 καίειν : κάειν SZ.   8 οὐ περιεῖχεν αὐτόν : sic corr. A.   ὑπερεῖχεν αὐτῶν pr. m.
ὑπῆρχεν αὐτῶν SZ.   10 ἀνῄει : ἀνῄειν SZ.

πάλιν ἀθρόον ἡδύ: and in 66 C we find
that this is just the effect produced on
the tongue by a pleasant taste: τὰ δὲ
παρὰ φύσιν ξυνεστῶτα ἢ κεχυμένα, τὰ μὲν
ξυνάγῃ, τὰ δὲ χαλᾷ, καὶ πάνθ' ὅ τι μάλιστα
ἱδρύῃ κατὰ φύσιν. For the use of διαχεῖν
compare 45 E, *Philebus* 46 E; and for
ξυνόδων see 58 B, 59 A, and 61 A. Compare also Theophrastos *de sensu* § 84 τὰ
δὲ σὺν τῇ ὑγρότητι τῇ ἐν τῇ γλώττῃ καὶ
διαχυτικὰ καὶ συστατικὰ εἰς τὴν φύσιν
γλυκέα.

3. ὀπός] This is another substance
which it seems impossible precisely to
identify. Martin understands opium;
but this in no wise agrees with the description. It rather is some powerful
vegetable acid, perhaps the juice of the
silphium, as in Hippokrates *de morbis
acutis* vol. II p. 92 Kühn. In Homer
Iliad V 902 it is a liquid used for curd-
ling milk, said to be the juice of the
wild fig: see Aristotle *historia animalium*
III xx 522b 2 πήγνυσι δὲ τὸ γάλα ὀπός τε
συκῆς καὶ πυετία: cf. *meteorologica* IV vii
384ª 20: see too Pliny *natural history*
XVI 72, XXIII 63. The name would
seem to have been applied to vegetable
acids in general, not confined to the sap
of one particular plant: wherefore, although I have acquiesced in the usual
explanation of ἐκ πάντων ἀφορισθὲν τῶν
χυμῶν, it is a question to my mind
whether Thomas Taylor is not more
correct in rendering these words 'is secreted from all liquors'. For ὀπός is no
more 'distinguished' from the other saps
than are wine, oil and honey; if anything, less so. I have adopted the term
'verjuice' as the nearest rendering I
could find, although this, I believe, is
properly confined to the juice of the wild
crab.

60 B—61 C, *c.* XXV. The chief forms
of earth are as follows: (1) *stone* is
formed when in a mixture of earth and
water the water is resolved into air and
issues forth; then the earth that remains
behind is strongly compressed by the
surrounding air and compacted into a
rocky substance: (2) *earthenware* or *pottery* is produced in a similar way, except
that the expulsion of the water is much
more violent and sudden through the action of fire, and therefore the substance
produced is more brittle than the former:
(3) the so-called 'black stone' is formed
when a certain portion of water is left

pores of the mouth to their natural condition, and by this property produces sweetness to the taste,—of this honey is the most general appellation ; lastly that which corrodes the flesh by burning, a sort of frothy substance, distinct from all the other saps, which has been named verjuice.

XXV. Of the different kinds of earth, that which is strained through water becomes a stony mass in the following way. When the commingled water is broken up in the mixing, it changes into the form of air; and having become air it darts up to its own region. Now there was no void surrounding it ; accordingly it gives a thrust to the neighbouring air. And the air, being weighty, when it is thrust and poured around the mass of earth, presses it hard and squeezes it into the spaces which the new-made air quitted. Thus earth, when compressed by air into a mass that will not dissolve in water, forms stone ; of which the transparent sort made of equal and uniform particles is fairer, while that of the opposite kind is less fair. But that

behind, rendering the stone fusible by fire : (4) *alkali* and *salt* are composed of a mixture of earth and water, consisting of fine saline particles of earth from which a large part of the water has been expelled, but which has never been thoroughly compacted, so that the substance is soluble in water: (5) there remain compounds of earth and water which are fusible by fire, but not soluble in water. The reason why this is so is as follows : Earth in its unmodified form is dissoluble by water alone ; for its interstices are large enough to give free passage to the particles of earth and fire: but the larger particles of water, forcing their way in, break up the mass. Earth highly compressed can only be dissolved by fire, for nothing else can find entrance. Water, when most compacted, can be dissolved by fire alone ; when in a less degree, by fire or air. The highest condensation of air can only be dissolved by conversion into another element; the less condensed forms are affected by fire only. Now into a compound of earth and water the particles of water from without can find

no entrance : but fire entering in dislocates the particles of water, and they dislocate the particles of earth, so that the whole compound is broken up and fused. Such substances are, if water predominates in the compound, glass and the like ; if earth, all kinds of wax.

7. κοπῇ] sc. κατὰ τὰ τρίγωνα. The water, becoming air, rushes to join the surrounding air; which then thrusts the earth together, exactly as described in the solidification of metals, 59 A.

11. ἀλύτως ὕδατι] There can be little doubt, I think, that these words are to be taken together, ' insoluble by water '. Martin joins ὕδατι with ξυνωσθεῖσα, 'forced into indissoluble union with water '. But Plato does not say that any of the water is left behind; and we find that when this takes place, the substance is fusible by fire, which is not here the case. Nor is it easy to see how such an inseparable conjunction could exist. The phrase seems pretty clearly contrasted with λυτῶ πάλιν ὑφ' ὕδατος in D.

12. ἡ τῶν ἴσων] i. e. precious stones and crystals. It is clear from this that

220   ΠΛΑΤΩΝΟΣ   [60 C—

ἐξαρπασθὲν καὶ κραυρότερον ἐκείνου ξυστάν, ᾧ γένει κέραμον ἐπω- D
νομάκαμεν, τοῦτο γέγονεν· ἔστι δὲ ὅτε νοτίδος ὑπολειφθείσης χυτὴ
γῆ γενομένη διὰ πυρός, ὅταν ψυχθῇ, γίγνεται τὸ μέλαν χρῶμα
ἔχων λίθος· τὼ δ' αὖ κατὰ ταὐτὰ μὲν ταῦτα ἐκ ξυμμίξεως ὕδατος
5 ἀπομονουμένω πολλοῦ, λεπτοτέρων δὲ ἐκ γῆς μερῶν ἁλμυρώ τε
ὄντε ἡμιπαγῆ γενομένω καὶ λυτὼ πάλιν ὑφ' ὕδατος, τὸ μὲν ἐλαίου
καὶ γῆς καθαρτικὸν γένος λίτρον, τὸ δ' εὐάρμοστον ἐν ταῖς κοινωνίαις ταῖς περὶ τὴν τοῦ στόματος αἴσθησιν ἁλῶν κατὰ λόγον E
νόμου θεοφιλὲς σῶμα ἐγένετο. τὰ δὲ κοινὰ ἐξ ἀμφοῖν ὕδατι μὲν

3 γίγνεται: γέγονε S.   4 ἔχων: ἔχον HSZ.   λίθος: εἶδος H e sua coniectura.
τὼ et cetera dualis numeri scripsi e Schneideri coniectura.   τῷ ceteraque concordantia HSZ.   τὰ A, qui tamen in sequentibus dativum habet.

ἀδάμας cannot be the diamond or any other crystal.

1. **ἐξαρπασθέν**] The construction with this verb seems unique, though it is of course common with ἐξαιρεῖσθαι. The rapid evaporation of the water by fire and the consequent sudden violence of the compression causes the pottery to be hard and brittle. For the rather elaborate form of expression ᾧ γένει ..τοῦτο γέγονεν cf. 40 B καθάπερ ἐν τοῖς πρόσθεν ἐρρήθη, κατ' ἐκεῖνα γέγονε.

2. **χυτὴ γῆ γενομένη**] The reason why the continuance of moisture in the stone renders it fusible by fire is explained below at 61 B.

3. **τὸ μέλαν χρῶμα ἔχων λίθος**] There is evidently some corruption in the text of the mss. The vulgate ἔχον cannot be construed at all: ἔχων is supported by A, but the article is not wanted with μέλαν χρῶμα. Hermann restores grammar by writing εἶδος for λίθος; yet this is not convincing. Nor yet can I acquiesce in the suggestion of the translator in the Engelmann edition, to read λίθου, supplying γένος from the previous sentence. Retaining ἔχων, we might perhaps insert ὁ before τὸ μέλαν χρῶμα. As to the nature of this μέλας λίθος, it would seem to be a substance of volcanic origin, probably lava. Compare Theophrastos de lapidibus § 14 ὁ δὲ λιπαραῖος ἐκφοροῦται τε τῇ καύσει καὶ γίνεται κισηροειδής, ὥσθ' ἅμα τε τὴν χρόαν μεταβάλλειν καὶ τὴν πυκνότητα, μέλας τε γὰρ καὶ λεῖός ἐστι καὶ πυκνὸς ἄκαυστος ὤν. This λιπαραῖος is a volcanic stone from the Lipari islands, which Theophrastos classes among the πυρὶ τηκτά: on being subjected to the action of fire it leaves a residuum which is light and porous like pumice stone. The description of it while still ἄκαυστος seems to agree very well with Plato's μέλας λίθος. Compare too Aristotle *meteorologica* IV vi 383[b] 5 τήκεται δὲ καὶ ὁ λίθος ὁ πυρίμαχος, ὥστε στάζειν καὶ ῥεῖν· τὸ δὲ πηγνύμενον ὅταν ῥυῇ, πάλιν γίγνεται σκληρόν, καὶ αἱ μύλαι τήκονται ὥστε ῥεῖν· τὸ δὲ ῥέον πηγνύμενον τὸ μὲν χρῶμα μέλαν. The μύλαι certainly were made of lava: see Strabo VI ii 3, where he says of the matter ejected from the Liparaean craters, ὕστερον δὲ παγῆναι καὶ γενέσθαι τοῖς μυλίταις λίθοις ἐοικότα τὸν πάγον. It is to be observed that Theophrastos assigns the same cause as Plato for the fusibility of some stones: see *de lapidibus* § 10 τὸ γὰρ τηκτὸν ἔνικμον εἶναι δεῖ καὶ ὑγρότητ' ἔχειν πλείω.

4. **τὼ δ' αὖ**] Schneider's correction seems indispensable: I can see no reasonable way of construing the dative: and why the Engelmann translator declares the emendation to be 'zum Nachtheil des Sinnes' I cannot understand. Soda and salt are compounds of earth and water only partially compacted and consequently soluble in water; which is

which is suddenly deprived of all its moisture by the rapid action of fire and is become more brittle than the first forms the class to which we have given the name of earthenware. Again when some moisture is left behind, earth, after having been fused by fire and again cooled, becomes a certain stone of a black colour. There are also two sorts which in the same manner after the admixture are robbed of a great part of the water, being formed of the finer particles of earth with a saline taste, and becoming only half solid and soluble again by water; of these what purifies from oil and earth is alkali; while that which easily blends with all the combinations of tastes on the palate is, in the words of the ordinance, the god-beloved substance of salt. The bodies which are composed of

not the case with bodies wherein the water and earth have been brought into a complete and stable union.

6. τὸ μὲν ἐλαῖον καὶ γῆς] I do not know that soda is specially applicable to the elimination of earth, and the words καὶ γῆς seem to me to be dubious. Lindau, imputing to Plato 'brevitatem prope similem Thucydidis', somehow extracts from the words the manufacture of soap and of glass: but such more than Pythian tenebricosity of diction, I think, even Thucydides would shrink from. By λίτρον we are to understand natron, or carbonate of soda.

7. τὸ δ' εὐάρμοστον ἐν ταῖς κοινωνίαις] By this Plato means that salt is an agreeable adjunct to many flavours and combinations of flavours.

8. κατὰ λόγον νόμου] This seems plainly to indicate, what would in any case be a natural supposition, that Plato quotes the expression θεοφιλὲς σῶμα from some well-known ordinance relating to sacrificial ceremonies or from some formula used therein: but I have not been able to trace the phrase to any such origin.

9. θεοφιλὲς σῶμα] The application of the epithet θεοφιλὲς to salt is, as aforesaid, probably due to its use for sacrificial and ceremonial purposes, though this is

not suggested by Plutarch in his curious little disquisition on the subject, *quaest. conv.* V 10. Salt was mixed with whole barley (οὐλοχύται) and sprinkled on the head of the victim. This appears to have been the only use of salt in sacrifice among the Greeks; but both in ancient and modern times it was held to be a potent preservative against witchcraft and evil spirits, and many curious customs connected with it are to be found in mediaeval folk-lore. It was likewise used in purifications—see Theokritos XXIV 94

καθαρῷ δὲ πυρώσατε δῶμα θεείῳ
πρᾶτον, ἔπειτα δ' ἄλεσσι μεμιγμένον, ὡς νενόμισται,
θαλλῷ ἐπιρραίνειν ἐστεμμένῳ ἀβλαβὲς ὕδωρ.

Homer terms it 'divine', *Iliad* IX 214 πάσσε δ' ἀλὸς θείοιο. According to a fable mentioned by Aristotle *meteorologica* II iii 359^a 27 it was a gift of Herakles to the Chaonians. In Tacitus *annals* XIII 57 we read that a spot where salt is found was held by the ancient Germans to be peculiarly sacred and in proximity to heaven. The passage of Athenion (apud Athenaeum XIV 79) which Stallbaum quotes as establishing the sacrificial use of salt has an opposite tendency:

οὐ λυτά, πυρὶ δέ, διὰ τὸ τοιόνδε οὕτω ξυμπήγνυται· γῆς ὄγκους
πῦρ μὲν ἀήρ τε οὐ τήκει· τῆς γὰρ ξυστάσεως τῶν διακένων αὐτῆς
σμικρομερέστερα πεφυκότα, διὰ πολλῆς εὐρυχωρίας ἰόντα, οὐ βια-
ζόμενα, ἄλυτον αὐτὴν ἐάσαντα ἄτηκτον παρέσχε· τὰ δὲ ὕδατος
5 ἐπειδὴ μείζω πέφυκε μέρη, βίαιον ποιούμενα τὴν διέξοδον, λύοντα
αὐτὴν τήκει. γῆν μὲν γὰρ ἀξύστατον ὑπὸ βίας οὕτως ὕδωρ μόνον 61 A
λύει, ξυνεστηκυῖαν δὲ πλὴν πυρὸς οὐδέν· εἴσοδος γὰρ οὐδενὶ πλὴν
πυρὶ λέλειπται. τὴν δὲ ὕδατος αὖ ξύνοδον τὴν μὲν βιαιοτάτην
πῦρ μόνον, τὴν δὲ ἀσθενεστέραν ἀμφότερα, πῦρ τε καὶ ἀήρ, δια-
10 χεῖτον, ὁ μὲν κατὰ τὰ διάκενα, τὸ δὲ καὶ κατὰ τὰ τρίγωνα. βίᾳ δὲ
ἀέρα ξυστάντα οὐδὲν λύει πλὴν κατὰ τὸ στοιχεῖον, ἀβίαστον δὲ
κατατήκει μόνον πῦρ. τὰ δὲ δὴ τῶν ξυμμίκτων ἐκ γῆς τε καὶ
ὕδατος σωμάτων, μέχριπερ ἂν ὕδωρ αὐτοῦ τὰ τῆς γῆς διάκενα καὶ B
βίᾳ ξυμπεπιλημένα κατέχῃ, τὰ μὲν ὕδατος ἐπιόντα ἔξωθεν εἴσοδον
15 οὐκ ἔχοντα μέρη περιρρέοντα τὸν ὅλον ὄγκον ἄτηκτον εἴασε, τὰ δὲ

1 ξυμπήγνυται : ξυμπηγνύναι A.   3 φαίνεται ante πεφυκότα habet A.
7 πυρός : πυρί A.

ὅθεν ἔτι καὶ νῦν τῶν προτέρων μεμνημένοι
τὰ σπλάγχνα τοῖς θεοῖσιν ὀπτῶσιν φλογί
ἅλας οὐ προσάγοντες· οὐ γὰρ ἦσαν οὐ-
δέπω
εἰς τὴν τοιαύτην χρῆσιν ἐξευρημένοι.

Originally, says the author, men both ate and sacrificed without salt; and even after they discovered that salt was good to eat, they went on sacrificing in the old way. Among some other nations, e.g. the Jews, salt was very extensively used for sacrificial purposes.

τὰ δὲ κοινὰ ἐξ ἀμφοῖν] We now come to compounds of earth and water. We have indeed had already one such combination, which is λυτὸν ὑφ' ὕδατος: but there the water is hardly a constituent of the solidified mass; the substance has parted with nearly all its moisture, but still remains ἡμιπαγές. Before explaining why these compounds are dissoluble by fire alone, Plato digresses a little to explain the mode in which the several elements are dissolved. Solution and dilatation alone are treated here, not the transmutation of one element into another.

1. **γῆς ὄγκους**] Earth in its normal condition, ἀξύστατον ὑπὸ βίας, is dissolved by water alone, for the interstices in its structure are so large that the minute particles of fire and air can pass in and out without obstruction and do not disturb the fabric: but those of water are too large to make their way without dislocating the particles of earth. When however earth is firmly compacted, ξυνεστηκυῖα, the interstices are so small that only fire can find an entrance.

8. **τὴν μὲν βιαιοτάτην**] Clearly metals are meant.

9. **τὴν δὲ ἀσθενεστέραν**] Ice, snow, hail, and hoar-frost: cf. 59 E. Air dissolves these κατὰ τὰ διάκενα, i.e. by separating the particles; for ice or snow exposed to the air above a certain temperature will melt; but it still retains the form of water. Fire on the other hand, may vaporise it; which means that the corpuscules of water are dissolved and recon-

earth and water combined cannot be dissolved by water, but by fire alone for the following reason. A mass of earth is resolved neither by fire nor by air, because their atoms are smaller than the interstices in its structure, so that they have abundant room to move in and do not force their way, wherefore instead of breaking it up they leave it undissolved: but whereas the parts of water are larger, they make their passage by force and dissolve the mass by breaking it up. Earth then, when it is not forcibly solidified, is thus dissolved by water only; but when it is solidified, only by fire, for no entrance is left except to fire. And of water the most forcible congelation is melted by fire alone, but the more feeble both fire and air break up; the latter by the interstices, the former by the triangles as well. Air, when forcibly condensed, can only be resolved into the elementary triangles, and when uncondensed fire alone dissolves it. In the case of a substance formed of water and earth combined, so long as water occupies the spaces in it that are forcibly compressed, the particles of water arriving from without find no entrance but simply flow round and leave the whole

---

stituted as corpuscules of air: this is dissolution κατὰ τὰ τρίγωνα.

10. βίᾳ δὲ ἀέρα ξυστάντα} Air in its highest condensation can only be resolved κατὰ τὰ τρίγωνα, that is by transmutation into another element. Stallbaum, not understanding this sentence, desires to corrupt it by altering πλὴν to πάλιν. But the text is perfectly sound and has been rightly explained by Martin. Condensed air means cloud: and cloud is ordinarily dissolved into a shower of rain; or, in the case of a thundercloud, lightning issues from it. Plato therefore, holding as he does that the cloud is a form of air, conceives it to be resolved κατὰ τὰ τρίγωνα, in the one case into water, in the other into fire. The agent which produces the metamorphosis is not specified in this instance.

11. ἀβίαστον δὲ κατατήκει] In its normal state air is subject to the influence of fire alone, which dilates it by insinuating its own particles between those of air. Plato must have observed the fact that air expands when heated. Of course it is κατὰ τὰ διάκενα that air yields to the influence of fire alone; for it may be resolved κατὰ τὰ τρίγωνα by either fire or water, on the principles laid down in 56 E.

12. τὰ δὲ δὴ τῶν ξυμμίκτων] Now we come to the reason why substances compounded by earth and water are fused by fire alone. So long as the interspaces between the earthy particles are occupied by the particles of water belonging to the ξύστασις, the particles of water external to it, supposing the body to be plunged in water, can find no entrance; consequently they can produce no effect upon it. But the particles of fire, finding their way in, force themselves between the particles of water and disturb them: and these in their turn, being thrust against the particles of earth, dislocate the latter, and so the structure of the whole mass is broken up and fused.

224   ΠΛΑΤΩΝΟΣ   [61 B—

πυρὸς εἰς τὰ τῶν ὑδάτων διάκενα εἰσιόντα, ὅπερ ὕδωρ γῆν, τοῦτο
ἀπεργαζόμενα, τηχθέντι τῷ κοινῷ σώματι ῥεῖν μόνα αἴτια ξυμβέβηκε. τυγχάνει δὲ ταῦτα ὄντα, τὰ μὲν ἔλαττον ἔχοντα ὕδατος
ἢ γῆς τό τε περὶ τὴν ὕαλον γένος ἅπαν ὅσα τε λίθων χυτὰ εἴδη
5 καλεῖται, τὰ δὲ πλέον ὕδατος αὖ πάντα ὅσα κηροειδῆ καὶ θυμιατικὰ C
σώματα ξυμπήγνυται.

XXVI. Καὶ τὰ μὲν δὴ σχήμασι κοινωνίαις τε καὶ μεταλλα-
γαῖς εἰς ἄλληλα πεποικιλμένα εἴδη σχεδὸν ἐπιδέδεικται· τὰ δὲ
παθήματα αὐτῶν δι᾽ ἃς αἰτίας γέγονε πειρατέον ἐμφανίζειν. πρῶτον
10 μὲν οὖν ὑπάρχειν αἴσθησιν δεῖ τοῖς λεγομένοις ἀεί· σαρκὸς δὲ καὶ
τῶν περὶ σάρκα γένεσιν, ψυχῆς τε ὅσον θνητόν, οὔπω διεληλύ-

1 post τοῦτο delevi πῦρ ἀέρα, quae dant codices omnes et HSZ.   τοῦτο δέ S.
7 σχήμασι : σχήματα HSZ.   8 εἴδη : ἤδη A.

1. ὅπερ ὕδωρ γῆν, τοῦτο ἀπεργαζόμενα] The words πῦρ ἀέρα, which in the mss. follow τοῦτο, I have rejected for more than one reason; the chief of which is that they are absolute nonsense. We have seen above that water acts upon earth by thrusting its particles between those of earth and forcing them asunder: likewise we have just seen that fire acts upon water by thrusting its particles between those of water and forcing them asunder. Therefore, as Plato says, fire has precisely the same action upon water that water has upon earth. But what conceivable sense is there in introducing air? Air neither is any constituent of the compound nor plays any part in its fusion: it is altogether beside the question. A minor, though still substantial, reason for rejecting the words is the grammar. If we retain πῦρ ἀέρα, not only is πῦρ out of all construction, but ἀπεργαζόμενα is left forlorn of any substantive wherewith to agree. On the other hand the rejection of those two words, which I conceive to have been inserted by a copyist in an over antithetical frame of mind, restores both sense and grammar. I suspect however that Plato's original words were τοῦθ᾽ ὕδωρ ἀπεργαζόμενα and that ὕδωρ was expelled by the two intruding elements, πῦρ ἀέρα: its insertion would be a gain to the sense.

4. λίθων χυτὰ εἴδη] For example the μέλαν χρῶμα ἔχων λίθος mentioned above, which we saw to have an admixture of water in its composition.

61 C—64 A, c. xxvi. In order to set forth thoroughly the properties of matter, we ought to explain the nature of their action upon our bodies and the nature of the bodies that are so affected. As both these subjects cannot be dealt with at once, let us first examine the sensible qualities of things. The sensation of heat is due to the penetrating power of fire, which enters and divides the flesh: cold is a contraction of the flesh under the influence of moisture. Hardness and softness depend on the form of the constituent corpuscule, the cube being most stable and therefore most resisting. Concerning heavy and light, it is necessary to clear away some popular misconceptions. It is common to speak as if the universe were divided into two regions, upper and lower, to the latter of which all heavy bodies naturally tend. But the truth is that, the universe being a sphere, there is no such thing as an upper and a lower region in it. For if one were to travel round the universe he

bulk undissolved; but those of fire enter into the interstices of the water, and acting upon it as water does upon earth, can alone cause the combined mass to melt and become liquid. In this class those which have less water than earth are all kinds of glass and all stones that are called fusible; and those which contain more water include all formations like wax and frankincense.

XXVI. Now all the manifold forms that arise from diverse shapes and combinations and changes from one to another have been pretty fully set forth; next we must try to explain their affections and the causes that lead to them. First we must assign to all the substances we have described the property of causing sensation. But the origin of flesh and all that belongs to it and of the mortal part of soul we have not yet discussed.

would be forced to call the same point successively above and below: since it would at one time be overhead, at another beneath him. The true explanation of gravity and attraction is as follows. Owing to the vibration of the universe, every element has its proper region in space; and every portion of any element which is in an alien sphere endeavours to escape to its own sphere. For this reason, if we raise portions of earth into the region of air, they tend to make their way back to earth again, and the larger portion strives more forcibly so to return than the smaller. Hence we say that earth is 'heavy' and tends 'downward'; while fire, because it seeks to fly away from earth to its own home, we say is 'light' and tends 'upward'. But could we reach the home of fire and raise portions of it into the air, we should find this condition reversed: fire would be 'heavy' and tend 'downwards' to its own home, and earth would be 'light' and tend 'upwards' to the home of earth. And so the gravitation of all bodies depends altogether upon their position in space relatively to their proper region; and the 'weight' of any body is simply the attraction which draws it towards its own home. Such is the nature of light and heavy: roughness is due to hardness and irregularity in the substance, smoothness to regularity and density.

7. καὶ τὰ μὲν δὴ σχήμασι] Having explained the structure of the various forms in which the four εἴδη appear and their combinations, our next task is to set forth the causes of the sensations they produce in us. For σχήμασι the editors from Stallbaum onwards, with the exception of Martin, read σχήματα *sub silentio*. This reading is not mentioned by Bekker, and no ms. testimony is by any one cited for it. It is by no means an improvement; and since I can find neither its origin nor its authority I have suffered it ἐρήμην ὀφλεῖν and reverted to the old reading. Ficinus translates 'eas species, quae figuris commutationibusque invicem variantur.'

8. τὰ δὲ παθήματα] The word πάθημα is here used in a rather peculiar manner. Elsewhere it denotes the impression sustained by the percipient subject from the external agent—see 64 B, C. But here πάθημα signifies a quality pertaining to the object which produces this impression on the subject. We have a similar unusual significance in ὑπάρχειν αἴσθησιν below; where αἴσθησις denotes the property of exciting sensation.

11. ψυχῆς τε ὅσον θνητόν] See 69 D, where the term is explained.

θαμεν. τυγχάνει δὲ οὔτε ταῦτα χωρὶς τῶν περὶ τὰ παθήματα ὅσα αἰσθητὰ οὔτ' ἐκεῖνα ἄνευ τούτων δυνατὰ ἱκανῶς λεχθῆναι, τὸ δὲ D ἅμα σχεδὸν οὐ δυνατόν· ὑποθετέον δὴ πρότερον θάτερα, τὰ δ' ὑποτεθέντα ἐπάνιμεν αὖθις. ἵνα οὖν ἑξῆς τὰ παθήματα λέγηται 5 τοῖς γένεσιν, ἔστω πρότερα ἡμῖν τὰ περὶ σῶμα καὶ ψυχὴν ὄντα. πρῶτον μὲν οὖν ᾗ πῦρ θερμὸν λέγομεν, ἴδωμεν ὧδε σκοποῦντες, τὴν διάκρισιν καὶ τομὴν αὐτοῦ περὶ τὸ σῶμα ἡμῶν γιγνομένην ἐννοηθέντες. ὅτι μὲν γὰρ ὀξύ τι τὸ πάθος, πάντες σχεδὸν αἰσθανόμεθα· E τὴν δὲ λεπτότητα τῶν πλευρῶν καὶ γωνιῶν ὀξύτητα τῶν τε μορίων 10 σμικρότητα καὶ τῆς φορᾶς τὸ τάχος, οἷς πᾶσι σφοδρὸν ὂν καὶ τομὸν ὀξέως τὸ προστυχὸν ἀεὶ τέμνει, λογιστέον ἀναμιμνησκομένοις τὴν τοῦ σχήματος αὐτοῦ γένεσιν, ὅτι μάλιστα ἐκείνη καὶ 62 A οὐκ ἄλλη φύσις διακρίνουσα ἡμῶν κατὰ σμικρά τε τὰ σώματα κερματίζουσα τοῦτο ὃ νῦν θερμὸν λέγομεν εἰκότως τὸ πάθημα καὶ 15 τοὔνομα παρέσχε. τὸ δ' ἐναντίον τούτων κατάδηλον μέν, ὅμως δὲ μηδὲν ἐπιδεὲς ἔστω λόγου. τὰ γὰρ δὴ τῶν περὶ τὸ σῶμα ὑγρῶν μεγαλομερέστερα εἰσιόντα, τὰ σμικρότερα ἐξωθοῦντα, εἰς τὰς ἐκείνων οὐ δυνάμενα ἕδρας ἐνδῦναι, ξυνωθοῦντα ἡμῶν τὸ νοτερὸν ἐξ B

2 αἰσθητά : αἰσθητικὰ AHSZ. 4 ὕστερα ante ὑποτεθέντα dat S.
15 τούτων : τούτῳ SZ.

1. **οὔτε ταῦτα χωρίς**] To explain the action of external objects upon the human body involves a description of the structure of the said body. But as two subjects cannot be expounded at once, we must assume (ὑποθετέον) one, and afterwards examine what we have assumed.

ὅσα αἰσθητά] I have taken upon me to make this correction of the ms. αἰσθητικί, which appears to me unmeaning. The two subjects to be handled are (1) the structure of flesh &c, how it is capable of receiving impressions, (2) the properties of objects, how they are capable of producing impressions. But this latter is expressed by αἰσθητά, not αἰσθητικά: how can the objects in this relation be termed sentient? The corruption has arisen, I doubt not, from failure to apprehend the peculiar significance of παθήματα. A similar confusion is found in 58 D, κινητικὸν for κινητόν.

5. **ἔστω πρότερα ἡμῖν**] That is to say, let us first assume their nature and construction; not let us first examine them. Plato, for the sake of continuity in his exposition, takes the παθήματα first, postponing the account of σαρκὸς γένεσις.

6. **ᾗ πῦρ θερμόν**] So then θερμὸν is the πάθημα of πῦρ: we have to inquire how fire acts, so as to possess this πάθημα.

τὴν διάκρισιν] Aristotle demurs to this explanation: see *de gen. et corr.* II ii 329[b] 26 θερμὸν γάρ ἐστι τὸ συγκρῖνον τὰ ὁμογενῆ (τὸ γὰρ διακρίνειν, ὅπερ φασὶ ποιεῖν τὸ πῦρ, συγκρίνειν ἐστὶ τὰ ὁμόφυλα· συμβαίνει γὰρ ἐξαιρεῖν τὰ ἀλλότρια), ψυχρὸν δὲ τὸ συνάγον καὶ συγκρῖνον ὁμοίως τά τε συγγενῆ καὶ τὰ μὴ ὁμόφυλα. Theophrastos also complains that Plato does not explain heat and cold on the same principle: *de sensu* § 87 ἄτοπον δὲ καὶ τούτου πρῶτον μὲν τὸ μὴ πάντα ὁμοίως

Now this cannot be adequately dealt with apart from the affections of sense, nor yet can the latter without the former; yet to treat them both at once is hardly possible. We must assume one side then, and afterwards we will return to examine what we assumed. In order then that the properties of the several elements may be discussed in due order, let us first assume the nature of body and soul. First then let us see what we mean by calling fire hot; which we must consider in the following way, remembering the power of dividing and cutting which fire exercises upon our body. That the sensation is a sharp one we are all well enough aware: and the fineness of the edges and sharpness of the angles, besides the smallness of its particles and the swiftness of its motion, all of which qualities combine to render it so vehement and piercing as keenly to cut whatever meets it—all this we must take into account, remembering the nature of its figure, that this more than any other kind penetrates our body and minutely divides it, whence the sensation that we now call heat justly derives its quality and its name. The opposite condition, though obvious enough, still must not lack an explanation. When the larger particles of moisture which surround the body enter into it, they displace the smaller, and because they are not able to pass into their places, they compress the moisture within

ἀποδοῦναι, μηδὲ ὅσα τοῦ αὐτοῦ γένους· ὁρίσας γὰρ τὸ θερμὸν σχήματι τὸ ψυχρὸν οὐχ ὡσαύτως ἀπέδωκεν. But it seems to me that the action of moisture in producing cold does, in Plato's account, depend on the form of the particles. It must at any rate be allowed that Plato's explanation has over Aristotle's, as propounded in the passage above cited, the advantage of clearness and simplicity.

11. λογιστέον ἀναμιμνησκομένοις] i.e. if we call to mind the form of its constituent particles, we cannot fail to see that fire must necessarily have a highly penetrating power.

14. ὃ νῦν θερμὸν λέγομεν] As is clearly indicated by νῦν, an etymology is intended; and the only possible reference is to κερματίζουσα. Plato would say that the word was originally κερμόν, 'cutting'.

πάθημα is again used as in 61 c.

16. τῶν περὶ τὸ σῶμα ὑγρῶν] Water then is for Plato the preeminently cold element: this view was shared by Aristotle; see *meteorologica* IV xi 389^b 15. Chrysippos said air: Plutarch in his treatise *de primo frigido* argues fantastically in favour of earth. Plato's theory of cold is this. The larger particles of moisture surrounding the body displace the smaller moist particles in the body, but owing to their size cannot occupy the place of the latter. Hence by the περίωσις the substance of the body is compressed to fill up the vacant spaces. This, in its extremest form, is freezing; and the mutual repulsion of the corporeal particles thus forced into unnatural con-

15—2

228   ΠΛΑΤΩΝΟΣ   [62 B—

ἀνωμάλου κεκινημένου τε ἀκίνητον δι' ὁμαλότητα καὶ τὴν ξύνωσιν
ἀπεργαζόμενα πήγνυσι. τὸ δὲ παρὰ φύσιν ξυναγόμενον μάχεται
κατὰ φύσιν αὐτὸ ἑαυτὸ εἰς τοὐναντίον ἀπωθοῦν. τῇ δὴ μάχῃ
καὶ τῷ σεισμῷ τούτῳ τρόμος καὶ ῥῖγος ἐτέθη, ψυχρόν τε τὸ πάθος
5 ἅπαν τοῦτο καὶ τὸ δρῶν αὐτὸ ἔσχεν ὄνομα. σκληρὸν δέ, ὅσοις ἂν
ἡμῶν ἡ σὰρξ ὑπείκῃ· μαλακὸν δέ, ὅσα ἂν τῇ σαρκί· πρὸς ἄλληλά
τε οὕτως. ὑπείκει δὲ ὅσον ἐπὶ σμικροῦ βαίνει· τὸ δὲ ἐκ τετραγώ-
νων ὂν βάσεων, ἅτε βεβηκὸς σφόδρα, ἀντιτυπώτατον εἶδος, ὅ τί τε C
ἂν εἰς πυκνότητα ξυνιὸν πλείστην ἀντίτονον ᾖ μάλιστα. βαρὺ δὲ
10 καὶ κοῦφον μετὰ τῆς τοῦ κάτω φύσεως ἄνω τε λεγομένης ἐξεταζό-
μενον ἂν δηλωθείη σαφέστατα. φύσει γὰρ δή τινας τόπους δύο
εἶναι διειληφότας διχῇ τὸ πᾶν ἐναντίους, τὸν μὲν κάτω, πρὸς ὃν
φέρεται πάνθ' ὅσα τινὰ ὄγκον σώματος ἔχει, τὸν δ' ἄνω, πρὸς ὃν
ἀκουσίως ἔρχεται πᾶν, οὐκ ὀρθὸν οὐδαμῇ νομίζειν· τοῦ γὰρ παντὸς

7 τε : γε A.   10 τοῦ ante κάτω omittunt SZ.

tiguity is trembling and shivering. Cf. *Philebus* 32 A.

2. **μάχεται κατὰ φύσιν**] Plutarch gives a somewhat different account of shivering: *de primo frigido* vi ὑφ' ὧν οὐκ ἀεὶ φεύγει καὶ ἀπολείπει τὸ θερμόν, ἀλλὰ πολλάκις ἐγκαταλαμβανόμενον ἀνθίσταται καὶ μάχεται, τῇ μάχῃ δ' αὐτῶν ὄνομα φρίκη καὶ τρόμος.

4. **τὸ πάθος...καὶ τὸ δρῶν**] i.e. we apply the term cold both to ice and to the sensation it produces in us.

6. **πρὸς ἄλληλά τε οὕτως**] i.e. the terms hard and soft are applied to them in relation to each other, as well as in relation to our flesh: thus lead, which yields to iron, is soft in relation to iron, though hard in relation to our flesh. Theophrastos takes exception to this definition also: *de sensu* § 87 ἐπεὶ δὲ μαλακὸν τὸ ὑπεῖκον, φανερὸν ὅτι τὸ ὕδωρ καὶ ὁ ἀὴρ καὶ τὸ πῦρ μαλακά· φησὶ γὰρ ὑπείκειν τὸ μικρὰν ἔχων βάσιν, ὥστε τὸ πῦρ ἂν εἴη μαλακώτατον. δοκεῖ δὲ τούτων οὐθὲν οὐδ' ὅλως τὸ μὴ μένον ἀλλὰ μεθιστάμενον εἶναι μαλακόν, ἀλλὰ τὸ εἰς τὸ βάθος ὑπεῖκον ἄνευ μεταστάσεως. Herein he follows Aristotle *meteorologica* IV iv 382ᵃ 12 μαλακὸν δὲ τὸ ὑπεῖκον τῷ μὴ ἀντι-περιίστασθαι· τὸ γὰρ ὕδωρ οὐ μαλακόν· οὐ γὰρ ὑπείκει τῇ θλίψει τὸ ἐπίπεδον εἰς βάθος ἀλλ' ἀντιπεριίσταται. This is of course merely a question of names.

9. **βαρὺ δὲ καὶ κοῦφον**] Here we have Plato's theory of attraction and gravitation, which is unquestionably by far the most lucid and scientific that has been propounded by any ancient authority. The popular notion was that the portion of the universe which we occupy is κάτω, and that above our heads ἄνω: βαρύ is that which has a tendency to move κάτω, κοῦφον that which has a tendency to move ἄνω, or at least a slighter tendency κάτω. Plato clearly saw the unscientific nature of this conception. The explanation he offered in its place was this. We have seen that the vibration of the ὑποδοχή tends to sift the four elements into separate regions in space; but owing to the πλῆσις portions of them are found scattered all over the universe. A mass of any element which finds itself in an alien sphere endeavours with all its might to escape to its proper region: and it is just this endeavour which constitutes its gravity: attraction is the effort of all matter to obey the sifting

c] ΤΙΜΑΙΟΣ. 229

us; and whereas it was irregular and mobile, they render it immovable owing to uniformity and contraction, and so it becomes rigid. And what is against nature contracted in obedience to nature struggles and thrusts itself apart; and to this struggling and quaking has been given the name of trembling and shivering: and both the effect and the cause of it are in all cases termed 'cold'.

'Hard' is the name given to all things to which our flesh yields; and 'soft' to those which yield to the flesh; and so also they are termed in their relation to each other. Those which yield are such as have a small base of support; and the figure with square surfaces, as it is most firmly based, is the most stubborn form; so too is whatever from the intensity of its compression offers the strongest resistance.

Of 'heavy' and 'light' we shall find the clearest explanation if we examine them together with the so-called 'below' and 'above'. That there are naturally two opposite regions, dividing the universe between them, one the lower, to which sink all things that have material bulk, the other upper, to which everything rises against its will, is altogether a false opinion. For

force which is in nature. So when we raise any substance of an earthy nature, the earthward impulse which we observe in it is not due to the fact that the earth is the downward region whither all heavy bodies tend to fall, but to this sifting force which causes the mass of earth to strive towards its own sphere.

Aristotle in his criticism of Plato's theory (*de caelo* IV ii 308ª 34 foll.) simply ignores the whole point of it from beginning to end. The extent to which he has done so may be gathered from the following citation: ὥστε οὐ δι' ὀλιγότητα τῶν τριγώνων ἐξ ὧν συνεστάναι φασὶν ἕκαστον αὐτῶν, τὸ πῦρ ἄνω φέρεσθαι πέφυκεν· τό τε γὰρ πλεῖον ἧττον ἂν ἐφέρετο καὶ βαρύτερον ἂν ἦν ἐκ πλειόνων ὂν τριγώνων. νῦν δὲ φαίνεται τοὐναντίον· ὅσῳ γὰρ ἂν ᾖ πλεῖον, κουφότερόν ἐστι καὶ ἄνω φέρεται θᾶττον. That is to say, Aristotle actually urges the fact that a larger body of flame has a stronger upward tendency than a smaller as an objection to Plato's theory; whereas it is precisely what Plato affirms must on his principles inevitably be the case. Aristotle's own doctrine differed but little from the vulgar notion on the subject: see *physica* IV v 212ª 24 ὥστ' ἐπεὶ τὸ μὲν κοῦφον τὸ ἄνω φερόμενόν ἐστι φύσει, τὸ δὲ βαρὺ τὸ κάτω, τὸ μὲν πρὸς τὸ μέσον περιέχον πέρας κάτω ἐστί, καὶ αὐτὸ τὸ μέσον, τὸ δὲ πρὸς τὸ ἔσχατον ἄνω, καὶ αὐτὸ τὸ ἔσχατον. Theophrastos in his statement of the Platonic theory (*de sensu* § 88) shows a clearer comprehension of it, though marred by a hankering after a ἁπλῶς βαρὺ καὶ κοῦφον. Anaxagoras divided space into ἄνω and κάτω: see Diogenes Laertius II § 8: but Aristotle says neither he nor Empedokles gave any definition of βαρὺ and κοῦφον: *de caelo* IV ii 309ª 20.

230   ΠΛΑΤΩΝΟΣ   [62 C—

οὐρανοῦ σφαιροειδοῦς ὄντος, ὅσα μὲν ἀφεστῶτα ἴσον τοῦ μέσου D
γέγονεν ἔσχατα, ὁμοίως αὐτὰ χρὴ ἔσχατα πεφυκέναι, τὸ δὲ μέσον
τὰ αὐτὰ μέτρα τῶν ἐσχάτων ἀφεστηκὸς ἐν τῷ καταντικρὺ νομίζειν
δεῖ πάντων εἶναι. τοῦ δὴ κόσμου ταύτῃ πεφυκότος τί τῶν εἰρημέ-
5 νων ἄνω τις ἢ κάτω τιθέμενος οὐκ ἐν δίκῃ δόξει τὸ μηδὲν προσῆκον
ὄνομα λέγειν; ὁ μὲν γὰρ μέσος ἐν αὐτῷ τόπος οὔτε κάτω πεφυκὼς
οὔτε ἄνω λέγεσθαι δίκαιος, ἀλλ' αὐτὸ ἐν μέσῳ· ὁ δὲ πέριξ οὔτε δὴ
μέσος οὔτ' ἔχων διάφορον αὐτοῦ μέρος ἕτερον θατέρου μᾶλλον πρὸς
τὸ μέσον ἤ τι τῶν καταντικρύ. τοῦ δὲ ὁμοίως πάντῃ πεφυκότος ποῖά
10 τις ἐπιφέρων ὀνόματα αὐτῷ ἐναντία καὶ πῇ καλῶς ἂν ἡγοῖτο λέγειν;
εἰ γάρ τι καὶ στερεὸν εἴη κατὰ μέσον τοῦ παντὸς ἰσοπαλές, εἰς
οὐδὲν ἄν ποτε τῶν ἐσχάτων ἐνεχθείη διὰ τὴν πάντῃ ὁμοιότητα 63 A
αὐτῶν· ἀλλ' εἰ καὶ περὶ αὐτὸ πορεύοιτό τις ἐν κύκλῳ, πολλάκις ἂν
στὰς ἀντίπους ταὐτὸν αὑτοῦ κάτω καὶ ἄνω προσείποι. τὸ μὲν γὰρ
15 ὅλον, καθάπερ εἴρηται νῦν δή, σφαιροειδὲς ὄν, τόπον τινὰ κάτω,
τὸν δὲ ἄνω λέγειν ἔχειν οὐκ ἔμφρονος· ὅθεν δὲ ὠνομάσθη ταῦτα
καὶ ἐν οἷς ὄντα εἰθίσμεθα δι' ἐκεῖνα καὶ τὸν οὐρανὸν ὅλον οὕτω
διαιρούμενοι λέγειν, ταῦτα διομολογητέον ὑποθεμένοις τάδε ἡμῖν. B
εἴ τις ἐν τῷ τοῦ παντὸς τόπῳ, καθ' ὃν ἡ τοῦ πυρὸς εἴληχε μάλιστα
20 φύσις, οὗ καὶ πλεῖστον ἂν ἠθροισμένον εἴη πρὸς ὃ φέρεται, ἐπεμβὰς

10 ἂν omittit A.   20 ἐπεμβάς: ἐπαναβὰς SZ.

3. **ἐν τῷ καταντικρύ**] The universe being a sphere, every point on the circumference (ἔσχατα) has precisely the same relation as every other to the centre, which is right opposite to each. There is therefore nothing whereby one portion of the circumference can be differentiated from another so as to justify us in terming one ἄνω and the other κάτω. Nor yet will Plato allow the correctness of terming the centre κάτω, as Aristotle subsequently did, nor ἄνω either: it is just 'the centre'—αὐτὸ ἐν μέσῳ. However in *Phaedo* 112 E the centre of the earth is regarded as the lowest point: but in that passage physics are largely tempered with mythology.

8. **μᾶλλον πρὸς τὸ μέσον**] That is, no part of the circumference has any difference in its relations towards the centre, as compared with any part on the opposite side.

11. **εἰ γάρ τι καὶ στερεὸν εἴη**] If there were a solid body at the centre of the universe (such as the earth in the Platonic cosmology actually was), such is the uniformity of the sphere in which it is, that it would have no tendency towards any one point in the circumference rather than any other: therefore for it there would be no ἄνω nor κάτω in any direction. Compare *Phaedo* 109 A ἰσόρροπον γὰρ πρᾶγμα ὁμοίου τινὸς ἐν μέσῳ τεθὲν οὐχ ἕξει μᾶλλον οὐδ' ἧττον οὐδαμόσε κλιθῆναι, ὁμοίως δ' ἔχον ἀκλινὲς μενεῖ.

13. **εἰ καὶ περὶ αὐτὸ πορεύοιτό τις**] A second illustration of the want of significance in the terms ἄνω and κάτω is this. If one were to travel round the circumference, he would be forced, if he used the words in the popular way, to call

since the form of the universe is spherical, all the extreme points, being equally distant from the centre, are by their very nature equally extreme; and the centre, being equally distant from all the extremes, ought to be regarded as opposite to all such points. This being the nature of the universe, how can one describe any of the said points as upper or lower, without justly being censured for using irrelevant terms? For the centre cannot properly be described as being above or below, but simply at the centre; while the circumference is neither itself central nor has any difference between the points on its surface, so that one has a different relation to the centre from an opposite point. Since then it is everywhere uniform, how and in what sense can we suppose we are speaking correctly if we use terms which imply opposition? For suppose in the midst of the universe there were a solid body in equilibrium, it would have no tendency towards any point in the circumference, owing to the absolute uniformity of the whole: indeed if we were to walk round the sphere, frequently, as we stood at the antipodes of our former position, we should call the same point on its surface successively 'above' and 'below'. For this universe being spherical, as we just now said, no rational man can speak of one region as upper, of another as lower: however whence these names were derived and under what conditions we use them to express this division of the entire universe, we may explain on the following hypothesis. In that region of the universe which is specially allotted to the element of fire, where indeed the greatest mass would be collected of that to which it is attracted, if one should attain to this place, and,

the same point both ἄνω and κάτω: for the point that now is κάτω will be ἄνω when he reaches the antipodes thereof. I think we must conceive the traveller to be moving round the inside of the circumference of the universe; not, as Stallbaum supposes, round the στερεόν. For were he walking round the latter, every point in it would always be κάτω in the vulgar sense.

19. καθ' ὅν] Stallbaum would expunge καθ'. But I think we may readily

supply an object with εἴληχε, 'in which fire has its allotted place.' Compare Aeschylus *Seven against Thebes* 423 Καπανεὺς δ' ἐπ' Ἠλέκτραισιν εἴληχεν πύλαις. See too 41 C above.

20. πλεῖστον ἂν ἠθροισμένον εἴη] Although detached portions of fire are to be found in all parts of the universe, yet, since all fire is perpetually struggling to reach its proper home, naturally the great bulk of the element will be accumulated in that region.

ἐπ' ἐκεῖνο καὶ δύναμιν εἰς τοῦτο ἔχων, μέρη τοῦ πυρὸς ἀφαιρῶν
ἰσταίη, τιθεὶς εἰς πλάστιγγας, αἴρων τὸν ζυγὸν καὶ τὸ πῦρ ἕλκων
εἰς ἀνόμοιον ἀέρα βιαζόμενος, δῆλον ὡς τοὔλαττόν που τοῦ μείζονος
ῥᾷον βιᾶται· ῥώμῃ γὰρ μιᾷ δυοῖν ἅμα μετεωριζομένοιν τὸ μὲν C
5 ἔλαττον μᾶλλον, τὸ δὲ πλέον ἧττον ἀνάγκη που κατατεινόμενον
ξυνέπεσθαι τῇ βίᾳ, καὶ τὸ μὲν πολὺ βαρὺ καὶ κάτω φερόμενον
κληθῆναι, τὸ δὲ σμικρὸν ἐλαφρὸν καὶ ἄνω. ταὐτὸν δὴ τοῦτο δεῖ
φωρᾶσαι δρῶντας ἡμᾶς περὶ τόνδε τὸν τόπον. ἐπὶ γὰρ γῆς βε-
βῶτες, γεώδη γένη διιστάμενοι καὶ γῆν ἐνίοτε αὐτὴν ἕλκομεν εἰς
10 ἀνόμοιον ἀέρα βίᾳ καὶ παρὰ φύσιν, ἀμφότερα τοῦ ξυγγενοῦς ἀντε-
χόμενα. τὸ δὲ σμικρότερον ῥᾷον τοῦ μείζονος βιαζομένοις εἰς τὸ D
ἀνόμοιον πρότερον ξυνέπεται· κοῦφον οὖν αὐτὸ προσειρήκαμεν καὶ
τὸν τόπον εἰς ὃν βιαζόμεθ' ἄνω, τὸ δ' ἐναντίον τούτοις πάθος βαρὺ
καὶ κάτω. ταῦτ' οὖν δὴ διαφόρως ἔχειν αὐτὰ πρὸς αὑτὰ ἀνάγκη
15 διὰ τὸ τὰ πλήθη τῶν γενῶν τόπον ἐναντίον ἄλλα ἄλλοις κατέχειν·
τὸ γὰρ ἐν ἑτέρῳ κοῦφον ὂν τόπῳ τῷ κατὰ τὸν ἐναντίον τόπον
ἐλαφρῷ καὶ τῷ βαρεῖ τὸ βαρὺ τῷ τε κάτω τὸ κάτω καὶ τῷ ἄνω τὸ
ἄνω πάντ' ἐναντία καὶ πλάγια καὶ πάντως διάφορα πρὸς ἄλληλα E
ἀνευρεθήσεται γιγνόμενα καὶ ὄντα· τόδε γε μὴν ἔν τι διανοητέον
20 περὶ πάντων αὐτῶν, ὡς ἡ μὲν πρὸς τὸ ξυγγενὲς ὁδὸς ἑκάστοις οὖσα

1. **πυρὸς ἀφαιρῶν ἰσταίη**] Our misconception about the nature of light and heavy is due to this cause. We are confined to this region of earth and water; and when we weigh masses of earth or water, we find that they always have a tendency in one direction. This tendency we call weight, and the direction in which they tend we call downward; and because earth and water resist our efforts to remove them from their own region, we conceive of them as absolutely heavy. Fire, on the other hand, so far from resisting any effort to lift it from the region which earth and water seek, has a natural impulse to fly from it; whence we conceive of fire as absolutely light. But this opinion is due to the limitation of our experience to one sphere. Could we reach the home of fire and endeavour to raise portions of it into the region of air, as we now do with earth and water, we should then find that fire resisted our efforts precisely as earth and water do now: it would have a similar tendency to revert to its proper region, and would be 'heavy'; while earth or water, so far from resisting the effort to remove it from the region of fire, would have a natural impulse to fly off in the direction of earth, and would be 'light'. Accordingly, whereas now we call the region of earth 'down', and things that tend towards it 'heavy', we should, in the supposed case, call the region of fire 'down' and things that tend towards fire 'heavy'. There is therefore no such thing as absolute lightness and heaviness; all things are light or heavy only relatively to the region in which they are situate.

4. **βιᾶται** is middle, as in Aeschylus *Agamemnon* 385 βιᾶται δ' ἁ τάλαινα πειθώ.

5. **ἧττον** is of course to be joined with ξυνέπεσθαι.

7. **ταὐτὸν δὴ τοῦτο δεῖ φωρᾶσαι**]

ΤΙΜΑΙΟΣ.

acquiring the needful power, should separate portions of fire and weigh them in scales, when he raises the balance and forcibly drags the fire into the alien air, evidently he overpowers the smaller portion more easily than the larger: for when two masses are raised at once by the same force, necessarily the smaller yields more readily to the force, the larger, owing to its resistance, less readily: hence the larger mass is said to be heavy and to tend downwards, the smaller to be light and to tend upwards. This is exactly what we ought to detect ourselves doing in our own region. Moving as we do on the earth, we separate portions of earthy substances or sometimes earth itself, and drag them into the alien air with unnatural force, for each portion clings to its own kind. Now the smaller mass yields more readily to our force than the larger and follows quicker into the alien element; therefore we call it 'light', and the place into which we force it 'above'; while to the opposite conditions we apply the terms 'heavy' and 'below'. Now that these mutual relations should vary is inevitable, because the bulk of the several elements occupy contrary positions in space. For as between a body that is light in one region and a body that is light in the opposite region, or as between two that are heavy, as well as upper and lower, all the lines of attraction will be found to become and remain relatively contrary and transverse and different in every possible way. But with all of them this one principle is to be borne in mind, that in every case it is the tendency towards the kindred element

What escapes our notice is that in lifting earth from earth, we are not lifting it 'up', but simply out of its own region. This we should realise if we tried the experiment on fire in the fire-home, because we should find our customary notions of up and down inverted.

10. ἀμφότερα] i.e. the earth in each scale.

14. ταῦτ' οὖν δὴ διαφόρως ἔχειν] These relations of 'light' and 'heavy' have no absolute fixity, because, as he goes on to explain, the same thing which is light in one region is heavy in another; and consequently the direction of 'up' and 'down' is reversed and altered in a variety of ways.

18. ἐναντία καὶ πλάγια] Different substances which are imprisoned in an alien region will have the lines of their attraction in some instances opposite, as in the case of masses of fire and of earth in the region of air, in others the lines may be inclined at any angle (πλάγια) one to another, according to the position occupied by the two bodies in relation to their proper regions. Plato is insisting that the lines of gravitation are not parallel.

20. ἡ μὲν πρὸς τὸ ξυγγενὲς ὁδός] Here we have the definite statement in so many words that gravity is just the attraction

234   ΠΛΑΤΩΝΟΣ   [63 E—

βαρὺ μὲν τὸ φερόμενον ποιεῖ, τὸν δὲ τόπον εἰς ὃν τὸ τοιοῦτον
φέρεται κάτω, τὰ δὲ τούτοις ἔχοντα ὡς ἑτέρως θάτερα. περὶ δὴ
τούτων αὖ τῶν παθημάτων ταῦτα αἴτια εἰρήσθω. λείου δ' αὖ καὶ
τραχέος παθήματος αἰτίαν πᾶς που κατιδὼν καὶ ἑτέρῳ δυνατὸς ἂν
5 εἴη λέγειν· σκληρότης γὰρ ἀνωμαλότητι μιχθεῖσα, τὸ δ' ὁμαλότης 64 A
πυκνότητι παρέχεται.

XXVII. Μέγιστον δὲ καὶ λοιπὸν τῶν κοινῶν περὶ ὅλον τὸ
σῶμα παθημάτων τὸ τῶν ἡδέων καὶ τῶν ἀλγεινῶν αἴτιον ἐν οἷς
διεληλύθαμεν, καὶ ὅσα διὰ τῶν τοῦ σώματος μορίων αἰσθήσεις
10 κεκτημένα καὶ λύπας ἐν αὑτοῖς ἡδονάς θ' ἅμα ἑπομένας ἔχει. ὧδ'
οὖν κατὰ παντὸς αἰσθητοῦ καὶ ἀναισθήτου παθήματος τὰς αἰτίας
λαμβάνωμεν, ἀναμιμῃσκόμενοι τὸ τῆς εὐκινήτου τε καὶ δυσκινή- B
του φύσεως ὅτι διειλόμεθα ἐν τοῖς πρόσθεν· ταύτῃ γὰρ δὴ μετα-
διωκτέον πάντα, ὅσα ἐπινοοῦμεν ἑλεῖν. τὸ μὲν γὰρ κατὰ φύσιν
15 εὐκίνητον, ὅταν καὶ βραχὺ πάθος εἰς αὐτὸ ἐμπίπτῃ, διαδίδωσι
κύκλῳ, μόρια ἕτερα ἑτέροις ταὐτὸν ἀπεργαζόμενα, μέχριπερ ἂν ἐπὶ

10 αὑτοῖς: αὐτοῖς A.

of a body towards its proper sphere; and for every substance the direction of its proper sphere, wherever that may be, is κάτω, and the opposite ἄνω. By τὰ δὲ τούτοις κ.τ.λ. Plato means that while in a given region we apply the term βαρὺ to a substance whose ὁδὸς πρὸς τὸ ξυγγενὲς is towards that region, we apply the term κοῦφον to a substance whose ὁδὸς πρὸς τὸ ξυγγενὲς is towards another. To adopt Martin's example, in the region of earth stones are heavy and vapour light; but in the region of air vapour is heavy and stones light.

5. σκληρότης γὰρ] With this clause τὸ μὲν has of course to be supplied.

64 A—65 B, c. xxvii. We have now to explain the nature and cause of pleasure and pain. Sensation is produced in the following way. If an impression from without lights upon a part of the body of which the particles are readily stirred, those particles which first received the impact transmit the motion to their neighbours; and so it is handed on until it reaches the seat of consciousness; at which point sensation is effected. If on the contrary the impression is received by a part of the body which is hard to stir, the motion is not transmitted, and no sensation ensues. This being so, the explanation of pleasure and pain is as follows. When any of the particles that constitute our body are suddenly and in considerable numbers forced out of their normal position, the result is pain; and when they in like manner return to their normal position, the result is pleasure. If however either process takes place on a very small scale or very gradually, it is imperceptible. When the corporeal particles yield to the external impact with extreme readiness, the process is accompanied by vivid perception, but neither by pleasure nor by pain. If the disturbance has been slow and gradual, and the restoration rapid and sudden, we experience pleasure without antecedent pain: but if these conditions are reversed, we feel pain in the disturbance, but the restoration affords no pleasure.

7. τῶν κοινῶν περὶ ὅλον τὸ σῶμα] An explanation of pleasure and pain will complete our account of the sensations

64 B]   ΤΙΜΑΙΟΣ.   235

that makes us call the falling body heavy, and the place to which it falls, below; while to the reverse relations we apply the opposite names. So much then for the causes of these conditions. Of the qualities of smooth and rough any one could perceive the cause and explain it to another: the latter is produced by a combination of hardness and irregularity, the former by a combination of uniformity and density.

XXVII. We have yet to consider the most important point relating to the affections which concern the whole body in common; that is, the cause of pleasure and pain accompanying the sensations we have discussed: and also the affections which produce sensation by means of the separate bodily organs and which involve attendant pains and pleasures. This then is how we must conceive the causes in the case of every affection, sensible or insensible, recollecting how we defined above the source of mobility and immobility: for this is the way we must seek the explanation we hope to find. When that which is naturally mobile is impressed by even a slight affection, it spreads abroad the motion, the particles one upon another producing the same effect, until coming to the sentient part it announces

which are not confined to any special organs, but affect the body as a whole: next we shall proceed to discuss the separate senses.

8. **ἐν οἷς διεληλύθαμεν**] i.e. in the perceptions treated in the preceding chapter.

11. **ἀναισθήτου παθήματος**] A πάθημα then, we see, is not always accompanied by αἴσθησις. The distinction is this. Every external influence affecting the body is a πάθημα, but, unless it is transmitted to the seat of consciousness, it does not produce αἴσθησις. Thus cutting the hair is a πάθημα, but not an αἴσθησις: or, to take another example, a deaf man has the πάθημα but not the αἴσθησις of sound; the air-vibrations are conveyed to his ear, but stop short there without being announced to the brain. The word πάθημα, it will be observed, being now applied to the subject, has a different significance from that in which we saw it used in the preceding chapter.

13. **ἐν τοῖς πρόσθεν**] See 55 B.

16. **μόρια ἕτερα ἑτέροις**] The word μόρια is usually considered as the object of διαδίδωσι. But this seems to me strained; since what the εὐκίνητον transmits is the πάθος, not its own particles. I should prefer to regard μόρια as placed in a kind of apposition, the construction being somewhat similar to that in Sophokles *Antigone* 259 λόγοι δ' ἐν ἀλλήλοισιν ἐρρόθουν κακοί, φύλαξ ἐλέγχων φύλακα: cf. Herodotus II cxxxiii (quoted by Prof. Campbell) ἵνα οἱ δυώδεκα ἔτεα ἀντὶ ἐξ ἐτέων γένηται, αἱ νύκτες ἡμέραι ποιεύμεναι. Just below the μόρια are spoken of as transmitting the πάθος, διαδιδόντων μορίων μορίοις ἄλλων ἄλλοις.

ταὐτὸν ἀπεργαζόμενα] i. e. affecting them with the same πάθος. The theory of sensation here enunciated is also set forth in *Philebus* 33 D: see too *Republic* 584 C αἵ γε διὰ τοῦ σώματος ἐπὶ τὴν

τὸ φρόνιμον ἐλθόντα ἐξαγγείλῃ τοῦ ποιήσαντος τὴν δύναμιν· τὸ δ᾽ ἐναντίον ἑδραῖον ὂν κατ᾽ οὐδένα τε κύκλον ἰὸν πάσχει μόνον, ἄλλο δὲ οὐ κινεῖ τῶν πλησίον, ὥστε οὐ διαδιδόντων μορίων μορίοις ἄλλων ἄλλοις τὸ πρῶτον πάθος ἐν αὑτοῖς ἀκίνητον εἰς τὸ πᾶν ζῷον γενόμενον ἀναίσθητον παρέσχε τὸ παθόν. ταῦτα δὲ περί τε ὀστᾶ καὶ τὰς τρίχας ἐστὶ καὶ ὅσ᾽ ἄλλα γήινα τὸ πλεῖστον ἔχομεν ἐν ἡμῖν μόρια· τὰ δὲ ἔμπροσθεν περὶ τὰ τῆς ὄψεως καὶ ἀκοῆς μάλιστα, διὰ τὸ πυρὸς ἀέρος τε ἐν αὐτοῖς δύναμιν ἐνεῖναι μεγίστην. τὸ δὴ τῆς ἡδονῆς καὶ λύπης ὧδε δεῖ διανοεῖσθαι. τὸ μὲν παρὰ φύσιν καὶ βίαιον γιγνόμενον ἀθρόον παρ᾽ ἡμῖν πάθος ἀλγεινόν, τὸ δ᾽ εἰς φύσιν ἀπιὸν πάλιν ἀθρόον ἡδύ, τὸ δὲ ἠρέμα καὶ κατὰ σμικρὸν ἀναίσθητον, τὸ δ᾽ ἐναντίον τούτοις ἐναντίως. τὸ δὲ μετ᾽ εὐπετείας γιγνόμενον ἅπαν αἰσθητὸν μὲν ὅ τι μάλιστα, λύπης δὲ καὶ ἡδονῆς οὐ μετέχον, οἷον τὰ περὶ τὴν ὄψιν αὐτὴν παθήματα, ᾗ δὴ σῶμα ἐν τοῖς πρόσθεν ἐρρήθη καθ᾽ ἡμέραν ξυμφυὲς ἡμῶν γίγνεσθαι. ταύτῃ γὰρ τομαὶ μὲν καὶ καύσεις καὶ ὅσα ἄλλα πάσχει λύπας οὐκ ἐμποιοῦσιν, οὐδὲ ἡδονὰς πάλιν ἐπὶ ταὐτὸν ἀπιούσης εἶδος, μέγισται δὲ αἰσθήσεις καὶ σαφέσταται καθότι τ᾽ ἂν πάθῃ καὶ ὅσων ἂν αὐτή πῃ προσβαλοῦσα ἐφάπτηται· βία γὰρ τὸ πάμπαν οὐκ ἔνι τῇ δια-

6 τὰς ante τρίχας omittunt SZ.   15 ἡμῖν S.   19 προσβαλοῦσα : προσβάλλουσα S.

ψυχὴν τείνουσαι: and compare Aristotle *de sensu* i 436ᵇ 6 ἢ δ᾽ αἴσθησις ὅτι διὰ τοῦ σώματος γίνεται τῇ ψυχῇ δῆλον καὶ διὰ τοῦ λόγου καὶ τοῦ λόγου χωρίς.

6. **ὀστᾶ καὶ τὰς τρίχας**] So says Aristotle *de anima* III xiii 435ᵃ 24 καὶ διὰ τοῦτο τοῖς ὀστοῖς καὶ ταῖς θριξὶ καὶ τοῖς τοιούτοις μορίοις οὐκ αἰσθανόμεθα, ὅτι γῆς ἐστίν.

9. **τὸ μὲν παρὰ φύσιν**] The first indication of this theory of pleasure and pain is to be found in *Republic* 583 C foll.: it is definitely set forth in *Philebus* 31 D foll. The Platonic theory is assailed by Aristotle, *nic. eth.* X iii 1173ᵃ 31. He objects (1) that a κίνησις involves the notion of speed, which pleasure does not; (2) if pleasure is a γένεσις, whereunto is it a γένεσις, and out of what constituents does it arise? (3) it cannot be an ἀποπλήρωσις, for that is a purely corporeal process, and it is not body but soul which perceives pleasure. As usual, Aristotle's objections miss the point. He is treating pleasure subjectively and psychologically; whereas Plato's theory is a purely physical one. There is no confusion in the latter's view between the subjective and objective aspects; but here he is only concerned with explaining the physical causes which give rise to pleasure and pain.

12. **τὸ δὲ μετ᾽ εὐπετείας**] We have seen that sensation is due to the corporeal particles being εὐκίνητα and transmitting the πάθος to the seat of consciousness. But pleasure and pain require a certain degree of resistance in the particles: for if they offer only the slightest possible opposition to the external influence, the perception is indeed acute, but is entirely unattended by physical pain or pleasure. An instance of this is furnished by the phenomena of sight.

the property of the agent: but a substance that is immobile is too stable to spread the motion round about, and thus merely receives the affection but does not stir any neighbouring part; so that as the particles do not pass on one to another the original impulse which affected them, they keep it untransmitted to the entire creature and thus leave the recipient of the affection without sensation. This takes place with our bones and hair and all the parts we have which are formed mostly of earth: while the former conditions apply in the highest degree to sight and hearing, because they contain the greatest proportion of fire and air. The nature of pleasure and pain must be conceived thus: an affection contrary to nature, when it takes place forcibly and suddenly within us, is painful; a sudden return to the natural state is pleasant; a gentle and gradual process is imperceptible; and one of an opposite character is perceptible. Now a process which takes place with perfect facility is perceptible in a high degree, but is accompanied neither by pleasure nor by pain. An example will be found in the affections of the visual current, which we said above was in the daytime a material body cognate with ourselves. In this cutting and burning and any other affection cause no pain; nor does pleasure ensue when it returns to its normal state: but its perceptions are most vivid and accurate of whatsoever impresses it or whatsoever itself meets and touches. For its dilation and contraction

The ὄψεως ῥεῦμα (which we must remember to be actually part of ourselves) is composed of extremely subtle and mobile particles, which yield without resistance to any external impulse. This may come in contact with fire or be divided by a sharp instrument, and yet, while the καῦσις and the τομή are clearly perceived, no pain is felt, notwithstanding that in either case the particles are very much dislocated. Plato is of course speaking merely of bodily pain and pleasure, not of the mental pleasure awakened by the sight of a beautiful object or of the disgust excited by a spectacle of contrary nature. The process of seeing, as such, is normally unattended by physical pain or pleasure.

14. ἐν τοῖς πρόσθεν] 45 B. By τὴν ὄψιν we are as before to understand the ὄψεως ῥεῦμα.

15. ξυμφυὲς ἡμῶν] Stallbaum is perhaps right in reading ἡμῖν. But as ξυγγενής is several times followed by the genitive (see 30 D) it seems possible that ξυμφυής might have the same construction. ξύμφυτος seems to have the same government in *Philebus* 51 D καὶ τούτων ξυμφύτους ἡδονὰς ἑπομένας.

18. καὶ ὅσων ἄν] A similar fulness of detail is in 45 C ὅτου τ' ἂν αὐτό ποτε ἐφάπτηται καὶ ὁ ἂν ἄλλο ἐκείνου.

19. διακρίσει τε αὐτῆς καὶ συγκρίσει] These terms are explained when Plato comes to treat of colours, 67 C foll.

238   ΠΛΑΤΩΝΟΣ   [64 E—

κρίσει τε αὐτῆς καὶ συγκρίσει. τὰ δ᾽ ἐκ μειζόνων μερῶν σώματα μόγις εἴκοντα τῷ δρῶντι, διαδιδόντα δὲ εἰς ὅλον τὰς κινήσεις, ἡδονὰς ἴσχει καὶ λύπας, ἀλλοτριούμενα μὲν λύπας, καθιστάμενα δὲ εἰς 65 A τὸ αὐτὸ πάλιν ἡδονάς. ὅσα δὲ κατὰ σμικρὸν τὰς ἀποχωρήσεις
5 ἑαυτῶν καὶ κενώσεις εἴληφε, τὰς δὲ πληρώσεις ἁθρόας καὶ κατὰ μεγάλα, κενώσεως μὲν ἀναίσθητα, πληρώσεως δὲ αἰσθητικὰ γιγνόμενα, λύπας μὲν οὐ παρέχει τῷ θνητῷ τῆς ψυχῆς, μεγίστας δὲ ἡδονάς· ἔστι δὲ ἔνδηλα περὶ τὰς εὐωδίας. ὅσα δὲ ἀπαλλοτριοῦται μὲν ἀθρόα, κατὰ σμικρὰ δὲ μόγις τε εἰς ταὐτὸ πάλιν ἑαυτοῖς καθί-
10 σταται, τοὐναντίον τοῖς ἔμπροσθεν πάντα ἀποδίδωσι· ταῦτα δ᾽ αὖ B περὶ τὰς καύσεις καὶ τομὰς τοῦ σώματος γιγνόμενά ἐστι κατάδηλα.

XXVIII. Καὶ τὰ μὲν δὴ κοινὰ τοῦ σώματος παντὸς παθήματα, τῶν τ᾽ ἐπωνυμιῶν ὅσαι τοῖς δρῶσιν αὐτὰ γεγόνασι, σχεδὸν εἴρηται· τὰ δ᾽ ἐν ἰδίοις μέρεσιν ἡμῶν γιγνόμενα, τά τε πάθη καὶ
15 τὰς αἰτίας αὖ τῶν δρώντων, πειρατέον εἰπεῖν, ἄν πῃ δυνώμεθα. C πρῶτον οὖν ὅσα τῶν χυμῶν πέρι λέγοντες ἐν τοῖς πρόσθεν ἀπελίπομεν, ἴδια ὄντα παθήματα περὶ τὴν γλῶτταν, ἐμφανιστέον ᾗ δυνατόν. φαίνεται δὲ καὶ ταῦτα, ὥσπερ οὖν καὶ τὰ πολλά, διὰ

4 τὸ αὐτὸ: ταὐτόν S.   9 καὶ post μόγις τε addit A.   ταὐτό: ταὐτὸν SZ.
10 ταῦτα: ταὐτά A.   15 αὖ omittit S, qui mox post δρώντων dedit αὐτά.   16 μὲν post πρῶτον addit S.   ἀπελίπομεν: ἀπελείπομεν A.

1. **ἐκ μειζόνων μερῶν**] It will be remembered that the visual stream consisted of very fine particles of fire; not the very finest, since the rays from some objects penetrate and divide the visual current: see 67 E.

7. **λύπας μὲν οὐ παρέχει**] When the dislocation has been very gradual and the restoration rapid, we have acute pleasure without any antecedent pain. Such pleasures are called in the *Republic* and *Philebus* καθαραὶ ἡδοναί, as distinguished from μικταί: see *Republic* 584 C and *Philebus* 51 B, where the example of sweet smells is given, as well as beautiful colours, shapes and sounds. In our present passage Plato adds a little to the explicitness of his statement: he shows that ὀσμαί are just as much καταστάσεις as the μικταί, only the κένωσις being insensible, we felt no preliminary pain. He seems to regard sweet odours as the natural nutriment of the nostrils, which suffer waste when those are absent: but the depletion is so imperceptible that it is only by sudden restoration of the natural state that we become conscious that there has been any lack. The statement in the *Philebus, l. l.*, though briefer, amounts to the same: ὅσα τὰς ἐνδείας ἀναισθήτοις ἔχοντα καὶ ἀλύποις τὰς πληρώσεις αἰσθητὰς καὶ ἡδείας καθαρὰς λυπῶν παραδίδωσιν. Aristotle tells us (*de sensu* V 445ᵃ 16) that certain Pythagoreans believed that some animals were nourished by smell.

8. **ἀπαλλοτριοῦται μὲν ἀθρόα**] On the other hand there are cases where the disturbance is violent and causes severe pain, but the restoration is too gradual to afford any pleasure. This is to be seen in wounds and burns and such like; the process of healing causes no pleasure.

65 E—66 C, c. xxviii. So much for the

are entirely free from violence. On the other hand bodies formed of larger particles, reluctantly yielding to the agent, and spreading the motions through the whole frame, cause pleasure and pain; when they are disturbed giving pain, and pleasure in being restored to their proper state. Those things which suffer a gradual withdrawing and emptying, but have their replenishment sudden and on a large scale, are insensible to the emptying but sensible of the replenishment; so that while they cause no pain to the mortal part of the soul, they produce very intense pleasure. This is to be observed in the case of sweet smells. But when the parts are disturbed suddenly, but gradually and laboriously restored to their former condition, they afford exactly the opposite result to the former: this may be seen in the case of burns and cuts on the body.

XXVIII. Now the affections common to the body as a whole and the names that have been given to the agents which produce them have been well-nigh expounded: next we must try to explain, if we can, what takes place in the separate parts of us, both as to the affections of them and the causes on the part of the agents. First then we must set forth to the best of our power all that we left unsaid concerning tastes, which are affections peculiar to the tongue. It appears that these,

sensations affecting the whole body and their causes; we have now to inquire into the separate sensory faculties. We will first take taste. This depends upon the contraction or dilatation of the pores of the tongue by substances that are dissolved in the mouth. Whatever powerfully contracts the small vessels of the tongue is harsh and astringent; that which has a detergent effect we call alkaline, or if its action is milder, saline. A substance which is volatile and inflames the vessels is called pungent; and one that produces a kind of fermentation or effervescence is acid. All the foregoing exercise a disturbing influence upon the substance of the tongue: that which mollifies it and restores the disturbed particles to their natural state, producing a pleasurable sensation, is named sweet.

13. τοῖς δρῶσιν αὐτά] i.e. the agents or forces which produce the παθήματα.

16. ἐν τοῖς πρόσθεν ἀπελίπομεν] The reference would seem to be to the enumeration of χυμοί in 60 A. Plato's statement is quoted by Theophrastos *de causis plantarum* VI 1: to the list of χυμοί given by Plato in the present passage he adds λιπαρός. Farther on he gives the views of Demokritos, who referred differences of taste to differences in the shape of the atoms: cf. *de sensu* §§ 65—69. Opinions not dissimilar to Plato's are ascribed to Alkmaion and to Diogenes of Apollonia by pseudo-Plutarch *de placitis philosophorum* IV 18.

17. περὶ τὴν γλῶτταν] The under surface of the soft palate is said by anatomists to share this function with the tongue.

συγκρίσεών τέ τινων καὶ διακρίσεων γίγνεσθαι, πρὸς δὲ αὐταῖς
κεχρῆσθαι μᾶλλόν τι τῶν ἄλλων τραχύτησί τε καὶ λειότησιν.
ὅσα μὲν γὰρ εἰσιόντα περὶ τὰ φλέβια, οἷόνπερ δοκιμεῖα τῆς
γλώττης τεταμένα ἐπὶ τὴν καρδίαν, εἰς τὰ νοτερὰ τῆς σαρκὸς καὶ D
5 ἁπαλὰ ἐμπίπτοντα γήϊνα μέρη κατατηκόμενα ξυνάγει τὰ φλέβια
καὶ ἀποξηραίνει, τραχύτερα μὲν ὄντα στρυφνά, ἧττον δὲ τραχύ-
νοντα αὐστηρὰ φαίνεται· τὰ δὲ τούτων τε ῥυπτικὰ καὶ πᾶν τὸ
περὶ τὴν γλῶτταν ἀποπλύνοντα, πέρα μὲν τοῦ μετρίου τοῦτο
δρῶντα καὶ προσεπιλαμβανόμενα, ὥστε ἀποτήκειν αὐτῆς τῆς φύ-
10 σεως, οἷον ἡ τῶν λίτρων δύναμις, πικρὰ πάνθ' οὕτως ὠνόμασται, E
τὰ δὲ ὑποδεέστερα τῆς λιτρώδους ἕξεως ἐπὶ τὸ μέτριόν τε τῇ ῥύψει
χρώμενα ἁλυκὰ ἄνευ πικρότητος τραχείας καὶ φίλα μᾶλλον ἡμῖν
φαντάζεται. τὰ δὲ τῇ τοῦ στόματος θερμότητι κοινωνήσαντα καὶ
λεαινόμενα ὑπ' αὐτοῦ, ξυνεκπυρούμενα καὶ πάλιν αὐτὰ ἀντικάοντα
15 τὸ διαθερμῆναν, φερόμενά τε ὑπὸ κουφότητος ἄνω πρὸς τὰς τῆς
κεφαλῆς αἰσθήσεις, τέμνοντά τε πάνθ' ὁπόσοις ἂν προσπίπτῃ, διὰ
ταύτας τὰς δυνάμεις δριμέα πάντα τοιαῦτα ἐλέχθη. τῶν δὲ αὐτῶν 66 A
προλελεπτυσμένων μὲν ὑπὸ σηπεδόνος, εἰς δὲ τὰς στενὰς φλέβας

3 δοκιμεῖα : δοκίμια HSZ.   14 λεαινόμενα : λειαινόμενα ASZ.

1. **διὰ συγκρίσεων**] Nearly all sense-perception is reduced by Plato to contraction and expansion, which however in different organs produce different classes of sensation. This is the agency by which taste is brought about, though the tongue is in a peculiar degree affected by the roughness or smoothness of the entering particles.

πρὸς δὲ αὐταῖς] sc. ταῖς συγκρίσεσι καὶ διακρίσεσι.

3. **οἷόνπερ δοκιμεῖα**] The word δοκιμεῖον or δοκίμιον signifies an instrument for testing, and is applied by Plato to the small blood-vessels of the tongue, which he holds to be both the cause of taste, through their contraction and expansion, and also the means of transmitting the πάθημα to the seat of consciousness. Of the nerves Plato, like Aristotle, understood nothing at all: their functions are attributed by him to the φλέβια.

5. **κατατηκόμενα**] Plato holds that all taste is produced by substances in a liquid state, whether liquefied before or after entering the mouth. In this opinion Aristotle coincides; see for instance *de anima* II x 422ᵃ 17 οὐδὲν δὲ ποιεῖ χυμοῦ αἴσθησιν ἄνευ ὑγρότητος, ἀλλ' ἔχει ἐνεργείᾳ ἢ δυνάμει ὑγρότητα. Aristotle's theory of taste will be found in that chapter.

6. **στρυφνά...αὐστηρά**] The first of these words evidently means 'astringent': αὐστηρὰ may be translated 'harsh'; but possibly it answers more to our 'bitter' than πικρά: at least we should hardly call soda bitter. The same word is applied to alkaline flavours by Aristotle *de sensu* iv 441ᵇ 6. πικρὸν is defined by Theophrastos *l. l.* as φθαρτικὸν τῆς ὑγρότητος ἢ πηκτικὸν ἢ δηκτικὸν ἢ ἁπλῶς τραχὺ ἢ μάλιστα τραχύν.

12. **φίλα μᾶλλον ἡμῖν φαντάζεται**] This is mentioned because all the substances hitherto enumerated, including salt, have a disturbing action upon the substance of the tongue, and are there-

66 Α]   ΤΙΜΑΙΟΣ.   241

like most other things, are brought about by contraction and dilation, besides which they have more to do than other sensations with roughness and smoothness in the agents. For whenever earthy particles enter in by the little veins which are a kind of testing instruments of the tongue, stretched to the heart, and strike upon the moist and soft parts of the flesh, these particles as they are being dissolved contract and dry the small veins; and if they are very rough, they are termed 'astringent'; if less so 'harsh'. Such substances again as are detergent and rinse the whole surface of the tongue, if they do this to an excessive degree and encroach so as to dissolve part of the structure of the flesh, as is the property of alkalies—all such are termed 'bitter': but those which fall short of the alkaline quality and rinse the tongue only to a moderate extent are saline without bitterness and seem to us agreeable rather than the reverse. Those which share the warmth of the mouth and are softened by it, being simultaneously inflamed and themselves in turn scorching that which heated them, and which owing to their lightness fly upward to the senses of the head, penetrating all that is in their path—owing to these properties all such substances are called 'pungent'. But sometimes these same substances, having been already refined by decomposition, enter into the narrow veins, being

fore presumably disagreeable. The irritation produced by salt is however so mild that it amounts to no more than a pleasant stimulation of the organ.

13. τὰ δὲ τῇ τοῦ στόματος θερμότητι] Compare the view assigned to Alkmaion by Theophrastos *de sensu* § 25: γλώττῃ δὲ τοὺς χυμοὺς κρίνειν· χλιαρὰν γὰρ οὖσαν καὶ μαλακὴν τήκειν τῇ θερμότητι· δέχεσθαι δὲ καὶ διαδιδόναι διὰ τὴν μανότητα τῆς ἁπαλότητος.

15. πρὸς τὰς τῆς κεφαλῆς αἰσθήσεις] A spoonful of strong mustard would probably produce very much the sort of experience which Plato describes. Theophrastos says δριμὺν δὲ τὸν πηκτικὸν ἢ δηκτικὸν ἢ ἐκκριτικὸν τῆς ἐν τῇ συμφύτῳ ὑγρότητι θερμότητος εἰς τὸν ἄνω τόπον ἢ ἁπλῶς χυμὸν καυτικὸν ἢ θερμαντικόν.

P. T.

There seems a lack of finish in his definition.

17. τῶν δὲ αὐτῶν προλελεπτυσμένων] In this portentous sentence it is quite probable that some corruptions may lurk. But no emendation suggests itself of sufficient plausibility to justify its admission into the text, although I have little doubt that ἐχόντων should be read for ἔχοντα. Stallbaum's proposed alterations are the result of his not understanding the construction: ὅσα ἀέρος is parallel to τοῖς γεώδεσι and equivalent to τοῖς ὅσα ἀέρος ἔνεστιν. As for the infinitives after ἃ δή, they are incurably ungrammatical: we must either suppose that the construction is carried on from ἐλέχθη in the previous sentence, or that it never recovers from the effects of ὥστε

16

242   ΠΛΑΤΩΝΟΣ   [66 A—

ἐνδυομένων, καὶ τοῖς ἐνοῦσιν αὐτόθι μέρεσι γεώδεσι καὶ ὅσα ἀέρος ξυμμετρίαν ἔχοντα, ὥστε κινήσαντα περὶ ἄλληλα ποιεῖν κυκᾶσθαι, κυκώμενα δὲ περιπίπτειν τε καὶ εἰς ἔτερα ἐνδυόμενα ἔτερα κοῖλα ἀπεργάζεσθαι περιτεινόμενα τοῖς εἰσιοῦσιν, ἃ δὴ νοτίδος περὶ ἀέρα B
5 κοίλης περιταθείσης, τοτὲ μὲν γεώδους, τοτὲ δὲ καὶ καθαρᾶς, νοτερὰ ἀγγεῖα ἀέρος ὕδατα κοῖλα περιφερῆ τε γενέσθαι, καὶ τὰ μὲν τῆς καθαρᾶς διαφανεῖς περιστῆναι κληθείσας ὄνομα πομφόλυγας, τὰ δὲ τῆς γεώδους ὁμοῦ κινουμένης τε καὶ αἱρομένης ζέσιν τε καὶ ζύμωσιν ἐπίκλην λεχθῆναι—τὸ δὲ τούτων αἴτιον τῶν παθημάτων
10 ὀξὺ προσρηθῆναι. ξύμπασι δὲ τοῖς περὶ ταῦτα εἰρημένοις πάθος ἐναντίον ἀπ' ἐναντίας ἐστὶ προφάσεως, ὁπόταν ἡ τῶν εἰσιόντων C ξύστασις ἐν ὑγροῖς, οἰκεία τῇ τῆς γλώττης ἕξει πεφυκυῖα, λεαίνῃ μὲν ἐπαλείφουσα τὰ τραχυνθέντα, τὰ δὲ παρὰ φύσιν ξυνεστῶτα ἢ κεχυμένα τὰ μὲν ξυνάγῃ, τὰ δὲ χαλᾷ, καὶ πάνθ' ὅ τι μάλιστα
15 ἱδρύῃ κατὰ φύσιν, ἡδὺ καὶ προσφιλὲς παντὶ πᾶν τὸ τοιοῦτον ἴαμα τῶν βιαίων παθημάτων γιγνόμενον κέκληται γλυκύ.

XXIX. Καὶ τὰ μὲν ταύτῃ ταῦτα· περὶ δὲ δὴ τὴν τῶν D μυκτήρων δύναμιν, εἴδη μὲν οὐκ ἔνι· τὸ γὰρ τῶν ὀσμῶν πᾶν ἡμιγενές, εἴδει δὲ οὐδενὶ ξυμβέβηκε ξυμμετρία πρὸς τό τινα ἔχειν

12 λεαίνῃ: λειαίνῃ ASZ.   17 δὴ post τὰ μὲν addit S.   19 ἔχειν: σχεῖν SZ.

early in the present one. However loose the syntax may be, the sense is not on the whole obscure. Acids are substances which have been refined by fermentation; these, when they enter the mouth, form a combination with the particles of earth and air which are therein, and stir and mix them up in such a way as to produce films of moisture enclosing air, in other words, bubbles: a kind of effervescence in fact is produced by the action of the acid on the substance of the tongue. The words εἰς ἕτερα ἐνδυόμενα ἔτερα κοῖλα ἀπεργάζεσθαι περιτεινόμενα τοῖς εἰσιοῦσιν are not clear: it would seem that the earthy particles within, by gathering round the entering particles of acid, vacate their former positions which are filled by air surrounded by the moisture attending the dissolution of the acid.

10. πάθος ἐναντίον] The χυμοί which act upon the tongue are thus divided into two classes, those which disturb the natural position of its constituent particles, and those which restore it. Of the former there are the six varieties herein before enumerated; of the latter there is but one, which we term sweet. This contracts what is unnaturally expanded and expands what is unnaturally contracted, and thus is 'a remedy of forcible affections', since by restoring the natural condition it produces a pleasant and soothing effect.

13. ξυνεστῶτα...κεχυμένα...ξυνάγῃ... χαλᾷ] Throughout this dialogue a distinct inclination to chiasmus may be observed.

66 D—67 C, c. xxix. Odours cannot be classified according to kinds. For no element in its normal state can be perceived by smell, because the vessels of the nostrils are too narrow to admit water or earth and too wide to be excited by air or fire. They can thus only perceive an element in process of disso-

duly proportioned to the earthy particles and the particles of air which are there, so that they set them in motion and mingle them together, and thereby cause them to jostle against one another and taking up other positions to form new hollows extended round the entering particles—which hollows consist of a film of moisture, sometimes earthy, sometimes pure, embracing a volume of air; and thus they form moist capsules containing air:—in some cases the films are of pure moisture and transparent and are called bubbles; in others they are of earthy liquid which effervesces and rises all together, when the name of seething and fermentation is given to it: and the cause of all these conditions is termed 'acid'. The opposite affection to all those which have been described is produced by an opposite cause: when the structure of the entering particles amid the moisture, having a natural affinity to the tongue's normal condition, smooths it by mollifying the roughened parts, and relaxes or contracts what is unnaturally contracted or expanded, and settles everything as much as possible in its natural state. Every such remedy of violent affections is to all of us pleasant and agreeable, and has received the name of 'sweet'.

XXIX. Enough of this subject. As regards the faculty of the nostrils no classification can be made. For smells are of a half-formed nature: and no class of figure has the adaptation requisite for producing any smell, but our veins in this

lution. The object of smell then is either vapour, which is water changing to air, or mist, which is air changing to water. That the object of smell is denser than air can be proved by placing some obstacle before the nostrils and then forcibly drawing breath: the air will pass in, but without any odour. The only classification we can make is that scents which disturb the substance of the nostrils are unpleasant, while those which restore the natural state are pleasant.

Sound is a vibration of the air, impinging upon the ear and thence transmitted first to the brain and finally to the liver: the pitch depends upon the rapidity, the quality upon the regularity, and the loudness upon the extent of the motion.

19. εἴδει δὲ οὐδενί] That is, it does not possess the structure of any of the four, fire, air, water, and earth. We were able to classify tastes, because we could point to a definite substance which caused the sensation in each case. Aristotle agrees with Plato that the sense of smell ἧττον εὐδιόριστόν ἐστι, *de anima* II ix 421ª 7: this he attributes to the fact that mankind possesses this sense in a very imperfect degree, being in this respect inferior to many animals. In the same chapter 421ᵇ 9 he says air or water is the medium of smell: ἔστι δὲ καὶ ἡ ὄσφρησις διὰ τοῦ μεταξύ, οἷον ἀέρος ἢ ὕδατος· καὶ γὰρ τὰ ἔνυδρα δοκοῦσιν ὀσμῆς αἰσθάνεσθαι. Elsewhere Aristotle denies that smells cannot be classified: *de sensu* v

244 ΠΛΑΤΩΝΟΣ [66 D—

ὀσμήν· ἀλλ' ἡμῶν αἱ περὶ ταῦτα φλέβες πρὸς μὲν τὰ γῆς ὕδατός
τε γένη στενότεραι ξυνέστησαν, πρὸς δὲ τὰ πυρὸς ἀέρος τε εὐρύ-
τεραι, διὸ τούτων οὐδεὶς οὐδενὸς ὀσμῆς πώποτε ᾔσθετό τινος, ἀλλὰ
ἢ βρεχομένων ἢ σηπομένων ἢ τηκομένων ἢ θυμιωμένων γίγνονταί
5 τινων. μεταβάλλοντος γὰρ ὕδατος εἰς ἀέρα ἀέρος τε εἰς ὕδωρ ἐν Ε
τῷ μεταξὺ τούτων γεγόνασιν, εἰσὶ δὲ ὀσμαὶ ξύμπασαι καπνὸς ἢ
ὁμίχλη· τούτων δὲ τὸ μὲν ἐξ ἀέρος εἰς ὕδωρ ἰὸν ὁμίχλη, τὸ δὲ ἐξ
ὕδατος εἰς ἀέρα καπνός· ὅθεν λεπτότεραι μὲν ὕδατος, παχύτεραι
δὲ ὀσμαὶ ξύμπασαι γεγόνασιν ἀέρος. δηλοῦνται δέ, ὁπόταν τινὸς
10 ἀντιφραχθέντος περὶ τὴν ἀναπνοὴν ἄγῃ τις βίᾳ τὸ πνεῦμα εἰς
αὑτόν· τότε γὰρ ὀσμὴ μὲν οὐδεμία ξυνδιηθεῖται, τὸ δὲ πνεῦμα
τῶν ὀσμῶν ἐρημωθὲν αὐτὸ μόνον ἕπεται. δι' οὖν ταῦτα ἀνώνυμα
τὰ τούτων ποικίλματα γέγονεν, οὐκ ἐκ πολλῶν οὐδ' ἁπλῶν εἰδῶν 67 A

2 στενότεραι : στενώτεραι AZ.   3 ἀλλὰ ἤ: ἀλλ' ἀεὶ S.   6 εἰσὶ δέ : εἰσί τε S.
12 δι' οὖν: δύ' οὖν ASZ.

443^b 17 οὐ γὰρ ὥσπερ τινές φασιν, οὐκ ἔστιν εἴδη τοῦ ὀσφραντοῦ, ἀλλ' ἔστιν: a little above he gives a list; καὶ γὰρ δριμεῖαι καὶ γλυκεῖαι εἰσὶν ὀσμαὶ καὶ αὐστηραὶ καὶ στρυφναὶ καὶ λιπαραί, καὶ τοῖς πικροῖς (sc. χυμοῖς) τὰς σαπρὰς ἄν τις ἀνάλογον εἴποι. Galen's opinion concerning this sense is similar to Plato's: see *de plac. Hipp. et Plat.* VII 628 πέμπτον γὰρ δὴ τοῦτό ἐστιν αἰσθητήριον, οὐκ ὄντων πέντε στοιχείων, ἐπειδὴ τῶν ὀσμῶν γένος ἐν τῷ μεταξὺ τὴν φύσιν ἐστὶν ἀέρος καὶ ὕδατος, ὡς καὶ Πλάτων εἶπεν ἐν Τιμαίῳ.

3. **ἀλλὰ ἢ βρεχομένων**] The sense of smell then perceives matter in an intermediate condition, as it is passing from one form to another. Herakleitos seems to have held some similar view: see Aristotle *de sensu* V 443^a 23 διὸ καὶ Ἡράκλειτος οὕτως εἴρηκεν, ὡς εἰ πάντα τὰ ὄντα καπνὸς γένοιτο, ῥῖνες ἂν διαγνοῖεν. Plato's doctrine of smell however, when considered in connexion with his corpuscular theory, has a striking peculiarity. Only ὁμίχλη and καπνὸς can be smelt, he says. But what are ὁμίχλη and καπνὸς? We cannot say simply that ὁμίχλη is the densest form of air and καπνὸς the rarest form of water, because Plato expressly tells us that they are transitional forms between air and water. Now the densest form of air is still formed of octahedrons, and the rarest form of water still formed of icosahedrons; so that no condensation of the one or rarefaction of the other constitutes any approach to a transition between the two. Now since ὁμίχλη and καπνὸς are not composed either of octahedrons or of icosahedrons, of what nature are the material particles which smell perceives? for no other regular solid figure beyond the five exists in nature. We are compelled to suppose that the agent which excites smell is actually unformed matter—matter, that is, which is dissolved out of one form, but not yet remoulded in another. It is evident that if the particles of water are dissolved and remoulded as particles of air, this is a physical process taking place in time : there is a time therefore when matter does exist in an unformed condition; and just in this time smell has the power of perceiving it. Aristotle, whose objections to the theory are stated in the chapter of the *de sensu* above cited, has nothing to say about this.

4. **γίγνονται**] sc. αἱ ὀσμαί.

7. **τὸ μὲν ἐξ ἀέρος**] Aristotle puts it rather differently: *meteorologica* I ix

ΤΙΜΑΙΟΣ.

part are formed too narrow for earth and water, and too wide for fire and air: for which cause no one ever perceived any smell of these bodies; but smells arise from substances which are being either liquefied or decomposed or dissolved or evaporated: for when water is changing into air and air into water, odours arise in the intermediate condition; and all odours are vapour or mist, mist being the conversion of air into water, and vapour the conversion of water into air; whence all smells are subtler than water and coarser than air. This is proved when any obstacle is placed before the passages of respiration, and then one forcibly inhales the air: for then no smell filters through with it, but the air bereft of all scent alone follows the inhalation. For this reason the complex varieties of odour are unnamed, and are ranked in classes neither numerous nor simple:

346ᵇ 32 ἔστι δ' ἡ μὲν ἐξ ὕδατος ἀναθυμίασις ἀτμίς, ἡ δ' ἐξ ἀέρος εἰς ὕδωρ νέφος· ὁμίχλη δὲ νεφέλης περίττωμα τῆς εἰς ὕδωρ συγκρίσεως.

8. ὕδατος εἰς ἀέρα] If the matter which is perceived by smell has no formed particles (as it cannot have), it is hard to see why it should not be so perceived when on the point of passing from water or air into fire, or the contrary: and in fact this seems actually suggested by θυμωμένων just above. However Plato presently affirms that the substances which excite smell, because they are in a transitional state between octahedrons and icosahedrons, are subtler than one and coarser than the other. This consequence seems equally hard to deduce from any interpretation of Plato's corpuscular theory.

9. ὀσμαί] i.e. the several substances which excite the olfactory organ.

τινὸς ἀντιφραχθέντος] When the air is filled with any odour, if a handkerchief, for instance, be pressed to the nostrils, and then a strong inhalation be taken, the air will force its way through the barrier, but the scent will not accompany it; whence Plato deduces the inference that the matter which excites the sensation of smell is less subtle than the particles of air. This led him to devise the theory of smell which we have been discussing. Martin curiously misunderstands this sentence, supposing that two people are concerned in the experiment: but τινὸς ἀντιφραχθέντος is of course neuter—'if an obstacle be placed'. It would seem then as if Plato conceived matter in its passage from air to water, or from water to air, to be made up of irregular figures intermediate in size between the particles of air and those of water: but how this comes about he does not explain. Theophrastos says curiously enough) in de sensu § 6 περὶ δὲ ὀσφρήσεως καὶ γεύσεως καὶ ἁφῆς ὅλως οὐδὲν εἴρηκεν [ὁ Πλάτων]: he means probably that Plato's account treats more of the αἰσθητὸν than the αἴσθησις: μᾶλλον ἀκριβολογεῖται περὶ τῶν αἰσθητῶν: still his statement cannot be considered accurate.

12. δι' οὖν ταῦτα] Although all the mss. agree in giving δύ' οὖν, it is impossible to retain it. For the δύο εἴδη could only refer to the two divisions specified below, which are not ἀνώνυμα, but ἡδὺ and λυπηρόν. It is the endless diversity of different scents that fall under these two heads—τὰ τούτων ποικίλματα—which are ἀνώνυμα.

13. οὐκ ἐκ πολλῶν] Tastes were divided into numerous species, which were

ὄντα, ἀλλὰ διχῇ τό θ' ἡδὺ καὶ τὸ λυπηρὸν αὐτόθι μόνω διαφανῆ λέγεσθον, τὸ μὲν τραχῦνόν τε καὶ βιαζόμενον τὸ κύτος ἅπαν, ὅσον ἡμῶν μεταξὺ κορυφῆς τοῦ τε ὀμφαλοῦ κεῖται, τὸ δὲ ταὐτὸν τοῦτο καταπραΰνον καὶ πάλιν ᾗ πέφυκεν ἀγαπητῶς ἀποδιδόν.

5 Τρίτον δὲ αἰσθητικὸν ἐν ἡμῖν μέρος ἐπισκοποῦσι τὸ περὶ τὴν ἀκοήν, δι' ἃς αἰτίας τὰ περὶ αὐτὸ ξυμβαίνει παθήματα, λεκτέον. B ὅλως μὲν οὖν φωνὴν θῶμεν τὴν δι' ὤτων ὑπ' ἀέρος ἐγκεφάλου τε καὶ αἵματος μέχρι ψυχῆς πληγὴν διαδιδομένην, τὴν δὲ ὑπ' αὐτῆς κίνησιν, ἀπὸ τῆς κεφαλῆς μὲν ἀρχομένην, τελευτῶσαν δὲ περὶ τὴν
10 τοῦ ἥπατος ἕδραν, ἀκοήν· ὅση δ' αὐτῆς ταχεῖα, ὀξεῖαν, ὅση δὲ βραδυτέρα, βαρυτέραν· τὴν δὲ ὁμοίαν ὁμαλήν τε καὶ λείαν, τὴν δὲ ἐναντίαν τραχεῖαν· μεγάλην δὲ τὴν πολλήν, ὅση δὲ ἐναντία, C σμικράν. τὰ δὲ περὶ ξυμφωνίας αὐτῶν ἐν τοῖς ὕστερον λεχθησομένοις ἀνάγκη ῥηθῆναι.

15 XXX. Τέταρτον δὴ λοιπὸν ἔτι γένος ἡμῖν αἰσθητικόν, ὃ

6 δι' ἃς: δι' ἃς δ' A.   11 βραδυτέρα: βραχυτέρα A.   13 τὰ δέ: τὰς δέ A.

ἁπλᾶ, because we could name the precise kind of substance which produced each and the mode of its action: smells are not ἁπλᾶ, because they do not proceed from any definite single substance, nor πολλά, because we can only classify them as agreeable or the reverse. Although a stricter classification than this can be made, Plato rightly regards taste as much more ἁπλοῦν than smell. For the more complex flavours which we 'taste' are really perceived by smell.

2. τὸ μὲν τραχῦνον] Plato's classification is based on his broad distinction between irritant and soothing agents.

3. μεταξὺ κορυφῆς τοῦ τε ὀμφαλοῦ] This must apply to extremely pungent and volatile scents, such as the fumes of strong ammonia: compare the description of δριμέα in 65 E.

7. τὴν δι' ὤτων] Plato's account of sound is in many respects consonant with modern acoustic science. He is correct in attributing it to vibrations which are propagated through the air until they strike upon the ear, and in saying that the loudness of the sound is proportionate to the amplitude of the sound-wave (μεγάλην δὲ τὴν πολλήν). He is also right in referring smoothness in the sound to regularity of the vibrations; for this is what constitutes the difference between a musical sound and mere noise; in the former case the vibrations are executed in regular periods, in the latter they are irregular. His explanation of the pitch is correct if by 'swiftness' he means the rapidity with which the vibrations are performed, but erroneous if he refers to the celerity of the sound's transmission through the air: from 80 A, B it would appear that he included both, supposing the more rapid vibrations to be propagated more swiftly through the atmosphere.

ἐγκεφάλου τε καὶ αἵματος] The construction of all these genitives is a little puzzling. Stallbaum constructs ἐγκεφάλου τε καὶ αἵματος with διά, but the interposition of ὑπ' ἀέρος surely renders this indefensible. I think we should join the words with πληγήν: 'a striking of the brain and blood by the air through the ears'. Plato conceives the vibrations, entering through the ears, to reach the brain and to be from thence transmitted

only two conspicuous kinds are in fact here distinguished, pleasant and unpleasant. The latter roughens and irritates all the cavity of the body that is between the head and the navel; the former soothes this same region and restores it with contentment to its own natural condition.

A third organ of sensation in us which we have to examine is that of hearing, and we must state the causes whence arise the affections connected with it. Let us in general terms define sound as a stroke transmitted through the ears by the air and passed through the brain and the blood to the soul; while the motion produced by it, beginning in the head and ending in the region of the liver, is hearing. A rapid motion produces a shrill sound, a slower one a deeper sound; regular vibration gives an even and smooth sound, and the opposite a harsh one; if the movement is large, the sound is loud; if otherwise, it is slight. Concerning accords of sound we must speak later on in our discourse.

XXX. A fourth faculty of sense yet remains, the intricate

through the blood-vessels to the liver. The liver appears to be selected because that region is the seat of the nutritive faculty of the soul, 70 D: and since the sensation of sound, as such, does not appeal to the intellectual organ, it is transmitted to that faculty which is specially concerned with sensation.

13. τὰ δὲ περὶ ξυμφωνίας] The account of concords is given in 80 A, where the transmission of sounds is explained. Aristotle's opinions concerning sound will be found in *de anima* II viii 419^b 4 foll., and scattered through the treatise *de sensu*.

67 C—69 A, *c*. xxx. The process of vision has already been explained: it only remains to give an account of colours. The particles which stream off from the objects perceived are some of them larger than those which compose the visual current, some smaller, and some of equal size. In case they are equal, the object whence they proceed is colourless and transparent; if they are smaller, they dilate the visual current;

if larger, they contract it. White is produced by dilation, black by contraction. Brightness and gleaming are the effects of a very swift motion of the particles, which divide the visual stream up to the very eyes themselves and draw forth tears. Red is the product of another kind of fire which penetrates the visual stream and mingles with the moisture of the eye. The other colours, yellow, violet, purple, chestnut, grey, buff, dark blue, pale blue, green, are produced by commixtures of the aforesaid, but in what proportions mingled God alone knows.

The physical processes we have been describing belong to the rank of subsidiary causes. For we must remember that there are in nature two classes of causes, the divine and the necessary; whereof we must search out the divine for the sake of happiness, and the necessary for the sake of the divine.

15. αἰσθητικόν] It is again a question whether we ought not to read αἰσθητόν, since colours are the object of investigation. Here however I think the

διελέσθαι δεῖ συχνὰ ἐν ἑαυτῷ ποικίλματα κεκτημένον, ἃ ξύμπαντα μὲν χρόας ἐκαλέσαμεν, φλόγα τῶν σωμάτων ἑκάστων ἀπορρέουσαν, ὄψει ξύμμετρα μόρια ἔχουσαν πρὸς αἴσθησιν. ὄψεως δ' ἐν τοῖς πρόσθεν αὖ τὸ περὶ τῶν αἰτίων τῆς γενέσεως ἐρρήθη. τῇ δ' D
5 οὖν τῶν χρωμάτων πέρι μάλιστα εἰκὸς πρέποι τ' ἂν τὸν ἐπιεικῆ λόγον διεξελθεῖν, τὰ φερόμενα ἀπὸ τῶν ἄλλων μόρια ἐμπίπτοντά τε εἰς τὴν ὄψιν τὰ μὲν ἐλάττω, τὰ δὲ μείζω, τὰ δ' ἴσα τοῖς αὐτῆς τῆς ὄψεως μέρεσιν εἶναι· τὰ μὲν οὖν ἴσα ἀναίσθητα, ἃ δὴ καὶ διαφανῆ λέγομεν, τὰ δὲ μείζω καὶ ἐλάττω, τὰ μὲν συγκρίνοντα, τὰ
10 δὲ διακρίνοντα αὐτήν, τοῖς περὶ τὴν σάρκα θερμοῖς καὶ ψυχροῖς καὶ τοῖς περὶ τὴν γλῶτταν στρυφνοῖς καὶ ὅσα θερμαντικὰ E ὄντα δριμέα ἐκαλέσαμεν ἀδελφὰ εἶναι, τά τε λευκὰ καὶ τὰ μέλανα, ἐκείνων παθήματα γεγονότα ἐν ἄλλῳ γένει τὰ αὐτά, φανταζόμενα δὲ ἄλλα διὰ ταύτας τὰς αἰτίας. οὕτως οὖν αὐτὰ
15 προσρητέον, τὸ μὲν διακριτικὸν τῆς ὄψεως λευκόν, τὸ δ' ἐναντίον αὐτοῦ μέλαν, τὴν δὲ ὀξυτέραν φορὰν καὶ γένους πυρὸς ἑτέρου προσ-

4 αὖ τό : αὐτό A. αὐτῶν HSZ. ὀλίγα post γενέσεως e margine codicis A dedit H. cieci cum SZ. 5 τὸν ἐπιεικῆ λόγον scripsi : τὸν ἐπιεικῆ λόγῳ AH. ἐπιεικεῖ λόγῳ SZ. sed forsitan melius legatur πρέπον τ' ἂν ἔτι εἴη λόγος.

ms. reading is defensible: we have, says Plato, to examine a fourth faculty of sense, which has various ποικίλματα : the ποικίλματα being the sensations we call colours. But he passes immediately from the subjective to the objective aspect of χρόαι, φλόγα τῶν σωμάτων ἑκάστων ἀπορρέουσαν.

3. ὄψει ξύμμετρα μόρια] i. e. particles of the right size to coalesce with the ὄψεως ῥεῦμα and form with it one sympathetic body. Stallbaum says Plato is following Empedokles, but this is incorrect: see Theophrastos de sensu § 7 Ἐμπεδοκλῆς δὲ περὶ ἁπασῶν ὁμοίως λέγει καὶ φησι τῷ ἐναρμόττειν εἰς τοὺς πόρους τοὺς ἑκάστης αἰσθάνεσθαι : cf. pseudo-Plutarch de placitis philosophorum I 15. The views of Aristotle concerning colour may be gathered from de sensu iii 439ᵃ 18 foll. and from the not very luminous treatise de coloribus. Aristotle considered the beauty of colours to depend upon numerical ratios: see de sensu iii 439ᵇ 31 τὰ μὲν γὰρ ἐν ἀριθμοῖς εὐλογίστοις χρώματα, καθάπερ ἐκεῖ τὰς συμφωνίας, τὰ ἥδιστα τῶν χρωμάτων εἶναι δοκοῦντα, οἷον τὸ ἁλουργὸν καὶ φοινικοῦν καὶ ὀλίγ' ἄττα τοιαῦτα, δι' ἥνπερ αἰτίαν καὶ αἱ συμφωνίαι ὀλίγαι, τὰ δὲ μὴ ἐν ἀριθμοῖς τἆλλα χρώματα, ἢ καὶ πάσας τὰς χρόας ἐν ἀριθμοῖς εἶναι, τὰς μὲν τεταγμένας τὰς δὲ ἀτάκτους, καὶ αὐτὰς ταύτας, ὅταν μὴ καθαραὶ ὦσι, διὰ τὸ μὴ ἐν ἀριθμοῖς εἶναι τοιαύτας γίνεσθαι. This has rather a Pythagorean sound.

6. τὰ φερόμενα ἀπὸ τῶν ἄλλων μόρια] i.e. the particles of fire which stream off from the object : it must be remembered that Plato's conception differs from the Demokritean or Empedoklean effluences, inasmuch as he does not hold that any image of the object is thrown off. τὴν ὄψιν again = τὸ τῆς ὄψεως ῥεῦμα.

8. τὰ μὲν οὖν ἴσα] Colours are then classified according to the relative size of the fiery particles from the object. If they are equal to those of the visual stream, we perceive no colour, but transparency alone : if smaller, so that they penetrate and dilate the ὄψεως ῥεῦμα, the

Ε] ΤΙΜΑΙΟΣ. 249

varieties of which it is our part to classify. To these we have given the name of *colours*, which consist of a flame streaming off from every object, having its particles so adjusted to those of the visual current as to excite sensation. We have already set forth the causes which gave origin to vision: thus therefore it will be most natural and fitting for a rational theory to treat of the question of colours. The particles which issue from outward objects and meet the visual stream are some of them smaller, some larger, and some equal in size to the particles of that stream. Those of equal size cause no sensation, and these we call *transparent*; but the larger and smaller, in the one case by contracting, in the other by dilating it, produce effects akin to the action of heat and cold on the flesh, and to the action on the tongue of astringent tastes and the heating sensations which we termed pungent. These are *white* and *black*, affections identical with those just mentioned, but occurring in a different class and seeming to be different for the causes aforesaid. We must then classify them as follows. What dilates the visual stream is white, and the opposite thereof is black. A swifter motion belonging to a different kind of fire, which meets and

colours produced are light and bright; if they are larger and compress the stream, the colours tend to be dark.

ἀναίσθητα] Since the particles are equal to those of the visual current, they do not affect the homogeneous structure of the latter.

10. τοῖς περὶ τὴν σάρκα] Plato merely means that the physical processes of contraction and dilation are the same in both instances; for in the other cases mentioned the sensations are pleasant or unpleasant, whereas the phenomena of vision are, physically regarded, unaccompanied either by pleasure or by pain.

13. ἐκείνων παθήματα] I take ἐκείνων to refer to τὰ συγκρίνοντα καὶ διακρίνοντα: the παθήματα belonging to the objects affecting the eye are the same as the παθήματα belonging to the objects of taste &c, namely σύγκρισις and διάκρισις. For the use of πάθημα compare 61 c, where παθήματα are the properties where-

by sensibles excite sensation. Stallbaum, following Stephanus, understands ἐκείνων to refer to θερμὰ and ψυχρά, στρυφνὰ and δριμέα, but this does not appear to me to give so good a sense. ἐν ἄλλῳ γένει = in another organ or mode of sensation. It is not generally recognised, Plato means, that the process is the same in the case of sight as in that of taste, because the sensible effect is so widely dissimilar.

14. διὰ ταύτας τὰς αἰτίας] i.e. because they are ἐν ἄλλῳ γένει and are not attended by pleasure or pain.

16. τὴν δ' ὀξυτέραν] Bright is distinguished from white (1) by dissimilarity between its fiery particles and those of white, (2) by its more rapid motion. It penetrates the ὄψεως ῥεῦμα right up to the eyes, the pores of which it displaces and dissolves, drawing forth a mixture of fire and water which we call tears. And so when the entering and issuing fires mingle and are quenched in the

πίπτουσαν καὶ διακρίνουσαν τὴν ὄψιν μέχρι τῶν ὀμμάτων, αὐτάς τε τῶν ὀφθαλμῶν τὰς διεξόδους βίᾳ διωθοῦσαν καὶ τήκουσαν, πῦρ μὲν ἀθρόον καὶ ὕδωρ, ὃ δάκρυον καλοῦμεν, ἐκεῖθεν ἐκχέουσαν, αὐτὴν δὲ οὖσαν πῦρ ἐξ ἐναντίας ἀπαντῶσαν, καὶ τοῦ μὲν ἐκπη-
5 δῶντος πυρὸς οἷον ἀπ' ἀστραπῆς, τοῦ δ' εἰσιόντος καὶ περὶ τὸ νοτερὸν κατασβεννυμένου, παντοδαπῶν ἐν τῇ κυκήσει ταύτῃ γιγνομένων χρωμάτων, μαρμαρυγὰς μὲν τὸ πάθος προσείπομεν, τὸ δὲ τοῦτο ἀπεργαζόμενον λαμπρόν τε καὶ στίλβον ἐπωνομάσαμεν. τὸ δὲ τούτων αὖ μεταξὺ πυρὸς γένος, πρὸς μὲν τὸ τῶν ὀμμάτων ὑγρὸν
10 ἀφικνούμενον καὶ κεραννύμενον αὐτῷ, στίλβον δὲ οὔ, τῇ δὲ διὰ τῆς νοτίδος αὐγῇ τοῦ πυρὸς μιγνυμένου χρῶμα ἔναιμον παρασχόμενον, τοὔνομα ἐρυθρὸν λέγομεν. λαμπρόν τε ἐρυθρῷ λευκῷ τε μιγνύμενον ξανθὸν γέγονε· τὸ δὲ ὅσον μέτρον ὅσοις, οὐδ' εἴ τις εἰδείη νοῦν ἔχει τὸ λέγειν, ὧν μήτε τινὰ ἀνάγκην μήτε τὸν εἰκότα
15 λόγον καὶ μετρίως ἄν τις εἰπεῖν εἴη δυνατός. ἐρυθρὸν δὲ δὴ μέλανι λευκῷ τε κραθὲν ἁλουργόν· ὄρφνινον δέ, ὅταν τούτοις μεμιγμένοις καυθεῖσί τε μᾶλλον συγκραθῇ μέλαν. πυρρὸν δὲ ξανθοῦ τε καὶ φαιοῦ κράσει γίγνεται, φαιὸν δὲ λευκοῦ τε καὶ μέλανος, τὸ δὲ ὠχρὸν λευκοῦ ξανθῷ μιγνυμένου. λαμπρῷ δὲ λευκὸν ξυνελθὸν
20 καὶ εἰς μέλαν κατακορὲς ἐμπεσὸν κυανοῦν χρῶμα ἀποτελεῖται,

3 ἀθρόον post ὕδωρ ponunt SZ. 10 τῇ: αὐτῇ A. 11 μιγνυμένου dedi cum S e Stephani correctione. μιγνυμένῃ AHZ. παρασχόμενον scripsi. παρασχομένῃ AHSZ. 19 μιγνυμένου: μεμιγμένου S. λευκόν: λαμπρόν A.

moisture, an agitation of the eyes is produced which we call 'dazzling'. As regards πῦρ ἀθρόον καὶ ὕδωρ, we must remember that, as Martin remarks, Plato considered all liquid water, and especially of course warm water, to be a mixture of fire and water; cf. 59 D.

8. τὸ δὲ τούτων αὖ μεταξύ] i.e. intermediate between the fire producing λευκόν and that producing στίλβον.

10. τῇ δὲ διὰ τῆς νοτίδος αὐγῇ] The reading of the ms. cannot be construed. I think it is necessary to receive μιγνυμένου and παρασχόμενον, agreeing with γένος. The sense will then be, the rays arriving at the eye, as their fire mingles with the gleam pervading the moisture which is there (i.e. with the fire residing in the eye itself), give it a blood-red colour. Stallbaum, accepting μιγνυμένου, oddly enough retains παρασχομένῃ.

13. τὸ δὲ ὅσον μέτρον] To give the exact proportions of the mixture is beyond the power of science and is not requisite κατὰ τὸν εἰκότα λόγον: cf. below, 68 D.

16. ὄρφνινον] This is probably a very deep shade of violet: compare Aristotle de coloribus ii 792[a] 25 ἐντεινόμενα γάρ πως πρὸς τὸ φῶς ἁλουργὲς ἔχει τὸ χρῶμα· ἐλάττονος δὲ τοῦ φωτὸς προσβάλλοντος ζοφερόν, ὃ καλοῦσιν ὀρφνιον. The word occurs again in the same form in chapter iv 794[b] 5. See too Xenophon Cyropaedia VIII iii 3 οὐδὲν φειδόμενος οὔτε πορφυρίδων οὔτε ὀρφνίνων οὔτε φοινικίδων οὔτε καρυκίνων (red-sauce-coloured) ἱματίων. It seems to have been an expensive

penetrates the visual stream quite up to the eyes, and forcibly displaces and decomposes the pores of the eyes themselves, draws from thence a combined body of fire and water, which we call a tear: and whereas this agent is itself fire, meeting the other from the opposite direction, and one fire leaps forth as lightning from the eyes, while the other enters in and is quenched in the moisture, all manner of colours arise in this commixture; and to the sensation we give the name of dazzling, and the agent which produces it we call bright and shining. A kind of fire which is intermediate between the two former, when it reaches the moisture of the eye and is mingled with it, but does not flash, produces a blood-like colour by the mixture of fire with the gleam of the moisture, and the name we give it is red. Bright combined with red and white makes yellow. In what proportion they are mingled, it were not reasonable to say, even if we knew; for there is neither any inevitable law nor any probable account thereof which we might properly declare. Red mingled with black and white becomes purple, which turns to dark violet when these ingredients are more burnt and a greater quantity of black is added. Chestnut arises from the mixture of yellow and grey, and grey from white and black: pale buff is from white mixed with yellow. When bright meets white and is steeped in intense black, a deep blue colour is the result; and

---

tint much in vogue among people who dressed handsomely: cf. Athenaeus XII 50, where it appears to represent the colour of the midnight star-lit sky. As regards ἀλουργόν, it may be noted that this is the same combination which is assigned by Demokritos to πορφυροῦν: Theophrastos *de sensu* § 77 τὸ δὲ πορφυροῦν ἐκ λευκοῦ καὶ μέλανος καὶ ἐρυθροῦ, πλείστην μὲν μοῖραν ἔχοντος τοῦ ἐρυθροῦ, μικρὰν δὲ τοῦ μέλανος, μέσην δὲ τοῦ λευκοῦ. A summary of the opinions of Demokritos concerning colour is given in §§ 73—78.

17. πυρρὸν δέ] This is a bright reddish brown, chestnut or auburn. φαιόν is a dusky grey: ὠχρὸν an ochreous yellow or buff.

20. **εἰς μέλαν κατακορές**] i.e. an intense, absolute black; the substance being, as it were, saturated with as much black as it can contain. This is a technical term to express vividness of colour: cf. Aristotle *de coloribus* v 795ᵃ 2 μᾶλλον μὲν οὖν τοῦ ὑγροῦ μελαινομένου τὸ ποῶδες γίνεται κατακορὲς ἰσχυρῶς καὶ πρασοειδές.

κυανοῦν χρῶμα] Dark blue. Demokritos gives a different account: Theophrastos *l. l.* τὸ δὲ κυανοῦν ἐξ ἰσάτιδος (the blue colour obtained from woad) καὶ πυρώδους. By γλαυκον a light blue is evidently meant. The elaborate distinctions of colour drawn in the present chapter certainly do not tend to support the theory which has been put forward that the Greeks were deficient in the colour-sense: indeed it is somewhat difficult to get a sufficient number of English terms to translate the Greek names.

252   ΠΛΑΤΩΝΟΣ   [68 C—

κυανοῦ δὲ λευκῷ κεραννυμένου γλαυκόν, πυρροῦ δὲ μέλανι πράσιον. τὰ δὲ ἄλλα ἀπὸ τούτων σχεδὸν δῆλα, αἷς ἂν ἀφομοιούμενα D
μίξεσι διασώζοι τὸν εἰκότα μῦθον. εἰ δέ τις τούτων ἔργῳ σκο-
πούμενος βάσανον λαμβάνοι, τὸ τῆς ἀνθρωπίνης καὶ θείας φύσεως
5 ἠγνοηκὼς ἂν εἴη διάφορον, ὅτι θεὸς μὲν τὰ πολλὰ εἰς ἓν ξυγκεραν-
νύναι καὶ πάλιν ἐξ ἑνὸς εἰς πολλὰ διαλύειν ἱκανῶς ἐπιστάμενος
ἅμα καὶ δυνατός, ἀνθρώπων δὲ οὐδεὶς οὐδέτερα τούτων ἱκανὸς οὔτε
ἔστι νῦν οὔτ' εἰσαῦθίς ποτ' ἔσται. ταῦτα δὴ πάντα τότε ταύτῃ E
πεφυκότα ἐξ ἀνάγκης ὁ τοῦ καλλίστου τε καὶ ἀρίστου δημιουργὸς
10 ἐν τοῖς γιγνομένοις παρελάμβανεν, ἡνίκα τὸν αὐτάρκη τε καὶ τὸν
τελεώτατον θεὸν ἐγέννα, χρώμενος μὲν ταῖς περὶ ταῦτα αἰτίαις
ὑπηρετούσαις, τὸ δὲ εὖ τεκταινόμενος ἐν πᾶσι τοῖς γιγνομένοις
αὐτός. διὸ δὴ χρὴ δύ' αἰτίας εἴδη διορίζεσθαι, τὸ μὲν ἀναγκαῖον,
τὸ δὲ θεῖον, καὶ τὸ μὲν θεῖον ἐν ἅπασι ζητεῖν κτήσεως ἕνεκα εὐδαί-
15 μονος βίου, καθ' ὅσον ἡμῶν ἡ φύσις ἐνδέχεται, τὸ δὲ ἀναγκαῖον 69 A
ἐκείνων χάριν, λογιζόμενον, ὡς ἄνευ τούτων οὐ δυνατὰ αὐτὰ ἐκεῖνα,
ἐφ' οἷς σπουδάζομεν, μόνα κατανοεῖν οὐδ' αὖ λαβεῖν οὐδ' ἄλλως
πως μετασχεῖν.

6 ἱκανῶς : ἱκανὸς ὡς SZ.   16 λογιζόμενον : λογιζομένους SZ.

1. **πυρροῦ δὲ μέλανι πράσιον**] This certainly seems an exceedingly odd combination. πράσιον is bright green, or leek-colour; and a mixture of chestnut and black appears very little likely to produce it. Aristotle more correctly classes green, along with red and violet, as a simple colour: see *meteorologica* III ii 372ᵃ 5 ἔστι δὲ τὰ χρώματα ταῦτα ἅπερ μόνα σχεδὸν οὐ δύνανται ποιεῖν οἱ γραφῆς· ἔνια γὰρ αὐτοὶ κεραννύουσι, τὸ δὲ φοινικοῦν καὶ πράσινον καὶ ἁλουργὸν οὐ γίγνεται κεραννύμενον. ἡ δὲ ἶρις ταῦτ' ἔχει τὰ χρώματα· τὸ δὲ μεταξὺ τοῦ φοινικοῦ καὶ πρασίνου φαίνεται πολλάκις ξανθόν. According to Demokritos πράσινον is ἐκ πορφυροῦ καὶ τῆς ἰσάτιδος, ἢ ἐκ χλωροῦ καὶ πορφυροειδοῦς: combinations which seem hardly better calculated than Plato's for producing the desired result.

5. **θεὸς μέν**] God, says Plato, can detect in the multifarious diversity of particulars one single form underlying them all; and again he can trace the development of that form through all the ramifications of its manifold appearances. Plato here probably has in view the problem of ἓν καὶ πολλὰ as presented by the methodical investigation of physical phenomena; the tendency of his later thought was however to the conclusion that the problem is one which can only approximately be grasped by finite intelligence. Compare 83 C.

11. **αἰτίαις ὑπηρετούσαις**] cf. *supra* 48 C, *Phaedo* 99 A, *Politicus* 281 D.

13. **τὸ μὲν ἀναγκαῖον, τὸ δὲ θεῖον**] The distinction between the two sorts of causes is obvious enough. The ἀναγκαῖον includes all the subsidiary causes, the physical forces and laws by means of which Nature carries on her work: the θεῖον is the final cause, the idea of τὸ βέλτιστον as existing in absolute intelligence. The operation of ἀνάγκη is to be studied either, as we were told at 59 C, for the sake of rational recreation, or more seriously, as we now

deep blue mingled with white produces pale blue; and chestnut with black makes green. And for the remaining colours, it is pretty clear from the foregoing to what combinations we ought to assign them so as to preserve the probability of our account: but if a man endeavour to make practical trial of these theories he will prove himself ignorant of the difference between divine and human intelligence: that God has sufficient understanding and power to blend the many into one and again to resolve the one into many; but no man is able to do either of these, now or henceforth for ever.

All these things being thus constituted by necessity, the creator of the most fair and perfect in the realm of becoming took them over, when he was generating the self-sufficing and most perfect god, using the forces in them as subservient causes, but himself working out the good in all things that come into being. Wherefore we must distinguish two kinds of causes, one of necessity and one of God: and the divine we must seek in all things for the sake of winning a happy life, so far as our nature admits of it; and the necessary for the sake of the divine, reflecting that without these we cannot apprehend by themselves the other truths, which are the object of our serious study, nor grasp them nor in any other way attain to them.

learn, as a stepping-stone to the knowledge of the θεῖον. This passage contains the strongest expression which is to be found in Plato in favour of the investigation of phenomena, when he says that it is necessary to study subsidiary causes as an aid to the study of the final cause. Particulars are nothing else but the form in which the ideas are made manifest to our bodily senses; therefore the study of particulars, in its highest aspect, is the study of ideas. But the sole value of this study lies in its bearing on the knowledge of the ideal world: the physical inquiry regarded as an end in itself Plato estimates quite as low in the *Timaeus* as in the *Republic*.

69 A—70 D, c. xxxi. Now therefore that we have completed our account of the accessory causes which God employed in carrying out his end, let us bring our story to a fitting close by setting forth how he thereafter fulfilled his design. God found all matter without form or law, obeying blind chance. He inspired into it form and order and made it to be a single universe, a living creature containing within it all things else that live. Of the divine he was himself the maker; but the creation of the mortal he committed to his children. And they, receiving from him the immortal essence, built for it a mortal body, bringing with it all the passions that belong to the flesh. And reason, which is immortal, they set in the head: but they made to dwell with it two mortal forms of soul, which they severed from the immortal by putting the neck to sunder them. And since the mortal form was twofold, they made the midriff for a wall to part the two: and they set emotion in the heart,

XXXI. Ὅτ' οὖν δὴ τὰ νῦν οἷα τέκτοσιν ἡμῖν ὕλη παράκειται τὰ τῶν αἰτίων γένη διυλασμένα, ἐξ ὧν τὸν ἐπίλοιπον λόγον δεῖ ξυνυφανθῆναι, πάλιν ἐπ' ἀρχὴν ἐπανέλθωμεν διὰ βραχέων, ταχύ τε εἰς ταὐτὸν πορευθῶμεν, ὅθεν δεῦρο ἀφικόμεθα, καὶ τελευ-
5 τὴν ἤδη κεφαλήν τε τῷ μύθῳ πειρώμεθα ἁρμόττουσαν ἐπιθεῖναι B τοῖς πρόσθεν. ὥσπερ οὖν καὶ κατ' ἀρχὰς ἐλέχθη, ταῦτα ἀτάκτως ἔχοντα ὁ θεὸς ἐν ἑκάστῳ τε αὐτῷ πρὸς αὑτὸ καὶ πρὸς ἄλληλα συμμετρίας ἐνεποίησεν, ὅσας τε καὶ ὅπῃ δυνατὸν ἦν ἀνάλογα καὶ σύμμετρα εἶναι. τότε γὰρ οὔτε τούτων ὅσον μὴ τύχῃ τι μετεῖχεν,
10 οὔτε τὸ παράπαν ὀνομάσαι τῶν νῦν ὀνομαζομένων ἀξιόλογον ἦν οὐδέν, οἷον πῦρ καὶ ὕδωρ καὶ εἴ τι τῶν ἄλλων· ἀλλὰ πάντα ταῦτα πρῶτον διεκόσμησεν, ἔπειτ' ἐκ τούτων πᾶν τόδε ξυνεστήσατο, ζῷον C ἓν ζῷα ἔχον τὰ πάντα ἐν αὑτῷ θνητὰ ἀθάνατά τε. καὶ τῶν μὲν θείων αὐτὸς γίγνεται δημιουργός, τῶν δὲ θνητῶν τὴν γένεσιν τοῖς
15 ἑαυτοῦ γεννήμασι δημιουργεῖν προσέταξεν· οἱ δὲ μιμούμενοι, παραλαβόντες ἀρχὴν ψυχῆς ἀθάνατον, τὸ μετὰ τοῦτο θνητὸν σῶμα αὐτῇ περιετόρνευσαν ὄχημά τε πᾶν τὸ σῶμα ἔδοσαν, ἄλλο τε εἶδος

2 διυλασμένα: διυλισμένα Π et ex correctione, ut videtur, Λ.
6 ταῦτα: αὐτὰ τά Λ.   13 ἔχον τὰ πάντα: ἔχοντα πάντα Λ.

and appetite they chained in the belly. This they did that the nobler part should hear the voice of the reason and pass its commands through all the swift channels of the blood, and so might aid it in subduing the rebellious swarm of lusts and passions. And knowing that the heart, excited by fear or passion, would leap and throb vehemently, they devised the cool soft structure of the lungs for a cushion to soothe and sustain it in the time of need.

1. ὕλη παράκειται] We have assorted our material by distinguishing the θεία αἰτία from the ἀναγκαία and by enumerating the manifold forms of the latter. The use of ὕλη is of course purely metaphorical, without any trace of the Aristotelian sense.

2. διυλασμένα] I can find no authority for using διυλισμένα, which Hermann keeps, in the sense here required. διυλίζειν is a late word signifying 'to filter'.

3. ἐπ' ἀρχὴν ἐπανέλθωμεν] We here resume our account, interrupted at 47 E, of the operation of intelligence, which now acts through the created gods in the generation of human beings. At the same time Plato fulfils the promise made in 61 D of expounding σαρκὸς καὶ τῶν περὶ σάρκα γένεσιν ψυχῆς τε ὅσον θνητόν.

4. τελευτὴν ἤδη κεφαλήν τε] Compare Phaedrus 264 C ἀλλὰ τόδε γε οἶμαί σε φάναι ἄν, δεῖν πάντα λόγον ὥσπερ ζῷον συνεστάναι σῶμά τι ἔχοντα αὐτὸν αὑτοῦ, ὥστε μήτε ἀκέφαλον εἶναι μήτε ἄπουν, ἀλλὰ μέσα τε ἔχειν καὶ ἄκρα, πρέποντ' ἀλλήλοις καὶ τῷ ὅλῳ γεγραμμένα: also Politicus 277 B ἀλλ' ἀτεχνῶς ὁ λόγος ἡμῖν ὥσπερ ζῷον τὴν ἔξωθεν μὲν περιγραφὴν ἔοικεν ἱκανῶς ἔχειν, τὴν δὲ οἷον τοῖς φαρμάκοις καὶ τῇ συγκράσει τῶν χρωμάτων ἐνάργειαν οὐκ ἀπειληφέναι πω.

6. κατ' ἀρχὰς ἐλέχθη] We have here a brief reference to the statements in 30 A, 42 D—43 A.

XXXI. Now therefore that the different kinds of causes lie ready sorted to our hand, like wood prepared for a carpenter, of which we must weave the web of our ensuing discourse, let us in brief speech return to the beginning and proceed once more to the spot whence we arrived at our present point; and so let us endeavour to add an end and a climax to our story conformable with what has gone before.

As was said then at the beginning, when God found these things without order either in the relation of each thing to itself or of one to another, he introduced proportion among them, in as many kinds and ways as it was possible for them to be proportionate and harmonious. For at that time neither had they any proportion, except by mere chance, nor did any of the bodies that now are named by us deserve the name, such as fire and water and the other elements: but first he ordered all these, and then out of them wrought this universe, a single living creature containing within itself all living creatures, mortal and immortal, that exist. And of the divine he himself was the creator; but the creation of mortals he delivered over to his own children to work out. And they, in imitation of him, having received from him the immortal principle of soul, fashioned round about her a mortal body and gave her all the body to

ἀτάκτως ἔχοντα] See note on 53 A. As to the construction, the accusative may be regarded as governed by the compound phrase συμμετρίας ἐνεποίησεν, as though Plato had written ξυνηρμόσατο. We had a somewhat similar sentence in 37 D, ἡμέρας γὰρ καὶ νύκτας καὶ μῆνας καὶ ἐνιαυτούς, οὐκ ὄντας πρὶν οὐρανὸν γενέσθαι, τότε ἅμα ἐκείνῳ ξυνισταμένῳ τὴν γένεσιν αὐτῶν μηχανᾶται.

9. τούτων] sc. τῶν συμμετριῶν.

10. οὔτε τὸ παράπαν ὀνομάσαι] Another shaft aimed at Demokritos: had fire and water received only just so much form as they might owe to τύχη, they could not even have been worthy of the names fire and water. The mere existence of such definite forms as fire air earth and water, even apart from their harmonisation into a single coherent κόσμος, could not have come to pass without the action of intelligence. Compare 53 B ἴχνη μὲν ἔχοντα αὐτῶν ἄττα.

12. ζῷον ἕν] cf. 30 C.

17. ὄχημα] Compare 44 E ὄχημα αὐτὸ τοῦτο καὶ εὐπορίαν ἔδοσαν. The notion of ὄχημα is not a vessel to contain the soul, but a means of her physical locomotion.

ἄλλο τε εἶδος...τὸ θνητόν] The nature of this θνητὸν εἶδος has been discussed in detail in my introduction to the *Phaedo*: a brief statement therefore of what I conceive it to mean may suffice here. The division into θεῖον and θνητὸν is obviously identical with the division into λογιστικὸν and ἄλογον in the *Republic*; and the subdivision of θνητὸν corresponds to the subdivision of ἄλογον in that dialogue into θυμοειδὲς and ἐπιθυμη-

ἐν αὐτῷ ψυχῆς προσῳκοδόμουν τὸ θνητόν, δεινὰ καὶ ἀναγκαῖα ἐν
ἑαυτῷ παθήματα ἔχον, πρῶτον μὲν ἡδονήν, μέγιστον κακοῦ δέλεαρ, D
ἔπειτα λύπας, ἀγαθῶν φυγάς, ἔτι δ᾿ αὖ θάρρος καὶ φόβον, ἄφρονε
ξυμβούλω, θυμὸν δὲ δυσπαραμύθητον, ἐλπίδα δ᾿ εὐπαράγωγον·
5 αἰσθήσει δὲ ἀλόγῳ καὶ ἐπιχειρητῇ παντὸς ἔρωτι ξυγκερασάμενοι
ταῦτα ἀναγκαίως τὸ θνητὸν γένος ξυνέθεσαν. καὶ διὰ ταῦτα δὴ
σεβόμενοι μιαίνειν τὸ θεῖον, ὅ τι μὴ πᾶσα ἦν ἀνάγκη, χωρὶς ἐκείνου
κατοικίζουσιν εἰς ἄλλην τοῦ σώματος οἴκησιν τὸ θνητόν, ἰσθμὸν E
καὶ ὅρον διοικοδομήσαντες τῆς τε κεφαλῆς καὶ τοῦ στήθους, αὐ-
10 χένα μεταξὺ τιθέντες, ἵνα εἴη χωρίς. ἐν δὴ τοῖς στήθεσι καὶ τῷ
καλουμένῳ θώρακι τὸ τῆς ψυχῆς θνητὸν γένος ἐνέδουν, καὶ ἐπειδὴ
τὸ μὲν ἄμεινον αὐτῆς, τὸ δὲ χεῖρον ἐπεφύκει, διοικοδομοῦσι τὸ τοῦ
θώρακος αὖ κύτος, διορίζοντες οἷον γυναικῶν, τὴν δὲ ἀνδρῶν χωρὶς 70 A
οἴκησιν, τὰς φρένας διάφραγμα εἰς τὸ μέσον αὐτῶν τιθέντες. τὸ

4 θυμὸν δέ: θυμόν τε et mox ἐλπίδα τ᾿ S.   5 αἰσθήσει δέ: αἰσθήσει τε SZ.
ξυγκερασάμενοι ταῦτα: ξυγκερασάμενοι τ᾿ αὐτά, facta post ἔρωτι interpunctione, SZ.
12 ἐπεφύκει: πεφύκει S.   τὸ τοῦ θώρακος αὖ: τὸ τοῦ θώρακος αὐτό A.   τοῦ θώρακος αὖ
τό SZ.

τικόν, and to the nobler and baser steed in the *Phaedrus*. It seems to me certain that these three εἴδη are but names for one and the same vital force manifesting itself in different relations. The intellect, seated in the head, is the soul acting by herself, performing her own proper function of thinking. But since she is brought into connexion with a material body, she must needs have πάθη which are concerned with that body. So then, if the θεῖον is her activity by herself, the θνητόν is her activity through the body; which activity Plato distributes into two classes of πάθη, one of which may be designated by the general term of emotions, the other by that of appetites. It will be noticed that this does not profess to give an exhaustive catalogue of the soul's activities through body: for sensuous perceptions are a mode of her action through body which does not fall under either head. For reasons in support of this view of the relation of the εἴδη I must refer to the introduction to the *Phaedo* aforesaid. The name θνητόν is applied by Plato to the lower

εἶδος, because, though soul is in herself and in her own activity eternal, her connexion with any particular body is temporary, and so must her action through such a body be also. Galen comments upon the term θνητόν as follows: *de plac. Hipp. et Plat.* IX 794 πότερον κυρίως ὀνομάζων εἴρηκεν ἐν Τιμαίῳ θνητὰ τὰ δύο μέρη τῆς ψυχῆς ἢ ταύτην αὐτοῖς ἐπήνεγκε τὴν προσηγορίαν ἀθανάτοις οὖσιν ὡς χείροσι τοῦ λογιστικοῦ καὶ ὡς κατὰ τὰ θνητὰ τῶν ζῴων ἐνεργοῦσι μόνον; Of this question he offers no determination, but that he raised the point is interesting.

1. δεινὰ καὶ ἀναγκαῖα] This and much more of the phraseology in the present passage is echoed from 42 A. ἀναγκαῖα = necessarily inherent in their nature.

3. ἄφρονε ξυμβούλω] Compare *Laws* 644 C, where pleasure and pain take the place of confidence and fear: δύο δὲ κεκτημένον ἐν ἑαυτῷ ξυμβούλω ἐναντίω τε καὶ ἄφρονε, ὣ προσαγορεύομεν ἡδονὴν καὶ λύπην.

6. τὸ θνητὸν γένος] sc. τῆς ψυχῆς.

7. σεβόμενοι μιαίνειν τὸ θεῖον] An-

ride in; and beside her they built in another kind of soul, even that which is mortal, having within itself dread and inevitable passions—first pleasure, the strongest allurement of evil, next pains, that scare good things away; confidence moreover and fear, a yoke of thoughtless counsellors; wrath hard to assuage and hope that lightly leads astray; and having mingled all these perforce with reasonless sensation and love that ventures all things, so they fashioned the mortal soul. And for this cause, in awe of defiling the divine, so far as was not altogether necessary, they set the mortal kind to dwell apart from the other in another chamber of the body, having built an isthmus and boundary between the head and the breast, setting the neck between them to keep them apart. So in the breast, or the thorax as it is called, they confined the mortal kind of soul. And whereas one part of it was nobler, the other baser, they built a party-wall across the hollow of the chest, as if they were marking off an apartment for women and another for men, and they put the midriff as a fence between them. That part of the

other reason why the intellect should be in the head is given in 90 A. Galen *de plac. Hipp. et Plat.* VI 505 says that Hippokrates agreed with Plato in making three ἀρχαί, the head heart and liver: this view Galen himself defends against that of Aristotle and Theophrastos, who made the heart the sole ἀρχή: cf. Aristotle *de iuventute* iii 469ᵃ 5. See note on 73 B οἱ γὰρ τοῦ βίου δεσμοὶ τῆς ψυχῆς τῷ σώματι ξυνδουμένης ἐν τούτῳ διαδούμενοι κατερρίζουν τὸ θνητὸν γένος.

ὅ τι μὴ πᾶσα ἦν ἀνάγκη] A certain loss of her divine nature is inseparable from the soul's differentiation and consequent material embodiment: all the gods could do was to reduce this to a minimum.

10. τῷ καλουμένῳ θώρακι] The epithet καλουμένῳ is inserted because the word θώραξ in this sense is a technical term of anatomy, the popular word being στέρνον or στῆθος. It occurs nowhere else in Plato, but is common in Aristotle, who sometimes, as *de partibus animalium* IV

xii 693ᵃ 25, uses the same expression, τὰ τοῦ καλουμένου θώρακος ἐπὶ τῶν τετραπόδων. Euripides has it once, *Hercules furens* 1095 νεανίαν θώρακα καὶ βραχίονα. Aristotle also uses the word in a more comprehensive sense than it bears nowadays, including the entire trunk: *historia animalium* I vii 491ᵃ 29.

13. οἷον γυναικῶν, τὴν δὲ ἀνδρῶν] This is no more than a mere simile: there is nothing in the words to warrant the titles which Martin bestows upon the two εἴδη—l'âme mâle and l'âme femelle; nor is there the slightest appropriateness in these names. It is not even said which division corresponds to the γυναικῶν, which to the ἀνδρῶν οἴκησις.

14. διάφραγμα] This word, which has since become specially appropriated to the midriff, is used in a general sense by Plato for a fence or partition: Aristotle applies it to the cartilaginous wall dividing the nostrils, *historia animalium* I xi 492ᵇ 16: the midriff he often calls διάζωμα.

P. T. 17

258   ΠΛΑΤΩΝΟΣ   [70 A—

μετέχον οὖν τῆς ψυχῆς ἀνδρείας καὶ θυμοῦ, φιλόνεικον ὄν, κατῴ-
κισαν ἐγγυτέρω τῆς κεφαλῆς μεταξὺ τῶν φρενῶν τε καὶ αὐχένος,
ἵνα τοῦ λόγου κατήκοον ὂν κοινῇ μετ᾽ ἐκείνου βίᾳ τὸ τῶν ἐπιθυμιῶν
κατέχοι γένος, ὁπότ᾽ ἐκ τῆς ἀκροπόλεως τῷ ἐπιτάγματι καὶ λόγῳ
5 μηδαμῇ πείθεσθαι ἑκὸν ἐθέλοι. τὴν δὲ δὴ καρδίαν ἅμμα τῶν φλεβῶν
καὶ πηγὴν τοῦ περιφερομένου κατὰ πάντα τὰ μέλη σφοδρῶς αἵ- B
ματος εἰς τὴν δορυφορικὴν οἴκησιν κατέστησαν, ἵνα, ὅτε ζέσειε τὸ
τοῦ θυμοῦ μένος, τοῦ λόγου παραγγείλαντος, ὥς τις ἄδικος περὶ
αὐτὰ γίγνεται πρᾶξις ἔξωθεν ἢ καί τις ἀπὸ τῶν ἔνδοθεν ἐπιθυμιῶν,
10 ὀξέως διὰ πάντων τῶν στενωπῶν πᾶν ὅσον αἰσθητικὸν ἐν τῷ
σώματι τῶν τε παρακελεύσεων καὶ ἀπειλῶν αἰσθανόμενον γίγνοιτο ἐπήκοον καὶ ἕποιτο πάντῃ, καὶ τὸ βέλτιστον οὕτως ἐν αὐτοῖς
πᾶσιν ἡγεμονεῖν ἐῷ. τῇ δὲ δὴ πηδήσει τῆς καρδίας ἐν τῇ τῶν C
δεινῶν προσδοκίᾳ καὶ τῇ τοῦ θυμοῦ ἐγέρσει, προγιγνώσκοντες ὅτι
15 διὰ πυρὸς ἡ τοιαύτη πᾶσα ἔμελλεν οἴδησις γίγνεσθαι τῶν θυ-
μουμένων, ἐπικουρίαν αὐτῇ μηχανώμενοι τὴν τοῦ πλεύμονος ἰδέαν
ἐνεφύτευσαν, πρῶτον μὲν μαλακὴν καὶ ἄναιμον, εἶτα σήραγγας
ἐντὸς ἔχουσαν οἷον σπόγγου κατατετρημένας, ἵνα τό τε πνεῦμα
καὶ τὸ πόμα δεχομένη, ψύχουσα, ἀναπνοὴν καὶ ῥᾳστώνην ἐν τῷ

1 ἀνδρείας: ἀνδρίας AZ.   5 ἅμμα: ἀρχὴν ἅμα S.   10 τῶν ante στενωπῶν omittunt AS.
13 ἐῷ: ἐῴη S.   15 οἴδησις: οἴκησις A.   19 πόμα: πῶμα A pr. m. SZ.

3. κατήκοον] Undoubtedly this means 'within hearing of': that was the object they had in view when they placed the θυμοειδὲς ἐγγυτέρω τῆς κεφαλῆς.

4. ἐκ τῆς ἀκροπόλεως] Compare Galen de placitis Hippocratis et Platonis II 230 καθάπερ ἐν ἀκροπόλει τῇ κεφαλῇ δίκην μεγάλου βασιλέως ὁ ἐγκέφαλος ἵδρυται.

5. ἅμμα] This reading has best ms. authority and gives the best sense: Stallbaum's ἀρχὴν ἅμα is comparatively feeble. It is true that Aristotle de iuventute iii 468ᵇ 31 has ἡ δὲ καρδία ὅτι ἐστὶν ἀρχὴ τῶν φλεβῶν: but that is no evidence that Plato wrote ἀρχήν here. Galen quotes this passage, de plac. II 292, and charges Chrysippos with plagiarising the Platonic doctrine.

6. σφοδρῶς] From this word Galen de plac. VI 573 infers that Plato makes the heart the ἀρχή of the arterial circulation only, not of the venous, the ἀρχή of which is the liver; τὸ μὲν γὰρ ἐξ ἥπατος ὁρμώμενον οὐ περιφέρεται σφοδρῶς. This seems however a slight basis on which to found the inference that Plato knew the difference between veins and arteries, which he nowhere else gives any sign of distinguishing. Compare pseudo-Hippokrates de alimentis vol. II p. 22 Kühn ῥίζωσις φλεβῶν ἧπαρ, ῥίζωσις ἀρτηριῶν καρδίη, ἐκ τουτέων ἀποπλανᾶται αἷμα καὶ πνεῦμα, καὶ θερμασίη διὰ τουτέων φοιτᾷ: the passage however has in it unmistakable marks of a date long subsequent to Plato's time or Aristotle's either. The distinction between veins and arteries seems also to have been unknown to Aristotle; and unquestionably he makes the heart the only ἀρχή.

9. τῶν ἔνδοθεν ἐπιθυμιῶν] Compare the functions of the φύλακες in protecting the city εἴτε τις ἔξωθεν ἢ καὶ τῶν ἔνδοθεν

c] ΤΙΜΑΙΟΣ. 259

soul which shares courage and anger, seeing that it is warlike, they planted nearer the head, between the midriff and the neck, that it might be within hearing of the reason and might join it in forcibly keeping down the tribe of lusts, when they would in no wise consent to obey the order and word of command from the citadel. And the heart, which is the knot of the veins and the fount of the blood which rushes vehemently through all the limbs, they made into the guardhouse, that whensoever the fury of anger boiled up at the message from the reason, that some unrighteous dealing is being wrought around them, either without, or, it may be, by the lusts within, swiftly through all the narrow channels all the sensitive power in the body might be aware of the admonitions and threats and be obedient to them and follow them altogether, and so permit the noblest part to be leader among them all.

For the throbbing of the heart in the anticipation of danger or the excitement of wrath, since they foreknew that all such swelling of passion should come to pass by means of fire, they devised a plan of relief, and framed within us the structure of the lungs, which in the first place is soft and void of blood, and next is perforated within with cavities like those of a sponge, in order that receiving the breath and the drink it might cause coolness and give rest and relief in the burning. Wherefore

ἴοι κακουργήσων 17 D.

10. διὰ πάντων τῶν στενωπῶν] i.e. through all the narrow blood-vessels; to which, as we have seen, Plato attributed the functions which are really discharged by the nerves.

11. τῶν τε παρακελεύσεων καὶ ἀπειλῶν] Cf. 71 B χαλεπὴ προσενεχθεῖσα ἀπειλῇ. τὸ βέλτιστον of course = τὸ λογιστικόν.

13. τῇ δὲ δὴ πηδήσει] The violent beating of the heart under the influence of strong emotion is due to its hot and fiery composition. So the lungs, a soft and bloodless structure, were placed beside it, partly to cool it, partly to provide a soft cushion to receive its bounding. Plato, as we shall see when we come to his account of respiration, was unaware

of the paramount importance of the lungs in the process of breathing and the purification of the blood: he is also of course quite wrong in calling them ἄναιμον. His view is impugned by Aristotle on grounds of comparative anatomy, de partibus animalium III vi 669ᵃ 18 τὸ δὲ πρὸς τὴν ἕλσιν εἶναι τὸν πλεύμονα τῆς καρδίας οὐκ εἴρηται καλῶς: further on, 669ᵇ 8, he says ὅλως μὲν οὖν ὁ πλεύμων ἐστὶν ἀναπνοῆς χάριν: but he does not seem to have had a very clear idea of the functions performed by the lungs.

18. τό τε πνεῦμα καὶ τὸ πόμα] In this curious error Plato is at one with all, or nearly all, the best medical science of the day. Plutarch de Stoicorum repugnantiis xxix says Πλάτων μὲν ἔχει τῶν ἰατρῶν τοὺς ἐνδοξοτάτους μαρτυροῦντας,

καύματι παρέχοι· διὸ δὴ τῆς ἀρτηρίας ὀχετοὺς ἐπὶ τὸν πλεύμονα D
ἔτεμον, καὶ περὶ τὴν καρδίαν αὐτὸν περιέστησαν οἷον ἅλμα μαλακόν, ἵν' ὁ θυμὸς ἡνίκα ἐν αὐτῇ ἀκμάζοι, πηδῶσα εἰς ὑπεῖκον καὶ
ἀναψυχομένη, πονοῦσα ἧττον, μᾶλλον τῷ λόγῳ μετὰ θυμοῦ δύ-
5 ναιτο ὑπηρετεῖν.

XXXII. Τὸ δὲ δὴ σίτων τε καὶ ποτῶν ἐπιθυμητικὸν τῆς
ψυχῆς καὶ ὅσων ἔνδειαν διὰ τὴν τοῦ σώματος ἴσχει φύσιν, τοῦτο
εἰς τὰ μεταξὺ τῶν τε φρενῶν καὶ τοῦ πρὸς τὸν ὀμφαλὸν ὅρου κατῴ- E
κισαν, οἷον φάτνην ἐν ἅπαντι τούτῳ τῷ τόπῳ τῇ τοῦ σώματος
10 τροφῇ τεκτηνάμενοι· καὶ κατέδησαν δὴ τὸ τοιοῦτον ἐνταῦθα ὡς
θρέμμα ἄγριον, τρέφειν δὲ ξυνημμένον ἀναγκαῖον, εἴπερ τι μέλλοι

2 ἅλμα μαλακόν: μάλαγμα H.   7 ὅσων: ὅσον H.

Ἱπποκράτην, Φιλιστίωνα, Διώξιππον τὸν Ἱπποκράτειον· καὶ τῶν ποιητῶν Εὐριπίδην, Ἀλκαῖον, Εὔπολιν, Ἐρατοσθένην, λέγοντας ὅτι τὸ ποτὸν διὰ τοῦ πνεύμονος διέξεισι. It is remarkable that Galen also held this view: cf. *de plac. Hipp. et Plat.* VIII 719 ἀλλὰ εἰ καὶ ζῷον, ὅ τι ἂν ἐθελήσῃς, διψῆσαι ποιήσεις, ὡς κεχρωσμένον ὕδωρ ὑπομεῖναι πιεῖν, εἰ δοίης εἴτε κυανῷ χρώματι χρώσας εἴτε μίλτῳ, εἶτα εὐθέως σφάξας ἀνατέμοις, εὑρήσεις κεχρωσμένον τὸν πνεύμονα. δῆλον οὖν ἐστὶν ὅτι φέρεταί τι τοῦ πόματος εἰς αὐτόν. Galen's observation is, I believe, correct, though his inference is not so. Aristotle, on the contrary, was aware that no fluid passes down the windpipe to the lungs: see *historia animalium* I xvi 495ᵇ 16 ἡ μὲν οὖν ἀρτηρία τοῦτον ἔχει τὸν τρόπον, καὶ δέχεται μόνον τὸ πνεῦμα καὶ ἀφίησιν, ἄλλο δ' οὐθὲν οὔτε ξηρὸν οὔθ' ὑγρόν, ἢ πόνον παρέχει, ἕως ἂν ἐκβήξῃ τὸ κατελθόν. See too *de partibus animalium* III iii 664ᵇ 9, where he gives divers demonstrations that the hypothesis is untenable. It is also denied by the writer of book IV of the Hippokratean treatise *de morbis*, vol. II pp. 373, 374 Kühn: but affirmed by the author of *de ossium natura*, a work of uncertain date, vol. I p. 515 Kühn. Galen *de plac.* VIII 715 points out that Plato conceives only a part of the fluid to pass down the trachea: οὐκ ἀθρόον οὐδὲ διὰ μέσης τῆς εὐ-ρυχωρίας τοῦ ὀργάνου φερόμενον, ἀλλὰ περὶ τὸν χιτῶνα αὐτοῦ δροσοειδῶς καταρρέον.

1. **τῆς ἀρτηρίας**] i.e. the windpipe: later it was designated ἡ τραχεῖα ἀρτηρία, whence trachea. This is the only usage of the word ἀρτηρία in Plato and Aristotle; it never means 'artery' in the modern sense. ὀχετούς is plural like ἀρτηρίας in 78 C, probably because of the bifurcation of the trachea into the bronchia before entering the lungs.

2. **ἅλμα μαλακόν**] There is certainly no reason for altering the text: Plato might very well say 'a soft leap' for 'a soft place to leap upon'. Martin's ἄγμα is a very unhappy suggestion, and Hermann's μάλαγμα is as inappropriate as arbitrary. μάλαγμα means a poultice or fomentation; but the function of the lungs is distinctly stated just below, πηδῶσα εἰς ὑπεῖκον: this is perfectly well expressed by the received reading. I believe that Aristotle had this word ἅλμα in his mind, when he wrote ἅλσις in the passage from *de partibus animalium* quoted above. The object of the lungs then, according to Plato, is to quiet down the agitation of the heart and thereby render the emotional faculty capable of taking sides with the reason against the ἐπιθυμητικόν.

4. **μετὰ θυμοῦ**] i.e. that the heart, along with the emotional faculty seated therein, may he enabled to obey the

ε]   ΤΙΜΑΙΟΣ.   261

they made the windpipe for a channel to the lungs, which they set around the heart, as it were a soft cushion to spring upon; so that when wrath was at its height therein, the heart might leap upon a yielding substance and become cooled, and thus being less distressed it might together with the emotions be better enabled to obey the reason.

XXXII. But that part of the soul which lusts after meat and drink and all things whereof it has need owing to the body's nature, this they set between the midriff and the navel as its boundary, constructing in all this region as it were a manger for the sustenance of the body: and here they chained it like a wild beast, which must yet be reared in conjunction with the rest, if a mortal race were to be at all. To the end

reason : that is to say, that the emotional faculty may not be hampered in its action by the physical agitation of the organ which it employs. From first to last, in this dialogue as in the *Republic*, Plato regards the emotions, if they are given fair play, as sure allies of the reason.

70 D—72 D, c. xxxii. But that part of the soul whereunto belongs the craving for meat and drink the gods placed in the belly, where they made, as it were, its stall: and so they kept it far away from the habitation of the intellect, that it might cause the least disquietude. And since they knew that it could not apprehend reason, but would be led by dreams and visions of the night, they devised for it the liver, which should copy off for it all the messages from the brain; either terrifying it by threats and pains and sickness, or soothing it by visions of peace. Here then they set up the oracular shrine in the body of man: and since the appetitive soul could not directly comprehend the precepts of reason, they thought to guide it by signs and tokens and dreams which might be comprehended of it. A proof that divination is a boon for human folly is this. No sane man in his waking senses is a true seer: only one that is asleep or delirious or in some way beside himself has this gift. The part of the sane man is to interpret the prophetic utterances of the distraught seer, for that the prophet cannot do. Whence the seer always has an interpreter to expound his sayings; who often, but wrongly, is himself termed a seer. So then the liver is the seat of prophecy: but it has this virtue only during life: after death it is blind.

Next to the liver is placed the spleen, which is as a sponge to purify it and carry off noxious humours.

7. διὰ τὴν τοῦ σώματος ἴσχει φύσιν] This clearly teaches that it is for the sake of the body alone that the appetitive soul desires meat and drink; for itself it needs no such thing. The inference thence is that the ἐπιθυμητικὸν detached from the body is just pure soul, the one and only soul; but *qua* ἐπιθυμητικὸν it is considered as working through and for the body, the nourishment of which it has to superintend.

9. οἷον φάτνην] This suggests a horse as the similitude, rather than a wild beast: compare *Phaedrus* 247 E.

10. ὡς θρέμμα ἄγριον] Compare *Republic* 588 C foll.

11. εἴπερ τι μέλλοι] If a mortal creature is to be, it must have a body; the body must be animated and sustained by

262   ΠΛΑΤΩΝΟΣ   [70 E—

τὸ θνητὸν ἔσεσθαι γένος. ἵν᾽ οὖν ἀεὶ νεμόμενον πρὸς φάτνῃ καὶ
ὅ τι πορρωτάτω τοῦ βουλευομένου κατοικοῦν, θόρυβον καὶ βοὴν
ὡς ἐλαχίστην παρέχον, τὸ κράτιστον καθ᾽ ἡσυχίαν περὶ τοῦ πᾶσι 71 A
κοινῇ ξυμφέροντος ἐῷ βουλεύεσθαι, διὰ ταῦτα ἐνταῦθ᾽ ἔδοσαν
5 αὐτῷ τὴν τάξιν. εἰδότες δὲ αὐτό, ὡς λόγου μὲν οὔτε ξυνήσειν
ἔμελλεν, εἴ τέ πῃ καὶ μεταλαμβάνοι τινὸς αὐτῶν αἰσθήσεως, οὐκ
ἔμφυτον αὐτῷ τὸ μέλειν τινῶν ἔσοιτο λόγων, ὑπὸ δὲ εἰδώλων καὶ
φαντασμάτων νυκτός τε καὶ μεθ᾽ ἡμέραν μάλιστα ψυχαγωγή-
σοιτο, τούτῳ δὴ θεὸς ἐπιβουλεύσας αὐτῷ τὴν ἥπατος ἰδέαν ξυνέ-
10 στησε καὶ ἔθηκεν εἰς τὴν ἐκείνου κατοίκησιν, πυκνὸν καὶ λεῖον B
καὶ λαμπρὸν καὶ γλυκὺ καὶ πικρότητα ἔχον μηχανησάμενος, ἵνα
ἐν αὐτῷ τῶν διανοημάτων ἡ ἐκ τοῦ νοῦ φερομένη δύναμις, οἷον ἐν

1 τὸ θνητόν : ποτὲ θνητὸν S.   6 αὐτῶν αἰσθήσεως : αὖ τῶν αἰσθήσεων SZ.

soul; hence there must be an ἐπιθυμητι-κόν, or, as Aristotle would say, a θρεπτι-κὸν εἶδος of soul. For, as has been said, the differentiation of souls into individuals involves materialisation and hence imperfection.

5. οὔτε ξυνήσειν ἔμελλεν] The lowest εἶδος would not have any comprehension of rational principles, or if haply it had some inkling of them, it would not care to pay any heed to them. Therefore they are expressed to this faculty in similitudes by means of the liver. It will be noticed that this symbolical representation of the dictates of the individual reason is exactly analogous to the symbolical manifestation of the ideas of universal reason by means of the sensible perception of particular objects.

6. αὐτῶν] This is doubtless right, referring to the τινῶν λόγων which follows. Stallbaum's reading is, as I think, weak in sense.

8. καὶ μεθ᾽ ἡμέραν] The phantasms of the daytime are the perceptions of the senses.

10. τὴν ἐκείνου κατοίκησιν] sc. τὴν τοῦ ἐπιθυμητικοῦ. In his account of the relations of the liver with the ἐπιθυμητι-κὸν Plato has by anticipation refined beyond the point made by Aristotle in nic. eth. I xiii 1102^b 23 foll. ἴσως δ᾽ οὐδὲν

ἧττον καὶ ἐν τῇ ψυχῇ νομιστέον εἶναί τι παρὰ τὸν λόγον, ἐναντιούμενον τούτῳ καὶ ἀντιβαῖνον. πῶς δ᾽ ἕτερον, οὐδὲν διαφέρει. λόγου δὲ καὶ τοῦτο φαίνεται μετέχειν, ὥσπερ εἴπομεν· πειθαρχεῖ γὰρ τῷ λόγῳ τὸ τοῦ ἐγκρατοῦς. ἔτι δ᾽ ἴσως εὐηκοώτερόν ἐστι τὸ τοῦ σώφρονος καὶ ἀνδρείου· πάντα γὰρ ὁμοφωνεῖ τῷ λόγῳ. φαίνεται δὴ καὶ τὸ ἄλογον διττόν. τὸ μὲν γὰρ φυτικὸν οὐδαμῶς κοινωνεῖ λόγου, τὸ δὲ ἐπιθυμητικὸν καὶ ὅλως ὀρεκτικὸν μετέχει πως, ᾗ κατήκοόν ἐστιν αὐτοῦ καὶ πειθαρχικόν. οὕτω δὴ τοῦ πατρὸς καὶ τῶν φίλων φαμὲν ἔχειν λόγον, καὶ οὐχ ὥσπερ τῶν μαθηματικῶν. ὅτι δὲ πείθεταί πως ὑπὸ τοῦ λόγου τὸ ἄλογον, μηνύει καὶ ἡ νουθέτησις καὶ πᾶσα ἐπιτίμησις καὶ παράκλησις. εἰ δὲ χρὴ καὶ τοῦτο φάναι λόγον ἔχειν, διττόν ἐστι καὶ τὸ λόγον ἔχον, τὸ μὲν κυρίως καὶ ἐν αὐτῷ, τὸ δὲ ὥσπερ τοῦ πατρὸς ἀκουστικόν τι. In Aristotle's analysis then the rational part is twofold, the one kind possessing reason absolutely, the other listening to its behests. The ἄλογον also is twofold, one kind being absolutely irrational, while the other μετέχει πῃ λόγου. It thus appears that the lower kind of λόγον ἔχον is identical with the higher kind of ἄλο-γον: that in fact they are the same thing viewed in different aspects. Comparing this with Plato's statement, we shall find that Aristotle's ἄλογον μετέχον πῃ λόγου

71 B] ΤΙΜΑΙΟΣ. 263

then that always feeding at its stall and dwelling as far as possible from the seat of counsel, it might produce the least possible tumult and uproar and allow the noblest part to consult in peace for the common weal, here they assigned it its place. And knowing that it would have no comprehension of reason, and that even if it did in some way gain any perception of rational thoughts, it was not in its nature to take heed to any such things, but that it would be entirely led away by images and shadows both by night and by day, God devised as a remedy for this the nature of the liver, which he constructed and set in its dwelling place: and he made it a body dense and smooth and bright and sweet with a share of bitterness. This he did to the end that the influence of thoughts proceeding

occupies the same position as Plato's θυμοειδὲς κατήκοον τοῦ λόγου. This *directly* hears and obeys the dictates of reason. If a man is betrayed by his friend, the declaration by the reason that such conduct is immoral is at once responded to by the θυμοειδὲς with a surge of indignation against the friend's baseness. But no such response would come from the ἐπιθυμητικόν, which is incapable of understanding the situation. The judgments of the reason must therefore be conveyed to it in the symbolic form which alone appeals to it, by signs and visions, by portents and presages and terrors. This indirect communication has no place in the statement of Aristotle, who would no doubt denounce it as πλασματώδες. It must of course not be forgotten that Aristotle's ἐπιθυμητικόν is not the same as Plato's.

A point worth noticing is a certain advance in the psychology of the *Timaeus* as compared with that of the *Phaedrus*. In the latter the lowest εἶδος is simply appetitive; but in the *Timaeus* it includes the functions of nutrition and growth. This is plain from 70 E οἷον φάτνην κ.τ.λ.; and also from the fact that the τρίτον εἶδος is assigned to plants. Aristotle then is in reality indebted to Plato for his θρεπτικὸν καὶ φυτικόν: though it must be confessed that the debt is by no means acknowledged.

11. ἵνα ἐν αὐτῷ] As this long sentence is very involved, a few words about the construction may not be amiss. The optatives belonging to ἵνα are φοβοῖ (the temporal clause after ὁπότε extending as far as παρέχοι) and the second ποιοῖ: while to ὁπότε belong ἐμφαίνοι, the first ποιοῖ, and παρέχοι; and to ὅτε belongs ἀποζωγραφοῖ only. The μὲν after φοβοῖ ought to have been answered by a δέ, when the soothing influence was first mentioned, but the length and intricacy of the sentence has interrupted the exact correspondence, so that the second member is introduced by καὶ instead of δέ. Again, it is not at first sight obvious, especially as the sentence is sometimes punctuated, to see where the apodosis to ὅτ' αὖ begins. I should without hesitation, putting a comma after ἀπευθύνουσα, make the beginning of the apodosis at ἵλεών τε: though, if we took the participles παρέχουσα and the rest in agreement with δύναμις instead of ἐπίπνοια, it would be possible to begin the apodosis at τῆς μὲν πικρότητος. But the former view seems to me in every way preferable. ἐν αὐτῷ is anticipative of the clause beginning οἷον ἐν κατόπτρῳ, from which we must supply the notion 'producing reflections in it'.

264 ΠΛΑΤΩΝΟΣ [71 B—

κατόπτρῳ δεχομένῳ τύπους καὶ κατιδεῖν εἴδωλα παρέχοντι, φοβοῖ
μὲν αὐτό, ὁπότε μέρει τῆς πικρότητος χρωμένη ξυγγενεῖ, χαλεπὴ
προσενεχθεῖσα ἀπειλῇ, κατὰ πᾶν ὑπομιγνῦσα ὀξέως τὸ ἧπαρ,
χολώδη χρώματα ἐμφαίνοι, ξυνάγουσά τε πᾶν ῥυσὸν καὶ τραχὺ
5 ποιοῖ, λοβὸν δὲ καὶ δοχὰς πύλας τε, τὰ μὲν ἐξ ὀρθοῦ κατακάμπ- C
τουσα καὶ ξυσπῶσα, τὰ δὲ ἐμφράττουσα συγκλείουσά τε, λύπας
καὶ ἄσας παρέχοι· καὶ ὅτ᾽ αὖ τἀναντία φαντάσματα ἀποζωγραφοῖ
πραότητός τις ἐκ διανοίας ἐπίπνοια, τῆς μὲν πικρότητος ἡσυχίαν
παρέχουσα τῷ μήτε κινεῖν μήτε προσάπτεσθαι τῆς ἐναντίας ἑαυτῇ
10 φύσεως ἐθέλειν, γλυκύτητι δὲ τῇ κατ᾽ ἐκεῖνο ξυμφύτῳ πρὸς αὐτὸ
χρωμένη καὶ πάντα ὀρθὰ καὶ λεῖα αὐτοῦ καὶ ἐλεύθερα ἀπευθύ- D
νουσα, ἵλεών τε καὶ εὐήμερον ποιοῖ τὴν περὶ τὸ ἧπαρ ψυχῆς
μοῖραν κατῳκισμένην, ἔν τε τῇ νυκτὶ διαγωγὴν ἔχουσαν μετρίαν,
μαντείᾳ χρωμένην καθ᾽ ὕπνον, ἐπειδὴ λόγου καὶ φρονήσεως οὐ
15 μετεῖχε. μεμνημένοι γὰρ τῆς τοῦ πατρὸς ἐπιστολῆς οἱ ξυστήσαντες
ἡμᾶς, ὅτε τὸ θνητὸν ἐπέστελλε γένος ὡς ἄριστον εἰς δύναμιν ποιεῖν,

5 τά: τὸ A.   10 αὐτό: ἑαυτὸ A.   15 ξυστήσαντες: ξυνιστάντες HS.

2. **μέρει τῆς πικρότητος χρωμένη ξυγγενεῖ**] Stallbaum understands τῷ ἥπατι after συγγενεῖ, saying 'ridicule enim quidam sic interpretantur, ac si rationis naturae cognatum intelligatur'. It appears to me that the 'ridiculous' interpretation is the only correct one: ξυγγενεῖ signifies, akin to the dark and gloomy nature of the thoughts which are conveyed by ἡ ἐκ τοῦ νοῦ φερομένη δύναμις: see below μήτε προσάπτεσθαι τῆς ἐναντίας ἑαυτῇ φύσεως ἐθέλειν. If the bitterness belonging to the liver is of a contrary nature to cheerful thoughts, it can hardly be very ridiculous to conceive that it is of kindred nature to thoughts that are gloomy. So Wagner, 'was seiner Natur (d. i. des Nachdenkens) entgegengesetzt ist'.

3. **ἀπειλῇ**] Hermann punctuates so as to join this word with κατὰ πᾶν ὑπομιγνῦσα κ.τ.λ., which surely gives it an intolerable situation. Cf. 70 B.

5. **λοβὸν δὲ καὶ δοχὰς πύλας τε**] The λοβὸς here meant is the lobe κατ᾽ ἐξοχήν, the large right lobe of the liver, in which the gall-bladder is situated; to which effect Stallbaum cites Rufus Ephesius: the δοχαὶ seem to be the small vessels in the liver: the πύλαι are the two entrances of the portal vein, which conveys blood to the liver; the plural is used because the vein divides into two branches immediately before entering the liver. That all these were of high importance in sacrificial divination is clear from Euripides *Electra* 827—829:

καὶ λοβὸς μὲν οὐ προσῆν
σπλάγχνοις, πύλαι δὲ καὶ δοχαὶ χολῆς
πέλας
κακὰς ἔφαινον τῷ σκοποῦντι προσβολάς.

Compare Aristotle *historia animalium* I xvii 496[b] 29 προσπέφυκε δὲ τῇ μεγάλῃ φλεβὶ τὸ ἧπαρ, τῇ δ᾽ ἀορτῇ οὐ κοινωνεῖ· διὰ γὰρ τοῦ ἥπατος διήκει ἡ ἀπὸ τῆς μεγάλης φλεβὸς φλέψ, ᾗ δὴ αἱ καλούμεναι πύλαι εἰσὶ τοῦ ἥπατος. The μεγάλη φλέψ is evidently the *vena cava*; see *de partibus animalium* III iv 666[b] 24 ὅτι δὲ πρῶτον ἐν τῇ καρδίᾳ γίνεται τὸ αἷμα πολλάκις εἰρήκαμεν, διὰ τὸ τὰς ἀρχηγοὺς φλέβας δύο εἶναι, τήν τε μεγάλην καλουμένην καὶ τὴν ἀορτήν· while ἡ ἀπὸ τῆς μεγάλης is as clearly the portal vein.

from the brain, when the liver received outlines of them, as if in a mirror, and exhibited reflections to view, might strike terror into the appetitive part, whenever making use of the bitter element akin to its own dark nature and threatening with stern approach, diffusing the bitterness swiftly throughout the whole liver it displayed a bilious colour, and contracting it made it all rough and wrinkled, and reaching the lobe and the vessels and the inlet, twisted the first from its right position and contorted it, while at the same time it obstructed and closed up the two latter, thereby producing pain and nausea: and on the other hand in order that, whenever a breath of mildness from the reason copied off on the liver visions of an opposite kind, giving relief from the bitterness, because it will not excite a nature opposite to its own nor have dealing with it, but using upon the liver the sweetness that exists therein and soothing everything till all is straight and smooth and free, it might render gentle and calm that part of the soul which is settled about the liver, and might enable it to secure a sober amusement at night, enjoying divination during sleep, in recompense for its deprivation of intelligence and wisdom. For our creators, because they remembered the behest of their father, when he commanded them to make the mortal race as perfect as they

τὰ μὲν] I suspect τὸν μὲν to be the right reading.

6. λύπας καὶ ἄσας] The effect is partly physical, partly moral: the pains and nausea would cause evil dreams, which served as portents and deterrents. Hermann, presumably by a typographical error, puts no stop at all after παρέχοι.

8. πραότητός τις...ἐπίπνοια] With this very striking expression compare the beautiful phrase in Aeschylus *Agamemnon* 740 φρόνημα νηνέμου γαλάνας. ἐπίπνοια is the regular word for divine inspiration: cf. *Phaedrus* 265 B, *Laws* 811 C.

10. γλυκύτητι τῇ κατ' ἐκεῖνο] sc. τὸ ἧπαρ: the ἐπίπνοια uses upon the liver (πρὸς αὐτὸ) the sweetness which permeates it. ξυμφύτῳ, i.e. akin to the ἐπίπνοια. Stallbaum understands πρὸς

αὐτὸ to refer to the ἐπιθυμητικόν: but this will not do. For αὐτὸ must surely have the same reference as αὐτοῦ, which necessarily means τοῦ ἤπατος.

12. ἵλεών τε καὶ εὐήμερον ποιοῖ] Aristotle (who must have been rather mystified by this passage) has a direct reference to these words in *de partibus animalium* IV ii 676ᵇ 22 διόπερ οἱ λέγοντες τὴν φύσιν τῆς χολῆς αἰσθήσεώς τινος εἶναι χάριν οὐ καλῶς λέγουσιν. φασὶ γὰρ εἶναι διὰ τοῦτο, ὅπως τῆς ψυχῆς τὸ περὶ τὸ ἧπαρ μόριον δάκνουσα μὲν συνιστῇ, λυόμενον δ' ἵλεων ποιῇ. Aristotle is himself decidedly sceptical concerning the prophetic character of dreams: see his exceedingly interesting treatise *de divinatione*.

13. ἐν τε τῇ νυκτί] The τε merely couples ἔχουσαν with ἵλεών τε καὶ εὐήμερον.

οὕτω δὴ κατορθοῦντες καὶ τὸ φαῦλον ἡμῶν, ἵνα ἀληθείας πῃ Ε
προσάπτοιτο, κατέστησαν ἐν τούτῳ τὸ μαντεῖον. ἱκανὸν δὲ σημεῖον, ὡς μαντικὴν ἀφροσύνῃ θεὸς ἀνθρωπίνῃ δέδωκεν· οὐδεὶς γὰρ
ἔννους ἐφάπτεται μαντικῆς ἐνθέου καὶ ἀληθοῦς, ἀλλ' ἢ καθ' ὕπνον
5 τὴν τῆς φρονήσεως πεδηθεὶς δύναμιν ἢ διὰ νόσον ἢ διά τινα ἐνθου-
σιασμὸν παραλλάξας. ἀλλὰ ξυννοῆσαι μὲν ἔμφρονος τά τε ῥηθέντα
ἀναμνησθέντα ὄναρ ἢ ὕπαρ ὑπὸ τῆς μαντικῆς τε καὶ ἐνθουσιασ-
τικῆς φύσεως, καὶ ὅσα ἂν φαντάσματα ὀφθῇ, πάντα λογισμῷ 72 A
διελέσθαι, ὅπῃ τι σημαίνει καὶ ὅτῳ μέλλοντος ἢ παρελθόντος ἢ
10 παρόντος κακοῦ ἢ ἀγαθοῦ· τοῦ δὲ μανέντος ἔτι τε ἐν τούτῳ μένοντος οὐκ ἔργον τὰ φανέντα καὶ φωνηθέντα ὑφ' ἑαυτοῦ κρίνειν, ἀλλ' εὖ
καὶ πάλαι λέγεται τὸ πράττειν καὶ γνῶναι τά τε αὑτοῦ καὶ ἑαυτὸν
σώφρονι μόνῳ προσήκειν. ὅθεν δὴ καὶ τὸ τῶν προφητῶν γένος ἐπὶ
ταῖς ἐνθέοις μαντείαις κριτὰς ἐπικαθιστάναι νόμος· οὓς μάντεις Β
15 αὐτοὺς ὀνομάζουσί τινες, τὸ πᾶν ἠγνοηκότες, ὅτι τῆς δι' αἰνιγμῶν
οὗτοι φήμης καὶ φαντάσεως ὑποκριταί, καὶ οὔ τι μάντεις, προφῆται δὲ
μαντευομένων δικαιότατα ὀνομάζοιντ' ἄν. ἡ μὲν οὖν φύσις ἥπατος
διὰ ταῦτα τοιαύτη τε καὶ ἐν τόπῳ ᾧ λέγομεν πέφυκε, χάριν μαν-
τικῆς· καὶ ἔτι μὲν δὴ ζῶντος ἑκάστου τὸ τοιοῦτον σημεῖα ἐναρ-

8 φαντάσματα: φάσματα SZ.   17 ἥπατος: τοῦ ἥπατος S.
19 ἐναργέστερα: ἐνεργέστερα Λ.

3. **ἀφροσύνῃ θεὸς ἀνθρωπίνῃ δέδωκεν**] The keen irony pervading the whole of this very curious and interesting passage is too evident to escape notice. Plato had no high opinion of μαντική and μάντεις: the μαντικὸς βίος comes low in order of merit in *Phaedrus* 248 E. See too the contemptuous reference to ἀγύρται καὶ μάντεις in *Republic* 364 B, and *Symposium* 203 A καὶ τὴν μαντείαν πᾶσαν καὶ γοητείαν. In *Politicus* 290 D he says with similar irony τὸ γὰρ δὴ τῶν ἱερέων σχῆμα καὶ τὸ τῶν μαντέων εὖ μάλα φρονήματος πληροῦται καὶ δόξαν σεμνὴν λαμβάνει διὰ τὸ μέγεθος τῶν ἐγχειρημάτων: but for all their assumption, they practise but a 'servile art', ἐπιστήμης διακόνου μόριον.

οὐδεὶς γὰρ ἔννους] Compare *Phaedrus* 244 A ἥ τε γὰρ δὴ ἐν Δελφοῖς προφῆτις αἵ τ' ἐν Δωδώνῃ ἱέρειαι μανεῖσαι μὲν πολλὰ δὴ καὶ καλὰ ἰδίᾳ τε καὶ δημοσίᾳ τὴν Ἑλλάδα εἰργάσαντο, σωφρονοῦσαι δὲ βραχέα ἢ οὐδέν. Presently follows the well-known derivation of μανική from μαντική. The most remarkable passage is at 244 D: ἀλλὰ μὴν νόσων γε καὶ πόνων τῶν μεγίστων, ἃ δὴ παλαιῶν ἐκ μηνιμάτων ποθὲν ἔν τισι τῶν γενῶν, ἡ μανία ἐγγενομένη καὶ προφητεύσασα οἷς ἔδει, ἀπαλλαγὴν εὕρετο, καταφυγοῦσα πρὸς θεῶν εὐχάς τε καὶ λατρείας, ὅθεν δὴ καθαρμῶν τε καὶ τελετῶν τυχοῦσα ἐξάντη ἐποίησε τὸν ἑαυτῆς ἔχοντα πρός τε τὸν παρόντα καὶ τὸν ἔπειτα χρόνον, λύσιν τῷ ὀρθῶς μανέντι καὶ κατασχομένῳ τῶν παρόντων κακῶν εὑρομένη: where see Thompson's note.

6. **παραλλάξας**] For this sense of the word see above, 27 C εἰ μή τι πάντασι παραλλάττομεν, and Euripides *Hippolytus* 935 λόγοι παραλλάσσοντες ἔξεδροι φρενῶν.

7. **ἀναμνησθέντα**] sc. ὑπὸ τοῦ ἔμφρονος: the order of words is somewhat peculiar.

13. **τὸ τῶν προφητῶν γένος**] The

were able, in this wise redeeming even the baser part of us, that it might have in some way a hold on the truth, placed in this region the seat of divination.

Now that divination is the gift of God to human folly, this is a sufficient proof. No man in his sound senses deals in true and inspired divination, but when the power of his understanding is fettered in sleep or by sickness, or if he has become distraught by some divine possession. The part of the sane man is to remember and interpret all things that are declared, dreaming or waking, by the prophetic and inspired nature; and whatsoever visions are beheld by the seer, to determine by reason in what way and to whom they betoken good or ill in the future or the present or the past: but it is not for him who has become mad and still is in that state to judge his own visions and utterances; the old saying remains true, that only for the sane man is it meet to act and to be the judge of his own actions and of himself. Whence has arisen the custom of setting up interpreters as judges of inspired prophecy: these are themselves called prophets by some who are altogether unaware that they are but the expounders of mystic speech and visions, and ought not in strict accuracy to be called prophets, but interpreters of the prophecies.

Such is the nature of the liver and its situation that we have described, for the purpose of prophecy as aforesaid. And while each body has life, this organ displays the signs clearly

function of the προφῆται is well illustrated by Euripides *Ion* 413—416:

ΞΟΥ. ἔσται τάδ'· ἀλλὰ τίς προφητεύει θεοῦ;
ΙΩΝ. ἡμεῖς τά γ' ἔξω, τῶν ἔσω δ' ἄλλοις μέλει
οἱ πλησίον θάσσουσι τρίποδος, ὦ ξένε,
Δελφῶν ἀριστῆς οὓς ἐκλήρωσεν πάλος.

This points to the existence at Delphi of two classes of προφῆται: one class, to which only high-born Delphians were admitted, heard the inspired utterances of the Pythia herself; the other and less exclusive class having to declare whatever was to be made known to the public without.

16. **οὔ τι μάντεις προφῆται δέ]** It must be confessed that Plato is himself guilty of a converse error, when in *Phaedrus* 244 B he applies the term προφῆτις to the Pythian priestess. This however is venial; for the Pythia may be regarded as the προφῆτις of Apollo, whereas her προφῆται are in no sense μάντεις.

18. **χάριν μαντικῆς]** Plato does not altogether ignore the physiological functions of the liver, as may be seen from the important part played by χολή, when this secretion is in a morbid condition, in his pathology. But he characteristically gives chief prominence to the final cause, which is to redeem the ἐπιθυμητικόν from complete irrationality.

268   ΠΛΑΤΩΝΟΣ   [72 B—

γέστερα ἔχει, στερηθὲν δὲ τοῦ ζῆν γέγονε τυφλὸν καὶ τὰ μαντεῖα ἀμυδρότερα ἔσχε τοῦ τι σαφὲς σημαίνειν. ἡ δ' αὖ τοῦ γείτονος C αὐτῷ ξύστασις καὶ ἕδρα σπλάγχνου γέγονεν ἐξ ἀριστερᾶς χάριν ἐκείνου, τοῦ παρέχειν αὐτὸ λαμπρὸν ἀεὶ καὶ καθαρόν, οἷον κατόπ-
5 τρῳ παρεσκευασμένον καὶ ἕτοιμον ἀεὶ παρακείμενον ἐκμαγεῖον· διὸ δὴ καὶ ὅταν τινὲς ἀκαθαρσίαι γίγνωνται διὰ νόσους σώματος περὶ τὸ ἧπαρ, πάντα ἡ σπληνὸς καθαίρουσα αὐτὰ δέχεται μανότης, ἅτε κοίλου καὶ ἀναίμου ὑφανθέντος· ὅθεν πληρούμενος τῶν ἀποκαθαιρομένων μέγας καὶ ὕπουλος αὐξάνεται, καὶ πάλιν, ὅταν D
10 καθαρθῇ τὸ σῶμα, ταπεινούμενος εἰς ταὐτὸν ξυνίζει.

XXXIII. Τὰ μὲν οὖν περὶ ψυχῆς, ὅσον θνητὸν ἔχει καὶ ὅσον θεῖον, καὶ ὅπῃ, καὶ μεθ' ὧν, καὶ δι' ἃ χωρὶς ᾠκίσθη, τὸ μὲν ἀληθὲς ὡς εἴρηται, θεοῦ ξυμφήσαντος, τότ' ἂν οὕτω μόνως διισχυριζοίμεθα· τό γε μὴν εἰκὸς ἡμῖν εἰρῆσθαι καὶ νῦν καὶ ἔτι μᾶλλον ἀνασκοποῦσι
15 διακινδυνευτέον τὸ φάναι, καὶ πεφάσθω. τὸ δ' ἑξῆς δὴ τούτοισι E

1. **στερηθὲν δὲ τοῦ ζῆν**] The function of the liver in divination is twofold, one mode being proper to man, the other to beasts. In the living man it is the means of warning him by dreams and visions; while the liver of the slaughtered beast gives omens of the future by its appearance when inspected. The efficacy in the first case Plato satirically allows, as a sop to human folly; to the second he will not allow even this.

5. **ἐκμαγεῖον**] Here we have a totally different use of the word from that in 50 C: it now means a sponge or napkin for wiping clean. The spleen then, according to Plato, exists solely for the sake of the liver, to purge it of superfluous and noxious humours, which it receives into itself and disposes of.

72 D—76 E, *c.* xxxiii. Now to assert that all we have said in the foregoing is certainly true were folly, wanting the assurance of some god, yet the account that seemed to us most likely, this we have given. On the same plan we have next to describe the remaining parts of the human body. First the intestines were devised as a precaution against gluttony and excess, in order that the food might not by passing through too rapidly leave a void that needed perpetual replenishment. Of bones and flesh the foundation is the marrow. This is made of the very finest and most perfect elements of fire air water and earth commingled. Part of this was moulded into a globe-like form and placed in the head; the rest, drawn out into a cylindrical shape, in the spinal column. And the marrow of the head, which we call the brain, is the habitation of the reason; while the lower forms of soul were attached to the spinal marrow. Bone is formed of fine earth kneaded with marrow and then tempered by being plunged alternately into fire and water; and of this was made a hard envelope to protect the vital marrow: and joints were inserted in the limbs for the sake of flexibility. And to prevent the structure of the bone decaying, the gods constructed flesh, and to impart the power of moving the limbs at will they made tendons. Flesh is a kind of ferment made with fire and water and earth, containing an acid and saline admixture; tendons, which are of a tougher and finer consistency, are made of unfermented flesh mingled with

enough; but when deprived of life, it is become blind and gives the token too dimly to afford any plain meaning. And the structure of the neighbouring organ and its position on the left has been planned for the sake of the liver, in order to keep it always bright and clean, as a napkin is prepared and laid ready for the cleansing of a mirror. Wherefore whenever any impurities arise in the region of the liver owing to sickness of the body, all is received and purified by the fine substance of the spleen, which is woven hollow and void of blood. This, when it is filled with the impurities from the liver, waxes swollen and festered; and again, when the body is purged, it is reduced and sinks again to its natural state.

XXXIII. Now as concerning soul, how far she has a mortal, how far a divine nature, and in what wise and with what conjunctions and for what causes she has her separate habitations, only when God has confirmed our statement can we confidently aver that it is true: nevertheless that we have given the probable account we may venture to say even now and still more on further meditation, and so let it be said. But what follows

bone. And such of the bones as contained the greatest amount of vital marrow the gods covered with the thinnest envelope of flesh; such as contained less, with a thicker envelope; to the end that the marrow in the former might not have its sensitiveness blunted by a thick covering. For this cause the head has but a slight covering, though a thicker one would have better protected it; since the gods deemed that a shorter and more intelligent life was preferable to a longer and less rational. In the construction of the mouth and neighbouring parts both the necessary cause and the divine cause were consulted: the necessary in view of the nutriment that must enter in, the divine in view of the speech that should issue forth. For the further protection of the head they devised the following. The surface of the flesh in drying formed a tough rind, which we call the skin: this is pierced by the internal fire of the head, and the moisture issuing through the punctures forms what we call hair. And the nails are formed by the skin at the end of the fingers, mixed with tendon and bone, being suddenly dried: for the gods knew that other creatures would arise out of mankind in future ages, which would need these defences.

14. τό γε μὴν εἰκός] It may be objected that soul is immaterial and eternal, and therefore we must not be satisfied with τὸ εἰκὸς concerning her. But here we are treating not of the nature of soul as she is in herself, but of her connexion with body: this belongs to the region of physics and consequently to that of the 'probable account'. Therefore Plato begins the chapter with a reiterated warning that we are dealing with matters where absolute certainty is impossible. But this does not apply to the exposition concerning the soul's own nature which we had in 34 B—37 C.

270   ΠΛΑΤΩΝΟΣ   [72 E—

κατὰ ταὐτὰ μεταδιωκτέον· ἦν δὲ τὸ τοῦ σώματος ἐπίλοιπον ᾗ γέγονεν. ἐκ δὴ λογισμοῦ τοιοῦδε ξυνίστασθαι μάλιστ᾽ ἂν αὐτὸ πάντων πρέποι. τὴν ἐσομένην ἐν ἡμῖν ποτῶν καὶ ἐδεστῶν ἀκολασίαν
ᾔδεσαν οἱ ξυντιθέντες ἡμῶν τὸ γένος, καὶ ὅτι τοῦ μετρίου καὶ ἀναγ-
5 καίου διὰ μαργότητα πολλῷ χρησοίμεθα πλέονι· ἵν᾽ οὖν μὴ φθορὰ
διὰ νόσους ὀξεῖα γίγνοιτο καὶ ἀτελὲς τὸ γένος εὐθὺς τὸ θνητὸν τελευτῷ, ταῦτα προορώμενοι τῇ τοῦ περιγενησομένου πόματος ἐδέ- 73 A
σματός τε ἕξει τὴν ὀνομαζομένην κάτω κοιλίαν ὑποδοχὴν ἔθεσαν,
εἵλιξάν τε πέριξ τὴν τῶν ἐντέρων γένεσιν, ὅπως μὴ ταχὺ διεκπε-
10 ρῶσα ἡ τροφὴ ταχὺ πάλιν τροφῆς ἑτέρας δεῖσθαι τὸ σῶμα ἀναγ-
κάζοι, καὶ παρέχουσα ἀπληστίαν διὰ γαστριμαργίαν ἀφιλόσοφον
καὶ ἄμουσον πᾶν ἀποτελοῖ τὸ γένος, ἀνυπήκοον τοῦ θειοτάτου τῶν
παρ᾽ ἡμῖν. τὸ δὲ ὀστῶν καὶ σαρκῶν καὶ τῆς τοιαύτης φύσεως
πέρι πάσης ὧδε ἔσχε. τούτοις ξύμπασιν ἀρχὴ μὲν ἡ τοῦ μυελοῦ B
15 γένεσις· οἱ γὰρ τοῦ βίου δεσμοὶ τῆς ψυχῆς τῷ σώματι ξυνδουμένης
ἐν τούτῳ διαδούμενοι κατερρίζουν τὸ θνητὸν γένος. αὐτὸς δὲ ὁ
μυελὸς γέγονεν ἐξ ἄλλων. τῶν γὰρ τριγώνων ὅσα πρῶτα ἀστραβῆ
καὶ λεῖα ὄντα πῦρ τε καὶ ὕδωρ καὶ ἀέρα καὶ γῆν δι᾽ ἀκριβείας
μάλιστα ἦν παρασχεῖν δυνατά, ταῦτα ὁ θεὸς ἀπὸ τῶν ἑαυτῶν
20 ἕκαστα γενῶν χωρὶς ἀποκρίνων, μιγνὺς δὲ ἀλλήλοις ξύμμετρα, C

6 τελευτῷ : τελευτῴη S.   7 πόματος : πώματος ASZ.

1. **ἦν δέ**] Referring back to 61 C σαρκὸς δὲ καὶ τῶν περὶ σάρκα γένεσιν, ψυχῆς τε ὅσον θνητόν, οὔπω διεληλύθαμεν.

8. **τὴν ὀνομαζομένην**] 'So-called', because ἡ κάτω κοιλία was a medical term: see Hippokrates *passim*: it denoted all the region of the body below the θώραξ strictly so called: cf. Aristotle *problemata* XXXIII ix 962ᵃ 35 τριῶν τόπων ὄντων, κεφαλῆς καὶ θώρακος καὶ τῆς κάτω κοιλίας, ἡ κεφαλὴ θειότατον. The θώραξ, though sometimes applied to the entire cavity of the body, was properly identical with ἡ ἄνω κοιλία, which included the stomach: cf. *de partibus animalium* III xiv 675ᵇ 29.

ὑποδοχήν] Plato does not seem to have understood very clearly the functions of this part of the human anatomy, merely regarding it as a safeguard against gluttony. Aristotle has a preciser conception: see *de partibus animalium* III xiv 674ᵃ 12 foll.

9. **ταχὺ διεκπερῶσα**] We should thus relapse into the life symbolised by the ἀγγεῖα τετρημένα καὶ σαθρά in *Gorgias* 493 E : cf. 494 B χαραδριοῦ τιν᾽ αὖ σὺ βίον λέγεις.

15. **οἱ γὰρ τοῦ βίου δεσμοί**] That is to say, it is through the marrow that the soul is linked to the body. Plato, though unacquainted with the nervous system, saw clearly that the spinal marrow and ultimately the brain was the centre of consciousness: a point wherein he is much ahead of Aristotle, who declared (1) that the brain and spinal marrow are essentially different substances, (2) that the function of the brain is merely to cool the region of the heart: see *de partibus*

upon the foregoing is the next object of our research: this was the manner wherein the rest of the body has come into being.

The following is the design on which it were most fitting to conceive that it is constructed. They who framed our race knew the intemperance in meat and drink that would prevail in us, and that for greed we should use far more than was moderate or necessary. In order then that swift destruction through sickness might not fall upon us, and that the mortal race might not perish out of hand before coming to completion, foreseeing the danger they made the abdomen, as it is called, a receptacle to contain the superfluity of food and drink, and coiled the bowels round about therein, lest the food passing speedily through should compel the body quickly to stand in need of a fresh supply, and thus producing an insatiable craving should render the whole race through gluttony devoid of philosophy and letters and disobedient to the highest part of our nature.

Concerning the bones and flesh and all such substances the case stands thus. The foundation of all these is the marrow: for the bonds of life whereby the soul is bound to the body were fastened in it throughout and planted therein the roots of human nature. But the marrow itself comes from other sources. Such of the primal triangles as were unwarped and smooth and thus able to produce fire and water and air and earth of the purest quality, these God selected and set apart, each from its own class, and mingling them in proportion one

animalium II vii 652ª 24 πολλοῖς γὰρ καὶ ὁ ἐγκέφαλος δοκεῖ μυελὸς εἶναι καὶ ἀρχὴ τοῦ μυελοῦ διὰ τὸ συνεχῆ τὸν ῥαχίτην αὐτῷ ὁρᾶν μυελόν. ἔστι δὲ πᾶν τοὐναντίον αὐτῷ τὴν φύσιν, ὡς εἰπεῖν· ὁ μὲν γὰρ ἐγκέφαλος ψυχρότατον τῶν ἐν τῷ σώματι μορίων, ὁ δὲ μυελὸς θερμὸς τὴν φύσιν. 652ᵇ 16 ἐπεὶ δ' ἅπαντα δεῖται τῆς ἐναντίας ῥοπῆς, ἵνα τυγχάνῃ τοῦ μετρίου καὶ τοῦ μέσου,...διὰ ταύτην τὴν αἰτίαν πρὸς τὸν τῆς καρδίας τόπον καὶ τὴν ἐν αὐτῇ θερμότητα μεμηχάνηται τὸν ἐγκέφαλον ἡ φύσις, καὶ τούτου χάριν ὑπάρχει τοῦτο τὸ μόριον τοῖς ζῴοις, τὴν φύσιν ἔχον κοινὴν

ὕδατος καὶ γῆς. Plato had considerably less knowledge of anatomy than Aristotle; but this is one of several cases where his superior scientific insight keeps him nearer to the truth.

16. **ἐν τούτῳ**] i.e. in the spinal marrow; for the brain was the seat of the θεῖον γένος.

17. **ἐξ ἄλλων**] sc. ἢ ὀστῶν καὶ σαρκῶν καὶ τῶν τοιούτων.

τῶν γὰρ τριγώνων] The triangles being the elements of the corpuscules of which matter is composed, Plato speaks of them as the elements of μυελός.

πανσπερμίαν παντὶ θνητῷ γένει μηχανώμενος, τὸν μυελὸν ἐξ αὐτῶν
ἀπειργάσατο, καὶ μετὰ ταῦτα δὴ φυτεύων ἐν αὐτῷ κατέδει τὰ τῶν
ψυχῶν γένη, σχημάτων τε ὅσα ἔμελλεν αὖ σχήσειν οἷά τε καθ᾽
ἕκαστα εἴδη, τὸν μυελὸν αὐτὸν τοσαῦτα καὶ τοιαῦτα διῃρεῖτο σχή-
5 ματα εὐθὺς ἐν τῇ διανομῇ τῇ κατ᾽ ἀρχάς. καὶ τὴν μὲν τὸ θεῖον
σπέρμα οἷον ἄρουραν μέλλουσαν ἕξειν ἐν αὑτῇ περιφερῆ πανταχῇ πλάσας ἐπωνόμασε τοῦ μυελοῦ ταύτην τὴν μοῖραν ἐγκέφαλον, D
ὡς ἀποτελεσθέντος ἑκάστου ζῴου τὸ περὶ τοῦτο ἀγγεῖον κεφαλὴν
γενησόμενον· ὃ δ᾽ αὖ τὸ λοιπὸν καὶ θνητὸν τῆς ψυχῆς ἔμελλε
10 καθέξειν, ἅμα στρογγύλα καὶ προμήκη διῃρεῖτο σχήματα, μυελὸν
δὲ πάντα ἐπεφήμισε, καὶ καθάπερ ἐξ ἀγκυρῶν βαλλόμενος ἐκ τούτων πάσης ψυχῆς δεσμοὺς περὶ τοῦτο ξύμπαν ἤδη τὸ σῶμα ἡμῶν
ἀπειργάζετο, στέγασμα μὲν αὐτῷ πρῶτον ξυμπηγνὺς περίβολον
ὀστέινον. τὸ δὲ ὀστοῦν ξυνίστησιν ὧδε· γῆν διαττήσας καθαρὰν Ε
15 καὶ λείαν ἐφύρασε καὶ ἔδευσε μυελῷ, καὶ μετὰ τοῦτο εἰς πῦρ αὐτὸ
ἐντίθησι, μετ᾽ ἐκεῖνο δὲ εἰς ὕδωρ βάπτει, πάλιν δὲ εἰς πῦρ αὖθίς τε
εἰς ὕδωρ· μεταφέρων δ᾽ οὕτω πολλάκις εἰς ἑκάτερον ὑπ᾽ ἀμφοῖν
ἄτηκτον ἀπειργάσατο. καταχρώμενος δὴ τούτῳ περὶ μὲν τὸν
ἐγκέφαλον αὐτοῦ σφαῖραν περιετόρνευσεν ὀστείνην, ταύτῃ δὲ στε-
20 νὴν διέξοδον κατελείπετο· καὶ περὶ τὸν διαυχένιον ἅμα καὶ νω- 74 A
τιαῖον μυελὸν ἐξ αὐτοῦ σφονδύλους πλάσας ὑπέτεινεν οἷον στρόφιγγας, ἀρξάμενος ἀπὸ τῆς κεφαλῆς, διὰ παντὸς τοῦ κύτους· καὶ
τὸ πᾶν δὴ σπέρμα διασῴζων οὕτω λιθοειδεῖ περιβόλῳ ξυνέφραξεν,

13 περίβολον: sic H e Valckenari coniectura.   περὶ ὅλον ASZ et codices omnes.
20 κατελείπετο: κατελίπετο SZ.   23 οὕτω: οὕτως A.

1. **πανσπερμίαν**] The marrow, being formed from all the four elements, was capable of supplying material for all parts of the human frame.

3. **ὅσα ἔμελλεν**] It is remarkable that, although Plato only mentions two σχήματα explicitly, his phraseology is so studiously vague concerning their number as to lead one to imagine that he may have suspected the existence of further ramifications of μυελός, such as in fact are the nerves.

καθ᾽ ἕκαστα εἴδη] sc. τῆς ψυχῆς: the shape of the different portions of marrow in the body was made to suit the nature of that particular function of soul which acted through it. There are however no special divisions of μυελός for the θυμοειδὲς and the ἐπιθυμητικὸν separately; the spinal cord serving for the θνητὸν as a whole.

5. **τῇ κατ᾽ ἀρχάς**] i. e. without waiting for the differentiation to be made in the course of evolution.

6. **περιφερῆ**] The brain is made approximately spherical, because, as we have seen, the action of reason is symbolised by the rotation of a sphere on its axis: cf. 44 D τὸ τοῦ παντὸς σχῆμα μιμούμενοι περιφερὲς ὂν εἰς σφαιροειδὲς σῶμα ἐνέδησαν.

8. **ὡς...γενησόμενον**] The construction is that which is known as the accusative absolute: compare *Protagoras* 342

with another, to make a common seed for all the race of mortals, he formed of them the marrow; and thereafter he implanted and fastened in it the several kinds of soul; and according to the number and fashion of the shapes that the soul should have corresponding to her kinds, into so many similar forms did he divide the marrow at the very outset of his distribution. And that which should be as it were a field to contain in it the divine seed he moulded in a spherical form all round; and this part of the marrow he called the brain, with the view that, when each animal was completed, the vessel containing it should be the head. But that which was to have the mortal part of soul which remained he distributed into moulds that were at once round and elongated: but he called all these forms marrow; and from these, as though from anchors, he put forth bonds to fasten all the soul, and then he wrought the entire body round about it, first building to fence it a covering of bone. And bone he formed in this way: having sifted out earth that was pure and smooth he kneaded and soaked it with marrow, and after that he placed it in fire; and next he set it in water, and again in fire, and once more in water: and thus having shifted it many times from one to another he made it indissoluble by either. Making use of this, he carved a bony sphere thereof to surround the brain, but on one side he left a narrow outlet; and around the marrow of the neck and back he made vertebrae of bone and set them to serve as pivots, beginning at the head and carrying them through the whole length of the body. Thus to preserve all the seed he enclosed it in a strong envelope, and he

C καὶ οἱ μὲν ὦτά τε κατάγνυνται μιμούμενοι αὐτούς, καὶ ἱμάντας περιειλίττονται καὶ φιλογυμναστοῦσι καὶ βραχείας ἀναβολὰς φοροῦσιν, ὡς δὴ τούτοις κρατοῦντας τῶν Ἑλλήνων τοὺς Λακεδαιμονίους.

10. στρογγύλα καὶ προμήκη] 'Round and elongated' is the same thing as 'cylindrical': this of course refers to the vertebral column.

12. πάσης ψυχῆς δεσμούς] The brain and spinal marrow serve as conductors of vital force; it is on them that the soul immediately acts—the λογιστικὸν working through the brain, the ἄλογον through the spinal marrow—and they transmit

her action to the rest of the body. The word δεσμοὺς does not refer to any ligament or the like, nor has it any physical significance: it is purely metaphorical. For the phrase καθάπερ ἐξ ἀγκυρῶν compare 85 E ἔλυσε τὰ τῆς ψυχῆς αὐτόθεν οἷον νεὼς πείσματα.

13. περίβολον] The ms. reading περὶ ὅλον will no doubt yield a reasonable sense. But Valckenaer's correction is so much more apt that I have not hesitated to follow Hermann in accepting it. Below in 74 A we have λιθοειδεῖ περιβόλῳ ξυνέφραξεν.

15. μετὰ τοῦτο εἰς πῦρ] The process

P. T.   18

274   ΠΛΑΤΩΝΟΣ   [74 A—

ἐμποιῶν ἄρθρα, τῇ θατέρου προσχρώμενος ἐν αὐτοῖς ὡς μέσῃ
ἐνισταμένῃ δυνάμει, κινήσεως καὶ κάμψεως ἕνεκα. τὴν δ' αὖ τῆς
ὀστεΐνης φύσεως ἕξιν ἡγησάμενος τοῦ δέοντος κραυροτέραν εἶναι
καὶ ἀκαμπτοτέραν, διάπυρόν τ' αὖ γιγνομένην καὶ πάλιν ψυχομένην
5 σφακελίσασαν ταχὺ διαφθερεῖν τὸ σπέρμα ἐντὸς αὑτῆς, διὰ ταῦτα
οὕτω τὸ τῶν νεύρων καὶ τὸ τῆς σαρκὸς γένος ἐμηχανᾶτο, ἵνα τῷ μὲν
ἅπαντα τὰ μέλη ξυνδήσας ἐπιτεινομένῳ καὶ ἀνιεμένῳ περὶ τοὺς
στρόφιγγας καμπτόμενον τὸ σῶμα καὶ ἐκτεινόμενον παρέχοι, τὴν
δὲ σάρκα προβολὴν μὲν καυμάτων πρόβλημα δὲ χειμώνων, ἔτι δὲ
10 πτωμάτων οἷον τὰ πιλητὰ ἔσεσθαι κτήματα, σώμασι μαλακῶς καὶ
πράως ὑπείκουσαν, θερμὴν δὲ νοτίδα ἐντὸς ἑαυτῆς ἔχουσαν θέρους
μὲν ἀνιδίουσαν καὶ νοτιζομένην ἔξωθεν ψῦχος κατὰ πᾶν τὸ σῶμα
παρέξειν οἰκεῖον, διὰ χειμῶνος δὲ πάλιν αὖ τούτῳ τῷ πυρὶ τὸν
προσφερόμενον ἔξωθεν καὶ περιιστάμενον πάγον ἀμυνεῖσθαι με-
15 τρίως. ταῦτα ἡμῶν διανοηθεὶς ὁ κηροπλάστης, ὕδατι μὲν καὶ πυρὶ
καὶ γῇ ξυμμίξας καὶ ξυναρμόσας, ἐξ ὀξέος καὶ ἁλμυροῦ ξυνθεὶς

12 ψῦχος : ψύχος SZ.

is obviously suggested by the tempering of metal.

1. **τῇ θατέρου προσχρώμενος**] This expression is very obscure; and no two interpreters agree as to its meaning. Stallbaum is entirely at sea: Lindau, at whom he scoffs, throws out a suggestion which is much more reasonable than anything in Stallbaum's note: 'eadem philosophum corpori et animo tribuere principia gravitatemque eum et expansionem comparare cum ratione sensibusque'. Martin's idea that ἡ θατέρου δύναμις means the synovial fluid is extremely far-fetched: could Plato possibly expect any one to understand him if he made such use of language? Dr Jackson has suggested to me an interpretation which is certainly much more natural and, I think, right. We know that θάτερον expresses plurality. Plato then, when he says that the gods used ἡ θατέρου δύναμις in the construction of the bones, simply signifies that by means of joints they divided the bones into a number of parts, κάμψεως καὶ κινήσεως ἕνεκα. ἐν μέσῃ I take to mean between the bones—the joints represent the principle of θάτερον, as being the cause of division and plurality.

4. **διάπυρόν τ' αὖ γιγνομένην**] That is to say, subjected to vicissitudes of temperature.

5. **σφακελίσασαν**] This is a medical term, signifying caries of the bones or gangrene of the flesh: it is also used of the blighting of plants; Aristotle *de iuventute* vi 470ᵃ 31 λέγεται σφακελίζειν καὶ ἀστρόβλητα γίνεσθαι τὰ δένδρα περὶ τοὺς καιροὺς τούτους.

τὸ σπέρμα] i.e. τὸν μυελόν: cf. 73 C.

6. **τὸ τῶν νεύρων**] By νεῦρα Plato always means tendons or ligaments, not nerves, which were entirely unknown to him. Aristotle always uses the word in the same sense: see *de partibus animalium* II ii 647ᵇ 16 τὰ δὲ ξηρὰ καὶ στερεὰ τῶν ὁμοιομερῶν ἐστιν, οἷον ὀστοῦν ἄκανθα νεῦρον φλέψ. The nature, almost the existence, of the nerves was not discovered till considerably after Plato's time: Erasistratos, who flourished in the next century, is said to have been the first who ascertained their functions. Aristotle seems to have had some sort of vague

c] ΤΙΜΑΙΟΣ. 275

made joints in it, using the power of the Other as an intermediary between the parts, for the sake of moving and bending them. But deeming that the structure of bone was too rigid and inflexible, and that should it be inflamed and cooled again, it would rot away and quickly destroy the seed within it, for this cause God devised the sinews and the flesh, that binding all the limbs together with the former he might by their tension and relaxation round their pivots enable the body to bend and extend itself; while the flesh he designed as a defence against heat and a shelter from cold; and moreover that it might be, like coverings of felt, a protection against falls, gently and easily yielding to external bodies; and containing a warm moisture within itself, in summer it might exude this, and spreading dampness on the surface might diffuse a natural coolness over all the body; but in winter on the other hand it might by its own fire afford a fair protection against the frost that assailed and surrounded it from without. Considering this, he that moulded us like wax made a mixture and blending of water and fire and earth; and compounding a ferment of acid and salt

knowledge of the optic and olfactory nerves, which he calls πόροι: cf. *de partibus animalium* II xii 656ᵇ 16 ἐκ μὲν οὖν τῶν ὀφθαλμῶν οἱ πόροι φέρουσιν εἰς τὰς περὶ τὸν ἐγκέφαλον φλέβας· πάλιν δ' ἐκ τῶν ὤτων ὡσαύτως πόρος εἰς τοὔπισθεν συνάπτει: also *historia animalium* I xvi 495ᵃ 11 φέρουσι δ' ἐκ τοῦ ὀφθαλμοῦ τρεῖς πόροι εἰς τὸν ἐγκέφαλον, ὁ μὲν μέγιστος καὶ ὁ μέσος εἰς τὴν παρεγκεφαλίδα ὁ δ' ἐλάχιστος εἰς αὐτὸν τὸν ἐγκέφαλον· ἐλάχιστος δ' ἐστὶν ὁ πρὸς τῷ μυκτῆρι μάλιστα. About the auditory nerve he gives a very confused statement, apparently, as Martin observes, mistaking for it the Eustachian tube: *ibid*. 492ᵃ 19 τοῦτο δ' εἰς μὲν τὸν ἐγκέφαλον οὐκ ἔχει πόρον, εἰς δὲ τὸν τοῦ στόματος οὐρανόν. Aristotle's notions concerning the brain are sufficient evidence that he did not really understand anything about the nature of the nerves. That Alkmaion was acquainted with the optic nerves, notwithstanding the statement of Kallisthenes adduced by Chalcidius, seems highly improbable: indeed the words of Kallisthenes, as there reported, hardly amount to this.

9. **προβολήν ... πρόβλημα**] There seems to be absolutely no difference in meaning between these two words, and the juxtaposition of two closely cognate forms without any distinction of sense is strange. Is it possible that we ought to read προβολὴν in both cases? Plato, like Sophokles, is given to repeating the same word with μέν and δέ; as in *Phaedrus* 247 D καθορᾷ μὲν αὐτὴν ·δικαιοσύνην, καθορᾷ δὲ σωφροσύνην, καθορᾷ δὲ ἐπιστήμην: see too below 87 A ποικίλλει μέν ...ποικίλλει δέ. And there is quite sufficient ornateness in the present passage to justify this rhetorical device. As to the construction, the future infinitives are substituted for the final clause: something like δ.ενοήθη must be mentally supplied.

13. **οἰκεῖον**] contrasted with τὸν περιφερόμενον ἔξωθεν.

16. **καὶ γῇ**] I see no sufficient reason

276   ΠΛΑΤΩΝΟΣ   [74 C—

ζύμωμα καὶ ὑπομίξας αὐτοῖς, σάρκα ἔγχυμον καὶ μαλακὴν ξυνέ- D
στησε· τὴν δὲ τῶν νεύρων φύσιν ἐξ ὀστοῦ καὶ σαρκὸς ἀζύμου κρά-
σεως μίαν ἐξ ἀμφοῖν μέσην δυνάμει ξυνεκεράσατο, ξανθῷ χρώματι
προσχρώμενος. ὅθεν συντονωτέραν μὲν καὶ γλισχροτέραν σαρκῶν,
5 μαλακωτέραν δὲ ὀστῶν ὑγροτέραν τε ἐκτήσατο δύναμιν νεῦρα. οἷς
ξυμπεριλαβὼν ὁ θεὶς ὀστᾶ καὶ μυελόν, δήσας πρὸς ἄλληλα νεύροις, μετὰ ταῦτα σαρξὶ πάντα αὐτὰ κατεσκίασεν ἄνωθεν. ὅσα μὲν E
οὖν ἐμψυχότατα τῶν ὀστῶν ἦν, ὀλιγίσταις συνέφραττε σαρξίν, ἃ δ᾽
ἀψυχότατα ἐντός, πλείσταις καὶ πυκνοτάταις. καὶ δὴ καὶ κατὰ
10 τὰς ξυμβολὰς τῶν ὀστῶν, ὅπῃ μή τινα ἀνάγκην ὁ λόγος ἀπέφαινε
δεῖν αὐτὰς εἶναι, βραχεῖαν σάρκα ἔφυσεν, ἵνα μήτε ἐμποδὼν ταῖς
καμπαῖσιν οὖσαι δύσφορα τὰ σώματα ἀπεργάζοιντο, ἅτε δυσκίνητα
γιγνόμενα, μήτ᾽ αὖ πολλαὶ καὶ πυκναὶ σφόδρα τε ἐν ἀλλήλαις
ἐμπεπιλημέναι, διὰ στερεότητα ἀναισθησίαν ἐμποιοῦσαι, δυσμνη-
15 μονευτότερα καὶ κωφότερα τὰ περὶ τὴν διάνοιαν ποιοῖεν. διὸ δὴ τό
τε τῶν μηρῶν καὶ κνημῶν καὶ τὸ περὶ τὴν τῶν ἰσχίων φύσιν τά τε 75 A
[περὶ τὰ] τῶν βραχιόνων ὀστᾶ καὶ τὰ τῶν πήχεων, καὶ ὅσα ἄλλα
ἡμῶν ἄναρθρα, ὅσα τε ἐντὸς ὀστᾶ δι᾽ ὀλιγότητα ψυχῆς ἐν μυελῷ
κενά ἐστι φρονήσεως, ταῦτα πάντα συμπεπλήρωται σαρξίν· ὅσα δ᾽
20 ἔμφρονα, ἧττον, εἰ μή πού τινα αὐτὴν καθ᾽ αὑτὴν αἰσθήσεων ἕνεκα

---

1 καὶ ante ὑπομίξας omittunt AHZ.   3 ἐξ ἀμφοῖν: συναμφοῖν supra scripto ἐξ A.
17 περὶ τὰ inclusi, quae retinet H.   omittunt SZ.

for abandoning the reading of all the mss., since σάρκα is readily supplied as the object of ξυμμίξας: and if γῆν be read, καὶ is positively bad. The insertion of καὶ before ὑπομίξας seems to me, in this accumulation of participles, almost necessary, although it is lacking in A.

1. ζύμωμα] This means a fermented mixture: it would seem to be intended thereby to explain the combined softness and elasticity of flesh. Flesh could also be made of unfermented materials, as we presently see: ἐξ ὀστοῦ καὶ σαρκὸς ἀζύμου: but the difference in the composition is not stated.

2. τὴν τῶν νεύρων φύσιν] The description of νεῦρα tallies closely with that given by Hippokrates *de locis in homine* vol. II. p. 107 Kühn τὰ δὲ νεῦρα ξηρά τέ ἐστι καὶ ἀκοίλια καὶ πρὸς τῷ ὀστέῳ πεφύκασι, καὶ τρέφονται δὲ τὸ πλεῖστον ἐκ τοῦ ὀστέου, τρέφονται δὲ καὶ ἀπὸ τῆς σαρκός, καὶ τὴν χροὴν καὶ τὴν ἰσχὺν μεταξὺ τῆς σαρκὸς καὶ τοῦ ὀστέου πεφύκασι. καὶ ὑγρότερα μέν εἰσι τοῦ ὀστέου καὶ σαρκοειδέστερα, ξηρότερα δὲ ἢ αἱ σάρκες καὶ ὀστοειδέστερα. This extract will explain the meaning of μέσην δυνάμει.

5. οἷς ξυμπεριλαβών] The reference of οἷς is to νεῦρα.

7. ὅσα μὲν οὖν ἐμψυχότατα] This rather curious expression denotes the bones which contain the greatest amount of marrow—marrow being the seat of life. By these are meant the bones of the skull and the vertebral process only; since it is clear from what Plato says a little below (διὸ δὴ τό τε τῶν μηρῶν κ.τ.λ.) that he entirely distinguished be-

he mingled it with them and produced soft flesh full of sap : the sinews he composed of bone and unfermented flesh, a separate substance having an intermediate function; and to this he added a yellow colour. Accordingly the sinews received a power more firm and tenacious than the flesh, but more soft and flexible than the bones.

With these God covered the bones and marrow; and after he had bound one part to another with sinews, he enveloped them over all with flesh. Those bones which were chiefly inhabited by soul, he enclosed with the smallest amount of flesh; but those wherein was least soul he covered most abundantly and densely with it: moreover at the joints of the bones, save where reason showed that it ought to be there, he put but little flesh, that neither it might render the body unwieldy by hindering its flexions and impeding its motions, nor again that a dense mass of flesh piled together, producing by its hardness a dulness of sensation, might render the faculties of the mind too slow of memory and hard of apprehension. Wherefore the thighs and the shins and the parts about the hips and the bones in the upper arms and the fore-arms and all parts of our limbs which are without joints, and all bones which are devoid of intelligence owing to the small amount of soul inhering in marrow within them, all these are abundantly furnished with flesh; but those which are the seat of intelligence have less: except in cases

tween the substance contained in the spinal column and what we call 'marrow' in other bones, which he does not account as μυελός at all. Aristotle, owing to his complete misconception of the functions belonging to the brain and spinal marrow, is much less clear on this point: see *de partibus animalium* II v 651[b] 32. It is true that Plato assigns as the reason for the fleshiness of the arms, thighs, &c, that these bones are ἄναρθρα: still, had they contained μυελός, that would have been a reason for giving them a thin covering of flesh.

11. αὐτάς] sc. τὰς σάρκας.

14. ἐμπεπιλημέναι] If from too much crowding the substance of the flesh became very stiff and solid, the free motions of its particles would be impeded, and consequently sensations would with difficulty make their way to the consciousness: cf. 64 B. This rather seems to apply to the density of the flesh than to its quantity; but doubtless the same effect might be produced by both.

20. εἰ μή που] The only instance in which an acutely sensitive part is of a fleshy nature is when the flesh itself is the instrument of perception; as in the case of the tongue, and that only. Of course in all cases the external πάθημα is conveyed through the flesh to the conscious centre; but in general the flesh is only the medium of transmission, and the less flesh there is to traverse, the more speedily and clearly will the sen-

278   ΠΛΑΤΩΝΟΣ   [75 A—

σάρκα ούτω ξυνέστησεν, οίον το της γλώττης είδος. τὰ δὲ πλεῖστα
ἐκείνως· ἡ γὰρ ἐξ ἀνάγκης γιγνομένη καὶ ξυντρεφομένη φύσις
οὐδαμῇ προσδέχεται πυκνὸν ὀστοῦν καὶ σάρκα πολλὴν ἅμα τε B
αὐτοῖς ὀξυήκοον αἴσθησιν. μάλιστα γὰρ ἂν αὐτὰ πάντων ἔσχεν ἡ
5 περὶ τὴν κεφαλὴν ξύστασις, εἴπερ ἅμα ξυμπίπτειν ἠθελησάτην, καὶ
τὸ τῶν ἀνθρώπων γένος σαρκώδη ἔχον ἐφ' ἑαυτῷ καὶ νευρώδη
κρατεράν τε κεφαλὴν βίον ἂν διπλοῦν καὶ πολλαπλοῦν καὶ ὑγιει-
νότερον καὶ ἀλυπότερον τοῦ νῦν κατεκτήσατο· νῦν δὲ τοῖς περὶ τὴν
ἡμετέραν γένεσιν δημιουργοῖς ἀναλογιζομένοις, πότερον πολυ-
10 χρονιώτερον χεῖρον ἢ βραχυχρονιώτερον βέλτιον ἀπεργάσαιντο C
γένος, συνέδοξε τοῦ πλείονος βίου, φαυλοτέρου δέ, τὸν ἐλάττονα
ἀμείνονα ὄντα παντὶ πάντως αἱρετέον· ὅθεν δὴ μανῷ μὲν ὀστῷ,
σαρξὶ δὲ καὶ νεύροις κεφαλήν, ἅτε οὐδὲ καμπὰς ἔχουσαν, οὐ ξυνε-
στέγασαν. κατὰ πάντα οὖν ταῦτα εὐαισθητοτέρα μὲν καὶ φρονιμω-
15 τέρα, πολὺ δὲ ἀσθενεστέρα παντὸς ἀνδρὸς προσετέθη κεφαλὴ
σώματι. τὰ δὲ νεῦρα διὰ ταῦτα καὶ οὕτως ὁ θεὸς ἐπ' ἐσχάτην τὴν
κεφαλὴν περιστήσας κύκλῳ περὶ τὸν τράχηλον ἐκόλλησεν ὁμοιό- D

9 ἀναλογιζομένοις: λογιζομένοις S.   12 τῷ ante μανῷ habet A.   13 οὐ delet A.

sation be registered in the consciousness. But in the case of the tongue, on the contrary, the fleshy structure is specifically adapted for the reception and discrimination of a particular class of sensations, and is no longer a mere passive medium. Hence Plato's distinction is sound.

2. **ἡ γὰρ ἐξ ἀνάγκης**] That is to say, the conditions of the material nature to which our soul is linked will not admit of the combination of a dense covering of flesh with acute sensitiveness. This would have seemed too obvious to need pointing out, but for Stallbaum's perverse comment 'intelligit animum'. Of course Plato does not mean anything so absurd as to deny that the flesh of the thigh, for instance, is acutely sensitive: he only means that the thigh is κενὸν φρονήσεως: it has no power of perceiving anything apart from the mere sense of touch residing in its nerves; whereas the parts containing μυελός are centres of consciousness, and the fleshy structure of the tongue is the organ of a special mode of sensation.

4. **μάλιστα γάρ**] Had such a combination been practicable, the gods would certainly have given the brain a more powerful protection than it now has: as it is, they sacrificed length of days and immunity from sickness to vividness of perception and power of reasoning. Aristotle attacks this doctrine because it does not fall in with his fantastic theory of the brain's functions: see *de partibus animalium* II xii 656a 15 οὐ γὰρ ὥσπερ τινές λέγουσιν, ὅτι εἰ σαρκώδης ἦν, μακροβιώτερον ἂν ἦν τὸ γένος· ἀλλ' εὐαισθησίας ἕνεκεν ἄσαρκον εἶναι φασιν· αἰσθάνεσθαι μὲν γὰρ τῷ ἐγκεφάλῳ, τὴν δ' αἴσθησιν οὐ προσίεσθαι τὰ μόρια τὰ σαρκώδη λίαν. τούτων δ' οὐδέτερόν ἐστιν ἀληθές, ἀλλὰ πολύσαρκος μὲν ὁ τόπος ὢν ὁ περὶ τὸν ἐγκέφαλον τοὐναντίον ἂν ἀπειργάζετο ὧν ἕνεκα ὑπάρχει τοῖς ζῴοις ὁ ἐγκέφαλος· οὐ γὰρ ἂν ἐδύνατο καταψύχειν ἀλεαίνων αὐτὸς λίαν· τῶν δ' αἰσθήσεων οὐκ αἴτιος οὐδεμιᾶς, ὅς γε ἀναίσθητος καὶ

where God has formed the flesh to be in itself an organ of sensation, as for instance the tongue: in most however it is as aforesaid; for this material nature which comes into being by the law of necessity and is reared with us does not allow dense bone and much flesh to be accompanied by ready and keen perception. For had these two conditions consented to combine, the structure of the head would have displayed them in the highest degree ; and the human being, bearing upon it a fleshy head, sinewy and strong, would have enjoyed a life twice, nay many times as long as now, besides being much more healthy and free from pain. But as it is, the creators who brought us to being considered whether they should make a long-lived race that was inferior, or one more short-lived which was nobler, and they agreed that every one must by all means choose a shorter and nobler life in preference to a longer but baser. Therefore they covered the head with thin bone, but not with flesh nor sinews; since it has no flexions. On all these grounds the head that is set upon the body of every man is much quicker of apprehension and understanding, but much weaker. For these reasons and in this manner God placed the sinews all round the base of the head about the neck and cemented them with

αὐτός ἐστιν ὥσπερ ὁτιοῦν τῶν περιττωμάτων. Aristotle is, I believe, to a certain extent right in his assertion respecting the ἀναισθησία of the brain; so that we have here again an instance of his drawing a false conclusion from correct data. One might have supposed that he who affirmed an ἀκίνητος ἀρχὴ κινήσεως need not have felt much difficulty about an ἀναίσθητος ἀρχὴ αἰσθήσεως.

αὐτά] i.e. a strong protective covering along with keenness of sensation.

13. σαρξὶ δὲ καὶ νεύροις] Hippokrates also denies that the head has νεῦρα: *de locis in homine* vol. II. p. 108 Kühn καὶ τὸ μὲν σῶμα πᾶν ἔμπλεον νεύρων, περὶ δὲ τὸ πρόσωπον καὶ τὴν κεφαλὴν οὐκ ἔστι νεῦρα.

14. εὐαισθητότερα] i.e. more sensitive than it would have been had the gods taken a different view.

16. ἐπ' ἐσχάτην τὴν κεφαλήν] Plato supposes the νεῦρα to pass up the neck and terminate at the base of the head, made fast to the jawbone.

17. ἐκόλλησεν ὁμοιότητι] It is impossible that ὁμοιότητι can simply stand for ὁμοίως, as Stallbaum asserts; nor is he justified by the passage he cites, *Republic* 555 A, ἔτι οὖν, ἦν δ' ἐγώ, ἀπιστοῦμεν μὴ κατὰ τὴν ὀλιγαρχουμένην πόλιν ὁμοιότητι τὸν φειδωλόν τε καὶ χρηματιστὴν τετάχθαι; there obviously the meaning is that the φειδωλὸς and χρηματιστὴ are ranked as corresponding to the oligarchical state because of their resemblance to it; and similarly in 576 C, ὁ γε τυραννικὸς κατὰ τὴν τυραννουμένην πόλιν ἂν εἴη ὁμοιότητι. In like manner I think we must take it here as an instrumental dative.

τητι, καὶ τὰς σιαγόνας ἄκρας αὐτοῖς ξυνέδησεν ὑπὸ τὴν φύσιν τοῦ
προσώπου· τὰ δ᾽ ἄλλα εἰς ἅπαντα τὰ μέλη διέσπειρε, ξυνάπτων
ἄρθρον ἄρθρῳ. τὴν δὲ δὴ τοῦ στόματος ἡμῶν δύναμιν ὀδοῦσι καὶ
γλώττῃ καὶ χείλεσιν ἕνεκα τῶν ἀναγκαίων καὶ τῶν ἀρίστων διε-
5 κόσμησαν οἱ διακοσμοῦντες, ᾗ νῦν διατέτακται, τὴν μὲν εἴσοδον
τῶν ἀναγκαίων μηχανώμενοι χάριν, τὴν δ᾽ ἔξοδον τῶν ἀρίστων· E
ἀναγκαῖον μὲν γὰρ πᾶν ὅσον εἰσέρχεται τροφὴν διδὸν τῷ σώματι,
τὸ δὲ λόγων νᾶμα ἔξω ῥέον καὶ ὑπηρετοῦν φρονήσει κάλλιστον καὶ
ἄριστον πάντων ναμάτων. τὴν δ᾽ αὖ κεφαλὴν οὔτε μόνον ὀστεΐνην
10 ψιλὴν δυνατὸν ἐᾶν ἦν διὰ τὴν ἐν ταῖς ὥραις ἐφ᾽ ἑκάτερον ὑπερ-
βολήν, οὔτ᾽ αὖ ξυσκιασθεῖσαν κωφὴν καὶ ἀναίσθητον διὰ τὸν τῶν
σαρκῶν ὄχλον περιιδεῖν γιγνομένην. τῆς δὴ σαρκοειδοῦς φύσεως
[οὐ] καταξηραινομένης λέμμα μεῖζον περιγιγνόμενον ἐχωρίζετο, 76 A
δέρμα τὸ νῦν λεγόμενον. τοῦτο δὲ διὰ τὴν περὶ τὸν ἐγκέφαλον
15 νοτίδα ξυνιὸν αὐτὸ πρὸς αὑτὸ καὶ βλαστάνον κύκλῳ περιημφίεννυε
τὴν κεφαλήν· ἡ δὲ νοτὶς ὑπὸ τὰς ῥαφὰς ἀνιοῦσα ἦρδε καὶ συνέ-
κλεισεν αὐτὸ ἐπὶ τὴν κορυφήν, οἷον ἅμμα ξυναγαγοῦσα· τὸ δὲ τῶν
ῥαφῶν παντοδαπὸν εἶδος γέγονε διὰ τὴν τῶν περιόδων δύναμιν καὶ

13 οὐ inclusi a tribus codicibus omissum. servant AHSZ.
ἐχωρίζετο : ἐχώριζε τό A.   14 δέρμα post τὸ νῦν λεγόμενον ponit S.

4. τῶν ἀναγκαίων καὶ τῶν ἀρίστων] This distinction differs from that of ἀναγκαῖα and θεῖα in 68 E; for here both ἀναγκαῖα and ἄριστα are an end, not a means.

8. λόγων νᾶμα] Compare the metaphor in Euripides *Hippolytus* 653 ἀγὼ ῥυτοῖς νασμοῖσιν ἐξομόρξομαι | εἰς ὦτα κλύζων. Somewhat similar is the metaphor in *Phaedrus* 243 D, ποτίμῳ λόγῳ οἷον ἁλμυρὰν ἀκοὴν ἀποκλύσασθαι.

10. ἐφ᾽ ἑκάτερον] sc. ἐπὶ πνῖγος καὶ ψῦχος.

11. τὸν τῶν σαρκῶν ὄχλον] cf. 42·C τὸν πολὺν ὄχλον καὶ ὕστερον προσφύντα ἐκ πυρὸς καὶ ὕδατος καὶ ἀέρος καὶ γῆς.

13. [οὐ] καταξηραινομένης] Notwithstanding the approximate unanimity of the mss., I do not see how it is possible to reconcile οὐ with the sense. Surely the λέμμα is formed by the drying of the surface of the flesh. The Engelmann translator indeed says it is 'durch den Sinn erfordert', and renders it 'welche nicht ausgetrocknet war': but obviously this would require καταξηρανθείσης. I suspect we ought to read αὖ.

λέμμα μεῖζον] λέμμα is a peel or rind: the skin, according to Plato's conception, is analogous to the membranous film which forms on the surface of boiled milk, for instance, when exposed to the air : cf. Aristotle *de generatione animalium* II vi 743[b] 5 τὸ δὲ δέρμα ξηραινομένης τῆς σαρκὸς γίνεται, καθάπερ ἐπὶ τοῖς ἑψήμασιν ἡ λεγομένη γραῦς. Aristotle's language, it may be observed by the way, supports the omission of οὐ before καταξηραινομένης. As to μεῖζον, I see nothing for it but to acquiesce in Lindau's 'dixit vero μεῖζον, quod cetera amplectitur': but I cannot believe that the word is genuine. That Plato should think it necessary to point out that the envelope is greater than that which it envelopes is altogether incredible : but

uniformity; and he fastened the extremities of the jaw-bones to them just under the face; and the rest he distributed over all the limbs, uniting joint to joint. And our framers ordained the functions of the mouth, furnishing it with teeth and tongue and lips, in the way it is now arranged, combining in their purpose the necessary and the best; for they devised the incoming with the necessary in view, but the outgoing with the most excellent. For all that enters in to give sustenance to the body is of necessity; but the stream of speech which flows out and ministers to understanding is of all streams the most noble and excellent. But as to the head, it was neither possible to leave it of bare bone, owing to the extremes of heat and cold in the seasons; nor yet by covering it over to allow it to become dull and senseless through the burden of flesh. Of the fleshy material as it was drying a larger film formed on the surface and separated itself; this is what is now called skin. This by the influence of the moisture of the brain combined and grew up and clothed the head all round: and the moisture rising up under the sutures saturated and closed it in on the crown, fastening it together like a knot. Now the form of the sutures is manifold, owing to the power of the soul's revolutions and of the aliment; if these

I cannot see my way to any satisfactory emendation.

14. δέρμα] Is this meant to be derived from λέμμα? The νῦν looks like it; and Plato's etymological audacity has adventured things κύντερα than this.

διὰ τὴν περὶ τὸν ἐγκέφαλον νοτίδα] Plato is explaining how it comes to pass that the skull is covered with skin, although, according to his account, there is no flesh upon it. He regards it as an extension of the skin on the face and neck, which grows up over the head from all sides, being nourished by the moisture belonging to the brain, and meets on the summit (ξυνιὸν αὐτὸ πρὸς αὑτό). Thereupon the moisture, issuing through the sutures, penetrates the skin and causes it to take root on the head and to grow firmly together where it meets in the middle, as it were fastened in a knot (οἷον ἅμμα ξυναγαγοῦσα).

17. τὸ δὲ τῶν ῥαφῶν] The number and diversity of the sutures depends upon the violence of the struggle described in 43 B foll. between the influx of aliment and the revolutions of the soul acting through the brain. There is a passage of Hippokrates which curiously falls in with Plato's connexion of the sutures with the soul's περίοδοι: *de capitis vulneribus* vol. III p. 347 Kühn ὅστις μηδετέρωθι μηδεμίαν προβολὴν ἔχει, οὗτος ἔχει τὰς ῥαφὰς τῆς κεφαλῆς ὡς γράμμα τὸ χῖ γράφεται: that is to say, the rounder the head the more nearly does the form of the sutures approximate to that of the letter X, which is the form of the intersection of the two circles. When the head is prominent in front, says Hippokrates, the sutures resemble T; when protuberent behind, the figure is reversed, ⊥; if protuberent both before and behind, the sutures form the figure H. Thus in

282   ΠΛΑΤΩΝΟΣ   [76 A—

τῆς τροφῆς, μᾶλλον μὲν ἀλλήλοις μαχομένων τούτων πλείους, ἧττον B
δὲ ἐλάττους. τοῦτο δὴ πᾶν τὸ δέρμα κύκλῳ κατεκέντει πυρὶ τὸ
θεῖον, τρωθέντος δὲ καὶ τῆς ἰκμάδος ἔξω δι' αὐτοῦ φερομένης τὸ μὲν
ὑγρὸν καὶ θερμὸν ὅσον εἰλικρινὲς ἀπῄειν, τὸ δὲ μικτὸν ἐξ ὧν καὶ τὸ
5 δέρμα ἦν, αἰρόμενον μὲν ὑπὸ τῆς φορᾶς ἔξω μακρὸν ἐτείνετο, λεπτότητα ἴσην ἔχον τῷ κατακεντήματι, διὰ δὲ βραδυτῆτα ἀπωθούμε-
νον ὑπὸ τοῦ περιεστῶτος ἔξωθεν πνεύματος πάλιν ἐντὸς ὑπὸ τὸ
δέρμα εἰλλόμενον κατερριζοῦτο, καὶ κατὰ ταῦτα δὴ τὰ πάθη τὸ C
τριχῶν γένος ἐν τῷ δέρματι πέφυκε, ξυγγενὲς μὲν ἱμαντῶδες ὂν
10 αὐτοῦ, σκληρότερον δὲ καὶ πυκνότερον τῇ πιλήσει τῆς ψύξεως,
ἣν ἀποχωριζομένη δέρματος ἑκάστη θρὶξ ψυχθεῖσα συνεπιλήθη.
τούτῳ δὴ λασίαν ἡμῶν ἀπειργάσατο τὴν κεφαλὴν ὁ ποιῶν, χρώμενος μὲν αἰτίοις τοῖς εἰρημένοις, διανοούμενος δὲ ἀντὶ σαρκὸς
αὐτὸ δεῖν εἶναι στέγασμα τῆς περὶ τὸν ἐγκέφαλον ἕνεκα ἀσφα-
15 λείας κοῦφον καὶ θέρους χειμῶνός τε ἱκανὸν σκιὰν καὶ σκέπην D
παρέχειν, εὐαισθησίας δὲ οὐδὲν διακώλυμα ἐμποδὼν γενησόμενον.
τὸ δὲ ἐν τῇ περὶ τοὺς δακτύλους καταπλοκῇ τοῦ νεύρου καὶ τοῦ
δέρματος ὀστοῦ τε, ξυμμιχθὲν ἐκ τριῶν, ἀποξηρανθὲν ἓν κοινὸν
ξυμπάντων σκληρὸν γέγονε δέρμα, τοῖς μὲν ξυναιτίοις τούτοις δη-
20 μιουργηθέν, τῇ δὲ αἰτιωτάτῃ διανοίᾳ τῶν ἔπειτα ἐσομένων ἕνεκα

3 τρωθέντος: τρηθέντος SZ.   7 ὑπὸ τοῦ: ἀπὸ τοῦ A.
8 τὸ τριχῶν: τὸ τῶν τριχῶν S.   10 πυκνότερον: πυκνώτερον S.

so far as the shape of the head departs from the spherical or normal shape, in the same degree the sutures depart from the figure X; and in the same degree we may suppose the struggle between the περίοδοι and the κῦμα τῆς τροφῆς to have been long and severe. The treatise concerning wounds on the head is one of those considered to be the genuine work of Hippokrates. In 92 A we find that in the lower animals the ἀργία τῶν περιφορῶν causes the head to assume an elongated shape.

2. τὸ θεῖον] i.e. the brain, which is the seat of τὸ θεῖον. Plato now passes to the growth of the hair, which he thus explains. The skin of the head is punctured all over by the fire issuing from the brain: through the punctures moisture escapes, of which so much as is pure evaporates and disappears; but that which contains an admixture of the substances composing the skin is forced outward in a cylindrical form fitting the size of the punctures. But owing to the slowness of its growth and the resistance of the surrounding atmosphere, the hair is pushed backwards, so that the end becomes rooted under the skin. Thus the hair is composed of the same substance as the skin, but by refrigeration and compression has become more hard and dense. As to its identity with the skin Aristotle agrees: cf. *de gen. anim.* II vi 745ᵃ 20 ὄνυχες δὲ καὶ τρίχες καὶ κέρατα καὶ τὰ τοιαῦτα ἐκ τοῦ δέρματος, διὸ καὶ συμμεταβάλλουσι τῷ δέρματι τὰς χρόας.

3. τρωθέντος] The suggestion τρηθέντος is certainly tempting: but the mss. are unanimous, and I retain their reading,

contend more vehemently one with another, the sutures are more in number; but if less so, they are fewer. Now the whole of this skin was pricked all about with fire by the divine part: and when it was pierced and the moisture issued forth through it, all the moisture and heat which was pure vanished away; but that which was mingled with the substances whereof the skin was formed, being lifted up by the impulse, stretched far outwards, in fineness equalling the size of the puncture; but owing to the slowness of its motion it was thrust back by the surrounding air, and being forced in and rolled up under the skin it took root there. Under these conditions hair grows up in the skin, being of similar nature but of threadlike appearance, and made harder and denser by the contraction of cooling: for every hair in being separated from the skin was cooled and contracted. Hereby has our creator made our head hairy, using the means aforesaid, and conceiving that this instead of flesh should be a covering for the protection of the brain, being light and capable of affording shade from heat and shelter from cold, while it would be no hindrance in the way of ready apprehension. The threefold combination of sinew skin and bone in the fabric of the fingers, when dried, forms out of all a single hard skin, for the construction of which these substances served as means, but the true cause and purpose of its formation was the welfare of races not

though with considerable hesitation.

4. ἀπῄειν] They at once departed in the course of nature to their own habitation: but the earthier substance, having no such impulse, was forced back by the pressure of the atmosphere.

8. εἱλλόμενον] 'rolled up': see note on 40 B.

13. αἰτίοις τοῖς εἰρημένοις] i. e. the subsidiary physical causes aforesaid: the final cause is given next..

16. γενησόμενον] Note the change of construction: the future participle stands in the place of δεῖν εἶναι in the prior cause.

17. καταπλοκῇ] That is to say, the three substances of tendon skin and bone are interwoven into one homogeneous body and completely dried; out of this are formed the nails. Plato's statement here differs somewhat from Aristotle's as cited above.

20. τῶν ἔπειτα ἐσομένων ἕνεκα] This is a very singular declaration. The nails, by this account, are formed solely for the development they will afterwards attain in the inferior animals, as though they were of no use whatsoever to mankind. The importance of them is no doubt more conspicuous in beasts and birds; but Plato's theory certainly appears rather paradoxically to ignore their value to the human race. There is however a curious approximation to Darwinism in his statement: the nails appeared first in a rudimentary form in the human race; and afterwards in course of evolution the claws of the lion and the talons of the

284   ΠΛΑΤΩΝΟΣ   [76 D—

εἰργασμένον. ὡς γάρ ποτε ἐξ ἀνδρῶν γυναῖκες καὶ τἆλλα θηρία
γενήσοιντο, ἠπίσταντο οἱ ξυνιστάντες ἡμᾶς, καὶ δὴ καὶ τῆς τῶν Ε
ὀνύχων χρείας ὅτι πολλὰ τῶν θρεμμάτων καὶ ἐπὶ πολλὰ δεήσοιτο
ᾔδεσαν, ὅθεν ἐν ἀνθρώποις εὐθὺς γιγνομένοις ὑπετυπώσαντο τὴν
5 τῶν ὀνύχων γένεσιν· τούτῳ δὴ τῷ λόγῳ καὶ ταῖς προφάσεσι
ταύταις δέρμα τρίχας <τ'> ὄνυχάς τε ἐπ' ἄκροις τοῖς κώλοις
ἔφυσαν.

XXXIV. Ἐπειδὴ δὲ πάντ' ἦν τὰ τοῦ θνητοῦ ζῴου ξυμπεφυ-
κότα μέρη καὶ μέλη, τὴν δὲ ζωὴν ἐν πυρὶ καὶ πνεύματι ξυνέβαινεν 77 Α
10 ἐξ ἀνάγκης ἔχειν αὐτῷ, καὶ διὰ ταῦτα ὑπὸ τούτων τηκόμενον κενούμενόν τ' ἔφθινε, βοήθειαν αὐτῷ θεοὶ μηχανῶνται. τῆς γὰρ
ἀνθρωπίνης ξυγγενῆ φύσεως φύσιν ἄλλαις ἰδέαις καὶ αἰσθήσεσι
κεραννύντες, ὥσθ' ἕτερον ζῷον εἶναι, φυτεύουσιν· ἃ δὴ νῦν ἥμερα

3 δεήσοιτο: δεήσοιντο Α.   6 τ' inserui.

eagle were developed from them. The notable point is that Plato evidently does not conceive that in the transmigrations any arbitrary change of form takes place, but that each successive organism is regularly developed out of its predecessors. Plato's notion rests on no zoological evidence, so far as we know; it is but a brilliant guess: none the less, perhaps all the more, seeing that such evidence was not at his command, it is a mark of his keen scientific insight.

6. **τρίχας <τ'> ὄνυχάς τε**] I have taken upon me to insert τε, since I do not believe δέρμα τρίχας ὄνυχάς τε can be Greek. It may be noticed that this correction almost restores a hexameter verse:
δέρμα τρίχας τ' ὄνυχάς τ' ἐπ' ἄκροις κώλοισιν ἔφυσαν.
Is Plato quoting from some old physical poet? Empedokles might have written such a line.

76 E—77 C, c. xxxiv. So when all the parts of the human frame had been combined in a body for ever suffering waste by fire and by air, the gods devised a means of its replenishment. They took wild plants and trained them by cultivation, so that they were fit for human sustenance. Plants are living and conscious beings; but they have the appetitive soul alone; they grow of their inborn vital force, without impulsion from without; they are stationary in one place, and cannot reflect upon their own nature.

9. **μέρη καὶ μέλη**] For this combination compare *Laws* 795 E τῶν τοῦ σώματος αὐτοῦ μελῶν τε καὶ μερῶν: and *Philebus* 14 E ὅταν τις ἑκάστου τὰ μέλη τε καὶ ἅμα μέρη διελὼν τῷ λόγῳ. The distinction between the terms is thus defined by Aristotle *historia animalium* I i 486[a] 8 τῶν δὲ τοιούτων ἔνια οὐ μόνον μέρη ἀλλὰ καὶ μέλη καλεῖται· τοιαῦτα δ' ἐστὶν ὅσα τῶν μερῶν ὅλα ὄντα ἕτερα μέρη ἔχει ἐν αὑτοῖς, οἷον κεφαλὴ καὶ σκέλος καὶ χεὶρ καὶ ὅλος ὁ βραχίων καὶ ὁ θώραξ· ταῦτα γὰρ αὐτά τέ ἐστι μέρη ὅλα, καὶ ἔστιν αὐτῶν ἕτερα μόρια. A μέλος then is that which is part of a whole, but is yet in itself a definite whole.

τὴν δὲ ζωὴν ἐν πυρὶ καὶ πνεύματι] Man's life is said to depend on fire and air because these are the agents of digestion and respiration, as we shall see in the next two chapters: cf. 78 D. These two elements in fact keep up the vital movement of the human body.

10. **τηκόμενον κενούμενόν τε**] Sc.

yet existing. For our creators were aware that men should pass into women, and afterwards into beasts; and they knew that many creatures would need the aid of nails for many purposes: wherefore at the very birth of the human race they fashioned the rudiments of nails. On such reasoning and with such purposes did they form skin and hair, and on the extremities of the limbs nails.

XXXIV. Now when all the parts and members of the mortal being were created in union, and since his life was made perforce dependent upon fire and air, and therefore his body suffered waste through being dissolved and left void by these, the gods devised succour for him. They engendered another nature akin to the nature of man, blending it with other forms and sensations, so as to be another kind of animal. These are

τηκόμενον ὑπὸ πυρός, κενούμενον ὑπ' ἀέρος. Plato enters more fully into this in 88 c foll.

12. ἄλλαις ἰδέαις καὶ αἰσθήσεσι] Plants are akin to the nature of mankind, inasmuch as they are animated by the same vital principle and are formed out of similar physical materials, so that they are able to repair the waste of the human structure. But the form of these organisms is diverse from man's, and their mode of sensation is peculiar to themselves. Whether Plato was a vegetarian or not, it is clear that he regards vegetables as the natural and primaeval food of man: see below 80 E, and *Epinomis* 975 A ἔστω δὴ πρῶτον μὲν ἡ τῆς ἀλληλοφαγίας τῶν ζῴων ἡμᾶς τῶν μέν, ὡς ὁ μῦθός ἐστι, τὸ παράπαν ἀποστήσασα, τῶν δὲ εἰς τὴν νόμιμον ἐδωδὴν καταστήσασα. We must of course allow for the possibility that the author of the *Epinomis* has overstated Plato's disapprobation of animal diet.

13. ἃ δὴ νῦν ἥμερα δένδρα] So then the device of the gods for the preservation of human life was not the invention of plants, but their cultivation: plants themselves existed as part of the general order of nature. It thus appears that in Plato's scheme plants do not, like the inferior animals, arise by degeneration from the human form. For as soon as man was first created, he would have need of plants to provide him with sustenance. It would appear then that in the Platonic mythology the erring soul in the course of her transmigrations does not enter any of the forms of plant-life; though the contrary was the belief of Empedokles—ἤδη γάρ ποτ' ἐγὼ γενόμην κοῦρός τε κόρη τε | θάμνος τ' οἰωνός τε καὶ εἰν ἀλὶ ἔλλοπος ἰχθύς. Martin however is mistaken in inferring this conclusion from the fact that plants possess only the third εἶδος of soul: this third εἶδος is simply the one vital force acting exclusively through matter—a degree of degeneracy to which any human soul, according to the theory of metempsychosis, might sink: indeed there are forms of what we call animal life, which are clearly within the limits of transmigration, but which possess little, if any, more independent activity of ψυχὴ than do plants. The simultaneous appearance of mankind and of plants in the world, while all intermediate forms of animal life are absent, is curious, and could hardly, I think, be defended upon ontological grounds.

δένδρα καὶ φυτὰ καὶ σπέρματα παιδευθέντα ὑπὸ γεωργίας τιθασῶς πρὸς ἡμᾶς ἔσχε, πρὶν δὲ ἦν μόνα τὰ τῶν ἀγρίων γένη, πρεσβύτερα τῶν ἡμέρων ὄντα. πᾶν γὰρ οὖν, ὅ τί περ ἂν μετάσχῃ B τοῦ ζῆν, ζῷον μὲν ἂν ἐν δίκῃ λέγοιτο ὀρθότατα· μετέχει γε μὴν
5 τοῦτο, ὃ νῦν λέγομεν, τοῦ τρίτου ψυχῆς εἴδους, ὃ μεταξὺ φρενῶν ὀμφαλοῦ τε ἱδρῦσθαι λόγος, ᾧ δόξης μὲν λογισμοῦ τε καὶ νοῦ μέτεστι τὸ μηδέν, αἰσθήσεως δὲ ἡδείας καὶ ἀλγεινῆς μετὰ ἐπιθυμιῶν. πάσχον γὰρ διατελεῖ πάντα, στραφέντι δ' αὐτῷ ἐν ἑαυτῷ περὶ ἑαυτό, τὴν μὲν ἔξωθεν ἀπωσαμένῳ κίνησιν, τῇ δ' οἰκείᾳ
10 χρησαμένῳ, τῶν αὑτοῦ τι λογίσασθαι κατιδόντι φύσιν οὐ παρα- C δέδωκεν ἡ γένεσις. διὸ δὴ ζῇ μὲν ἔστι τε οὐχ ἕτερον ζῴου, μόνιμον

10 αὑτοῦ : αὐτοῦ A. φύσιν : φύσει A.

2. **ἔσχε**] i.e. attained the condition in which now they are.

3. **πᾶν γὰρ οὖν**] This passage is of the highest importance, as proving beyond controversy that Plato in the fullest degree maintained the unity of all life. He drew no arbitrary line between 'animal' and 'vegetable' life: all things that live are manifestations of the same eternal essence: only as this evolved itself through countless gradations of existence, the lower ranks of organisms possess less and less of the pure activity of soul operating by herself, until in plants and the lowest forms of animal life the vital force only manifests itself in the power of sensation and growth.

Aristotle agrees with Plato in ascribing to plants ζωή and ψυχή, but he does not allow them αἴσθησις: see *de anima* I v 410^b 23 φαίνεται γὰρ τὰ φυτὰ ζῆν οὐ μετέχοντα φορᾶς καὶ αἰσθήσεως: cf. II ii 413^a 25, and *de partibus animalium* I i 641^b 6. They had according to him the θρεπτικὴ ψυχή alone: *de anima* II ii 413^b 7 θρεπτικὸν δὲ λέγομεν τὸ τοιοῦτον μόριον τῆς ψυχῆς οὗ καὶ τὰ φυτὰ μετέχει. This coincides with Plato's statement. Aristotle however draws the distinction between ζῷα and φυτὰ that the former possess αἴσθησις, the latter possess it not: *de iuventute* i 467^b 24 τὰ μὲν φυτὰ ζῇ μέν, οὐκ ἔχει δ' αἴσθησιν· τῷ δ' αἰσθάνεσθαι τὸ ζῷον πρὸς τὸ μὴ ζῷον διορίζομεν. See however *hist. anim.* VIII i.

In the pseudo-Aristotelian treatise *de plantis* i 815^b 16 it is affirmed that Anaxagoras Empedokles and Demokritos attributed thought and knowledge to plants: ὁ δὲ Ἀναξαγόρας καὶ ὁ Δημόκριτος καὶ ὁ Ἐμπεδοκλῆς καὶ νοῦν καὶ γνῶσιν εἶπον ἔχειν τὰ φυτά: they of course assigned them ἐπιθυμία and αἴσθησις also: *ibid.* 815^a 15 Ἀναξαγόρας μὲν οὖν καὶ Ἐμπεδοκλῆς ἐπιθυμίᾳ ταῦτα κινεῖσθαι λέγουσιν, αἰσθάνεσθαί τε καὶ λυπεῖσθαι καὶ ἥδεσθαι διαβεβαιοῦνται. ὧν ὁ μὲν Ἀναξαγόρας καὶ ζῷα εἶναι καὶ ἥδεσθαι καὶ λυπεῖσθαι εἶπε, τῇ τε ἀπορροῇ τῶν φύλλων καὶ τῇ αὐξήσει τοῦτο ἐκλαμβάνων· ὁ δὲ Ἐμπεδοκλῆς γένος ἐν τούτοις κεκραμένον εἶναι ἐδόξασεν. Sextus Empiricus *adv. math.* VIII 286 confirms the statement that Empedokles allowed reason to plants: πάντα γὰρ ἴσθι φρόνησιν ἔχειν καὶ νώματος αἶσαν. Diogenes of Apollonia was of a contrary opinion: Theophrastos *de sensu* § 44 τὰ δὲ φυτὰ διὰ τὸ μὴ εἶναι κοῖλα μηδὲ ἀναδέχεσθαι τὸν ἀέρα παντελῶς ἀφῃρῆσθαι τὸ φρονεῖν. In our estimate of such statements however we must allow for the fact that these early philosophers only very imperfectly distinguished between αἰσθάνεσθαι and φρονεῖν: Theophrastos says of Parmenides τὸ γὰρ αἰσθάνεσθαι καὶ τὸ φρονεῖν ὡς ταὐτὸ λέγει:

ΤΙΜΑΙΟΣ.

the cultivated trees and plants and seeds, which are now trained by culture and domesticated with us; but formerly there existed only the wild kinds, which are older than the cultivated. For indeed everything which partakes of life may with perfect justice and fitness be termed an animal; but the kind of which we are now speaking shares only the third form of soul, which our theory says is seated between the midriff and the navel, and which has nothing to do with opinion and reasoning and thought, but only with sensation, pleasant or painful, with appetites accompanying. For it ever continues passively receptive of all sensations, and having its circulation in itself about its own centre, it rejects all motion from without and uses only its own; but its nature has not bestowed upon it any power of observing its own being and reflecting thereon. Wherefore it is indeed alive and in no wise differs from an animal, but it is

and this is no doubt still more true of others.

7. αἰσθήσεως δέ] The θρεπτικὴ δύναμις, though not explicitly mentioned here, is of course included, as we see from the account of the τρίτον εἶδος in 70 D foll.

8. πάσχον γὰρ διατελεῖ πάντα] i.e. it passively submits to the influences which work upon it: since it does not possess the two more active forms of soul, the passive conditions of nutrition growth and decay, together with sensation, are all that belong to it.

στραφέντι δ' αὐτῷ ἐν ἑαυτῷ] That is to say, its motions, e.g. the circulation of the sap, take place within it: its movement is not κατὰ τόπον, but ἐν ταὐτῷ.

9. τὴν μὲν ἔξωθεν ἀπωσαμένῳ] It rejects motion from without and avails itself of its own innate force: that is, its growth is not due to any external compulsion, but the development of its own impulse. As Aristotle would put it, a plant has its proper motion κατὰ φύσιν, the motion ἔξωθεν only κατὰ συμβεβηκός. Plato means that it αὐτὸ ἑαυτὸ κινεῖ and therefore must possess ψυχή, which alone is self-moved.

10. τῶν αὐτοῦ τι λογίσασθαι κατιδόντι φύσιν] i.e. it is conscious, but not self-conscious. Man can look into his own consciousness and realise his own identity and personality: he can speculate upon his relation to other personalities and to the sensible objects around him. The plant can do none of this: it can but take its sensations as they come, without inquiring what they are, what it is that feels them, what is the line of continuity that binds them together. The meaning of this phrase is plain enough; but the expression of it is a little strange. There is an overwhelming preponderance of mss. evidence in favour of φύσει, and I am not sure that it ought not to be restored: Schneider however is alone, I believe, in adopting it.

11. ἔστι τε οὐχ ἕτερον ζῴου] It would seem a necessary consequence that a thing which ζῇ is ζῷον: and Aristotle is perhaps somewhat inconsistent in allowing plants ζῆν, while refusing them the title of ζῷα. Also Plato seems more scientific than Aristotle in attributing αἴσθησις to plants. What manner of αἴσθησις belongs to plants may or may not be dis-

288   ΠΛΑΤΩΝΟΣ   [77 C—

δὲ καὶ κατερριζωμένον πέπηγε διὰ τὸ τῆς ὑφ' ἑαυτοῦ κινήσεως ἐστερῆσθαι.

XXXV. Ταῦτα δὴ τὰ γένη πάντα φυτεύσαντες οἱ κρείττους τοῖς ἥττοσιν ἡμῖν τροφήν, τὸ σῶμα αὐτὸ ἡμῶν διωχέτευσαν τέμ-
5 νοντες οἷον ἐν κήποις ὀχετούς, ἵνα ὥσπερ ἐκ νάματος ἐπιόντος ἄρδοιτο. καὶ πρῶτον μὲν ὀχετοὺς κρυφαίους ὑπὸ τὴν ξύμφυσιν D τοῦ δέρματος καὶ τῆς σαρκὸς δύο φλέβας ἔτεμον νωτιαίας διδύμους, ὡς τὸ σῶμα ἐτύγχανε δεξιοῖς τε καὶ ἀριστεροῖς ὄν· ταύτας δὲ καθῆκαν παρὰ τὴν ῥάχιν καὶ τὸν γόνιμον μεταξὺ λαβόντες
10 μυελόν, ἵνα οὗτός τε ὅ τι μάλιστα θάλλοι, καὶ ἐπὶ τἆλλα εὔρους ἐντεῦθεν ἅτε ἐπὶ κάταντες ἡ ἐπίχυσις γιγνομένη παρέχοι τὴν ὑδρείαν ὁμαλήν. μετὰ δὲ ταῦτα σχίσαντες περὶ τὴν κεφαλὴν τὰς φλέβας καὶ δι' ἀλλήλων ἐναντίας πλέξαντες διεῖσαν, τὰς μὲν E ἐκ τῶν δεξιῶν ἐπὶ τἀριστερὰ τοῦ σώματος, τὰς δ' ἐκ τῶν ἀριστε-

6 κρυφαίους : κρυφαίως Λ.   7 διδύμους : δίδυμον SZ.   14 τἀριστερά : τὰ ἀριστερά S.

covered or discoverable by science; but it seems at least improbable that anywhere a hard and fast line can be drawn between the αἴσθησις of animals, from man down to the zoophyte, and the corresponding πάθος in plants. Plato here as everywhere in his system preserves the principle of continuity, the germ of which he inherited from Herakleitos, and which attained so astonishing a development in his hands. Brief as is Plato's treatment of the subject, the union of poetical imagination and scientific grasp which it displays renders this short chapter on plants singularly interesting. And but for it, we should have been forced inferentially to fill up a space in his theory, for which we now have the authority of his explicit statement.

1. **τῆς ὑφ' ἑαυτοῦ κινήσεως ἐστερῆσθαι.**] This is not inconsistent, though at first sight it may appear so, with τῇ οἰκείᾳ χρησαμένῳ above. For there the question was of motion ἐν τῷ αὐτῷ, now it is of motion from place to place. The plant is free to carry on all its natural movements within its own structure, but it is incapable of transferring itself from place to place. Yet this stationary condition is no reason for refusing it the name of ζῷον: for indeed the κόσμος itself has its motion only ἐν τῷ αὐτῷ. Galen evidently had τῆς ἐξ ἑαυτοῦ, for he proposes to read ἔξω: ἐνενόησα λείπειν τὸ ω στοιχεῖον, γράψαντος τοῦ Πλάτωνος διὰ τὸ τῆς ἔξω ἑαυτοῦ. The emendation does him credit: but there is no reason for interfering with our present text.

77 C—79 A, c. xxxv. Then the gods made two channels down the body, embedded in the flesh, one on either side of the spine, to irrigate it with blood: and at the head they cleft the veins and caused them to cross each other transversely, that the head might be firmly fixed on the neck, and that communication might be preserved between both sides of the body. This scheme for the irrigation of the body we shall best understand, if we reflect that all substances composed of finer particles exclude those of coarser, while the coarser are easily penetrated by the finer. So then when food and drink enter the belly, they are retained; but fire and air are too subtle to be confined therein. Therefore the gods wove a web of fire and of air spread over the cavity of the body and

stationary and rooted fast, because it has been denied the power of self-motion.

XXXV. Thus did the higher powers create all these kinds as sustenance for us who were feebler; and next they made canals in the substance of our body, as though they were cutting runnels in a garden, that it might be irrigated as by an inflowing stream. And first they carried like hidden rills, under the place where the skin and the flesh are joined, two veins down the back, following the twofold division of the body into right and left. These they brought down on either side of the spine and the seminal marrow, first in order that this might be most vigorous, next that the current might have an easy flow downwards and render the irrigation regular. After that, they cleft the veins around the head, and interweaving them crossed them in opposite directions, carrying these from the right side of the body to the left and those from the left to the right. This

placed therein two lesser webs opening into the mouth and nostrils. And they made alternately the great web to flow towards the lesser webs, and again the lesser towards the greater. In the former case the airy envelope of the greater web penetrated through the porous substance of the body to the cavity within, in the latter the lesser webs passed through the body outwards; and in either case the fire followed with the air. This alternation is kept up perpetually so long as a man lives, and we give it the name of respiration. And so when the fire, passing to and fro, encounters food and drink in the stomach, it dissolves them and driving them onwards forces them to flow through the veins, like water drawn into pipes from a fountain.

3. οἱ κρείττους] Plato several times applies this phrase to supernal powers: cf. *Sophist* 216 B τάχ' οὖν ἂν καὶ σοί τις οὗτος τῶν κρειττόνων συνέποιτο, φαύλους ἡμᾶς ὄντας ἐν τοῖς λόγοις ἐποψόμενός τε καὶ ἐλέγξων, θεός τις ὢν ἐλεγκτικός: *Symposium* 188 D τοῖς κρείττοσιν ἡμῶν θεοῖς: *Euthydemus* 291 A μή τις τῶν κρειττόνων παρὼν αὐτὰ ἐφθέγξατο: the

last passage being ironical.

4. τέμνοντες...ὀχετούς] cf. 70 D τῆς ἀρτηρίας ὀχετοὺς ἐπὶ τὸν πλεύμονα ἔτεμον.

7. δύο φλέβας] The two 'veins' are, according to Martin, the aorta and the vena cava.

8. δεξιοῖς τε καὶ ἀριστεροῖς ὄν] i.e. with right and left sides: I doubt whether μέρεσιν is to be supplied, any more than μέρη with the phrases ἐπὶ δεξιά, ἐπ' ἀριστερά.

9. τὸν γόνιμον...μυελόν] cf. 73 C.

11. ἐπὶ κάταντες] As Galen objects, this seems to leave out of sight the circulation of the blood in the head and neck, which would be ἄναντες.

14. ἐκ τῶν δεξιῶν ἐπὶ τἀριστερά] Plato makes the blood-vessels belonging to the right side of the head pass to the left side of the body and *vice versa* for two reasons: first that the consequent interlacing of the veins might fasten the head (which we have seen to be destitute of νεῦρα) firmly on the trunk; secondly that the sensations might be conveyed from either side of the brain to the opposite side of the body, and so all parts of the body might be kept in communica-

290   ΠΛΑΤΩΝΟΣ   [77 E—

ρῶν ἐπὶ τὰ δεξιὰ κλίναντες, ὅπως δεσμὸς ἅμα τῇ κεφαλῇ πρὸς
τὸ σῶμα εἴη μετὰ τοῦ δέρματος, ἐπειδὴ νεύροις οὐκ ἦν κύκλῳ
κατὰ κορυφὴν περιειλημμένη, καὶ δὴ καὶ τὸ τῶν αἰσθήσεων πάθος
ἵν' ἀφ' ἑκατέρων τῶν μερῶν εἰς ἅπαν τὸ σῶμα εἴη διαδιδόμενον.
5 τὸ δ' ἐντεῦθεν ἤδη τὴν ὑδραγωγίαν παρεσκεύασαν τρόπῳ τινὶ
τοιῷδε, ὃν κατοψόμεθα ῥᾷον προδιομολογησάμενοι τὸ τοιόνδε, ὅτι 73 Α
πάντα, ὅσα ἐξ ἐλαττόνων ξυνίσταται, στέγει τὰ μείζω, τὰ δ' ἐκ
μειζόνων τὰ σμικρότερα οὐ δύναται· πῦρ δὲ πάντων γενῶν σμικρο-
μερέστατον, ὅθεν δι' ὕδατος καὶ γῆς ἀέρος τε καὶ ὅσα ἐκ τούτων
10 ξυνίσταται διαχωρεῖ καὶ στέγειν οὐδὲν αὐτὸ δύναται. ταὐτὸν δὴ
καὶ περὶ τῆς παρ' ἡμῖν κοιλίας διανοητέον, ὅτι σιτία μὲν καὶ
ποτὰ ὅταν εἰς αὐτὴν ἐμπέσῃ στέγει, πνεῦμα δὲ καὶ πῦρ σμικρο- Β
μερέστερα ὄντα τῆς αὐτῆς ξυστάσεως οὐ δύναται. τούτοις οὖν
κατεχρήσατο ὁ θεὸς εἰς τὴν ἐκ τῆς κοιλίας ἐπὶ τὰς φλέβας
15 ὑδρείαν, πλέγμα ἐξ ἀέρος καὶ πυρὸς οἷον οἱ κύρτοι ξυνυφηνάμενος,

4 διαδιδόμενον : διαδιδὸν^{πλ} Α.

tion. The notion that the blood-vessels are wanted to fasten the head is of course erroneous; the latter part of his theory, had nerves but been substituted for veins, is a nearer guess at the truth.

5. **τὸ δ' ἐντεῦθεν ἤδη**] cf. Galen *de plac. Hipp. et Plat.* VIII 706 τὸ μὲν οὖν ἀέρι καὶ πυρὶ χρῆσθαι τὴν φύσιν πρὸς πέψιν τροφῆς αἱματώσιν τε καὶ ἀνάδοσιν ὀρθῶς εἴρηται, τὸ δὲ ἐξ αὐτῶν πλέγμα γεγονέναι καὶ μὴ διὰ ὅλων κρᾶσιν οὐκέτι ἐπαινῶ, καθάπερ οὐδὲ τὸ πῦρ ὀνομάζειν αὐτόν [? αὐτό], ἐνόν, ὥς Ἱπποκράτης, ἔμφυτον θερμόν. The principle that smaller particles can pass through the interstices of larger ones, while the larger cannot penetrate the smaller, is thus applied by Plato to explain the process of digestion: the nutriment swallowed must on the one hand have a receptacle provided which is able to contain it, while on the other hand it must be subjected to the action of fire. The walls of the receptacle are therefore constructed of material sufficiently fine to retain the food, but not fine enough to arrest the passage of fire and air: the two latter therefore are enabled to circulate freely through the substance and lining of the body and to act upon the food contained within it. It will thus be seen that Plato conceives respiration solely as subsidiary to digestion: an opinion which is perhaps peculiar to him alone among ancient thinkers: the ordinary view being that its function was to regulate the temperature of the body, as thought Aristotle: cf. *de respiratione* xvi 478^{a} 28 καταψύξεως μὲν οὖν ὅλως ἡ τῶν ζῴων δεῖται φύσις, διὰ τὴν ἐν τῇ καρδίᾳ τῆς ψυχῆς ἐμπύρωσιν. ταύτην δὲ ποιεῖται διὰ τῆς ἀναπνοῆς. Demokritos thought it served to keep up the supply of ψυχή in the body: *ibid.* iv 471^{b} 30 foll.: not, Aristotle observes, that Demokritos conceived that Nature designed it for that end; ὅλως γάρ, ὥσπερ καὶ οἱ ἄλλοι φυσικοί, καὶ οὗτος οὐθὲν ἅπτεται τῆς τοιαύτης αἰτίας.

8. **πῦρ δὲ πάντων γενῶν**] Air seems more concerned with the process of respiration; but we must remember that in Plato's view fire was the actual instrument of assimilating the food, and also that it was the agent which started the

78 B] ΤΙΜΑΙΟΣ. 291

they did, partly in order that together with the skin they might form a bond to fasten the head to the body, seeing that it was not set round with sinews on the crown; and also that this might be a means of distributing from each side throughout the whole body the sensation due to the perceptions. And next to this they designed the irrigation on a kind of plan which we shall better discern by assuming the following premises. All bodies which are composed of smaller particles exclude the larger, but the larger cannot exclude the smaller. Fire is composed of finer particles than any other element, whence it penetrates through water and earth and air and whatever is composed of them, and nothing can keep it out. This rule must also be applied to the human belly; when food and drink enter into it, it keeps them in; but air and fire, being finer than its own structure, it cannot keep in. Accordingly God used these two elements for the conveyance of liquid from the belly to the veins, weaving of air and fire a network

air in its oscillations, cf. 79 D. Air then plays a part only subsidiary to fire.

13. τούτοις οὖν κατεχρήσατο] He used fire and air (1) for the conversion of the food into blood, (2) for its conveyance into the blood-vessels.

15. πλέγμα ἐξ ἀέρος καὶ πυρός] This theory of respiration is by far the most obscure and perplexing of Plato's physiological lucubrations, partly owing to the enigmatical form in which it is expressed, partly to actual gaps in the exposition. An important light however is thrown upon it by a fragment of Galen's treatise on the *Timaeus*, which deals with this passage. This fragment, which was previously known only in an imperfect Latin translation, was found by M. Daremberg in the Paris library and published by him in 1848. On Galen's commentary the ensuing explanation is based: I cannot however persuade myself that it fully clears up statements which Galen himself declares to be δυσνόητά τε καὶ δύσρητα.

First we must determine the meaning of κύρτος and ἐγκύρτιον. The first was a fishing-trap, or weel, woven of reeds; it seems to have had a narrow funnel-shaped neck, through which the fish entered, but was unable to return, owing to the points of the reeds being set against it. (Martin conceives it to consist of two baskets, one fitting into the other; but Galen says it is ἁπλοῦν.) The ἐγκύρτιον—a word which is only found in the present passage—is explained by Stallbaum (whom Liddell and Scott follow) to mean the entrance or neck of the κύρτος. But on this point Galen is explicit: he says it is ὅμοιον μὲν τῷ μεγάλῳ, μικρὸν δέ. We must therefore conceive the ἐγκύρτια to be two smaller κύρτοι similar to the larger, contained within it and opening into its neck.

Applying these premises, we shall find that the κύρτος or large πλέγμα consists of two layers, one of fire, one of air. The outer layer (τὸ κύρτος) is the stratum of air in contact with all the outer surface of the body; the inner layer (τὰ ἔνδον τοῦ πλοκάνου) is the vital heat contained in the blood and pervading all the substance of the body between the skin and the cavity within. The two ἐγκύρτια, which are formed entirely of air, represent re-

19—2

292   ΠΛΑΤΩΝΟΣ   [78 B—

διπλᾶ κατὰ τὴν εἴσοδον ἐγκύρτια ἔχον, ὧν θάτερον αὖ πάλιν
διέπλεξε δίκρουν· καὶ ἀπὸ τῶν ἐγκυρτίων δὴ διετείνατο οἷον
σχοίνους κύκλῳ διὰ παντὸς πρὸς τὰ ἔσχατα τοῦ πλέγματος. τὰ
μὲν οὖν ἔνδον ἐκ πυρὸς συνεστήσατο τοῦ πλοκάνου ἅπαντα, τὰ δ᾽ C
5 ἐγκύρτια καὶ τὸ κύτος ἀεροειδῆ, καὶ λαβὼν αὐτὸ περιέστησε τῷ
πλασθέντι ζῴῳ τρόπον τοιόνδε. τὸ μὲν τῶν ἐγκυρτίων εἰς τὸ
στόμα μεθῆκε· διπλοῦ δὲ ὄντος αὐτοῦ κατὰ μὲν τὰς ἀρτηρίας εἰς
τὸν πλεύμονα καθῆκε θάτερον, τὸ δ᾽ εἰς τὴν κοιλίαν παρὰ τὰς
ἀρτηρίας. τὸ δ᾽ ἕτερον σχίσας τὸ μέρος ἑκάτερον κατὰ τοὺς
10 ὀχετοὺς τῆς ῥινὸς ἀφῆκε κοινόν, ὥσθ᾽ ὅτε μὴ κατὰ στόμα ἴοι
θάτερον, ἐκ τούτου πάντα καὶ τὰ ἐκείνου ῥεύματα ἀναπληροῦσθαι. D
τὸ δ᾽ ἄλλο κύτος τοῦ κύρτου περὶ τὸ σῶμα ὅσον κοῖλον ἡμῶν
περιέφυσε, καὶ πᾶν δὴ τοῦτο τοτὲ μὲν εἰς τὰ ἐγκύρτια ξυρρεῖν
μαλακῶς, ἅτε ἀέρα ὄντα, ἐποίησε, τοτὲ δὲ ἀναρρεῖν μὲν τὰ ἐγ-
15 κύρτια, τὸ δὲ πλέγμα, ὡς ὄντος τοῦ σώματος μανοῦ, δύεσθαι εἴσω
δι᾽ αὐτοῦ καὶ πάλιν ἔξω, τὰς δὲ ἐντὸς τοῦ πυρὸς ἀκτῖνας διαδε-

spectively the thoracic and abdominal cavities of the body: the first having a double outlet, one by the larynx, the other by the orifices of the nostrils: the second has one outlet only, through the oesophagus into the mouth. These preliminaries laid down, we shall be able to understand more or less precisely the remaining statements in the chapter. Martin's interpretation, which is most lucidly stated, would probably have been modified had the commentary of Galen in the original been before him.

I give a diagram, which, without aiming at anatomical accuracy, may perhaps help to elucidate Plato's meaning.

1. **διπλᾶ κατὰ τὴν εἴσοδον**] i.e. having two separate entrances, the windpipe and the oesophagus, one to each ἐγκύρτιον.

2. **διέπλεξε δίκρουν**] The ἐγκύρτιον occupying the cavity of the thorax he constructed with a double outlet, one by the larynx through the mouth, the other through the nostrils.

διετείνατο οἷον σχοίνους] Here Plato has departed somewhat from his analogy of the fishing-trap. The σχοῖνοι of course represent the arteries and veins which permeate the structure of the body.

3. **τὰ μὲν οὖν ἔνδον ἐκ πυρός**] This is the inner layer of the κύρτος, which, as we have seen, consisted of the vital heat contained in the solid part of the body lying between the surrounding air and the ἐγκύρτια, or cavities within.

6. **τὸ μὲν τῶν ἐγκυρτίων**] Galen warns us against taking this 'one of the ἐγκύρτια', in which case, as he justly remarks,

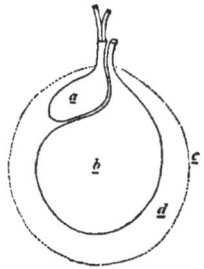

a. upper ἐγκύρτιον, opening into the mouth and bifurcating in the passages of the nostrils.
b. lower ἐγκύρτιον, opening into the mouth only.
c. κύτος τοῦ πλοκάνου, or stratum of air surrounding the body.
d. τὰ ἔνδον τοῦ πλοκάνου, or the heat residing in the solid part of the body.

D]   ΤΙΜΑΙΟΣ.   293

like a fish-trap or weel, having two lesser weels within with a double inlet; one of which inlets he again wove with two passages; and from the lesser weels he stretched as it were cords on all sides to the extremities of the network. All the inner part of the net he constructed of fire, but the lesser weels and the envelope he made of airy substance; and he took the net and wrapped it in manner following about the animal he had moulded. The structure of the lesser weels he carried into the mouth: and, these being twofold, he let down one of them by the windpipe into the lungs, the other past the windpipe into the belly. The one weel he split in two, and let both inlets meet by the passages of the nostrils, so that when the first inlet was not in action by way of the mouth, all its currents also might be replenished from the second. But with the general surface of the network he enveloped all the hollow part of our body; and all this, seeing it was air, he now caused to flow gently into the lesser weels, now made them flow back upon it; and since the body is of porous texture, the network passes through it inward and again outward, and the beams of fire

Plato would have gone on 'τὸ δὲ εἰς τόδε τι τοῦ σώματος'. He understands πλόκανον, in which he is probably right. The subdivision of the πλόκανον into the two ἐγκύρτια begins at διπλοῦ δὲ ὄντος αὐτοῦ.

7. τὰς ἀρτηρίας]' See the note on 70 C.

8. τὸ δὲ εἰς τὴν κοιλίαν] The other ἐγκύρτιον, occupying the abdominal cavity, had its outlet past the windpipe by way of the oesophagus: this had only one opening.

9. τὸ δ' ἕτερον] The ἐγκύρτιον which occupied the chest had a twofold outlet, one through the mouth, the other through the nose; and this latter was again divided into the two channels of the nostrils. The object of this double outlet was to allow respiration to be carried on through the nostrils when the passage by way of the mouth was not working, that we might not always have to open our mouths in order to breathe.

12. τὸ δ' ἄλλο κύτος] i.e. the stratum of air in contact with the body. This first, penetrating through the porous substance of the flesh, flows through it into the cavity of the ἐγκύρτια, the airy contents of which have passed up through the passages of respiration: presently the ἐγκύρτια flow down again into the body, and the air that had come in through the flesh passes forth again by the way that it came. The inner layer of the κύρτος, which was formed of fire, also oscillates to and fro, accompanying the motions of the airy envelope. And this oscillation must ceaselessly continue so long as we live. There are then two modes by which the air effects an entrance into the interior of the body: one by way of the tubes and orifices constructed for that purpose; the other through the substance of the body, which is too porous to bar its ingress, seeing that the flesh is partly constructed out of the coarser elements of water and earth.

16. τὰς δὲ ἐντὸς τοῦ πυρὸς ἀκτῖνας]

294 ΠΛΑΤΩΝΟΣ [78 D—

δεμένας ἀκολουθεῖν ἐφ' ἑκάτερα ἰόντος τοῦ ἀέρος, καὶ τοῦτο,
ἕωσπερ ἂν τὸ θνητὸν ξυνεστήκῃ ζῷον, μὴ διαπαύεσθαι γιγνόμε- E
νον· τούτῳ δὲ δὴ τῷ γένει τὸν τὰς ἐπωνυμίας θέμενον ἀναπνοὴν
καὶ ἐκπνοὴν λέγομεν θέσθαι τοὔνομα. πᾶν δὲ δὴ τό τ' ἔργον
5 καὶ τὸ πάθος τοῦθ' ἡμῶν τῷ σώματι γέγονεν ἀρδομένῳ καὶ
ἀναψυχομένῳ τρέφεσθαι καὶ ζῆν· ὁπόταν γὰρ εἴσω καὶ ἔξω τῆς
ἀναπνοῆς ἰούσης τὸ πῦρ ἐντὸς ξυνημμένον ἕπηται, διαιωρούμενον
δὲ ἀεὶ διὰ τῆς κοιλίας εἰσελθὸν τὰ σιτία καὶ ποτὰ λάβῃ, τήκει 79 A
δή, καὶ κατὰ σμικρὰ διαιροῦν, διὰ τῶν ἐξόδων ᾗπερ πορεύεται
10 διάγον, οἷον ἐκ κρήνης ἐπ' ὀχετοὺς ἐπὶ τὰς φλέβας ἀντλοῦν αὐτά,
ῥεῖν ὥσπερ αὐλῶνος διὰ τοῦ σώματος τὰ τῶν φλεβῶν ποιεῖ ῥεύματα.
XXXVI. Πάλιν δὲ τὸ τῆς ἀναπνοῆς ἴδωμεν πάθος, αἷς χρώ-

7 ἰούσης : οὔσης A.   11 αὐλῶνος διά : δι' αὐλῶνος S.

This is the same as τὰ ἔνδον τοῦ πλοκάνου above: i.e. the ἔμφυτον θερμόν, or vital heat residing in the substance of the body.

3. ἀναπνοὴν καὶ ἐκπνοήν] Plato uses the word ἀναπνοή for what was later termed εἰσπνοή, ἀναπνοή being reserved for the whole process of εἰσπνοή + ἐκπνοή. Aristotle uses ἀναπνοή similarly: *de respiratione* xxi 480b 9 καλεῖται δ' ἡ μὲν εἴσοδος τοῦ ἀέρος ἀναπνοή, ἡ δ' ἔξοδος ἐκπνοή. The dynamical cause of inspiration and expiration is explained in the next chapter.

5. ἀρδομένῳ καὶ ἀναψυχομένῳ] It would appear from this that Plato did regard respiration as serving the purpose of tempering the vital heat of the body: but this is a merely secondary object; its chief end being to effect the digestion of the food.

6. τῆς ἀναπνοῆς] Here ἀναπνοή is simply equivalent to the breath.

8. διὰ τῆς κοιλίας εἰσελθόν] The air and the fire which accompanies it, in the course of its oscillation to and fro, encounter the food which has been received into the body; and since it is composed of much finer particles than the latter, they penetrate and divide the food, converting it into blood (the red colour is due to the tinge imparted by fire as we find at 80 E); and then they drive the now fluid substance through the small vessels which they themselves permeate, and so pump it into the veins.

11. ὥσπερ αὐλῶνος] The body is compared to an aqueduct through which the veins pass as pipes or conduits irrigating all parts of it. The metaphor has become a little mixed here; above the body was likened to the κῆποι which had to be watered.

79 A—E, c. xxxvi. Let us more closely examine the conditions of the process described in the foregoing chapter. The cause of it is that there is no void space in the nature of things. Therefore when the breath issues forth of the mouth it thrusts against the neighbouring air, which transmits the impulse till it is received by the air in immediate contact with the body: this then forces its way in through the pores and replenishes the space within which the departing air leaves. Again this newly entered air, passing out once more through the pores of the body, in its turn thrusts the outside air and forces it to pass inward again through the passages of respiration to replenish the deserted space: and this process goes on continually, like a wheel turning to and fro. The cause of this oscillation is the vital heat which re-

which are confined within follow the air as it moves in either direction: and this never ceases to go on so long as the mortal creature holds together. To this process he who appointed names gave, we say, the titles of inspiration and expiration: and from this condition, both active and passive, it has come about that our body, deriving moisture and coolness, has its sustenance and life. For when, as the respiration passes in and out, the interwoven fire within follows it and entering the belly swings up and down and meets the food and drink, it dissolves them, and reducing them to small particles, drives them along the channels through which it flows, pumping them into the veins like spring-water into conduits, and so it makes the current of the veins flow through the body as through an aqueduct.

XXXVI. Let us once more examine the process of respira-

sides in the body. For the air within the body, being warmed thereby, rushes upward through the mouth and nose, and the cool air surrounding the body rushes in through the pores. Then this in its turn, becoming heated, rushes out through the pores, and the cool external air comes in through the passages of the breath. And thus a perpetual alternation of inspiration and expiration is kept up for the preservation of life.

Plato's theory then depends (1) upon his principle of περίωσις, by which he has explained the melting of metals &c, and by which in the next chapter he explains a variety of natural phenomena; (2) upon the vibration of the ὑποδοχή, which causes every element to strive towards its proper situation in space.

12. πάλιν δέ] Plato's account of respiration falls into two parts; in the first he simply describes the process, in the second he points out the physical causes of it. His theory bears a certain resemblance to that of Empedokles, which will be found in a passage quoted by Aristotle *de respiratione* vi 473^b 9, 275—299 Karsten. According to his statement, which is not very clear, the blood-vessels are only partially filled with blood; and when the blood rushes one way, the air follows through the pores into the body; when the blood moves in the other direction, the air is again expelled through the pores: this he illustrates by the analogy of a girl playing with a clepsydra; she covers the mouth with her hand and then plunges the instrument in water: the air, detained in the vessel by her hand, will not suffer the water to enter through the perforations; when she removes her hand the water enters at the bottom and expels the air through the mouth: similarly if the vessel is full of water, the air is unable to find entrance, but passes in as the water flows out.

Aristotle criticises Plato's theory in *de respiratione* v 472^b 6 foll.: it does not explain, he says, why only land animals breathe, or if fishes &c do so also, how they do it; again it assumes that ἐκπνοή is prior to εἰσπνοή, the contrary being the case; γίνεται μὲν γὰρ ταῦτα παρ' ἄλληλα, τελευτῶντες δὲ ἐκπνέουσιν, ὥστ' ἀναγκαῖον εἶναι τὴν ἀρχὴν εἰσπνοήν. Aristotle's own mechanical explanation is given in *de resp.* xxi 480^a 16. More

μενον αἰτίαις τοιοῦτον γέγονεν, οἷόνπερ τὰ νῦν ἐστίν. ὧδ' οὖν.
ἐπειδὴ κενὸν οὐδέν ἐστιν, εἰς ὃ τῶν φερομένων δύναιτ' ἂν εἰσελθεῖν
τι, τὸ δὲ πνεῦμα φέρεται παρ' ἡμῶν ἔξω, τὸ μετὰ τοῦτο ἤδη παντὶ
δῆλον, ὡς οὐκ εἰς κενόν, ἀλλὰ τὸ πλησίον ἐκ τῆς ἕδρας ὠθεῖ· τὸ
5 δ' ὠθούμενον ἐξελαύνει τὸ πλησίον ἀεί, καὶ κατὰ ταύτην τὴν
ἀνάγκην πᾶν περιελαυννόμενον εἰς τὴν ἕδραν, ὅθεν ἐξῆλθε τὸ
πνεῦμα, εἰσιὸν ἐκεῖσε καὶ ἀναπληροῦν αὐτὴν ξυνέπεται τῷ πνεύματι, καὶ τοῦτο ἅμα πᾶν οἷον τροχοῦ περιαγομένου γίγνεται διὰ
τὸ κενὸν μηδὲν εἶναι. διὸ δὴ τὸ τῶν στηθῶν καὶ τοῦ πλεύμονος
10 ἔξω μεθιὲν τὸ πνεῦμα πάλιν ὑπὸ τοῦ περὶ τὸ σῶμα ἀέρος, εἴσω
διὰ μανῶν τῶν σαρκῶν δυομένου καὶ περιελαυνομένου, γίγνεται
πλῆρες· αὖθις δὲ ἀποτρεπόμενος ὁ ἀὴρ καὶ διὰ τοῦ σώματος ἔξω
ἰὼν εἴσω τὴν ἀναπνοὴν περιωθεῖ κατὰ τὴν τοῦ στόματος καὶ τὴν
τῶν μυκτήρων δίοδον. τὴν δὲ αἰτίαν τῆς ἀρχῆς αὐτῶν θετέον
15 τήνδε· πᾶν ζῷον ἑαυτοῦ τἀντὸς περὶ τὸ αἷμα καὶ τὰς φλέβας
θερμότατα ἔχει, οἷον ἐν ἑαυτῷ πηγήν τινα ἐνοῦσαν πυρός· ὃ δὴ
καὶ προσεικάζομεν τῷ τοῦ κύρτου πλέγματι, κατὰ μέσον διατε-
ταμένον ἐκ πυρὸς πεπλέχθαι πᾶν, τὰ δὲ ἄλλα, ὅσα ἔξωθεν, ἀέρος.
τὸ θερμὸν δὴ κατὰ φύσιν εἰς τὴν αὑτοῦ χώραν ἔξω πρὸς τὸ
20 ξυγγενὲς ὁμολογητέον ἰέναι· δυοῖν δὲ ταῖν διεξόδοιν οὔσαιν, τῆς
μὲν κατὰ τὸ σῶμα ἔξω, τῆς δὲ αὖ κατὰ τὸ στόμα καὶ τὰς ῥῖνας,
ὅταν μὲν ἐπὶ θάτερα ὁρμήσῃ, θάτερα περιωθεῖ· τὸ δὲ περιωσθὲν

9 τὸ ante τοῦ πλεύμονος dant SZ.   15 ἑαυτοῦ: αὑτοῦ SZ.
τἀντός : πάντως A.   16 θερμότατα : θερμότητα A.   20 δυοῖν : δυεῖν S.

cogent arguments against the Platonic account are adduced by Galen *de plac. Hipp. et Plat.* VIII 708 foll.; his chief objection being that Plato ignores respiration as a voluntary action; also Galen prefers ὁλκή to περίωσις as its cause.

6. **περιελαυννόμενον**] The outside air receives as a whole an impulse from the breath essaying to issue forth. Now the only region in which it is possible for it to yield to this impulse is that which is being vacated by the issuing air. It matters not therefore in what direction the originating impulse is given: if room is to be found outside the body for the breath as it comes forth, it must be by an equal quantity of air entering the cavity which it quits.

8. **τροχοῦ περιαγομένου**] The 'wheel' does not move in continuous revolution, but alternately describes first a semicircle forward then a semicircle backward *usque ad infinitum*: cf. Galen *de plac.* VIII 711.

14. **τὴν δὲ αἰτίαν τῆς ἀρχῆς**] Hitherto the περίωσις has been the physical law alleged; now comes in the other principle, the vibration of the ὑποδοχή, which is the primary motive power producing respiration. The original motion is due to the fire within the body which constitutes its vital heat. The air within the ἐγκύρτια, coming in contact with this fire, becomes heated; that is, is mingled

tion and the causes which have led to its present conditions. These are as follows. Since there is no void into which any moving body could enter, and since the breath issues forth from us, the consequence is clear to every one: instead of entering into a void space it thrusts the neighbouring matter out of its place. And this, yielding to the thrust, drives before it that which is immediately nearest; and all being driven round by this compulsion enters into the place whence the breath came forth, and replenishing the same follows after the breath; and this whole process goes on like the rotation of a wheel, because there is no void. Therefore when the cavity of the chest and the lungs send forth the breath, they are again replenished by the air surrounding the body, which penetrates inwards through the flesh, seeing it is porous, and is forced round in a circuit. And again when the air returns and passes forth through the body, it thrusts the breath back again inwards through the passages of the mouth and nostrils. The cause which sets this principle in action we may describe thus. In every animal the inner parts about the blood and veins are the hottest, as if there were a fount of fire contained in it. This is what we compare to the network of the weel, supposing that all the part extending from the middle to the sides is woven of fire, but the outer part of air. Now we must admit that the heat naturally tends outwards to its own region and its own kin. And whereas there are two means of egress, one out through the body, the other by way of the mouth and nostrils, when it makes for one exit, it impels the air round towards the other. And the air so impelled falling into the fire

with fire. Now fire, as we know, ever seeks to escape upwards to its own region; therefore the mixture of air and fire is impelled to quit the body in search of its own kind. This it may do by either of two outlets—by penetrating through the porous substance of the body, or by passing upward through the respiratory passages. Whichever of these passages it selects, it thrusts against the air outside, and each particle of air pressing upon its neighbour, the air nearest the body is forced into the body by the other entrance. The original impulse then is given by the fire in the body seeking to escape to its own kindred element.

17. **προσεικάζομεν τῷ τοῦ κύρτου πλέγματι**] This seems sufficiently to confirm the explanation of the κύρτος given above, and the identification of the inner layer thereof with the vital heat which by means of the blood-vessels pervades all the substance of the body.

298 ΠΛΑΤΩΝΟΣ [79 E—

εἰς τὸ πῦρ ἐμπῖπτον θερμαίνεται, τὸ δ' ἐξιὸν ψύχεται. μεταβαλλούσης δὲ τῆς θερμότητος καὶ τῶν κατὰ τὴν ἑτέραν ἔξοδον θερμοτέρων γιγνομένων πάλιν ἐκείνῃ ῥέπον αὖ τὸ θερμότερον μᾶλλον, πρὸς τὴν αὑτοῦ φύσιν φερόμενον, περιωθεῖ τὸ κατὰ
5 θάτερα· τὸ δὲ τὰ αὐτὰ πάσχον καὶ τὰ αὐτὰ ἀνταποδιδὸν ἀεί, κύκλον οὕτω σαλευόμενον ἔνθα καὶ ἔνθα ἀπειργασμένον ὑπ' ἀμφοτέρων τὴν ἀναπνοὴν καὶ ἐκπνοὴν γίγνεσθαι παρέχεται.

XXXVII. Καὶ δὴ καὶ τὰ τῶν περὶ τὰς ἰατρικὰς σικύας παθημάτων αἴτια καὶ τὰ τῆς καταπόσεως τά τε τῶν ῥιπτουμένων, 80 A
10 ὅσα ἀφεθέντα μετέωρα καὶ ὅσα ἐπὶ γῆς φέρεται, ταύτῃ διωκτέον, καὶ ὅσοι φθόγγοι ταχεῖς τε καὶ βραδεῖς ὀξεῖς τε καὶ βαρεῖς

6 κύκλον: κύκλῳ S.

1. **μεταβαλλούσης δὲ τῆς θερμότητος**] So far as the theory has yet been set forth, no reason has been assigned why the heated air escapes alternately through the respiratory passages and through the pores of the body; the wheel might always turn in the same direction. Plato now endeavours to supply a cause for this: but it must be confessed that, if I rightly apprehend his meaning, it is a very inadequate one: however it seems to be as follows. Let us suppose the process to be at this point, that the heated air in the ἐγκύρτια has just passed up through the trachea into the outer atmosphere; accordingly the cool stratum of air surrounding the body has passed in through the pores to supply its place. Now why should this newly entered air, when it in its turn is heated and endeavours to escape, return through the body instead of following its predecessor up the trachea? The reason assigned is this: the warm air on passing forth out of the mouth or nostrils finds itself plunged in the cool atmosphere without; at the same time the air newly arrived in the body is heated. The preponderance of warmth is now in the neighbourhood of the outlet through the flesh: the heated air therefore seeks the nearest and easiest way of escape by passing outward through the pores of the body, as it had entered; whereupon the περίωσις sends a current of air down the respiratory passages. Then precisely the same process takes place at the other entrance: the air that entered through the trachea is warmed, and likewise seeks to escape by the nearest outlet, viz. the trachea. Thus the air that passes into the body by either entrance is always impelled to return by that same entrance and not by the other. But this part of the theory is both obscure and unsatisfactory, unless some better interpretation of it can be found. Plato's hypothesis, it will be observed, renders the process entirely independent of any muscular action of the body; and Galen's criticism is pertinent: ἐν οὐδετέρᾳ δὲ αὐτῶν ὁ Πλάτων προσχρῆται τῇ προαιρέσει, καίτοι φανερῶς ἐν ἡμῖν ὄντος καὶ τὸ θᾶττον καὶ βραδύτερον ἔλαττόν τε καὶ πλέον καὶ πυκνότερον εἰσπνεῦσαί τε καὶ ἐκπνεῦσαι.

79 A—80 C, c. xxxvii. The same principle of circular impulsion will account for the action of cupping-glasses, for the process of swallowing, for the motion of projected bodies, whether through the air or along the ground, and for the consonance of high and deep notes, which is produced by the gradual retardation of the swifter sound until it coincides with the motion of the slower. To the same cause is due the flowing of water, the falling of the thunderbolt, and the force

is heated, but that which passes out is cooled. So the heat changes its position and the parts about the other outlet become warmer; therefore the heat now has a stronger tendency in the new direction, seeking its own affinity, and impels the air by the other passage: and this, undergoing the same change and reproducing the same process, is thus by these two impulses converted into a wheel swaying backwards and forwards, and so it gives rise to respiration.

XXXVII. In the same direction are we to look for the explanation of the phenomena of medical cupping-glasses and of swallowing and of projected bodies, whether cast through the air or moving along the ground; and of sounds too, which from their swiftness and slowness seem to us shrill or deep,

of attraction exercised by amber and the loadstone. All these diverse phenomena are due to the manifold interaction of these two principles—the absence of void, which is the cause of the circular impulsion, and the vibratory motion which causes every substance to strive towards its own peculiar region in space.

8. περὶ τὰς ἰατρικὰς σικύας] Plato now applies his two great dynamical principles to the explanation of various natural phenomena. He does not work out the mode of their operation in detail, but leaves that to be done by the reader. A full commentary on the present chapter will be found in Plutarch *quaestiones platonicae* vii. The explanation of the cupping instruments is this. When the cup is applied to the flesh, the air within it becomes warmed and consequently dilated; and escaping through the pores of the metal, it thrusts the surrounding air, which in its turn, pressing on the surface of the body, forces the humours to exude into the cup: cf. *Timaeus Locrus* 102 A.

9. τὰ τῆς καταπόσεως] The food, propelled downwards by the muscles of the throat, thrusts the air in front of it: this, escaping through the pores, thrusts the air outside, which by the περίωσις presses upon the food from behind and pushes it downward: and since at every moment of its progress more air is displaced to set the περίωσις in motion, the downward impulse is continually maintained.

τά τε τῶν ῥιπτουμένων] The process is the same here as in the preceding instance: if a stone is hurled through the air, the air displaced in front of the stone sets up a περίωσις which impels it behind and keeps it going. The problem which seemed to the ancient thinkers to demand solution was, when the stone has left the hand of the thrower and consequently is no longer directly receiving any propulsion from it, what is it that keeps the stone moving? what enables it to withstand the force of gravitation which would otherwise cause it to fall perpendicularly earthward? A clear understanding of the point of view from which this question was regarded will be gained from Aristotle *physica* VIII x 266[b] 27 foll. Aristotle, who seems to adopt Plato's explanation, remarks that the propelling hand communicates to the stone not only passive motion, but an active power of moving the air before it: it ceases to be κινούμενον at the moment it leaves the hand (relatively to the hand, Aristotle should have added), but remains κινοῦν so long as it is in motion.

11. καὶ ὅσοι φθόγγοι] It is not at

ΠΛΑΤΩΝΟΣ [80 A—

φαίνονται, τοτὲ μὲν ἀνάρμοστοι φερόμενοι δι' ἀνομοιότητα τῆς
ἐν ἡμῖν ὑπ' αὐτῶν κινήσεως, τοτὲ δὲ ξύμφωνοι δι' ὁμοιότητα. τὰς
γὰρ τῶν προτέρων καὶ θαττόνων οἱ βραδύτεροι κινήσεις ἀποπαυο-
μένας ἤδη τε εἰς ὅμοιον ἐληλυθυίας, αἷς ὕστερον αὐτοὶ προσφερό- B
5 μενοι κινοῦσιν ἐκείνας, καταλαμβάνουσι, καταλαμβάνοντες δὲ οὐκ
ἄλλην ἐπεμβάλλοντες ἀνετάραξαν κίνησιν, ἀλλ' ἀρχὴν βραδυτέρας
φορᾶς κατὰ τὴν τῆς θάττονος ἀπoληγούσης δὲ ὁμοιότητα προσ-
άψαντες μίαν ἐξ ὀξείας καὶ βαρείας ξυνεκεράσαντο πάθην· ὅθεν
ἡδονὴν μὲν τοῖς ἄφροσιν, εὐφροσύνην δὲ τοῖς ἔμφροσι διὰ τὴν
10 τῆς θείας ἁρμονίας μίμησιν ἐν θνηταῖς γενομένην φοραῖς παρέσχον.
καὶ δὴ καὶ τὰ τῶν ὑδάτων πάντα ῥεύματα, ἔτι δὲ τὰ τῶν κεραυνῶν C
πτώματα καὶ τὰ θαυμαζόμενα ἠλέκτρων περὶ τῆς ἕλξεως καὶ

first obvious how the principle of περίωσις applies here. But I think it is clear
that Plato does not mean the περίωσις
to account for the consonance of different
sounds, but only for their propagation
from the sounding body to the ear. This
is effected in exactly the same way as the
projection of a stone through the air.
Sound is produced by the vibration of a
certain body of air, or of some other conducting medium: it is propagated by the
transmission of this vibration, or rather,
on Plato's theory, of this vibrating body
of air through the atmosphere; for it,
like the stone, displaces the air in front,
which keeps perpetually rushing in and
propelling it behind. This interpretation differs from that given by Plutarch
quaestiones platonicae vii 9, which is, I
think, unquestionably erroneous. He
supposes the περίωσις to account for the
consonance of high and deep notes, and
explains it thus: the acuter sound, travelling faster than the deeper, strikes first
upon the ear; then passing round by the
περίωσις, but with gradually diminishing
speed, it overtakes the slower, and assimilating its motion to that of the latter
reaches the ear again along with it: ὁ δὴ
σφόδρα καὶ συντόνως πληγεὶς προσμίγνυσι
τῇ ἀκοῇ πρῶτος, εἶτα περιιὼν πάλιν καὶ
καταλαμβάνων τὸν βραδύτερον συνέπεται
καὶ συμπαραπέμπει τὴν αἴσθησιν. But

there are grave objections to be brought
against this: (1) it is a totally illegitimate
use of the περίωσις: it is as if a stone
hurled in the air should describe a circular
orbit; (2) Plutarch makes the swifter
sound overtake the slower; but Plato distinctly speaks of the slower overtaking
the swifter, when the latter is relaxing its
speed. If however we suppose the περίωσις to be accountable merely for the
transmission of the sounds, the explanation as above is quite plain and simple;
and for the consonance it is not wanted.
Compare Aristotle *de audibilibus* 804[a] 4
foll.

2. τὰς γὰρ τῶν προτέρων] The cause
of consonance, according to Plato, is this.
If a high and a low note be sounded together, the high note, which travels more
swiftly through the air, will reach the
ear first and communicate its vibrations
to it. Presently the deeper note arrives.
But by that time the vibrations of the
higher note, which have been gradually
becoming slower, are synchronous with
the vibrations added by the deeper note,
and a consonance ensues. If the vibrations of the higher note have not slackened down to the speed of the lower, discord is the result instead of concord:
thus if we strike simultaneously two notes
at the interval of a semitone, a sharp discord is produced, because the two sounds

sometimes having no harmony in their movements owing to the irregularity of the vibrations they produce in us, sometimes being harmonious through regularity. For the slower sounds overtake the motions of the first and swifter sounds, when these are already beginning to die away and have become assimilated to the motions which the slower on their arrival impart to them: and on overtaking them they do not produce discord by the intrusion of an alien movement, but adding the commencement of a slower motion, which corresponds to that of the swifter now that the latter is beginning to cease, they form one harmonious sensation by the blending of shrill and deep. Thereby they afford pleasure to the foolish, but to the wise joy, through the imitation of the divine harmony which is given by mortal motions. And the flowing of all waters, the fall of thunderbolts, and the wonderful attracting power of

are so nearly of the same pitch that the lower reaches the ear before the higher has had time to slacken at all. It is evident from Plato's language that he conceived the acuter sound both to travel more swiftly through the air and to have more rapid vibrations: he thus comes very near the correct explanation of pitch, but falls into the not unnatural error of supposing that the more rapid vibration causes a swifter progress through the air. His theory of consonance is entirely unsatisfactory: apart from any other objection, the process he describes could only produce unison, not concord. For he cannot mean merely that the swifter vibrations slackened down so as to produce a due numerical ratio to the slower, since such a numerical ratio might have as well existed at first. It is strange that Plato, with his fondness for ἀναλογία, should not have based harmony of accords upon this. It will be observed that the principle of περίωσις is in no way concerned with the present hypothesis.

9. ἡδονὴν μὲν τοῖς ἄφροσιν] See note on 47 D. The ἔμφρονες enjoy music because they recognise that it is based on the same harmonic ratios as are found in the soul: in plainer language, because it expresses to the ear truths of the unseen world. For εὐφροσύνην compare *Cratylus* 419 D παντὶ γὰρ δῆλον ὡς ἀπὸ τοῦ εὖ ἐν τοῖς πράγμασιν τὴν ψυχὴν ξυμφέρεσθαι τοῦτο ἔλαβε τὸ ὄνομα, εὐφεροσύνην. The word expresses a calm enjoyment, different from the undisciplined pleasure of the multitude, the ἄπειρος ἡδονὴ beloved of Philebus.

11. τὰ τῶν ὑδάτων πάντα ῥεύματα] The cause of the flowing of water is pretty much the same as that alleged in 58 E for the flowing of molten metal, except that here we have to assume the original impulse, which there is explained. It seems strange that Plato makes no use here of the force of gravitation: perhaps that is assumed as obviously auxiliary; and this chapter is but an exceedingly brief summary.

τῶν κεραυνῶν πτώματα] The action in this instance is precisely identical with that in the case of the projection of a stone through the air.

12. τὰ θαυμαζόμενα ἠλέκτρων] The explanation given by Plutarch is as follows. Amber contains within it something φλογοειδὲς ἢ πνευματικόν, a rare and

τῶν Ἡρακλείων λίθων, πάντων τούτων ὁλκὴ μὲν οὐκ ἔστιν οὐδενί
ποτε, τὸ δὲ κενὸν εἶναι μηδὲν περιωθεῖν τε αὐτὰ ταῦτα εἰς ἄλληλα,
τό τε διακρινόμενα καὶ συγκρινόμενα πρὸς τὴν αὑτῶν διαμειβόμενα
ἕδραν ἕκαστ᾽ ἰέναι πάντα, τούτοις τοῖς παθήμασι πρὸς ἄλληλα
5 συμπλεχθεῖσι τεθαυματουργημένα τῷ κατὰ τρόπον ζητοῦντι
φανήσεται.

XXXVIII. Καὶ δὴ καὶ τὸ τῆς ἀναπνοῆς, ὅθεν ὁ λόγος ὥρμησε, D
κατὰ ταῦτα καὶ διὰ τούτων γέγονεν, ὥσπερ ἐν τοῖς πρόσθεν
εἴρηται, τέμνοντος μὲν τὰ σιτία τοῦ πυρός, αἰωρουμένῳ δὲ ἐντὸς
10 τῷ πνεύματι ξυνεπομένου, τὰς φλέβας τε ἐκ τῆς κοιλίας τῇ
ξυναιωρήσει πληροῦντος τῷ τὰ τετμημένα αὐτόθεν ἐπαντλεῖν·
καὶ διὰ ταῦτα δὴ καθ᾽ ὅλον τὸ σῶμα πᾶσι τοῖς ζῴοις τὰ τῆς

4 ἕκαστ᾽: ἕκαστα S.   8 ταῦτα: ταὐτὰ AH.
9 αἰωρουμένῳ coniecit H. αἰωρουμένου ΛSZ.   10 τῇ: τε Λ.

subtle substance, which is released by friction, the pores of the amber being expanded. This substance on escaping and coming into collision with the adjacent air sets up a περίωσις: and the air impinging from behind drives before it any light object in the vicinity, until it reaches the electrified piece of amber. Theophrastos seems to confound amber with the loadstone: *de lapidibus* § 29 ἐπεὶ δὲ καὶ τὸ ἤλεκτρον λίθος...μάλιστα δ᾽ ἐπίδηλος καὶ φανερωτάτη ἡ τὸν σίδηρον ἄγουσα.

1. τῶν Ἡρακλείων λίθων] This name is said to have been given to the loadstone from the town of Herakleia in Lydia. Plato's theory of the magnet is very much the same as in the case of the amber. There stream off from the magnet large and heavy particles of air, which, in the περίωσις that they occasion, themselves strike upon the iron and drive it towards the magnet. The reason why iron alone is so influenced is, according to Plutarch, that iron, being more dense than wood but less so than gold and other metals, has its pores of exactly the right size to retain the particles of air, which thus, instead of slipping off as they do in the case of other substances, propel the iron before them.

A peculiarity in this theory is that the air which escapes from the magnet itself is returned to it by the περίωσις: this is necessitated by the fact that iron and nothing else is attracted, iron being amenable to that particular kind of air alone. It is possible however that Plutarch may not have exactly represented Plato's meaning. On the subject of the loadstone compare *Ion* 533 D ὥσπερ ἐν τῇ λίθῳ, ἣν Εὐριπίδης μὲν Μαγνῆτιν ὠνόμασεν, οἱ δὲ πολλοὶ Ἡρακλείαν. καὶ γὰρ αὕτη ἡ λίθος οὐ μόνον αὐτοὺς τοὺς δακτυλίους ἄγει τοὺς σιδηροῦς, ἀλλὰ καὶ δύναμιν ἐντίθησι τοῖς δακτυλίοις, ὥστ᾽ αὖ δύνασθαι ταὐτὸν τοῦτο ποιεῖν ὅπερ ἡ λίθος, ἄλλους ἄγειν δακτυλίους, ὥστ᾽ ἐνίοτε ὁρμαθὸς μακρὸς πάνυ σιδηρῶν δακτυλίων ἐξ ἀλλήλων ἤρτηται· πᾶσι δὲ τούτοις ἐξ ἐκείνης τῆς λίθου ἡ δύναμις ἀνήρτηται. Compare also Lucretius VI 998—1064.

ὁλκὴ μὲν οὐκ ἔστιν] It is this denial of ὁλκή which Galen chiefly complains of in Plato's physics: *de plac.* VIII 708 ἀναιρεῖ γὰρ ὁλκήν, ᾗ πρὸς πολλὰ τῶν φυσικῶν ἔργων ὁ Ἱπποκράτης χρῆται. διὰ τοῦτο ἠναγκάσθη τῶν ἐνεργειῶν ἐνίας οὐκ ἄνευ τῆς ὁλκῆς γινομένας εἰς περίωσιν ἀναφέρειν.

3. τό τε διακρινόμενα] i.e. under the pressure of the πλῆσις the various bodies

ΤΙΜΑΙΟΣ.

amber and of the loadstone—all these are due to no drawing power, but to two causes: first there is no void, and the atoms jostle one upon another; secondly when they are divided or contracted they change places and move severally towards their own region; and by the complication of such conditions all these wonders arise, as will be plain to him who examines them by the proper method.

XXXVIII. The process of respiration then, whence this discussion arose, rests on the principles and causes which have been set forth: fire divides the food, following the air as it sways up and down within; and through this oscillation it replenishes the veins from the belly by pumping into them from thence the comminuted food. In this way throughout the whole body of all animals the streams of nourishment are kept con-

are constantly changing their form and their appropriate region in space. The text can hardly be sound here.

5. τεθαυματουργημένα] Owing to the endless complexity and intricacy of the interaction which these two forces exert upon one another, many of their effects appear to us marvellous, because we have not the means of tracing the conditions which gave rise to them. Compare *Laws* 893 D διὸ δὴ τῶν θαυμαστῶν ἀπάντων πηγὴ γέγονεν, ἅμα μεγάλοις καὶ σμικροῖς κύκλοις βραδυτῆτάς τε καὶ τάχη ὁμολογούμενα πορεύουσα, ἀδύνατον ὡς ἄν τις ἐλπίσειε γίγνεσθαι πάθος.

80 D—81 E, *c.* xxxviii. Respiration then is subsidiary to digestion: the fire which accompanies the oscillation of the air comminutes the food, which is then pumped into the blood-vessels and distributed throughout the body. The nutriment, consisting as it does of different kinds of vegetables, has naturally a variety of hues; but the action of the fire reduces it all to a predominant red colour. Now the microcosm of the human body has its motions conformable to those of the great universe: the law that like seeks to like holds good of it also. So as the substance of our bodies is continually being dissolved and evaporated by the action of the external elements, the food that is assimilated by virtue of this natural law proceeds to replenish the void left by that which is lost: and the body increases or diminishes according as the replenishment exceeds or falls short of the waste. In a young child the substance of the body, though soft, has its triangles true and sharp: therefore they readily overpower and assimilate the blunter triangles of the nutriment; but as time goes on, the triangles are blunted and cannot so well subdue the others; whence is old age and decay. Finally when the triangles of the vital marrow can no more hold out, the bonds of the soul are loosed, and she flies away rejoicing: for though death which comes by wounds or sickness is painful, when it is the result of natural decay it is painless and brings pleasure rather than distress.

7. ὅθεν ὁ λόγος ὥρμησε] i.e. the exposition of the law of περίωσις. Plato now passes from respiration to the processes of nutrition, growth, decay and death. It seems to me that κατὰ ταῦτα is clearly to be preferred over κατὰ ταὐτὰ for the sake of symmetry.

11. ἐπαντλεῖν] See above, 79 A.

304   ΠΛΑΤΩΝΟΣ   [80 E—

τροφῆς νάματα οὕτως ἐπίρρυτα γέγονε. νεότμητα δὲ καὶ ἀπὸ
ξυγγενῶν ὄντα, τὰ μὲν καρπῶν, τὰ δὲ χλόης, ἃ θεὸς ἐπ' αὐτὸ Ε
τοῦθ' ἡμῖν ἐφύτευσεν εἶναι τροφήν, παντοδαπὰ μὲν χρώματα ἴσχει
διὰ τὴν ξύμμιξιν, ἡ δ' ἐρυθρὰ πλείστη περὶ αὐτὸ χρόα διαθεῖ,
5 τῆς τοῦ πυρὸς τομῆς τε καὶ ἐξομόρξεως ἐν ὑγρῷ δεδημιουργημένη
φύσις· ὅθεν τοῦ κατὰ τὸ σῶμα ῥέοντος τὸ χρῶμα ἔσχεν οἵαν ὄψιν
διεληλύθαμεν. ὃ καλοῦμεν αἷμα, νομὴν σαρκῶν καὶ ξύμπαντος
τοῦ σώματος, ὅθεν ὑδρευόμενα ἕκαστα πληροῖ τὴν τοῦ κενουμένου 81 Α
βάσιν. ὁ δὲ τρόπος τῆς πληρώσεως ἀποχωρήσεώς τε γίγνεται,
10 καθάπερ ἐν τῷ παντὶ παντὸς ἡ φορὰ γέγονεν, ἣν τὸ ξυγγενὲς
πᾶν φέρεται πρὸς ἑαυτό. τὰ μὲν γὰρ δὴ περιεστῶτα ἐκτὸς ἡμᾶς
τήκει τε ἀεὶ καὶ διανέμει πρὸς ἕκαστον εἶδος τὸ ὁμόφυλον ἀποπέμποντα, τὰ δὲ ἔναιμα αὖ, κερματισθέντα ἐντὸς παρ' ἡμῖν καὶ
περιειλημμένα ὥσπερ ὑπ' οὐρανοῦ ξυνεστῶτος ἑκάστου τοῦ ζῴου,
15 τὴν τοῦ παντὸς ἀναγκάζεται μιμεῖσθαι φοράν· πρὸς τὸ ξυγγενὲς Β
οὖν φερόμενον ἕκαστον τῶν ἐντὸς μερισθέντων τὸ κενωθὲν τότε
πάλιν ἀνεπλήρωσεν. ὅταν μὲν δὴ πλέον τοῦ ἐπιρρέοντος ἀπίῃ,
φθίνει πᾶν, ὅταν δὲ ἔλαττον, αὐξάνεται. νέα μὲν οὖν ξύστασις

1 γέγονε: γεγονέναι ASZ.   12 ἀποπέμποντα: ἀποπέμπον ASZ.
15 τοῦ ante παντὸς delet A.

1. **ἐπίρρυτα γέγονε**] cf. 43 A ἐπίρρυτον σῶμα καὶ ἀπόρρυτον.
ἀπὸ ξυγγενῶν] i.e. composed of the
same elements. On the subject of vegetable diet see note on 77 A.
5. **τῆς τοῦ πυρὸς τομῆς τε καὶ ἐξομόρξεως**] See the account of the γένεσις
of red in 68 B. The colour of the blood
is due to the commingling of fire and
moisture: the fire, as it were, prints off
(ἐξομόργνυται) its own colour on the blood,
effacing the other hues.
8. **τὴν τοῦ κενουμένου βάσιν**] i.e.
the place left vacant by the particles
flying off in the natural process of waste.
βάσιν = τὸ ἐφ' ᾧ βέβηκε, the spot in which
it rests.
9. **ὁ δὲ τρόπος τῆς πληρώσεως**] Plato
conceives the human microcosm to work
on just the same principles as the οὐρανὸς
in which it has its being. The vibration
of the ὑποδοχὴ is the force which governs
the circulation of the blood. By the action of the elements which surround us
the substance of the body is perpetually
undergoing transmutation and depletion.
This body is to the blood within it as it
were an enclosing οὐρανός; and as changes
take place in its substance, the blood is
drawn to and fro according to the affinities
of its particles. Each change that takes
place in any part of the body affects the
affinity of the blood towards that part,
and consequently its tendency to flow in
that direction. Accordingly, as changes
are continually going on in all parts of
the body, the blood is constantly being
hurried to and fro throughout its whole
extent. This action is further supple-
mented by the principle of περίωσις. For
as fast as any vacancy is created by the
waste of the particles which are absorbed
by the surrounding elements, the blood
must rush in to take its place: whence
arises the necessity for a continual supply
of aliment. Such seems to be Plato's

81 B]   ΤΙΜΑΙΟΣ.   305

stantly supplied. And the particles of food, being freshly severed and from kindred substances—some from fruits and some from herbs, which God planted just to be our sustenance,—have all manner of colours owing to their intermixture; but a red hue pervades them most of all, through the natural contrivance whereby the fire divides the food and imprints its own hue upon it: whence the colour of the fluid that circulates through the body has the appearance we have described. This we call blood, which is the sustenance of the flesh and of all the body, and from which all parts draw moisture to fill up the places that are left void. And the mode of replenishment and evacuation is like the motion of all things in the universe, whereby all kindred substances seek each other. The elements that surround us without are constantly dissolving our substance and distributing it to its several kinds, returning each to its own kindred: and again the particles of blood, being minutely divided within us and enveloped in every creature by the body, as though by a heaven surrounding them, are forced to copy the universal motion. Therefore each of the divided particles within us is carried to its own kind and thus replenishes again what was left void. Now when the loss is greater than the replenishment, everything diminishes, but when less, it increases. The young

general meaning: but the exact part played respectively by the two principles of 'like seeks to like' and the περίωσις is not very clearly indicated.

11. τὰ μὲν γὰρ δὴ περιεστῶτα] The surrounding elements are conceived to have a solvent effect upon the body: they convert icosahedrons into octahedrons, and so forth. Consequently these particles, on changing their forms, change their natural homes, and flying off πρὸς τὸ ὁμόφυλον, leave a deficiency in the substance of the body.

15. πρὸς τὸ ξυγγενές] i.e. the particles of the blood which are akin to those of any special portion of the body flow thither so soon as room is made for them by the efflux of any particles from that spot.

18. νέα μὲν οὖν ξύστασις] Now fol-

lows the account of αὔξησις and φθίσις. When the human frame is still young, the particles of which it is composed, and especially those of the vital fire, have all their angles true and keen. The particles whereof the nutriment is formed are, on the contrary, comparatively blunt through age; hence the fiery particles have no difficulty in dividing them and performing the work described at 79 A. Consequently the food is very thoroughly assimilated and dispersed throughout the body, and the child grows apace. Notwithstanding the minute elaboration of this and several previous chapters, we read in Aristotle *de gen. et corr.* I ii 315ᵃ 29 Πλάτων μὲν οὖν μόνον περὶ γενέσεως ἐσκέψατο καὶ φθορᾶς, ὅπως ὑπάρχει τοῖς πράγμασι, καὶ περὶ γενέσεως οὐ πάσης, ἀλλὰ τῆς τῶν στοιχείων· πῶς δὲ σάρκες

P. T.   20

τοῦ παντὸς ζῴου, καινὰ τὰ τρίγωνα οἷον ἐκ δρυόχων ἔτι ἔχουσα τῶν γενῶν, ἰσχυρὰν μὲν τὴν ξύγκλεισιν αὐτῶν πρὸς ἄλληλα κέκτηται, ξυμπέπηγε δὲ ὁ πᾶς ὄγκος αὐτῆς ἁπαλός, ἅτ᾽ ἐκ μυελοῦ μὲν νεωστὶ γεγονυίας, τεθραμμένης δὲ ἐν γάλακτι. τὰ δὴ περιλαμ-
5 βανόμενα ἐν αὐτῇ τρίγωνα ἔξωθεν ἐπεισελθόντα, ἐξ ὧν ἂν ᾖ τά τε σιτία καὶ ποτά, τῶν ἑαυτῆς τριγώνων παλαιότερα ὄντα καὶ ἀσθενέστερα καινοῖς ἐπικρατεῖ τέμνουσα, καὶ μέγα ἀπεργάζεται τὸ ζῷον τρέφουσα ἐκ πολλῶν ὁμοίων. ὅταν δ᾽ ἡ ῥίζα τῶν τριγώνων χαλᾷ διὰ τὸ πολλοὺς ἀγῶνας ἐν πολλῷ χρόνῳ πρὸς πολλὰ
10 ἠγωνίσθαι, τὰ μὲν τῆς τροφῆς εἰσιόντα οὐκέτι δύναται τέμνειν εἰς ὁμοιότητα ἑαυτοῖς, αὐτὰ δὲ ὑπὸ τῶν ἔξωθεν ἐπεισιόντων εὐπετῶς διαιρεῖται· φθίνει δὴ πᾶν ζῷον ἐν τούτῳ κρατούμενον, γῆράς τε ὀνομάζεται τὸ πάθος. τέλος δέ, ἐπειδὰν τῶν περὶ τὸν μυελὸν τριγώνων οἱ ξυναρμοσθέντες μηκέτι ἀντέχωσι δεσμοὶ τῷ πόνῳ
15 διιστάμενοι, μεθιᾶσι τοὺς τῆς ψυχῆς αὖ δεσμούς, ἡ δὲ λυθεῖσα κατὰ φύσιν μεθ᾽ ἡδονῆς ἐξέπτατο. πᾶν γὰρ τὸ μὲν παρὰ φύσιν ἀλγεινόν, τὸ δ᾽ ᾗ πέφυκε γιγνόμενον ἡδύ· καὶ θάνατος δὴ κατὰ ταῦτα ὁ μὲν κατὰ νόσους καὶ ὑπὸ τραυμάτων γιγνόμενος ἀλγεινὸς καὶ βίαιος, ὁ δὲ μετὰ γήρως ἰὼν ἐπὶ τέλος κατὰ φύσιν ἀπονώτατος
20 τῶν θανάτων καὶ μᾶλλον μεθ᾽ ἡδονῆς γιγνόμενος ἢ λύπης.

5 ἐν αὐτῇ: ἑαυτῆς A.   δὲ post ἐπεισελθόντα inserit A.
15 διιστάμενοι: διεσταμένοι A. διεσταμένοι HSZ.   19 γήρως: γῆρας SZ.

ἢ ὀστᾶ ἢ τῶν ἄλλων τι τῶν τοιούτων, οὐδέν. ἔτι οὔτε περὶ ἀλλοιώσεως οὔτε περὶ αὐξήσεως, τίνα τρόπον ὑπάρχουσι τοῖς πράγμασιν.

1. **οἷον ἐκ δρυόχων**] i.e. new-made, like a ship from the stocks, and tightly fitting. τῶν γενῶν is construed with τρίγωνα.

3. **ὁ πᾶς ὄγκος**] As a whole the infantine body is soft, but this of course does not mean that the particles whereof it is composed, taken individually, are soft.

8. **ἡ ῥίζα τῶν τριγώνων**] This phrase is somewhat obscure. Stallbaum supposes it to mean simply the radical triangles. But as no other triangles can possibly be in question, this is utterly pointless. Martin renders it 'la pointe'; but this seems to restrict the meaning too much. I conceive ῥίζα to mean the fundamental structure of the triangles: the outlines composing it, its sides and angles, from long wear and tear, are no longer so true in form as once they were.

10. **τὰ μὲν τῆς τροφῆς**] Compare Hippokrates *de prisca medicina* vol. I p. 27 Kühn ὅσα μὲν ἰσχυρότερα ᾖ οὐ δυνήσεται κρατέειν ἡ φύσις, ἢν ἐσβάληται, ἀπὸ τουτέων δ᾽ αὐτῶν πόνους τε καὶ νόσους καὶ θανάτους ἔσεσθαι· ὅσων δ᾽ ἂν δύνηται ἐπικρατέειν, ἀπὸ τουτέων τροφήν τε καὶ αὔξησιν καὶ ὑγιείην.

11. **αὐτὰ δὲ ὑπὸ τῶν ἔξωθεν**] Instead of dividing and assimilating the particles of the food, the particles of the body are themselves divided; and the constitution being thus generally enfeebled, the condition ensues which we call old age. Plato has not expressly distinguished be-

frame then of the entire creature, having the triangles of its elements still as if fresh from the workshop, has them firmly linked one to another; but the whole mass is soft in substance, seeing that it has been newly formed out of marrow and nurtured upon milk. Now forasmuch as the triangles of the substances composing the food and drink, which enter from without and are received within the young creature, are older and feebler than those of the latter, it divides and subdues them with its new triangles, and by the assimilation of a large number nourishes and increases the animal: but when the exact outline of the triangles is blunted, because they have been for a long time struggling with many others, they are not able as of old to comminute and assimilate the entering aliment, but are themselves easily divided by the incoming particles. At such a time every living thing is enfeebled and wastes away; and this condition is termed old age. Finally when the bonds of the triangles belonging to the marrow no longer hold to their fastenings but snap asunder with the stress, they loose in their turn the bonds of the soul; and she, being in the course of nature released, flies away with gladness. For all that is contrary to nature is painful; but whatsoever takes place in the natural way is pleasant. On the same principle death which ensues upon sickness or wounds is painful and violent; but that which draws to the natural end in the course of old age is of all deaths the least distressing and is accompanied rather by pleasure than by pain.

tween the fiery particles, which do the work of digestion, and those which enter into the composition of the body at large. We must suppose that when the pyramids of the vital fire become too much blunted to perform their duty properly, the incoming aliment makes war upon and weakens the general structure of the body.

14. μηκέτι ἀντέχωσι δεσμοί] Finally, when the triangles of the marrow itself become blunted, the bonds of the soul are loosed, and she flies forth with joy: or translating into plain prose, the brain and spinal marrow become no longer a fit medium for the soul to act upon. For δεσμοί see 73 D.

15. διιστάμενοι] The form διεστάμενοι, adopted by the more recent editors from A, seems to me very suspicious. The only parallel quoted, so far as I can find, is κατεστέαται in Herodotus 1 196, where there is a variant κατεστᾶσι, which Abicht reads. Altogether the word appears to need more support than it has yet received.

16. κατὰ φύσιν] The doctrine that death in the course of nature is painless, if not pleasurable, is conformable to Plato's general theory of pleasure and pain. Pain is the result of a condition which is παρὰ φύσιν: therefore death, which is κατὰ φύσιν, cannot be painful.

308   ΠΛΑΤΩΝΟΣ   [81 E—

XXXIX. Τὸ δὲ τῶν νόσων ὅθεν ξυνίσταται, δῆλόν που καὶ παντί. τεττάρων γὰρ ὄντων γενῶν, ἐξ ὧν συμπέπηγε τὸ σῶμα, 82 A γῆς πυρὸς ὕδατός τε καὶ ἀέρος, τούτων ἡ παρὰ φύσιν πλεονεξία καὶ ἔνδεια καὶ τῆς χώρας μετάστασις ἐξ οἰκείας ἐπ᾽ ἀλλοτρίαν
5 γιγνομένη, πυρός τε αὖ καὶ τῶν ἑτέρων ἐπειδὴ γένη πλείονα ἑνὸς ὄντα τυγχάνει, τὸ μὴ προσῆκον ἕκαστον ἑαυτῷ προσλαμβάνειν καὶ πάνθ᾽ ὅσα τοιαῦτα στάσεις καὶ νόσους παρέχει· παρὰ φύσιν γὰρ ἑκάστου γιγνομένου καὶ μεθισταμένου θερμαίνεται μὲν ὅσα ἂν πρότερον ψύχηται, ξηρὰ δὲ ὄντα εἰς ὕστερον γίγνεται νοτερά, B
10 καὶ κοῦφα δὴ καὶ βαρέα, καὶ πάσας πάντῃ μεταβολὰς δέχεται. μόνως γὰρ δή, φαμέν, ταὐτὸν ταὐτῷ κατὰ ταὐτὸ καὶ ὡσαύτως καὶ ἀνὰ λόγον προσγιγνόμενον καὶ ἀπογιγνόμενον ἐάσει ταὐτὸν ὂν αὑτῷ σῶν καὶ ὑγιὲς μένειν· ὃ δ᾽ ἂν πλημμελήσῃ τι τούτων ἐκτὸς ἀπιὸν ἢ προσιόν, ἀλλοιότητας παμποικίλας καὶ νόσους
15 φθοράς τε ἀπείρους παρέξεται. δευτέρων δὴ ξυστάσεων αὖ κατὰ

8 ὅσα ἂν : ὅσαπερ ἂν S.   11 μόνως : μόνον S.   ταὐτόν : ταὐτό S.
15 δευτέρων δὴ : δευτέρων δέ S.

81 E—84 C, c. xxxix. A classification of diseases now follows. These arise (1) from excess or deficiency of any of the primary substances of which the body is formed, viz. fire air water and earth; this causes disturbance of the natural conditions and consequently pain and sickness: (2) from disorder in the secondary structures of the body and reversal of their natural relations. For naturally the blood feeds the flesh, and the flesh secretes a fluid which nourishes the bones and marrow: but in disease the flesh degenerates and dissolves into the blood, forming bile of divers kinds and phlegm. But if the evil affects the flesh alone, the danger is not so great; more serious is it when the cement which unites the flesh to the bones is attacked; for then the very roots of the flesh are severed, and it is loosed from the bones and tendons. Yet graver is the case when the mischief seizes upon the bones themselves; but most deadly of all, if the malady is in the marrow; for then the whole course of the body's nature is reversed from the very beginning.

2. τεττάρων] Plato distinguishes between the primary and the secondary structures of the body. The first are simply the fire air earth and water whereof it is composed: the second are structures formed out of these; blood, flesh, tendons, bone, and marrow. The maladies arising from disorders of the first class are not here specified; but in 86 A we have continued and intermittent fevers referred hereto; and probably most minor ailments would be assigned to this cause. These πρῶται ξυστάσεις are termed in the *Timaeus Locrus* 102 C ταὶ ἁπλαῖ δυνάμιες, θερμότας ἢ ψυχρότας ἢ ὑγρότας ἢ ξηρότας.

5. πυρός τε αὖ καὶ τῶν ἑτέρων] Stallbaum, joining these words with the preceding, gives a very unsatisfactory account of this passage. There is no difficulty in it, if we expunge the comma which he places after ἑτέρων and take the genitives after γένη. Plato is giving two causes of sickness; the first is the excess or defect or unnatural situation of some element; the second (introduced by αὖ) is that, whereas diverse kinds exist of each element (cf. 57 C), the wrong sort is

XXXIX. Now the cause whence sicknesses arise is doubtless evident to all. For seeing there are four elements of which the body is composed, earth fire water and air, any unnatural excess or defect of these or change of position from their own to an alien region, and also—since there are more than one kind of fire and the other elements—the reception by each of an unfitting kind, and other such causes, all combine to produce discord and disease. For when any of them changes its nature and position, the parts that formerly were cool are heated, and those that were dry become afterwards moist, and the light become heavy, and all undergo every kind of change. The only way we allow in which one and the same substance can remain whole and unchanged and sound is that the same element should be added to it or taken away from it on the same principle and in the same manner and proportion; and whatsoever errs in any of these points in its outgoings or incomings causes a vast diversity of vicissitudes and diseases and destructions. Next in the secondary structures which are in a

present. The subject of παρέχει is the sentence τὸ μὴ προσῆκον...τοιαῦτα.

7. στάσεις καὶ νόσους] Compare *Sophist* 228 A νόσον ἴσως καὶ στάσιν οὐ ταὐτὸν νενόμικας.

8. θερμαίνεται μέν] Compare Hippokrates *de natura hominis* vol. 1 p. 350 Kühn πολλὰ γάρ εἰσιν ἐν τῷ σώματι ἐόντα, ἃ ὁκόταν ὑπ' ἀλλήλων παρὰ φύσιν θερμαίνηταί τε καὶ ψύχηται, καὶ ξηραίνηταί τε καὶ ὑγραίνηται, νούσους τίκτει. This refers, as appears a little further on, to the four vital fluids enumerated by Hippokrates p. 352 τὸ δὲ σῶμα τοῦ ἀνθρώπου ἔχει ἐν ἑαυτῷ αἷμα καὶ φλέγμα καὶ χολὴν διττήν, ἤγουν ξανθήν τε καὶ μέλαιναν, καὶ ταῦτ' ἐστὶν αὐτέῳ ἡ φύσις τοῦ σώματος, καὶ διὰ ταῦτα ἀλγέει καὶ ὑγιαίνει. ὑγιαίνει μὲν οὖν μάλιστα, ὁκόταν μετρίως ἔχῃ ταῦτα τῆς πρὸς ἄλληλα κρήσιος καὶ δυνάμιος καὶ τοῦ πλήθεος, καὶ μάλιστα ἢν μεμιγμένα ᾖ· ἀλγέει δέ, ὁκόταν τι τουτέων ἔλασσον ἢ πλέον ᾖ, ἢ χωρισθῇ ἐν τῷ σώματι καὶ μὴ κεκρημένον ᾖ τοῖσι ξύμπασι. This statement of Hippokrates is approved by Galen as more correct than Plato's, *de plac. Hipp. et Plat.* VIII 677, 678. Compare a statement attributed to Alkmaion by Stobaeus *florilegium* 100 λέγει δὲ τὰς νόσους συμπίπτειν, ὡς μὲν ὑφ' οὗ, δι' ὑπερβολὴν θερμότητος ἢ ξηρότητος, ὡς δὲ ἐξ οὗ, διὰ πλῆθος τροφῆς ἢ ἐνδείας, ὡς δὲ ἐν οἷς, αἷμα ἢ μυελὸν ἢ ἐγκέφαλον: and again 101 Ἀλκμαίων ἔφη τῆς μὲν ὑγιείας εἶναι συνεκτικὴν τὴν ἰσονομίαν τῶν δυνάμεων ὑγροῦ ξηροῦ ψυχροῦ θερμοῦ πικροῦ γλυκέος καὶ τῶν λοιπῶν· τὴν δ' ἐν αὐτοῖς μοναρχίαν νόσοις παρασκευαστικὴν εἶναι.

11. μόνως γὰρ δή] i.e. each several part must have a continuous and unchanging supply in due proportion of the elements which contribute to its substance.

15. δευτέρων δὴ ξυστάσεων] The δεύτεραι ξυστάσεις are the various ὁμοιομερῆ, in Aristotelian terminology, of which the body is constructed; blood, flesh, bones &c. Galen *de plac.* VIII 680 is wrong in blaming Plato for making blood a δευτέρα ξύστασις, since his πρῶται

310   ΠΛΑΤΩΝΟΣ   [82 B—

φύσιν ξυνεστηκυιῶν δευτέρα κατανόησις νοσημάτων τῷ βουλομένῳ C
γίγνεται ξυννοῆσαι. μυελοῦ γὰρ ἐξ ἐκείνων ὀστοῦ τε καὶ σαρκὸς
καὶ νεύρου ξυμπαγέντος, ἔτι τε αἵματος ἄλλον μὲν τρόπον, ἐκ δὲ
τῶν αὐτῶν γεγονότος, τῶν μὲν ἄλλων τὰ πλεῖστα ᾗπερ τὰ πρόσθεν,
5 τὰ δὲ μέγιστα τῶν νοσημάτων τῇδε χαλεπὰ ξυμπέπτωκεν, ὅταν
ἀνάπαλιν ἡ γένεσις τούτων πορεύηται, τότε ταῦτα διαφθείρεται.
κατὰ φύσιν γὰρ σάρκες μὲν καὶ νεῦρα ἐξ αἵματος γίγνεται, νεῦρον
μὲν ἐξ ἰνῶν διὰ τὴν ξυγγένειαν, σάρκες δὲ ἀπὸ τοῦ παγέντος, D
ὃ πήγνυται χωριζόμενον ἰνῶν· τὸ δὲ ἀπὸ τῶν νεύρων καὶ σαρκῶν
10 ἀπιὸν αὖ γλίσχρον καὶ λιπαρὸν ἅμα μὲν τὴν σάρκα κολλᾷ πρὸς
τὴν τῶν ὀστῶν φύσιν αὐτό τε τὸ περὶ τὸν μυελὸν ὀστοῦν τρέφον
αὔξει, τὸ δ' αὖ διὰ τὴν πυκνότητα τῶν ὀστῶν διηθούμενον καθαρώ-
τατον γένος τῶν τριγώνων λειότατόν τε καὶ λιπαρώτατον,
λειβόμενον ἀπὸ τῶν ὀστῶν καὶ στάζον, ἄρδει τὸν μυελόν· καὶ E
15 κατὰ ταῦτα μὲν γιγνομένων ἑκάστων ὑγίεια ξυμβαίνει τὰ πολλά·
νόσοι δέ, ὅταν ἐναντίως. ὅταν γὰρ τηκομένη σὰρξ ἀνάπαλιν εἰς
τὰς φλέβας τὴν τηκεδόνα ἐξιῇ, τότε μετὰ πνεύματος αἷμα πολύ
τε καὶ παντοδαπὸν ἐν ταῖς φλεψὶ χρώμασι καὶ πικρότησι ποικίλ-

3 ἔτι : ἐπί A.

ξυστάσεις differed from those of Hippokrates and Galen. His distinction is that each of the πρῶται ξυστάσεις consists of one element only, a single geometrical form; whereas a δευτέρα ξύστασις is composite, being formed of two or more πρῶται ξυστάσεις.

2. ἐξ ἐκείνων] sc. ἐκ τῶν τεττάρων.

3. ἄλλον μὲν τρόπον] That is to say, the blood is prepared by a process peculiar to itself, being formed directly from the aliment by the action of the internal fire, as described at 79 A: cf. 73 B—74 D.

4. τὰ πλεῖστα ᾗπερ τὰ πρόσθεν] i.e. the majority of ailments are due to defects of the πρῶται ξυστάσεις, but the most serious to those of the δεύτεραι.

6. ἀνάπαλιν ἡ γένεσις] In disease the order of nature's process is reversed: the natural γένεσις is from blood, which is the sustenance of the whole body, successively to flesh, tendons, and the oily fluid which nourishes the bones and marrow. But sickness causes flesh to degenerate and liquefy and pass into the blood, contrary to the order of nature; and in severe cases this degeneration begins higher up, with the bones or even the vital marrow itself.

8. ἐξ ἰνῶν] That is, from the fibrine of the blood, which both Plato and Aristotle distinguished from the serum, ἰχώρ, though the globules were unknown to them. In 84 A ἰνῶν appears to mean the fibrine of the flesh, not of the blood. Compare Aristotle *historia animalium* III vi 515^b 27 αἱ δὲ ἶνές εἰσι μεταξὺ νεύρου καὶ φλεβός. ἔνιαι δ' αὐτῶν ἔχουσιν ὑγρότητα τὴν τοῦ ἰχῶρος, καὶ διέχουσιν ἀπό τε τῶν νεύρων πρὸς τὰς φλέβας καὶ ἀπ' ἐκείνων πρὸς τὰ νεῦρα. ἔστι δὲ καὶ ἄλλο γένος ἰνῶν, ὃ γίνεται μὲν ἐν αἵματι, οὐκ ἐν ἅπαντος δὲ ζῴου αἵματι· ὧν ἐξαιρουμένων ἐκ τοῦ αἵματος οὐ πήγνυται τὸ αἷμα, ἐὰν δὲ μὴ ἐξαιρεθῶσι, πήγνυται: cf. III xvi 519^b 32, *de partibus animalium* II ix 654^b 28, and II iv 651^a 1 αἱ δ' ἶνες στερεὸν καὶ γεῶδες, ὥστε γίνονται οἷον πυρίαι ἐν τῷ αἵματι καὶ ζέσιν ποιοῦσιν ἐν τοῖς θυμοῖς: he compares

natural state of union a second class of diseases may be discerned by one who would scrutinise them. For whereas marrow and bone and flesh and sinew are composed of the four elements, and blood is formed of the same though in a different way, most of the diseases arise in the manner before explained, but the gravest afflict them with especial severity in the following way: that is to say, when the order of their generation is reversed, these structures are then destroyed. For in the course of nature flesh and sinews arise from blood, the sinews from the fibrine, owing to their affinity; the flesh from the clots which are formed when the fibrine is separated. From the sinews and flesh again proceeds a glutinous and oily fluid, which not only cements the flesh to the structure of the bones and itself gives nourishment and growth to the bone which encloses the marrow, but also so much of it as filters through the dense substance of the bones, being formed of the purest and smoothest and most slippery kind of the triangles, as it distils and oozes from the bones, irrigates the marrow. When these structures are produced in this order, health is the result as a rule; but when this is reversed, sickness ensues. For when the flesh decomposes and returns the deliquescent matter to the veins, then is mingled with air in the veins much blood of manifold kinds, with diverse hues and bitter qualities, as well as acid and saline

them to the earthy element in mud.

9. ὃ πήγνυται χωριζομένων ἰνῶν] This is a curious statement: he conceives the flesh to be formed by the concretion of what is left of this blood after the ἶνες have gone to form νεῦρα.

10. γλίσχρον καὶ λιπαρόν] This glutinous and oily secretion of the flesh and tendons is perhaps identical with the synovial fluid, which lubricates the joints. Plato supposes it to form by coagulation the periosteum, or membrane enclosing the bones, and therefore to cement together flesh and bones: it also penetrates the bony envelope of the spinal column and nourishes the vital marrow, as well as the bones which protect it.

12. τὸ δ' αὖ answers ἅμα μέν: while part of the oily fluid is employed as above, another part, the finest and smoothest, filters through to the marrow.

17. μετὰ πνεύματος αἷμα] This indicates that Plato regarded the veins as ducts for air as well as blood. Aristotle also held that air passed through the blood-vessels: see *historia animalium* I xvii 496[a] 30 ἐπάνω δ' εἰσὶν οἱ ἀπὸ τῆς καρδίας πόροι· οὐδεὶς δ' ἐστὶ κοινὸς πόρος, ἀλλὰ διὰ τὴν σύναψιν δέχονται τὸ πνεῦμα καὶ τῇ καρδίᾳ διαπέμπουσιν. The word πόρος is elsewhere applied by Aristotle to a nerve; but here he is clearly speaking of a blood-vessel. It was supposed by some authorities after his time that the arteries, as distinguished from the veins, were filled with air alone: see Cicero *de natura deorum* § 138 coque modo ex his partibus et sanguis per venas in omne cor-

λόμενον, ἔτι δὲ ὀξείαις καὶ ἁλμυραῖς δυνάμεσι, χολὰς καὶ ἰχῶρας καὶ φλέγματα παντοῖα ἴσχει. παλιναίρετα γὰρ πάντα γεγονότα καὶ διεφθαρμένα τό τε αἷμα αὐτὸ πρῶτον διόλλυσι, καὶ αὐτὰ οὐδεμίαν τροφὴν ἔτι τῷ σώματι παρέχοντα φέρεται πάντη διὰ 83 A
5 τῶν φλεβῶν, τάξιν τῶν κατὰ φύσιν οὐκέτ' ἴσχοντα περιόδων, ἐχθρὰ μὲν αὐτὰ αὑτοῖς διὰ τὸ μηδεμίαν ἀπόλαυσιν ἑαυτῶν ἔχειν, τῷ ξυνεστῶτι δὲ τοῦ σώματος καὶ μένοντι κατὰ χώραν πολέμια, διολλύντα καὶ τήκοντα. ὅσον μὲν οὖν ἂν παλαιότατον ὂν τῆς σαρκὸς τακῇ, δύσπεπτον γιγνόμενον μελαίνει μὲν ὑπὸ παλαιᾶς
10 ξυγκαύσεως, διὰ δὲ τὸ πάντη διαβεβρῶσθαι πικρὸν ὂν παντὶ χαλεπὸν προσπίπτει τοῦ σώματος, ὅσον ἂν μήπω διεφθαρμένον ᾖ· B καὶ τοτὲ μὲν ἀντὶ τῆς πικρότητος ὀξύτητα ἔσχε τὸ μέλαν χρῶμα, ἀπολεπτυνθέντος μᾶλλον τοῦ πικροῦ· τοτὲ δὲ ἡ πικρότης αὖ βαφεῖσα αἵματι χρῶμα ἔσχεν ἐρυθρώτερον, τοῦ δὲ μέλανος τούτου
15 ξυγκεραννυμένου χλοῶδες· ἔτι δὲ ξυμμίγνυται ξανθὸν χρῶμα μετὰ τῆς πικρότητος, ὅταν νέα ξυντακῇ σὰρξ ὑπὸ τοῦ περὶ τὴν φλόγα πυρός. καὶ τὸ μὲν κοινὸν ὄνομα πᾶσι τούτοις ἤ τινες ἰατρῶν που χολὴν ἐπωνόμασαν ἢ καί τις ὢν δυνατὸς εἰς πολλὰ C μὲν καὶ ἀνόμοια βλέπειν, ὁρᾶν δὲ ἐν αὐτοῖς ἓν γένος ἐνὸν ἄξιον
20 ἐπωνυμίας πᾶσι· τὰ δ' ἄλλα ὅσα χολῆς εἴδη λέγεται, κατὰ τὴν χρόαν ἔσχε λόγον αὐτῶν ἕκαστον ἴδιον. ἰχὼρ δέ, ὁ μὲν αἵματος ὀρὸς πρᾷος, ὁ δὲ μελαίνης χολῆς ὀξείας τε ἄγριος, ὅταν ξυμμιγνύηται διὰ θερμότητα ἁλμυρᾷ δυνάμει· καλεῖται δὲ ὀξὺ φλέγμα

5 οὐκέτ' ἴσχοντα : οὐκέτι σχόντα A. οὐκέτ' ἔχοντα S. 14 ἐρυθρώτερον : ἐρυθρότερον S. 15 χλοῶδες dedi ex Cornari correctione et nonnullis codicibus. χολῶδες AHSZ.

pus diffunditur et spiritus per arterias. Cicero uses the word 'arteria' in the modern sense.

1. **χολὰς καὶ ἰχῶρας καὶ φλέγματα**] The decomposition of the flesh produces bile and serum and phlegm. By χυλὰς we must understand morbid conditions or excessive abundance of that fluid: since in 71 B, C Plato expressly recognises that χολή is a normal and necessary constituent of the body; which is more than Aristotle did: cf. *de partibus animalium* IV ii 676^b 31, 677^a 11—22. The same applies to ἰχῶρας, viz. that an abnormal condition is to be understood.

2. **παλιναίρετα**] i.e. ἀνάπαλιν τὴν γένεσιν ἔχοντα.

5. **τάξιν τῶν κατὰ φύσιν**] Although Plato was of course ignorant concerning the circulation of the blood, he conceived it to have regular periodic motions.

6. **μηδεμίαν ἑαυτῶν ἀπόλαυσιν**] i.e. they do not contribute to each other's nourishment.

9. **δύσπεπτον**] Being old firm flesh, it yields reluctantly to the decomposing agent.

μελαίνει μέν] i.e. it is blackened by long-standing inflammation and corrosion. The degeneration of flesh produces a morbid kind of χολή; of which are enumerated four classes, (1) black,

properties; and this contains all kinds of bile and serum and phlegm. For as all these are going the wrong way and have become corrupt, first they ruin the blood itself, and furnishing no nutriment to the body rush in all directions through the veins, paying no heed to the periods appointed by nature, but at war one with another, because they have no good of each other; at war also with all that is established and fixed in the body, which they corrupt and dissolve. Now when the oldest part of the flesh is decomposed, being hard to soften, it turns black through long-continued burning, and through being everywhere corroded it is bitter and dangerous to whatever part of the body it attacks which is not yet corrupted. Sometimes this black sort is acid instead of bitter, when the bitterness is more refined away; and again the bitter sort being steeped in blood gains a redder hue; and when black is mingled with this, it is greenish: sometimes too a yellow colour is added to the bitterness, when new flesh is decomposed by the fire of the inflammation. To all these symptoms the general name of *bile* has been given, either by physicians, or by some one who in looking at many dissimilar appearances was able to see one universal quality pervading them all which deserved a name. All other kinds of bile which are reckoned have their several descriptions according to their colour. Of lymph, one kind is the mild serum of blood,—the other is an acrid secretion of black and acid bile, when that is blended through inflammation with a saline property: this kind is called acid phlegm. But that

either bitter or acid, produced by the degeneration of old flesh, (2) reddish, where there is an admixture of blood, (3) green, apparently a combination of the two former, (4) yellow, from the corrosion of newly-formed flesh.

15. χλοώδες] This reading is clearly right: when Plato is classifying χολαί according to colour, it were absurd to call one class χολώδες. It will be remembered too that at 68 c green is derived from a mixture of red and black. χλοῶδες is found in one ms. and the margin of another, and is also confirmed by Galen.

16. τοῦ περὶ τὴν φλόγα πυρός] If φλόγα is right it must signify 'the inflammation'; but it is curiously abrupt, and I am disposed to agree with Lindau in suspecting it to be corrupt, though I cannot approve of his suggested alteration.

17. καὶ τὸ μὲν κοινὸν ὄνομα] All these different forms have received the general name of χολή, bestowed either by medical men (and presumably somewhat at hap-hazard), or more scientifically by a philosopher skilled in discerning ἓν ἐπὶ πολλοῖς. Compare 68 D.

23. καλεῖται δὲ ὀξὺ φλέγμα] Of

314   ΠΛΑΤΩΝΟΣ   [83 C—

τὸ τοιοῦτον. τὸ δ' αὖ μετ' ἀέρος τηκόμενον ἐκ νέας καὶ ἀπαλῆς
σαρκός, τούτου δὲ ἀνεμωθέντος καὶ ξυμπεριληφθέντος ὑπὸ ὑγρό- D
τητος, καὶ πομφολύγων ξυστασῶν ἐκ τοῦ πάθους τούτου καθ'
ἑκάστην μὲν ἀοράτων διὰ σμικρότητα, ξυναπασῶν δὲ τὸν ὄγκον
5 παρεχομένων ὁρατόν, χρῶμα ἐχουσῶν διὰ τὴν τοῦ ἀφροῦ γένεσιν
ἰδεῖν λευκόν, ταύτην πᾶσαν τηκεδόνα ἀπαλῆς σαρκὸς μετὰ πνεύματος ξυμπλακεῖσαν λευκὸν εἶναι φλέγμα φαμέν. φλέγματος δὲ
αὖ νέου ξυνισταμένου ὀρὸς ἱδρὼς καὶ δάκρυον, ὅσα τε ἄλλα τοιαῦτα E
σῶμα τὸ καθ' ἡμέραν χεῖται καθαιρόμενον· καὶ ταῦτα μὲν δὴ
10 πάντα νόσων ὄργανα γέγονεν, ὅταν αἷμα μὴ ἐκ τῶν σιτίων καὶ
ποτῶν πληθύσῃ κατὰ φύσιν, ἀλλ' ἐξ ἐναντίων τὸν ὄγκον παρὰ
τοὺς τῆς φύσεως λαμβάνῃ νόμους. διακρινομένης μὲν οὖν ὑπὸ
νόσων τῆς σαρκὸς ἑκάστης, μενόντων δὲ τῶν πυθμένων αὐταῖς
ἡμίσεια τῆς ξυμφορᾶς ἡ δύναμις· ἀνάληψιν γὰρ ἔτι μετ' εὐπετείας
15 ἴσχει· τὸ δὲ δὴ σάρκας ὀστοῖς ξυνδοῦν ὁπότ' ἂν νοσήσῃ, καὶ μηκέτι 84 A
αὐτὸ ἐξ ἰνῶν ἅμα καὶ νεύρων ἀποχωριζόμενον ὀστῷ μὲν τροφή,
σαρκὶ δὲ πρὸς ὀστοῦν γίγνηται δεσμός, ἀλλ' ἐκ λιπαροῦ καὶ λείου
καὶ γλίσχρου τραχὺ καὶ ἁλμυρὸν αὐχμῆσαν ὑπὸ κακῆς διαίτης
γένηται, τότε ταῦτα πάσχον πᾶν τὸ τοιοῦτον καταψήχεται μὲν
20 αὐτὸ πάλιν ὑπὸ τὰς σάρκας καὶ τὰ νεῦρα, ἀφιστάμενον ἀπὸ τῶν
ὀστῶν, αἱ δ' ἐκ τῶν ῥιζῶν ξυνεκπίπτουσαι τά τε νεῦρα γυμνὰ B

7 ξυμπλακεῖσαν: ξυμπλεκεῖσαν A.   16 αὐτὸ scripsi: αὖ τὸ AHSZ.
ἅμα, quod suadente Lindavio recepi, probavit nec tamen admisit S.   αἷμα AHSZ.

φλέγμα two sorts are distinguished, ὀξὺ
and λευκόν. The first is the serum of
μέλαινα χολή, and a morbid humour: the
second, formed by the dissolution of new-
formed flesh and highly aerated, is in its
normal state a natural and healthy secretion, viz. perspiration or tears; but if
produced to excess, it is a source of disease.
  1. **ἐκ νέας καὶ ἀπαλῆς σαρκός**] Galen, while approving Plato's description
of φλέγμα, dissents from his account of
its origin: see *de plac.* VIII 699 τὸ δὲ ἐκ
συντήξεως ἀπαλῆς σαρκὸς γενέσθαι ποτὲ
φλέγμα τῶν ἀτοπωτάτων ἐστί: his own
statement is δέδεικται γὰρ ἢ γε τοῦ φλέγματος γένεσις ἐκ τροφῆς φύσει ψυχροτέρας
ἐνδεῶς ὑπὸ τῆς ἐμφύτου θερμασίας κατερ-
γασθείσης ἀποτελουμένη.
  2. **ξυμπεριληφθέντος ὑπὸ ὑγρότητος**]
This seems to be a loose way of expressing that the air-bubbles are enclosed
in the moisture of the φλέγμα.
  9. **τὸ καθ' ἡμέραν**] i.e. in the normal
healthy course of life.
  11. **ἀλλ' ἐξ ἐναντίων**] i.e. when it
feeds upon the flesh or other structures of
the body, instead of the food: see above,
82 E.
  13. **μενόντων δὲ τῶν πυθμένων**] That
is, if the mischief is comparatively superficial, and the fundamental structure of
the flesh is unhurt, recovery is still easy.
  15. **τὸ δὲ δὴ σάρκας ὀστοῖς ξυνδοῦν**]
sc. the γλίσχρον καὶ λιπαρόν, which by
coagulation forms the periosteum, as ex-

which is formed in conjunction with air by the liquefaction of new and tender flesh,—when it is inflated with air enveloped by moisture, and through this condition bubbles are formed, invisible separately because of their smallness, but all together becoming visible in the mass and presenting a white colour to view by the formation of froth—this liquefaction of tender flesh in combination with air we term white phlegm. And the serum of freshly formed phlegm is sweat and tears, and whatever other secretions purify the body from day to day. All these become a means of disease, when the blood is not replenished from the food and drink in the natural way, but receives its volume in the contrary manner in despite of nature's laws. Now when the flesh is anywhere pierced by disease, but the foundations of it remain intact, the malady has only half its power; for there is still the prospect of ready recovery. But when that which unites the flesh to the bones is diseased, and in turn no longer by distilling both from the fibres and sinews nourishes the bones and cements the flesh to them, but instead of being oily and smooth and glutinous becomes harsh and saline and shrivelled through an unhealthy habit of life, under these conditions all that substance crumbles away under the flesh and the sinews and separates from the bones; while the flesh, falling away from its foundations, leaves the sinews

plained in 82 D.

16. **αὐτὸ ἐξ ἰνῶν ἅμα**] The reading of the mss. seems here unquestionably corrupt. The passage obviously refers to the substance mentioned immediately above, the cement which joins flesh and bones together. But this substance is not blood, nor is the blood ἐξ ἰνῶν καὶ νεύρων ἀποχωριζόμενον; which the cement however is, provided we understand ἰνῶν here as signifying the fibrine of the flesh, not of the blood: see note on 82 D. It is plain then that αἷμα is wrong; and Lindau's suggestion ἅμα seems to me a good one. But furthermore αὖ τὸ surely cannot be right; for αὖ introduces an antithesis where none exists, and the article seems to mark the mention of some new substance, whereas Plato is still speaking of the cement of flesh and bones. I have therefore made the slight alteration to αὐτό, which may, I think, be justified as setting off the fluid against the bones which it nourishes and the flesh which it fastens to them: it is itself no longer secreted and it therefore fails to nourish the bones and cement the flesh: cf. 82 E τό τε αἷμα αὐτὸ πρῶτον διόλλυσι, καὶ αὐτὰ οὐδεμίαν τροφὴν ἔτι τῷ σώματι παρέχοντα φέρεται.

19. **καταψήχεται μὲν αὐτό**] The periosteum dries up and crumbles, and the flesh, no longer cemented to the bones, falls away from them: cf. Aristotle *historia animalium* III xiii 519[b] 5 ψιλούμενά τε τὰ ὀστᾶ τῶν ὑμένων σφακελίζει.

21. **ἐκ τῶν ῥιζῶν**] The ῥίζαι are the πυθμένες mentioned above in 83 E.

καταλείπουσι καὶ μεστὰ ἄλμης, αὐταὶ δὲ πάλιν εἰς τὴν αἵματος φορὰν ἐμπεσοῦσαι τὰ πρόσθεν ῥηθέντα νοσήματα πλείω ποιοῦσι. χαλεπῶν δὲ τούτων περὶ τὰ σώματα παθημάτων γιγνομένων μείζω ἔτι γίγνεται τὰ πρὸ τούτων, ὅταν ὀστοῦν διὰ πυκνότητα σαρκὸς
5 ἀναπνοὴν μὴ λαμβάνον ἱκανήν, ὑπ' εὐρῶτος θερμαινόμενον, σφακελίσαν μήτε τὴν τροφὴν καταδέχηται πάλιν τε αὐτὸ εἰς ἐκείνην C ἐναντίως ἴῃ ψηχόμενον, ἡ δ' εἰς σάρκας, σὰρξ δὲ εἰς αἷμα ἐμπίπτουσα τραχύτερα πάντα τῶν πρόσθεν τὰ νοσήματα ἀπεργάζηται· τὸ δ' ἔσχατον πάντων, ὅταν ἡ τοῦ μυελοῦ φύσις ἀπ' ἐνδείας ἤ τινος
10 ὑπερβολῆς νοσήσῃ, τὰ μέγιστα καὶ κυριώτατα πρὸς θάνατον τῶν νοσημάτων ἀποτελεῖ, πάσης ἀνάπαλιν τῆς τοῦ σώματος φύσεως ἐξ ἀνάγκης ῥυείσης.

XL. Τρίτον δ' αὖ νοσημάτων εἶδος τριχῇ δεῖ διανοεῖσθαι γιγνόμενον, τὸ μὲν ὑπὸ πνεύματος, τὸ δὲ φλέγματος, τὸ δὲ χολῆς. D
15 ὅταν μὲν γὰρ ὁ τῶν πνευμάτων τῷ σώματι ταμίας πλεύμων μὴ καθαρὰς παρέχῃ τὰς διεξόδους ὑπὸ ῥευμάτων φραχθείς, ἔνθα μὲν οὐκ ἰόν, ἔνθα δὲ πλεῖον ἢ τὸ προσῆκον πνεῦμα εἰσιὸν τὰ μὲν οὐ τυγχάνοντα ἀναψυχῆς σήπει, τὰ δὲ τῶν φλεβῶν διαβιαζόμενον καὶ ξυνεπιστρέφον αὐτὰ τῆκόν τε τὸ σῶμα εἰς τὸ μέσον αὐτοῦ
20 διάφραγμά τ' ἴσχον ἐναπολαμβάνεται, καὶ μυρία δὴ νοσήματα E

18 διαβιαζόμενον: διαβιαζομένων A. 20 τ' ἴσχον: τί σχόν A.

2. τὰ πρόσθεν ῥηθέντα νοσήματα] sc. the χολαί and φλέγματα.

4. τὰ πρὸ τούτων] i.e. when the degeneration begins further back; the bones being regarded as posterior in the order of γένεσις to the flesh.

διὰ πυκνότητα σαρκός] Perhaps then, after all, if the gods had given our heads a thick covering of flesh, we might not have lived any the longer for it.

5. ἀναπνοήν] cf. 85 A, C: 'ventilation' seems to be the meaning here.

6. τὴν τροφήν] i.e. the oily fluid which nourishes them. The bones decompose and mingle with this fluid, the fluid with the flesh, and the flesh with the blood.

11. πάσης ἀνάπαλιν] The μυελός is the very citadel of life; so that when the disease assails that, the foundations of health are sapped: the course of nature flows backward from its utmost fount.

84 C—86 A, c. xl. A third class of maladies remains for consideration: those engendered by air, by phlegm, and by bile. When an excessive amount of air passes into the veins and penetrating their sides finds its way into the flesh and is there imprisoned, various evil results follow; in some cases convulsions and tetanus, which will hardly yield to treatment, and diseases of the lungs. By phlegm are produced leprosies and all manner of skin-diseases; and when in conjunction with bile it attacks the head, epilepsy ensues, which is called the 'sacred disease', because it affects the divinest part. All kinds of inflammatory disorders, accompanied by pustules and eruptions, arise from bile; which also

E]   ΤΙΜΑΙΟΣ.   317

bare and full of brine, and itself falling back into the current of the blood aggravates the diseases that have been described. But distressing as are these symptoms which affect the body, yet more serious are those which are prior in order; when the bones, owing to denseness of the flesh, cannot get sufficient air and becoming mouldy and heated decay away, and while they will not receive their nourishment, crumble down and return by a reversed process into their nourishing fluid, and that in its turn passing into flesh, and the flesh into blood, they render all the diseases more virulent than those already mentioned. The most desperate case of all is when the substance of the marrow becomes diseased by any defect or excess: this produces the most serious and fatal disorders, seeing that the whole nature of the body is forced to proceed in a backward course.

XL. A third class of diseases we must conceive as occurring in three ways: one by the agency of air, the second of phlegm, the third of bile. For when the lungs, which are the dispensers of air to the body, do not keep their passages clear, because they are impeded by catarrhs, the air, failing to pass through some, and in others entering with a volume unduly great, causes the decomposition of the parts which lack their supply of air, and forces its way through the channels of the veins and dislocates them, and dissolving the body it is confined amid its substance, occupying the midriff; and so countless painful diseases are produced from these causes, accompanied by

seizes upon the fibrine of the blood, and preventing its due circulation causes chills and shuddering; and sometimes penetrating to the vital marrow sets free the soul: but if its fury be less violent, it gives rise to diarrhoea and dysentery. Continuous, quotidian, tertian, and quartan fevers are caused by a superabundance of fire, air, water, and earth respectively predominating in the composition of the body.

14. τὸ μὲν ὑπὸ πνεύματος] This class of diseases is distinct from those caused by a mere superfluity of air entering into the composition of the body. We are at present concerned with the maladies arising from the confinement of large quantities of air in places where it has no right to be.

18. τὰ δὲ τῶν φλεβῶν] Here again the veins are considered as passages for air: the ingress of air is normal; it is the excessive amount which gives rise to disease: see note on 82 E.

19. εἰς τὸ μέσον αὐτοῦ] These words are best taken with ἐναπολαμβάνεται. But the sentence does not run smoothly, and I suspect that something has gone amiss with it. διάφραγμα ἴσχον, if the words are sound, means taking possession of the midriff, pressing against it.

ἐκ τούτων ἀλγεινὰ μετὰ πλήθους ἱδρῶτος ἀπείργασται. πολλάκις δ' ἐν τῷ σώματι διακριθείσης σαρκὸς πνεῦμα ἐγγενόμενον καὶ ἀδυνατοῦν ἔξω πορευθῆναι τὰς αὐτὰς τοῖς ἐπεισεληλυθόσιν ὠδῖνας παρέσχε, μεγίστας δέ, ὅταν περὶ τὰ νεῦρα καὶ τὰ ταύτῃ φλέβια 5 περιστὰν καὶ ἀνοιδῆσαν τούς τε ἐπιτόνους καὶ τὰ ξυνεχῆ νεῦρα οὕτως εἰς τὸ ἐξόπισθεν κατατείνῃ τούτοις· ἃ δὴ καὶ ἀπ' αὐτοῦ τῆς συντονίας τοῦ παθήματος τὰ νοσήματα τέτανοί τε καὶ ὀπισθότονοι προσερρήθησαν. ὧν καὶ τὸ φάρμακον χαλεπόν· πυρετοὶ γὰρ οὖν δὴ τὰ τοιαῦτα ἐγγιγνόμενοι μάλιστα λύουσι. τὸ δὲ λευκὸν φλέγμα 85 A 10 διὰ τὸ τῶν πομφολύγων πνεῦμα χαλεπὸν ἀποληφθέν, ἔξω δὲ τοῦ σώματος ἀναπνοὰς ἴσχον, ἠπιώτερον μέν, καταποικίλλει δὲ τὸ σῶμα λεύκας ἀλφούς τε καὶ τὰ τούτων ξυγγενῆ νοσήματα ἀποτίκτον· μετὰ χολῆς δὲ μελαίνης κερασθὲν ἐπὶ τὰς περιόδους τε τὰς ἐν τῇ κεφαλῇ θειοτάτας οὔσας ἐπισκεδαννύμενον καὶ ξυνταράττον 15 αὐτάς, καθ' ὕπνον μὲν ἰὸν πρᾳότερον, ἐγρηγορόσι δὲ ἐπιτιθέμενον B δυσαπαλλακτότερον· νόσημα δὲ ἱερᾶς ὂν φύσεως ἐνδικώτατα ἱερὸν λέγεται. φλέγμα δ' ὀξὺ καὶ ἁλμυρὸν πηγὴ πάντων νοσημάτων, ὅσα γίγνεται καταρροϊκά· διὰ δὲ τοὺς τόπους, εἰς οὓς ῥεῖ, παντοδαποὺς ὄντας παντοῖα ὀνόματα εἴληφεν. ὅσα δὲ φλεγμαίνειν

1 πολλάκις post ἱδρῶτος inserunt AS. 9 ἐγγιγνόμενοι: ἐπιγιγνόμενοι S.

1. **μετὰ πλήθους ἱδρῶτος**] Plato evidently has in view consumption and kindred maladies.

2. **διακριθείσης σαρκός**] In the former case the air entered from without: an equally bad, though different, result is produced when the imprisoned air has been produced within the body by dissolution of the flesh.

5. **τούς τε ἐπιτόνους**] The ἐπίτονοι are the great tendons of the shoulders and arms.

7. **τέτανοί τε καὶ ὀπισθότονοι**] The first is the generic term for diseases the symptoms of which are spasmodic contraction of the muscles : ὀπισθότονος was a special form in which the muscles are drawn violently backwards: see Hippokrates *de morbis* vol. II p. 303 Kühn: the opposite form was ἐμπροσθότονος. Aristotle also attributes these disorders to the action of air: *meteorologica* II viii 366ᵇ 25

οἵ τε γὰρ τέτανοι καὶ οἱ σπασμοὶ πνεύματος μὲν εἰσὶ κινήσεις.

8. **πυρετοὶ γὰρ οὖν δή**] Compare Hippokrates *aphorisms* vol. III p. 735 Kühn ὑπὸ σπασμοῦ ἢ τετάνου ἐνοχλουμένῳ πυρετὸς ἐπιγενόμενος λύει τὸ νόσημα. Plato means that in cases which do not end fatally it is this natural relief, rather than medical treatment, which saves the patient's life.

10. **διὰ τὸ τῶν πομφολύγων πνεῦμα**] The diseases produced by the λευκὸν φλέγμα are ultimately to be traced to πνεῦμα, since they are due to the air which is enclosed in the former ; they are less dangerous however, because they are thrown off at the surface.

12. **λεύκας ἀλφούς τε**] These are diseases of the skin described by Celsus V xxviii 19.

15. **καθ' ὕπνον μὲν ἰὸν πρᾳότερον**] 'In many epileptics the fits occur during the

excessive sweat. Often too when the flesh is broken up, air is formed in the body, and being unable to find an exit it produces the same torments as are caused by the air which enters in; the most severe of all, when gathering and swelling up around the sinews and the blood-vessels in these parts it strains the tendons of the shoulders and the muscles attached to them in a backward direction: and owing to the intense strain produced in this condition these affections are called tetanus and opisthotonus. For these the remedy is severe: for in fact fevers supervening chiefly give relief in such cases. The white phlegm when intercepted is dangerous owing to the air in the bubbles: but when it finds an escape to the surface of the body it is more mild; yet it disfigures the person by engendering scabs and leprosies and kindred maladies. Sometimes it is mingled with black bile and is shed upon the revolutions in the head, which are the most divine, and confounds them; and if this occurs during sleep, the effects are milder, but if in the waking hours, it is harder to relieve. This, as affecting the sacred part, is justly called the sacred disease. Acid and saline phlegm is the source of all diseases that take the form of catarrh: and these have received manifold names according to the diverse places in which the discharge takes place. Inflammations in various parts

night as well as during the day, but in some instances they are entirely nocturnal, and it is well known that in such cases the disease may long exist and yet remain unrecognised either by the patient or the physician.' Dr Affleck in the *Encyclopaedia Britannica*, article *Epilepsy.*

16. ἐνδικώτατα ἱερὸν λέγεται] The name ἱερὰ νόσος was given to epilepsy because, owing to the suddenness of the attack and its appalling symptoms, it seemed like the direct visitation of some divine power, which without warning struck down its victim. Hippokrates in the true scientific spirit protests against this superstition: see *de morbo sacro* vol. I p. 587 Kühn οὐδέν τί μοι δοκέει τῶν ἄλλων θειοτέρη εἶναι νούσων οὐδὲ ἱερωτέρη, ἀλλὰ φύσιν μὲν ἔχει ἣν καὶ τὰ λοιπὰ νουσήματα

ὅθεν γίνεται. φύσιν δὲ αὐτῇ καὶ πρόφασιν οἱ ἄνθρωποι ἐνόμισαν θεῖον εἶναι ὑπὸ ἀπειρίης καὶ θαυμασιότητος, ὅτι οὐδὲν ἔοικε ἑτέρῃσι νούσοισι. καὶ κατὰ μὲν τὴν ἀπορίην αὐτοῖσι τοῦ μὴ γινώσκειν τὸ θεῖον αὐτῇ διασώζεται, κατὰ δὲ τὴν εὐπορίην τοῦ τρόπου τῆς ἰήσιος ἰῶνται· ἀπολύονται γὰρ ᾗ καθ ιρμοῖσι ἢ ἐπαοιδῇσι. Plato, as his manner is, adopts the popular appellation, but gives it a new and higher significance of his own: it is the sacred disease because peculiarly affecting the divinest part of us.

18. καταρροϊκά] i.e. catarrhs, in whatever part of the body they may occur.

19. φλεγμαίνειν λέγεται] Notwithstanding the name φλεγμαίνειν, Plato would say, inflammations are not owing to φλέγμα at all, but to χολή.

λέγεται τοῦ σώματος, ἀπὸ τοῦ κάεσθαί τε καὶ φλέγεσθαι, διὰ
χολὴν γέγονε πάντα. λαμβάνουσα μὲν οὖν ἀναπνοὴν ἔξω παντοῖα C
ἀναπέμπει φύματα ζέουσα, καθειργνυμένη δ' ἐντὸς πυρίκαυτα
νοσήματα πολλὰ ἐμποιεῖ, μέγιστον δέ, ὅταν αἵματι καθαρῷ ξυγκε-
5 ρασθεῖσα τὸ τῶν ἰνῶν γένος ἐκ τῆς ἑαυτῶν διαφορῇ τάξεως, αἳ
διεσπάρησαν μὲν εἰς αἷμα, ἵνα συμμέτρως λεπτότητος ἴσχοι καὶ
πάχους καὶ μήτε διὰ θερμότητα ὡς ὑγρὸν ἐκ μανοῦ τοῦ σώματος
ἐκρέοι, μήτ' αὖ πυκνότερον δυσκίνητον ὂν μόλις ἀναστρέφοιτο ἐν
ταῖς φλεψί. καιρὸν δὴ τούτων ἶνες τῇ τῆς φύσεως γενέσει φυλάτ- D
10 τουσιν· ἃς ὅταν τις καὶ τεθνεῶτος αἵματος ἐν ψύξει τε ὄντος πρὸς
ἀλλήλας συναγάγῃ, διαχεῖται πᾶν τὸ λοιπὸν αἷμα, ἐαθεῖσαι δὲ
ταχὺ μετὰ τοῦ περιεστῶτος αὐτὸ ψύχους ξυμπηγνύασι. ταύτην
δὴ τὴν δύναμιν ἐχουσῶν ἰνῶν ἐν αἵματι χολὴ φύσει παλαιὸν αἷμα
γεγονυῖα καὶ πάλιν ἐκ τῶν σαρκῶν εἰς τοῦτο τετηκυῖα, θερμὴ καὶ
15 ὑγρὰ κατ' ὀλίγον τὸ πρᾶτον ἐμπίπτουσα πήγνυται διὰ τὴν τῶν E
ἰνῶν δύναμιν, πηγνυμένη δὲ καὶ βίᾳ κατασβεννυμένη χειμῶνα
καὶ τρόμον ἐντὸς παρέχει· πλείων δ' ἐπιρρέουσα, τῇ παρ' αὐτῆς
θερμότητι κρατήσασα, τὰς ἶνας εἰς ἀταξίαν ζέσασα διέσεισε· καὶ
ἐὰν μὲν ἱκανὴ διὰ τέλους κρατῆσαι γένηται, πρὸς τὸ τοῦ μυελοῦ
20 διαπεράσασα γένος καίουσα ἔλυσε τὰ τῆς ψυχῆς αὐτόθεν οἷον
νεὼς πείσματα μεθῆκέ τε ἐλευθέραν· ὅταν δ' ἐλάττων ᾖ τό τε
σῶμα ἀντίσχῃ τηκόμενον, αὐτὴ κρατηθεῖσα ἢ κατὰ πᾶν τὸ σῶμα
ἐξέπεσεν, ἢ διὰ τῶν φλεβῶν εἰς τὴν κάτω ξυνωσθεῖσα ἢ τὴν ἄνω
κοιλίαν, οἷον φυγὰς ἐκ πόλεως στασιασάσης ἐκ τοῦ σώματος
25 ἐκπίπτουσα, διαρροίας καὶ δυσεντερίας καὶ τὰ τοιαῦτα νοσήματα 86 A
πάντα παρέσχετο. τὸ μὲν οὖν ἐκ πυρὸς ὑπερβολῆς μάλιστα

8 μόλις: μόγις SZ.   9 τούτων: τοῦτον A.   17 αὐτῆς: αὐτῆι AHS.
22 αὐτή: αὕτη A.

1. **διὰ χολὴν γέγονε πάντα**] This was, according to Aristotle, the opinion of Anaxagoras and his school: cf. *de partibus animalium* IV ii 677^a 5 οὐκ ὀρθῶς δὲ ἐοίκασιν οἱ περὶ Ἀναξαγόραν ὑπολαμβάνειν ὡς αἰτίαν οὖσαν [sc. τὴν χολὴν] τῶν ὀξέων νοσημάτων· ὑπερβάλλουσαν γὰρ ἀπορραίνειν πρὸς τε τὸν πλεύμονα καὶ τὰς φλέβας καὶ τὰ πλευρά.

2. **λαμβάνουσα μὲν οὖν ἀναπνοήν**] i.e. when it is thrown off in an eruption: Plato is aware that the suppressed inflammation is much more dangerous.

7. **ἐκ μανοῦ τοῦ σώματος ἐκρέοι**] i.e. percolate through the substance of the body.

11. **διαχεῖται πᾶν τὸ λοιπὸν αἷμα**] Hence we see that although Plato conceived that flesh was formed by condensation of the ἰχώρ (82 D), he did not suppose that blood deprived of the ἶνες would coagulate on exposure to the air.

13. **παλαιὸν αἷμα γεγονυῖα**] The flesh is formed of the blood, and χολή (that is,

of the body, so called from the heat and burning that occurs, are all due to bile. When they have egress, they seethe up and send forth all kinds of pustules; but if they are suppressed within, they cause many inflammatory diseases; of which the worst is when the inflammation entering into pure blood carries away from its proper place the fibrine which was distributed through the blood in order that it might preserve a due measure of thinness and thickness and neither be so much liquefied by heat as to flow out through the porous texture of the body, nor become sluggish from excessive density and circulate with difficulty in the veins. Now the fibrine by the nature of its composition preserves the due mean in these respects. For if from blood that is dead and beginning to cool the fibrine be gathered apart, the rest of the blood is dissipated; but if the fibrine be allowed to remain, by the help of the cold air surrounding, it quickly congeals it. The fibrine then in the blood having this property, bile which is naturally formed of old blood and is dissolved again into blood out of the flesh, enters warm and liquid into the blood, at first gradually, and is condensed by the power of the fibrine; and as it is condensed and forced to cool, it produces internal chill and shivering. But when a greater quantity flows in, it subdues the fibrine with its heat, and boiling up scatters it abroad; and if it is able to obtain the mastery to the end, it penetrates to the substance of the marrow, and consuming it looses from thence the bonds of the soul, as it were the moorings of a ship, and sets her free. But when the bile is too feeble for this, and the body holds out against the dissolution, itself is vanquished, and either is expelled by an eruption over the whole body, or is driven through the veins into the lower or upper belly, like an exile banished from a city that has been at civil war; and as it issues forth from the body, it causes diarrhoea and dysentery and all diseases of that kind.

When a body has been stricken with sickness chiefly through

χολὴ of a morbid nature) is formed by degeneration of the flesh, and hence is παλαιὸν αἷμα.

16. χειμῶνα καὶ τρόμον] The solidification of the χολή causes tremor and shivering on the principle enunciated in

62 A, B: τὸ παρὰ φύσιν ξυναγόμενον μάχεται κατὰ φύσιν αὐτὸ ἑαυτὸ εἰς τοὐναντίον ἀπωθοῦν.

20. οἷον νεὼς πείσματα] Compare 73 D καθάπερ ἐξ ἀγκυρῶν βαλλόμενος ἐκ τούτων πάσης ψυχῆς δεσμούς.

322   ΠΛΑΤΩΝΟΣ   [86 A—

νοσῆσαν σῶμα ξυνεχῆ καύματα καὶ πυρετοὺς ἀπεργάζεται, τὸ
δ' ἐξ ἀέρος ἀμφημερινούς, τριταίους δ' ὕδατος διὰ τὸ νωθέστερον
ἀέρος καὶ πυρὸς αὐτὸ εἶναι· τὸ δ' ἐκ γῆς, τετάρτως ὂν νωθέστατον
τούτων, ἐν τετραπλασίαις περιόδοις χρόνου καθαιρόμενον, τεταρ-
5 ταίους πυρετοὺς ποιῆσαν ἀπαλλάττεται μόγις.
    XLI. Καὶ τὰ μὲν περὶ τὸ σῶμα νοσήματα ταύτῃ ξυμβαίνει Β
γιγνόμενα, τὰ δὲ περὶ ψυχὴν διὰ σώματος ἕξιν τῇδε. νόσον μὲν
δὴ ψυχῆς ἄνοιαν ξυγχωρητέον, δύο δ' ἀνοίας γένη, τὸ μὲν μανίαν,
τὸ δὲ ἀμαθίαν. πᾶν οὖν ὅ τι πάσχων τις πάθος ὁπότερον αὐτῶν
10 ἴσχει, νόσον προσρητέον, ἡδονὰς δὲ καὶ λύπας ὑπερβαλλούσας τῶν
νόσων μεγίστας θετέον τῇ ψυχῇ· περιχαρὴς γὰρ ἄνθρωπος ὢν
ἢ καὶ τἀναντία ὑπὸ λύπης πάσχων, σπεύδων τὸ μὲν ἑλεῖν ἀκαίρως, C
τὸ δὲ φυγεῖν, οὔθ' ὁρᾶν οὔτε ἀκούειν ὀρθὸν οὐδὲν δύναται, λυττᾷ
δὲ καὶ λογισμοῦ μετασχεῖν ἥκιστα τότε δὴ δυνατός ἐστι. τὸ δὲ

3 τὸ δ' ἐκ : τὸ δὲ SZ.   5 μόγις, ut videtur, A.   μόλις H.

2. ἀμφημερινούς] i.e. cases in which there is a period of fever and a period of relaxation in every twenty-four hours. As Martin observes, the names given to these recurrent fevers denote, not their period, but the number of days necessary for determining the period: thus in a τριταῖος there is a day of fever and a day of relief; the fever returning on the third day marks the period as comprising two days: similarly in a τεταρταῖος there is a day of fever and two days of relief, the fever returning on the fourth day. Galen de plac. Hipp. et Plat. VIII 697 disputes Plato's account of fever, which he ascribes not to the four elements, but to the four primary fluids of the body. The ancient medical writers also mention a species of tertian fever called ἡμιτριταῖος, the period of which was thirty-six hours of fever (more or less) and twelve hours of comparative relaxation; see Celsus III 3, III 8.

86 B—87 B, c. xli. Maladies of the soul arise from morbid conditions of the body. Now the sickness of the soul is foolishness; and of this there are two kinds, madness and ignorance. Pleasure and pain in excess are the most calamitous of mental disorders, for they lead a man vehemently to seek one thing and eschew another without reflection or understanding. Whenever the seminal marrow is abundant and vigorous, it prompts to indulgence in bodily pleasures which enfeeble the soul. But the profligate are unjustly reproached as criminals: in truth they are sick in soul. For no one is willingly evil; this comes to a man against his will through derangement. For when the vicious humours of the body are pent up therein and find no vent, the vapours of them rise up and choke the movements of the soul at all her seats, causing moroseness and melancholy, rashness and cowardice, forgetfulness and dulness. And these evils are further aggravated by bad institutions and teaching and lack of wholesome training. Wherefore the teachers are more to blame than the sinners themselves, whom we ought to strive to bring into a healthier habit of mind.

7. διὰ σώματος ἕξιν] The corporeal ἕξεις which cause sickness to the soul may be classified in two divisions. (1) susceptibility to pleasures and pains (these arise from σώματος ἕξεις, because, although it is the soul, not the body that

c]   ΤΙΜΑΙΟΣ.   323

excess of fire, it exhibits continued inflammations and fevers; excess of air causes quotidian fevers; excess of water tertian, because it is more sluggish than air or fire; excess of earth, which is by four measures most sluggish of all, being purged in a fourfold period of time, gives rise to quartan fevers, and is with difficulty banished.

XLI. Such are the conditions connected with diseases of the body; those of the soul depend upon bodily habit in the following way. We must allow that disease of the soul is senselessness; and of this there are two forms, madness and stupidity. Every condition then in which a man suffers from either of these must be termed a disease. We must also affirm that the gravest maladies of the soul are excessive pleasures or pains. For if a man is under the influence of excessive joy, or, on the other hand, of extreme pain, and is eager unduly to grasp the one or shun the other, he is able neither to see nor to hear anything aright; he is delirious, and at that moment entirely unable to obey reason.

perceives them, yet they affect the soul through the body), which blind a man to his real interest and highest happiness: (2) physical ill health, which, by enfeebling the parts through which the soul acts upon the body, impedes her actions and stifles her intelligence. Compare *Phaedo* 66 B μυρίας μὲν γὰρ ἡμῖν ἀσχολίας παρέχει τὸ σῶμα διὰ τὴν ἀναγκαίαν τροφήν· ἔτι δ' ἄν τινες νόσοι προσπέσωσιν, ἐμποδίζουσιν ἡμῶν τὴν τοῦ ὄντος θήραν.

8. τὸ μὲν μανίαν, τὸ δὲ ἀμαθίαν] This classification, though not discordant, is not identical with that given in *Sophist* 228 A foll. In that passage we have two εἴδη of κακία in the soul, one being a νόσος or στάσις, the other αἶσχος or ἀμετρία. The νόσος is πονηρία, the αἶσχος is ἄγνοια. Further ἄγνοια is subdivided into ἀμαθία, defined as τὸ μὴ κατειδότα τι δοκεῖν εἰδέναι, and τὰ ἄλλα μέρη ἀγνοίας, which are left unnamed. In the *Timaeus* the distinction between νόσος and αἶσχος is sunk: for all that belongs to πονηρία in the *Sophist* here falls under ἄνοια, whereof ἀμαθία also is a form. This does not mean any ethical discrepancy between the two dialogues; rather the minuter διαίρεσις of the *Sophist* is made in furtherance of the dialectical ends of that dialogue, but is needless for the ethical object of the present passage. ἀμαθία can hardly be translated by any English word: it signifies ignorance combined with dulness which hinders the ἀμαθής from perceiving his ignorance. It must also be observed that μανία is not simply 'madness' in the ordinary sense of the word: as ἀμαθία is a defect of the θεῖον εἶδος τῆς ψυχῆς, a failure of reason, so is μανία a defect of the θνητὸν εἶδος, a want of due subordination to the θεῖον, leading to incontinence and the supremacy of the passions.

11. περιχαρὴς γὰρ ἄν] i.e. excessive sensitiveness either to bodily pleasure or pain is a species of madness which distracts the soul and prevents her from exercising the reason, impelling a man blindly to seek the pleasant and shun the painful without consideration of τὸ βέλτιστον.

21—2

σπέρμα ὅτῳ πολὺ καὶ ῥυῶδες περὶ τὸν μυελὸν γίγνηται καὶ καθαπερεὶ δένδρον πολυκαρπότερον τοῦ ξυμμέτρου πεφυκὸς ᾖ, πολλὰς μὲν καθ' ἕκαστον ὠδῖνας, πολλὰς δ' ἡδονὰς κτώμενος ἐν ταῖς ἐπιθυμίαις καὶ τοῖς περὶ τὰ τοιαῦτα τόκοις, ἐμμανὴς τὸ
5 πλεῖστον γιγνόμενος τοῦ βίου διὰ τὰς μεγίστας ἡδονὰς καὶ λύπας, D νοσοῦσαν καὶ ἄφρονα ἴσχων ὑπὸ τοῦ σώματος τὴν ψυχήν, οὐχ ὡς νοσῶν ἀλλ' ὡς ἑκὼν κακὸς δοξάζεται· τὸ δὲ ἀληθὲς ἡ περὶ τὰ ἀφροδίσια ἀκολασία κατὰ τὸ πολὺ μέρος διὰ τὴν ἑνὸς γένους ἕξιν ὑπὸ μανότητος ὀστῶν ἐν σώματι ῥυώδη καὶ ὑγραίνουσαν νόσος
10 ψυχῆς γέγονε. καὶ σχεδὸν δὴ πάντα, ὁπόσα ἡδονῶν ἀκράτεια καὶ ὄνειδος ὡς ἑκόντων λέγεται τῶν κακῶν, οὐκ ὀρθῶς ὀνειδίζεται· κακὸς μὲν γὰρ ἑκὼν οὐδείς, διὰ δὲ πονηρὰν ἕξιν τινὰ τοῦ σώματος E

1 γίγνηται scripsi: γίγνεται AHSZ codicesque omnes. καὶ inclusit H. 7 κακῶς post κακὸς cum A omisi. servant HSZ. in nonnullis codicibus, qui κακῶς tuentur, abest κακός. 10 ἀκράτεια: ἀκρατία S.

1. περὶ τὸν μυελόν] Compare 73 C, 91 C.

γίγνηται] I believe this slight alteration restores Plato's sentence. The vulgate γίγνεται καὶ cannot possibly stand; and Hermann's excision of καὶ leaves a construction sorely needing defence. Of the omission of ἂν with the relative instances are to be found in Attic prose: see Thucydides IV xvii 2 ἐπιχώριον ὂν ἡμῖν, οὗ μὲν βραχεῖς ἀρκῶσι, μὴ πολλοῖς χρῆσθαι. And above in 57 B we have the very similar construction πρὶν... ἐκφύγῃ: and so Laws 873 A πρὶν...κοιμίσῃ.

4. τοῖς περὶ τὰ τοιαῦτα τόκοις] i.e. pleasure exists (1) in desire, (2) in the gratification of desire. Note that Plato says, not that pleasure is ἐπιθυμία, which would be contrary to his principles, but that it is ἐν ταῖς ἐπιθυμίαις: it is pleasure of anticipation. See Philebus 35 E foll. τόκοις of course signifies the realising of the anticipation.

8. τὴν ἑνὸς γένους ἕξιν] sc. τοῦ μυελοῦ.

10. ἀκράτεια καὶ ὄνειδος] The text seems hitherto to have escaped suspicion; but certainly the phraseology is very extraordinary: I see however no plausible correction.

12. κακὸς μὲν γὰρ ἑκὼν οὐδείς] This passage is one of the most important ethical statements in Plato's writings. Plato's position, which he maintains consistently from first to last, that all vice and error are involuntary, is clearly to be distinguished from the Sokratic identification of ἀρετή with ἐπιστήμη, κακία with ἀμαθία. In the Platonic doctrine ἐπιστήμη is the indispensable condition to true ἀρετή (not to δημοτικὴ καὶ πολιτικὴ ἀρετή), and his teaching on this point is part of a comprehensive theory of determinism. No man, he says, wilfully and wittingly prefers bad to good. In making choice between two courses of action the determining motive is the real or apparent preponderance of good in one: if a man chooses the worse course, it is because either from physical incapacity or faulty training, or both combined, his discernment of good has been dimmed or distorted. We ought not then to rail upon him as a villain, but to pity him as one grievously afflicted and needing succour: compare Laws 731 C, D πᾶς δ' ἄδικος οὐχ ἑκὼν ἄδικος...ἀλλὰ ἐλεεινὸς μὲν πάντως ὅ γε ἄδικος καὶ ὁ τὰ κακὰ ἔχων, ἐλεεῖν δὲ τὸν μὲν ἰάσιμα ἔχοντα ἐγχωρεῖ καὶ ἀνείργοντα τὸν θυμὸν πραΰνειν καὶ μὴ

In whomsoever the seed in the region of the marrow is abundant and fluid and like a tree that is fruitful beyond due measure, he feels from time to time many a sore pang and many a delight amid his passions and their fruits; and he becomes mad for the greater part of his life owing to the intensity of pleasures and pains, keeping his soul in a state of disease and derangement through the power of the body; he is not however regarded as sick, but as willingly vicious. But the truth is that incontinence in sensual pleasures is a disease of the soul for the most part arising from the fluid and moist condition of one element in the body owing to porousness of the bones. So it is too with nearly all intemperance in pleasure; and the reproach attaching thereto, as if men were willingly vicious, is incorrectly brought against them. For no one is willingly wicked; but it is owing

ἀκραχολοῦντα. He admits however that θυμὸς is a useful ally in desperate cases: τῷ δ' ἀκράτως καὶ ἀπαραμυθήτως πλημμελεῖ καὶ κακῷ ἐφιέναι δεῖ τὴν ὀργήν· διὸ δὴ θυμοειδῆ πρέπειν καὶ πρᾷόν φαμεν ἑκάστοτε εἶναι δεῖν τὸν ἀγαθόν. Hence it necessarily follows that all punishment is either curative or deterrent, never vindictive or retributive; of this there are many explicit statements; see *Laws* 854 D, 862 D, E, and especially 934 A; *Phaedo* 113 D, E, *Gorgias* 477 A, 505 C, 525 B. The greatest benefit we can confer upon the wicked is to punish them and so deliver them from their wickedness. Even the punishment of death inflicted upon incurable criminals is regarded not only as a protection to society and as a warning to the evil-disposed, but also as a deliverance to the offender himself from a life of guilt and misery : cf. *Laws* 958 A οἷσι δὲ ὄντως ἐπικεκλωσμέναι, θάνατον ἅμα ταῖς οὕτω διατεθείσαις ψυχαῖς διανέμοντες, also 854 C.

Now this view of vice, that it is an involuntary affection of the soul, will be seen to be an inevitable inference from Plato's ontology; and it well illustrates how admirably the various parts of his system fit together. Soul, as such, is good entirely. Absolute being, absolute thought, and absolute goodness are one and the same. Therefore from the absolute or universal soul can come no evil. The particular soul is derived from the universal soul, whence she has her essence: therefore her nature, *qua* soul, is entirely good. No evil therefore can arise from the voluntary choice of the soul. Evil then must of necessity arise from the conditions of her limitation, which takes the form of bodily environment. And it is clear that all defects in this respect are due either to physical aberrations or faulty treatment. Therefore Plato's ethical is necessitated by his ontological theory. And the Interpreter's declaration in the *Republic* αἰτία ἑλομένου, θεὸς ἀναίτιος not only is not inconsistent with the maxim κακὸς ἑκὼν οὐδείς, but is inevitably implied in it : each statement in fact involves the other and could not be true without it.

In the region of sensibles ugliness and deformity are due to the imperfect manner in which the senses convey to us representations of the ideas: a perfect symbol of an idea would be perfectly beautiful; all imperfection being due to divergence from the type. So also moral deformity is due to divergence from the type; and the choice of evil arises from

ΠΛΑΤΩΝΟΣ

καὶ ἀπαίδευτον τροφὴν ὁ κακὸς γίγνεται κακός, παντὶ δὲ ταῦτα
ἐχθρὰ καὶ ἄκοντι προσγίγνεται. καὶ πάλιν δὴ τὸ περὶ τὰς λύπας
ἡ ψυχὴ κατὰ ταὐτὰ διὰ σῶμα πολλὴν ἴσχει κακίαν. ὅπου γὰρ
ἂν οἱ τῶν ὀξέων καὶ τῶν ἁλυκῶν φλεγμάτων καὶ ὅσοι πικροὶ καὶ
5 χολώδεις χυμοὶ κατὰ τὸ σῶμα πλανηθέντες ἔξω μὲν μὴ λάβωσιν
ἀναπνοήν, ἐντὸς δὲ εἰλλόμενοι τὴν ἀφ᾽ αὑτῶν ἀτμίδα τῇ τῆς ψυχῆς 87 Α
φορᾷ ξυμμίξαντες ἀνακερασθῶσι, παντοδαπὰ νοσήματα ψυχῆς
ἐμποιοῦσι μᾶλλον καὶ ἧττον καὶ ἐλάττω καὶ πλείω, πρός τε τοὺς
τρεῖς τόπους ἐνεχθέντα τῆς ψυχῆς, πρὸς ὃν ἂν ἕκαστ᾽ αὐτῶν
10 προσπίπτῃ, ποικίλλει μὲν εἴδη δυσκολίας καὶ δυσθυμίας παντο-
δαπά, ποικίλλει δὲ θρασύτητός τε καὶ δειλίας, ἔτι δὲ λήθης ἅμα
καὶ δυσμαθίας. πρὸς δὲ τούτοις, ὅταν οὕτω κακῶς παγέντων
πολιτεῖαι κακαὶ καὶ λόγοι κατὰ πόλεις ἰδίᾳ τε καὶ δημοσίᾳ Β
λεχθῶσιν, ἔτι δὲ μαθήματα μηδαμῇ τούτων ἰατικὰ ἐκ νέων μαν-
15 θάνηται, ταύτῃ κακοὶ πάντες οἱ κακοὶ διὰ δύο ἀκουσιώτατα
γιγνόμεθα· ὧν αἰτιατέον μὲν τοὺς φυτεύοντας ἀεὶ τῶν φυτευομένων
μᾶλλον καὶ τοὺς τρέφοντας τῶν τρεφομένων, προθυμητέον μήν,
ὅπῃ τις δύναται, καὶ διὰ τροφῆς καὶ δι᾽ ἐπιτηδευμάτων μαθημάτων
τε φυγεῖν μὲν κακίαν, τοὐναντίον δὲ ἑλεῖν. ταῦτα μὲν οὖν δὴ
20 τρόπος ἄλλος λόγων.

2 ἄκοντι : κακόν τι ASZ.   4 οἱ : ἢ Α.   12 δυσμαθίας : δυσμαθείας H.

imperfect apprehension of the type. All men necessarily desire what is good: but many causes combine to distort their apprehension of the good: whence arises vice.

2. ἐχθρὰ καὶ ἄκοντι] Cornarius' correction of κακόν τι into ἄκοντι seems nearly as certain as an emendation can be; and I can only wonder at Stallbaum's defence of the old reading. Perhaps Plato wrote the words as a crasis, κἄκοντι: this would readily become κακόν τι, after which the insertion of καὶ before it would follow as a matter of course.

τὸ περὶ τὰς λύπας] Here then we see what Plato means by calling pains ἀγαθῶν φυγαί in 69 D.

8. μᾶλλον καὶ ἧττον] I apprehend that these words apply to the intensity of the attack, ἐλάττω καὶ πλείω to the gravity of the disorder. There is a similar combi-nation of μᾶλλον καὶ ἐπὶ πλέον with ἧττον καὶ ἐπ᾽ ἔλαττον in Phaedo 93 Β.

πρός τε τοὺς τρεῖς τόπους] i.e. the seats of the three εἴδη of the soul, the liver, heart, and head: attacking the first, the vapours produce δυσκολία and δυσθυμία, attacking the heart, θρασύτης and δειλία, attacking the brain, they cause λήθη and δυσμαθία. The view that mental deficiencies are frequently due to bodily infirmity can be traced back to Sokrates: cf. Xenophon memorabilia III xii 6 ἐν πάσαις δὲ ταῖς τοῦ σώματος χρείαις πολὺ διαφέρει ὡς βέλτιστα τὸ σῶμα ἔχειν· καὶ γὰρ ἐν ᾧ δοκεῖς ἐλαχίστην σώματος χρείαν εἶναι, ἐν τῷ διανοεῖσθαι, τίς οὐκ οἶδεν ὅτι καὶ ἐν τούτῳ πολλοὶ μεγάλα σφάλλονται διὰ τὸ μὴ ὑγιαίνειν τὸ σῶμα; καὶ λήθη δὲ καὶ ἀθυμία καὶ δυσκολία καὶ μανία πολλάκις πολλοῖς διὰ τὴν τοῦ σώματος καχεξίαν εἰς τὴν διάνοιαν ἐμπίπτουσιν οὕτως, ὥστε καὶ τὰς ἐπιστήμας ἐκβάλλειν.

ΤΙΜΑΙΟΣ.

to some bad habit of body and unenlightened training that the wicked man becomes wicked; and these are always unwelcome and imposed against his will. And where pains are concerned, the soul likewise derives much evil from the body. For where the humours of acid and salt phlegms and those that are bitter and bilious roam about the body and find no outlet to the surface, but being pent up within and blending their own exhalations with the movement of the soul are mingled therewith, they induce all kinds of mental diseases, more or less violent and serious: and rushing to the three regions of the soul, in the part which each attacks they multiply manifold forms of moroseness and melancholy, of rashness and timidity, of forgetfulness and dulness. And when, besides these vicious conditions, there are added bad governments and bad principles maintained in public and private speech; when moreover no studies to be an antidote are pursued from youth up, then it is that all of us who are wicked become so, owing to two causes entirely beyond our own control. The blame must lie rather with those who train than with those who are trained, with the educators than the educated: however we must use our utmost zeal by education, pursuits, and studies to shun vice and embrace virtue. This subject however belongs to a different branch of inquiry.

10. ποικίλλει μὲν εἴδη] This comes to the same thing as γεννᾷ ποικίλα εἴδη.

14. λεχθῶσιν] There is an obvious zeugma here: with πολιτεῖαι we must mentally supply something like ξυστῶσιν. ἰδίᾳ καὶ δημοσίᾳ is used as in 88 A.

16. ὧν αἰτιατέον] Compare the famous passage in *Republic* 492 A ἦ καὶ σὺ ἡγεῖ ὥσπερ οἱ πολλοί, διαφθειρομένους τινὰς εἶναι ὑπὸ σοφιστῶν νέους, διαφθείροντας δέ τινας σοφιστὰς ἰδιωτικούς, ὅ τι καὶ ἄξιον λόγου, ἀλλ' οὐκ αὐτοὺς τοὺς ταῦτα λέγοντας μεγίστους μὲν εἶναι σοφιστάς, παιδεύειν δὲ τελεώτατα καὶ ἀπεργάζεσθαι οἴους βούλονται εἶναι καὶ νέους καὶ πρεσβυτέρους καὶ ἄνδρας καὶ γυναῖκας; Of course, on the other hand, the same allowance must be made to the teachers also, that they do not educate badly from preference of the bad, but because they know no better.

19. ταῦτα μὲν οὖν δή] i.e. the evil results of physical imperfection and bad training. The discussion of this subject, Plato says, is τρόπος ἄλλος λόγων, that is to say, it belongs to an ethical treatise.

87 C—89 D, c. xlii. But it is a pleasanter task to describe the means whereby the body is preserved and strengthened. All that is good is fair, all that is fair is symmetrical. Now we take great heed to lesser symmetries, but the most important of all, the symmetry of soul and body, we utterly neglect. Neither should the body be too weak for the soul, nor the soul too weak for the body. Just as some bodily disproportion is the cause of pain and fatigue to the sufferer, so it is here: either the soul wears out the body in her pursuit of knowledge, or the body hampers and stifles the soul. The only safeguard is to give due exercise both to

XLII. Τὸ δὲ τούτων ἀντίστροφον αὖ, τὸ περὶ τὰς τῶν C σωμάτων καὶ διανοήσεων θεραπείας αἷς αἰτίαις σῴζεται, πάλιν εἰκὸς καὶ πρέπον ἀνταποδοῦναι· δικαιότερον γὰρ τῶν ἀγαθῶν πέρι μᾶλλον ἢ τῶν κακῶν ἴσχειν λόγον. πᾶν δὴ τὸ ἀγαθὸν καλόν, 5 τὸ δὲ καλὸν οὐκ ἄμετρον· καὶ ζῷον οὖν τὸ τοιοῦτον ἐσόμενον ξύμμετρον θετέον. ξυμμετριῶν δὲ τὰ μὲν σμικρὰ διαισθανόμενοι ξυλλογιζόμεθα, τὰ δὲ κυριώτατα καὶ μέγιστα ἀλογίστως ἔχομεν. πρὸς γὰρ ὑγιείας καὶ νόσους ἀρετάς τε καὶ κακίας οὐδεμία ξυμμε- D τρία καὶ ἀμετρία μείζων ἢ ψυχῆς αὐτῆς πρὸς σῶμα αὐτό· ὧν 10 οὐδὲν σκοποῦμεν οὐδ' ἐννοοῦμεν, ὅτι ψυχὴν ἰσχυρὰν καὶ πάντῃ μεγάλην ἀσθενέστερον καὶ ἔλαττον εἶδος ὅταν ὀχῇ, καὶ ὅταν αὖ τοὐναντίον ξυμπαγῆτον τούτω, οὐ καλὸν ὅλον τὸ ζῷον· ἀξύμμετρον γὰρ ταῖς μεγίσταις ξυμμετρίαις· τὸ δὲ ἐναντίως ἔχον πάντων θεαμάτων τῷ δυναμένῳ καθορᾶν κάλλιστον καὶ ἐρασμιώτατον. 15 οἷον οὖν ὑπερσκελὲς ἢ καί τινα ἑτέραν ὑπέρεξιν ἄμετρον ἑαυτῷ E τι σῶμα ὂν ἅμα μὲν αἰσχρόν, ἅμα δ' ἐν τῇ κοινωνίᾳ τῶν πόνων πολλοὺς μὲν κόπους, πολλὰ δὲ σπάσματα καὶ διὰ τὴν παραφορότητα πτώματα παρέχον μυρίων κακῶν αἴτιον ἑαυτῷ· ταὐτὸν δὴ διανοητέον καὶ περὶ τοῦ ξυναμφοτέρου, ζῷον ὃ καλοῦμεν, ὡς ὅταν 20 τε ἐν αὐτῷ ψυχὴ κρείττων οὖσα σώματος περιθύμως ἴσχῃ, διασείουσα πᾶν αὐτὸ ἔνδοθεν νόσων ἐμπίπλησι, καὶ ὅταν εἴς τινας 88 A μαθήσεις καὶ ζητήσεις συντόνως ἴῃ, κατατήκει, διδαχάς τ' αὖ καὶ μάχας ἐν λόγοις ποιουμένη δημοσίᾳ καὶ ἰδίᾳ δι' ἐρίδων καὶ φιλο-

10 σκοποῦμεν: ἐσκοποῦμεν A.   11 ὀχῇ: ἐχῇ A.   15 ὑπέρεξιν: ὑπὲρ ἕξιν A.
22 συντόνως: εὐτόνως A.

body and to soul: the student must practise gymnastic, the athlete must cultivate his mind. We must in this matter follow the law of the universe. For the human body is subject to external influences, which, if left to themselves, quickly destroy it: but if it be exercised on the plan of the universal movement, it will be enabled to resist them; for by exercise the cognate and congenial particles are brought together, and the unlike and discordant are prevented from preying on each other. The best kind of exercise is when the body is moved by its own agency; it is less good if the agent is some external force, especially if only part of the body is moved: similarly of purifications the best is wrought by gymnastic, the next best by conveyance in vehicles; while that by drugs should only be employed in case of positive necessity. For every malady has its own natural period, which it is best not to disturb with medicine; and so has every individual and every species. Nature then should be suffered to take her own course and not be vexed by leechcraft.

3. δικαιότερον] We are endeavouring to trace how νοῦς ordered all things ἐπὶ τὸ βέλτιστον: therefore it is more appropriate to set forth ἀγαθά than κακά.

5. τὸ δὲ καλὸν οὐκ ἄμετρον] So the good is resolved into the beautiful, and beauty into proportion and symmetry, in

XLII. The counterpart to what has been said, the treatment of body and mind and the principles by which they are preserved, were the proper and fitting complement of our discourse: for it is more just to dwell upon good than upon evil. All that is good is fair, and what is fair is not disproportionate. Accordingly an animal that is to be fair must, we affirm, be well-proportioned. Now the smaller proportions we discern and reason upon them; but of the greatest and most momentous we take no account. For in view of health and sickness and virtue and vice no proportion or disproportion is more important than that existing between body and soul themselves: yet we pay no heed to these, nor do we reflect that if a feebler and smaller frame be the vehicle of a soul that is strong and mighty in all respects; or if the relation between the two be reversed, then the entire creature is not fair; for it is defective in the most essential proportions. But the opposite condition is to him who can discern it of all sights the fairest and loveliest. For example, a body which possesses legs of excessive length or which is unsymmetrical owing to any other disproportion, is not only ugly, but in taking its share of labour brings infinite distress on itself, suffering frequent fatigue and spasms, and often falling in consequence of inability to control its motions: the same then we must suppose to hold good of the combination of soul and body which we call an animal; when the soul in it is more powerful than the body and of ardent temperament, she agitates it and fills it from within with sickness; and when she impetuously pursues some study or research, she wastes the body away: and in giving instruction and conducting discussions private or

Philebus 64 E νῦν δὴ καταπέφευγεν ἡμῖν ἡ τοῦ ἀγαθοῦ δύναμις εἰς τὴν τοῦ καλοῦ φύσιν· μετριότης γὰρ καὶ ξυμμετρία κάλλος δήπου καὶ ἀρετὴ πανταχοῦ ξυμβαίνει γίγνεσθαι.

τὸ τοιοῦτον] sc. καλόν.

11. ὅταν ὀχῇ] Cf. 69 C ὄχημά τε πᾶν τὸ σῶμα ἔδοσαν.

12. ἀξύμμετρον γὰρ ταῖς μεγίσταις ξυμμετρίαις] The expression is remarkable. I cannot cite an instance which seems to me exactly parallel.

18. ταὐτὸν δὴ διανοητέον] Compare

Republic 535 D φιλοπονίᾳ οὐ δεῖ χωλὸν εἶναι τὸν ἁψόμενον, τὰ μὲν ἡμίσεα φιλόπονον, τὰ δ' ἡμίσεα ἄπονον, ἔστι δὲ τοῦτο, ὅταν τις φιλογυμναστὴς μὲν καὶ φιλόθηρος ᾖ καὶ πάντα τὰ διὰ τοῦ σώματος φιλοπονῇ, φιλομαθὴς δὲ μή, μηδὲ φιλήκοος μηδὲ ζητητικός, ἀλλ' ἐν πᾶσι τούτοις μισοπονῇ· χωλὸς δὲ καὶ ὁ τἀναντία τούτου μεταβεβληκὼς τὴν φιλοπονίαν.

20. περιθύμως ἴσχῃ] This simply means impetuous or masterful, without any special reference to the θυμοειδές.

23. δημοσίᾳ καὶ ἰδίᾳ] Plato evi-

νεικίας γιγνομένων διάπυρον αὐτὸ ποιοῦσα λύει, καὶ ῥεύματα ἐπάγουσα, τῶν λεγομένων ἰατρῶν ἀπατῶσα τοὺς πλείστους, τἀναντία αἰτιᾶσθαι ποιεῖ· σῶμά τε ὅταν αὖ μέγα καὶ ὑπέρψυχον σμικρᾷ ξυμφυὲς ἀσθενεῖ τε διανοίᾳ γένηται, διττῶν ἐπιθυμιῶν οὐσῶν φύσει B
5 κατ' ἀνθρώπους, διὰ σῶμα μὲν τροφῆς, διὰ δὲ τὸ θειότατον τῶν ἐν ἡμῖν φρονήσεως, αἱ τοῦ κρείττονος κινήσεις κρατοῦσαι καὶ τὸ μὲν σφέτερον αὔξουσαι, τὸ δὲ τῆς ψυχῆς κωφὸν καὶ δυσμαθὲς ἀμνῆμόν τε ποιοῦσαι, τὴν μεγίστην νόσον ἀμαθίαν ἐναπεργάζονται. μία δὴ σωτηρία πρὸς ἄμφω, μήτε τὴν ψυχὴν ἄνευ σώματος κινεῖν μήτε
10 σῶμα ἄνευ ψυχῆς, ἵνα ἀμυνομένω γίγνησθον ἰσορρόπω καὶ ὑγιῆ. τὸν δὴ μαθηματικὸν ἤ τινα ἄλλην σφόδρα μελέτην διανοίᾳ κατερ- C γαζόμενον καὶ τὴν τοῦ σώματος ἀποδοτέον κίνησιν, γυμναστικῇ προσομιλοῦντα, τόν τε αὖ σῶμα ἐπιμελῶς πλάττοντα τὰς τῆς ψυχῆς ἀνταποδοτέον κινήσεις, μουσικῇ καὶ πάσῃ φιλοσοφίᾳ προσ-
15 χρώμενον, εἰ μέλλει δικαίως τις ἅμα μὲν καλός, ἅμα δὲ ἀγαθὸς ὀρθῶς κεκλήσεσθαι. κατὰ δὲ ταὐτὰ ταῦτα καὶ τὰ μέρη θεραπευτέον, τὸ τοῦ παντὸς ἀπομιμούμενον εἶδος. τοῦ γὰρ σώματος ὑπὸ τῶν

dently means forensic oratory on the one hand and eristic discussions on the other, cf. *Sophist* 225 B, 268 B: dialectic seems to be excluded by δι' ἐρίδων, perhaps because the calm and dispassionate temper in which the true philosopher conducts his arguments is less likely to lead to injury of his health.

2. **τἀναντία αἰτιᾶσθαι**] The physicians set down to purely physical causes what is really due to the action of a vigorous mind upon a body which is too feeble for it. Martin falls into a strange error in imagining that Plato would actually sacrifice the vigour and excellence of the soul in order to preserve due proportion with the body—'les qualités de l'âme ne sauraient jamais être ni devenir trop belles'. What Plato says is that the model ζῷον is the union of a fair and vigorous soul with a fair and vigorous body; and if the body is too weak for the soul, unfortunate results are likely to happen. For this reason the body ought to receive due attention and training that it may be preserved in such health and vigour as to render it a fitting vehicle for the soul. But nothing can be more alien to the whole spirit of Plato's thought than the notion that the soul is not to be cultivated to the highest degree, even though she have the misfortune to be united to an inferior body. We can never make the soul 'trop belle'; but we must not neglect to keep her corporeal habitation fit for her residence.

3. **ὑπέρψυχον**] i.e. too great for the soul. This reading is indubitably right, although according to the general analogy the word would mean 'having an excess of soul', like ὑπέρθυμος, and ὑπερσκελὲς above. The old reading was ὑπέρψυχρον, which is found in some mss.

7. **τὸ δὲ τῆς ψυχῆς**] Compare the passage of the *Phaedo* 66 B quoted above. The teaching of the present passage is not in any way at variance with the doctrine of the *Phaedo* that the soul should withdraw herself so far as she can from the company of the body. However completely the body may be in

public in a spirit of contention and rivalry, she inflames and weakens its fabric and brings on chills; and thus deceiving most of the so-called physicians induces them to assign causes for the malady which are really in no way concerned with it. When on the other hand a large body, too great for the soul, is joined with a small and feeble mind,—two kinds of appetites being natural to mankind, on account of the body a craving for nourishment and on account of the divinest part of us for knowledge—the motions of the stronger prevail and strengthen their faculty, but that of the soul they render dull and slow of learning and of recollection, and so produce stupidity, the most grievous of maladies. There is but one safeguard against both these misfortunes: neither should the soul be exercised without the body nor the body without the soul, in order that they may be a match for each other and attain balance and health. So the mathematician, or whosoever is intensely absorbed in any intellectual study, must allow corresponding exercise to his body, submitting to athletic training; while he who is careful in forming his body must in turn give due exercise to his soul, calling in the aid of art and of all philosophy, if he is justly to be called at once fair and in the true sense good. The same treatment too should be applied to the separate parts, in imitation of the fashion of the All. For as the body is inflamed and cooled

---

subjection to the soul, it must be kept as healthy as possible, else it impedes the activity of the intellect: neglect of the body actually hinders the withdrawal of the soul, since her companion is perpetually forcing itself upon her notice with its maladies. At the same time when Plato is, as here, treating physically of the perfection of the ζῷον, he naturally lays more stress than in the *Phaedo* upon the attention due to the body. For the *Phaedo* gives us a 'study of death', the *Timaeus* a theory of life.

13. τόν τε αὖ σῶμα ἐπιμελῶς πλάττοντα] This sentence is, I think, sufficient to show the superfluity of the diverse emendations that have been proposed in *Phaedo* 82 D ἀλλὰ μὴ σῶμά τι (or σώματα) πλάττοντες. Compare *Re-*

public 377 C τοὺς δ' ἐγκριθέντας πείσομεν τὰς τροφούς τε καὶ μητέρας λέγειν τοῖς παισί, καὶ πλάττειν τὰς ψυχὰς αὐτῶν τοῖς μύθοις πολὺ μᾶλλον ἢ τὰ σώματα ταῖς χερσίν.

17. τὸ τοῦ παντὸς ἀπομιμούμενον εἶδος] i. e. imitating the vibration of the ὑποδοχή, which sifts the elements into their appropriate regions.

ὑπὸ τῶν εἰσιόντων] This seems to refer to the action of fire and air upon the nutriment received into the body: see 79 A. It is notable that Plato makes the temperature dependent upon internal agencies, assigning to the external merely variation of dryness and moisture: did he know, for instance, that the temperature of the blood is normally almost unaffected by the temperature of the air?

332 ΠΛΑΤΩΝΟΣ [88 D—

εἰσιόντων καομένου τε ἐντὸς καὶ ψυχομένου, καὶ πάλιν ὑπὸ τῶν D
ἔξωθεν ξηραινομένου καὶ ὑγραινομένου καὶ τὰ τούτοις ἀκόλουθα
πάσχοντος ὑπ' ἀμφοτέρων τῶν κινήσεων, ὅταν μέν τις ἡσυχίαν
ἄγον τὸ σῶμα παραδιδῷ ταῖς κινήσεσι, κρατηθὲν διώλετο, ἐὰν δὲ
5 ἥν τε τροφὸν καὶ τιθήνην τοῦ παντὸς προσείπομεν μιμῆταί τις, καὶ
τὸ σῶμα μάλιστα μὲν μηδέποτε ἡσυχίαν ἄγειν ἐᾷ, κινῇ δὲ καὶ
σεισμοὺς ἀεί τινας ἐμποιῶν αὐτῷ διὰ παντὸς τὰς ἐντὸς καὶ ἐκτὸς E
ἀμύνηται κατὰ φύσιν κινήσεις, καὶ μετρίως σείων τά τε περὶ τὸ
σῶμα πλανώμενα παθήματα καὶ μέρη κατὰ ξυγγενείας εἰς τάξιν
10 κατακοσμῇ πρὸς ἄλληλα, κατὰ τὸν πρόσθεν λόγον, ὃν περὶ τοῦ
παντὸς ἐλέγομεν, οὐκ ἐχθρὸν παρ' ἐχθρὸν τιθέμενον ἐάσει πολέμους
ἐντίκτειν τῷ σώματι καὶ νόσους, ἀλλὰ φίλον παρὰ φίλον τεθὲν
ὑγίειαν ἀπεργαζόμενον παρέξει. τῶν δ' αὖ κινήσεων ἡ ἐν ἑαυτῷ 89
ὑφ' αὑτοῦ ἀρίστη κίνησις· μάλιστα γὰρ τῇ διανοητικῇ καὶ τῇ τοῦ
15 παντὸς κινήσει ξυγγενής· ἡ δὲ ὑπ' ἄλλου χείρων· χειρίστη δὲ ἡ
κειμένου τοῦ σώματος καὶ ἄγοντος ἡσυχίαν δι' ἑτέρων αὐτὸ κατὰ
μέρη κινοῦσα. διὸ δὴ τῶν καθάρσεων καὶ ξυστάσεων τοῦ σώματος
ἡ μὲν διὰ τῶν γυμνασίων ἀρίστη, δευτέρα δὲ ἡ διὰ τῶν αἰωρήσεων
κατά τε τοὺς πλοῦς καὶ ὅπῃ περ ἂν ὀχήσεις ἄκοποι γίγνωνται·
20 τρίτον δὲ εἶδος κινήσεως σφόδρα ποτὲ ἀναγκαζομένῳ χρήσιμον,
ἄλλως δὲ οὐδαμῶς τῷ νοῦν ἔχοντι προσδεκτέον, τὸ τῆς φαρμα- B
κευτικῆς καθάρσεως γιγνόμενον ἰατρικόν. τὰ γὰρ νοσήματα,
ὅσα μὴ μεγάλους ἔχει κινδύνους, οὐκ ἐρεθιστέον φαρμακείαις.

5 τε post ἦν delet S.   11 ἐλέγομεν : λέγομεν A.

1. ὑπὸ τῶν ἔξωθεν] i.e. by the circumfluent elements: see 81 A.

6. μάλιστα μέν] These words suggest a δεύτερος πλοῦς, implied but not expressed—'if possible keep the body in constant activity, or at least as nearly so as may be'.

7. σεισμοὺς ἀεί τινας] Plato's meaning is that the natural and voluntary motions of the body will do for it what the vibration of the ὑποδοχὴ does for the universe; that is to say, it will sift things into their right places. The various forces which act upon the body tend to dissolve its substance and confuse it at random, and thus produce sickness and discomfort by the juxtaposition of uncongenial and incongruous particles. This is counteracted by the natural movement of the body, which restores the due relative position of the particles: thus if ὑπὸ τῶν ἔξωθεν a particle of water is changed into one of air, and so we have air where water ought to be, the motion of the body sends the air where it ought to be and supplies its former place with water. In such manner equilibrium and health are preserved.

9. παθήματα καὶ μέρη] A somewhat curious collocation. The παθήματα are roaming about the body seeking ἀναπνοή, which the σεισμοί enable them to find: the μέρη are the elemental particles, which are thus shifted each into its proper place.

13. τῶν δ' αὖ κινήσεων] The modes

within by the particles that enter, and again is dried and moistened by those that are outside, and by the agency of these two forces suffers all that ensues upon these conditions, if we submit the body passively to the forces aforesaid, it is overcome and destroyed: but if we imitate what we have called the fostress and nurse of the All, and allow the body, if possible, never to be inactive, but keep it astir and, exciting continual vibrations in it, furnish it with the natural defence against the motions from without and within; and by moderately exercising it bring into orderly relation with each other according to their affinities the affections and particles that are going astray in the body; then, as we have already described in speaking of the universe, we shall not suffer mutually hostile particles to be side by side and to engender discord and disease in the body, but we shall set friend beside friend so as to bring about a healthy state. Of all motions that which arises in any body by its own action is the best (for it is most nearly allied to the motion of thought and of the All), but that which is brought about by other agency is inferior; and the worst of all is that which, while the body is lying still, is produced by other agents which move it piecemeal. Accordingly of all modes of purifying and restoring the body gymnastic is the best; the next best is any swinging motion such as of sailing or any other conveyance of the body which does not tire it: a third kind is useful sometimes under absolute necessity, but in no other circumstances should be employed by a judicious person, I mean medical purgation effected by drugs. No disease, not involving imminent danger, should be irritated by drugs. For

in which the body may be exercised are threefold: (1) when it moves itself as a whole; (2) when it is moved as a whole by some external agency; (3) when parts are moved by external agency, the rest remaining stationary. The first and best is gymnastic; the second travelling in a boat or any other means of conveyance; the third includes the action of medical cathartics, which are to be avoided, unless absolutely necessary. Compare *Laws* 789 C τὰ σώματα πάντα ὑπὸ τῶν σεισμῶν τε καὶ κινήσεων κινούμενα ἄκοπα ὀνίναται

πάντων, ὅσα τε ὑπὸ ἑαυτῶν ἢ καὶ ἐν αἰώραις ἢ καὶ κατὰ θάλατταν καὶ ἐφ' ἵππων ὀχούμενα καὶ ὑπ' ἄλλων ὁπωσοῦν δὴ φερομένων τῶν σωμάτων κινεῖται.

18. αἰωρήσεων] This refers probably to a gymnastic machine called αἰώρα, a kind of swing.

23. οὐκ ἐρεθιστέον φαρμακείαις] Compare Hippokrates *aphorisms*, vol. III p. 711 Kühn τὰ κρινόμενα καὶ τὰ κεκριμένα ἀρτίως μὴ κινέειν μηδὲ νεωτερ·ποιέειν μήτε φαρμακίοισι μήτε ἄλλοισι ἐρεθισμοῖσι, ἀλλ' ἐᾶν.

334   ΠΛΑΤΩΝΟΣ   [89 B—

πᾶσα γὰρ ξύστασις νόσων τρόπον τινὰ τῇ τῶν ζῴων φύσει προσέοικε. καὶ γὰρ ἡ τούτων ξύνοδος ἔχουσα τεταγμένους τοῦ βίου γίγνεται χρόνους τοῦ τε γένους ξύμπαντος καὶ κατ' αὐτὸ τὸ ζῷον εἱμαρμένον ἕκαστον ἔχον τὸν βίον φύεται, χωρὶς τῶν ἐξ ἀνάγκης παθημάτων· C
5 τὰ γὰρ τρίγωνα εὐθὺς κατ' ἀρχὰς ἑκάστου δύναμιν ἔχοντα ξυνίσταται μέχρι τινὸς χρόνου δυνατὰ ἐξαρκεῖν, οὗ βίου οὐκ ἄν ποτέ τις εἰς τὸ πέραν ἔτι βιῴη. τρόπος οὖν ὁ αὐτὸς καὶ τῆς περὶ τὰ νοσήματα ξυστάσεως· ἣν ὅταν τις παρὰ τὴν εἱμαρμένην τοῦ χρόνου φθείρῃ φαρμακείαις, ἅμα ἐκ σμικρῶν μεγάλα καὶ πολλὰ ἐξ ὀλίγων
10 νοσήματα φιλεῖ γίγνεσθαι. διὸ παιδαγωγεῖν δεῖ διαίταις πάντα τὰ τοιαῦτα, καθ' ὅσον ἂν ᾖ τῳ σχολή, ἀλλ' οὐ φαρμακεύοντα κακὸν D δύσκολον ἐρεθιστέον.

XLIII. Καὶ περὶ μὲν τοῦ κοινοῦ ζῴου καὶ τοῦ κατὰ τὸ σῶμα αὐτοῦ μέρους, ᾗ τις ἂν καὶ διαπαιδαγωγῶν καὶ διαπαιδαγωγούμενος
15 ὑφ' αὑτοῦ μάλιστ' ἂν κατὰ λόγον ζῴη, ταύτῃ λελέχθω· τὸ δὲ δὴ παιδαγωγῆσον αὐτὸ μᾶλλόν που καὶ πρότερον παρασκευαστέον εἰς δύναμιν ὅ τι κάλλιστον καὶ ἄριστον εἰς τὴν παιδαγωγίαν εἶναι. δι' ἀκριβείας μὲν οὖν περὶ τούτων διελθεῖν ἱκανὸν ἂν γένοιτο αὐτὸ καθ' αὑτὸ μόνον ἔργον· τὸ δ' ἐν παρέργῳ κατὰ τὰ πρόσθεν ἑπόμενος E

3 κατ' αὐτό : καθ' αὐτὸ SZ.   19 τὰ πρόσθεν : τὸ πρόσθεν A.

1. **πᾶσα γὰρ ξύστασις**] Every form of disease has a certain correspondence with the constitution of animals. For as there are fixed periods for which both the individual and the species will endure, but no longer, seeing that the elementary triangles are calculated to hold out a certain definite time against the forces of dissolution, even so every disease has its fixed period to run; and if this be rashly interfered with by medicine, a slight ailment may easily be converted into a dangerous sickness. Compare the discussion on medical treatment in *Republic* 405 D foll.

2. **ἡ τούτων ξύνοδος**] Their conjunction, i.e. their composition or constitution.

3. **τοῦ τε γένους ξύμπαντος**] Plato's statement that the species wears out as well as the individual is very notable. Although he does not explain the cause why a species becomes extinct, we may well suppose him to conceive that in course of generations the triangles transmitted by the parent to the offspring are no longer fresh and accurate; so that every succeeding generation becomes more feeble, and finally the race disappears.

4. **χωρὶς τῶν ἐξ ἀνάγκης παθημάτων**] i.e. apart from accidents or illness. This use of the word ἀνάγκη falls in with the explanation of it offered above on p. 166.

10. **διὸ παιδαγωγεῖν**] That is, we should guide the disease, not drive it; and by suitable diet and mode of life suffer it to run its course in the easiest and safest way.

11. **καθ' ὅσον ἂν ᾖ τῳ σχολή**] i.e. he must not pay exclusive attention to it so as to leave no time for mental culture.

89 D—90 D, c. xliii. Man then being formed of body and soul united, his guide is the soul: therefore must he diligently take heed to her well-being. And where-

ΤΙΜΑΙΟΣ.

every form of sickness has a certain correspondence to the nature of living creatures. Their constitution is so ordered as to have definite periods of life both for the kind and for the individual, which has its own fixed span of existence, always excepting inevitable accidents. For the triangles of each creature are composed at the very outset with the capacity of holding out for a certain definite time; beyond which its life cannot be prolonged. The same applies also to the constitution of diseases; if these are interfered with by medicine to the disregard of their appointed period, it often happens that a few slight maladies are rendered numerous and grave. Wherefore we should guide all such sicknesses by careful living, so far as we have time to attend to it, and not provoke a troublesome mischief by medical treatment.

XLIII. Now so far as concerns the animate creature and the bodily part of it, how a man, guiding the latter and by himself being guided, should live a most rational life, let this discussion suffice. But the part which is to guide the body must beforehand be trained with still greater care to be most perfect and efficient for education. To deal with this subject minutely would in itself be a sufficient task: but if we may merely touch upon it in conformity with our previous discourse, we should

as there are three forms of soul existing in man, that form will be the most powerful which is most fully exercised. Wherefore he must be careful to give freest activity to that divinest part, which is his guiding genius, and which lifts him up towards his birthplace in the heavens. For he whose care is for earthly lusts and ambitions will become, so far as that may be, mortal altogether, he and all his thoughts; but whoso sets his heart upon knowledge and truth, he, so far as man may attain to immortality, will be immortal and supremely blessed. And this he must ensue by dwelling in thought upon the eternal truth; and making his soul like to that she contemplates, so he will fulfil the perfect life.

13. τοῦ κοινοῦ ζῴου] i.e. the living creature consisting of soul and body united, the ξυναμφότερον.

14. διαπαιδαγωγῶν καὶ διαπαιδαγωγούμενος] Stallbaum gives a strange perversion of this passage. He desires to read ὑπ' αὐτοῦ for ὑφ' αὐτοῦ, giving the truly remarkable result that man must be guided by his body! 'Cette monstrueuse altération du texte', as Martin not too forcibly terms it, is unworthy of discussion. The vulgate is obviously right; the sense being that a man must train his bodily part and be trained by himself, that is, by his true self, the soul.

15. τὸ δὲ δὴ παιδαγωγῆσον] This is of course ψυχή.

18. δι' ἀκριβείας] Such an exposition does in fact occupy nearly a whole book, the seventh, of the *Republic*; where we have the following programme laid down: (1) arithmetic, (2) plane geometry, (3) solid geometry, (4) astronomy, (5) harmony, (6) dialectic.

336 ΠΛΑΤΩΝΟΣ [89 E—

ἄν τις οὐκ ἄπο τρόπου τῇδε σκοπῶν ὧδε τῷ λόγῳ διαπεράναιτ' ἄν.
καθάπερ εἴπομεν πολλάκις, ὅτι τρία τριχῇ ψυχῆς ἐν ἡμῖν εἴδη
κατῴκισται, τυγχάνει δὲ ἕκαστον κινήσεις ἔχον, οὕτω κατὰ ταὐτὰ
καὶ νῦν ὡς διὰ βραχυτάτων ῥητέον, ὅτι τὸ μὲν αὐτῶν ἐν ἀργίᾳ
5 διάγον καὶ τῶν ἑαυτοῦ κινήσεων ἡσυχίαν ἄγον ἀσθενέστατον ἀνάγκη
γίγνεσθαι, τὸ δ' ἐν γυμνασίοις ἐρρωμενέστατον· διὸ φυλακτέον,
ὅπως ἂν ἔχωσι τὰς κινήσεις πρὸς ἄλληλα συμμέτρους. τὸ δὲ περὶ 90 A
τοῦ κυριωτάτου παρ' ἡμῖν ψυχῆς εἴδους διανοεῖσθαι δεῖ τῇδε, ὡς
ἄρα αὐτὸ δαίμονα θεὸς ἑκάστῳ δέδωκε, τοῦτο ὃ δή φαμεν οἰκεῖν μὲν
10 ἡμῶν ἐπ' ἄκρῳ τῷ σώματι, πρὸς δὲ τὴν ἐν οὐρανῷ ξυγγένειαν ἀπὸ
γῆς ἡμᾶς αἴρειν ὡς ὄντας φυτὸν οὐκ ἔγγειον ἀλλὰ οὐράνιον, ὀρθότατα λέγοντες· ἐκεῖθεν γάρ, ὅθεν ἡ πρώτη τῆς ψυχῆς γένεσις ἔφυ,
τὸ θεῖον τὴν κεφαλὴν καὶ ῥίζαν ἡμῶν ἀνακρεμαννὺν ὀρθοῖ πᾶν τὸ B
σῶμα. τῷ μὲν οὖν περὶ τὰς ἐπιθυμίας ἢ περὶ φιλονεικίας τετευτα-
15 κότι καὶ ταῦτα διαπονοῦντι σφόδρα πάντα τὰ δόγματα ἀνάγκη
θνητὰ ἐγγεγονέναι, καὶ παντάπασι καθ' ὅσον μάλιστα δυνατὸν
θνητῷ γίγνεσθαι, τούτου μηδὲ σμικρὸν ἐλλείπειν, ἅτε τὸ τοιοῦτον
ηὐξηκότι· τῷ δὲ περὶ φιλομαθίαν καὶ περὶ τὰς ἀληθεῖς φρονήσεις
ἐσπουδακότι καὶ ταῦτα μάλιστα τῶν αὐτοῦ γεγυμνασμένῳ φρονεῖν C
20 μὲν ἀθάνατα καὶ θεῖα, ἄνπερ ἀληθείας ἐφάπτηται, πᾶσα ἀνάγκη

7 δὴ post τὸ δὲ addit S. 14 ἐπιθυμίας: προθυμίας A. περὶ ante φιλονεικίας omittunt SZ. 18 φιλομαθίαν: φιλομάθειαν SZ. τὰς ἀληθεῖς: τὰς τῆς ἀληθείας S.

2. **τρία τριχῇ**] This seems a favourite phrase with Plato; see above, 52 D, ὄν τε καὶ χώραν καὶ γένεσιν εἶναι, τρία τριχῇ. Compare too *Sophist* 266 D τίθημι δύο διχῇ ποιητικῆς εἴδη.

7. **πρὸς ἄλληλα συμμέτρους**] Not in equal measure, but properly proportioned to their relative merits, so that the highest εἶδος may be supreme, and the two lower in due subordination.

8. **ὡς ἄρα αὐτὸ δαίμονα**] Compare Plutarch *de genio Socratis* § 22 τὸ μὲν οὖν ὑποβρύχιον ἐν τῷ σώματι ψυχὴ λέγεται· τὸ δὲ φθορᾶς λειφθὲν οἱ πολλοὶ νοῦν καλοῦντες ἐντὸς εἶναι νομίζουσιν αὐτῶν, ὥσπερ ἐν τοῖς ἐσόπτροις τὰ φαινόμενα κατ' ἀνταύγειαν· οἱ δ' ὀρθῶς ὑπονοοῦντες ὡς ἐκτὸς ὄντα δαίμονα προσαγορεύουσι. Plutarch here deviates in more than one point from Plato's doctrine. Plato, in calling the intellect δαίμων, does not of course mean that it is ἐκτός. Also Plutarch, like many of the later, especially neoplatonist, writers, draws an unplatonic distinction between νοῦς and ψυχή, although a little above he has used correcter language. In Plato νοῦς is simply ψυχή exercising her own unimpeded functions. Plato gives us to understand that the true δαίμων ὃν ἕκαστος εἴληχεν is our own mind: we are to look for guidance not to any external source, but to ourselves, to the divinest part of our nature.

10. **πρὸς δὲ τὴν ἐν οὐρανῷ ξυγγένειαν**] See 41 D, E. The affinity of the highest part of the soul to the skies is poetically assigned as the cause why man alone of all animals walks upright: compare 91 E foll. It is amusing to compare the prosaic and matter-of-fact treatment of the same

find a consistent answer to the question from the following reflections. As we have often said, three forms of soul with threefold functions are implanted in us, and each of these has its proper motions. Accordingly we may say as briefly as possible that whichever of these continues in idleness and keeps its own motions inactive, this must needs become the weakest; but that which is in constant exercise waxes strongest: wherefore we must see that they exercise their motions in due proportion. As to the supreme form of soul that is within us, we must believe that God has given it to each of us as a guiding genius— even that which we say, and say truly, dwells in the summit of our body and raises us from earth towards our celestial affinity, seeing we are of no earthly, but of heavenly growth: since to heaven, whence in the beginning was the birth of our soul, the diviner part attaches the head or root of us and makes our whole body upright. Now whoso is busied with appetites or ambitions and labours hard after these, all the thoughts of his heart must be altogether mortal; and so far as it is possible for him to become utterly mortal, he falls no whit short of this; for this is what he has been fostering. But he whose heart has been set on the love of learning and on true wisdom, and has chiefly exercised this part of himself, this man must without fail have thoughts that are immortal and divine, if he lay hold upon

subject by Sokrates: Xenophon *memorabilia* I iv 11.

13. τὴν κεφαλὴν καὶ ῥίζαν ἡμῶν] The significance of this bold and beautiful metaphor is that, as a plant draws its sustenance through its roots from its native earth, so does the soul draw her spiritual sustenance through the head from her native heavens. Very different is the spirit of Aristotle's comparison, *de anima* II iv 416ᵃ 4 ὡς ἡ κεφαλὴ τῶν ζῴων οὕτως αἱ ῥίζαι τῶν φυτῶν: the analogy only refers to physical nutriment, cf. II i 412ᵇ 3 αἱ δὲ ῥίζαι τῷ στόματι ἀνάλογον· ἄμφω γὰρ ἕλκει τὴν τροφήν: and similarly Galen *de plac. Hipp. et Plat.* v 524 ὁποῖον γάρ τι τοῖς ζῴοις ἐστὶ τὸ στόμα, τοιοῦτον τοῖς φυτοῖς τὸ πέρας τῆς ῥιζώσεως ἀτεχνῶς φάναι δοκεῖ στοματίων πολλῶν ἑλκόντων ἐκ τῆς γῆς τροφὴν ὑπὸ τῆς φύσεως δεδημιουργημένην.

16. καθ' ὅσον μάλιστα δυνατόν] Do what he will, he cannot become altogether θνητός, because, to whatever degraded form of organic life he may descend, he always has the ἀθάνατος·ἀρχὴ which the δημιουργὸς delivered to the gods. τὸ τοιοῦτον = τὸ θνητόν.

19. φρονεῖν μὲν ἀθάνατα] Compare *Symposium* 212 A ἢ οὐκ ἐνθυμεῖ, ἔφη, ὅτι ἐνταῦθα αὐτῷ μοναχοῦ γενήσεται, ὁρῶντι ᾧ ὁρατὸν τὸ καλόν, τίκτειν οὐκ εἴδωλα ἀρετῆς, ἅτε οὐκ εἰδώλου ἐφαπτομένῳ, ἀλλ' ἀληθῆ, ἅτε τοῦ ἀληθοῦς ἐφαπτομένῳ· τεκόντι δὲ ἀρετὴν ἀληθῆ καὶ θρεψαμένῳ ὑπάρχει θεοφιλεῖ γενέσθαι, καὶ εἴπερ τῳ ἄλλῳ ἀνθρώπων, ἀθανάτῳ καὶ ἐκείνῳ; see too Aristotle *nicomachean ethics* X vii

που, καθ' ὅσον δ' αὖ μετασχεῖν ἀνθρωπίνη φύσις ἀθανασίας ἐνδέχεται, τούτου μηδὲν μέρος ἀπολείπειν, ἅτε δὲ ἀεὶ θεραπεύοντα τὸ
θεῖον ἔχοντά τε αὐτὸν εὖ κεκοσμημένον τὸν δαίμονα ξύνοικον ἐν
αὑτῷ, διαφερόντως εὐδαίμονα εἶναι. θεραπεία δὲ δὴ παντὶ πάντως
5 μία, τὰς οἰκείας ἑκάστῳ τροφὰς καὶ κινήσεις ἀποδιδόναι· τῷ δ' ἐν
ἡμῖν θείῳ ξυγγενεῖς εἰσὶ κινήσεις αἱ τοῦ παντὸς διανοήσεις καὶ D
περιφοραί· ταύταις δὴ ξυνεπόμενον ἕκαστον δεῖ τὰς περὶ τὴν γένεσιν
ἐν τῇ κεφαλῇ διεφθαρμένας ἡμῶν περιόδους ἐξορθοῦντα διὰ τὸ
καταμανθάνειν τὰς τοῦ παντὸς ἁρμονίας τε καὶ περιφορὰς τῷ
10 κατανοουμένῳ τὸ κατανοοῦν ἐξομοιῶσαι κατὰ τὴν ἀρχαίαν φύσιν,
ὁμοιώσαντα δὲ τέλος ἔχειν τοῦ προτεθέντος ἀνθρώποις ὑπὸ θεῶν
ἀρίστου βίου πρός τε τὸν παρόντα καὶ τὸν ἔπειτα χρόνον.

XLIV. Καὶ δὴ καὶ τὰ νῦν ἡμῖν ἐξ ἀρχῆς παραγγελθέντα E
διεξελθεῖν περὶ τοῦ παντὸς μέχρι γενέσεως ἀνθρωπίνης σχεδὸν
15 ἔοικε τέλος ἔχειν. τὰ γὰρ ἄλλα ζῷα ᾗ γέγονεν αὖ, διὰ βραχέων
ἐπιμνηστέον, ὃ μή τις ἀνάγκη μηκύνειν· οὕτω γὰρ ἐμμετρώτερός
τις ἂν αὑτῷ δόξειε περὶ τοὺς τούτων λόγους εἶναι. τῇδ' οὖν τὸ
τοιοῦτον ἔστω λεγόμενον. τῶν γενομένων ἀνδρῶν ὅσοι δειλοὶ καὶ
τὸν βίον ἀδίκως διῆλθον, κατὰ λόγον τὸν εἰκότα γυναῖκες μετε-
20 φύοντο ἐν τῇ δευτέρᾳ γενέσει. καὶ κατ' ἐκεῖνον δὴ τὸν χρόνον διὰ 91 A
ταῦτα θεοὶ τὸν τῆς ξυνουσίας ἔρωτα ἐτεκτήναντο, ζῷον τὸ μὲν ἐν

3 μάλα post εὖ addit S.   4 πάντως: παντὸς S.
16 ἐμμετρώτερος: ἐμμετρότερος HS.

1177ᵇ 30 εἰ δὴ θεῖον ὁ νοῦς πρὸς τὸν ἄνθρωπον, καὶ ὁ κατὰ τοῦτον βίος θεῖος πρὸς
τὸν ἀνθρώπινον βίον· οὐ χρὴ δὲ κατὰ τοὺς
παραινοῦντας ἀνθρώπινα φρονεῖν ἄνθρωπον
ὄντα οὐδὲ θνητὰ τὸν θνητόν, ἀλλ' ἐφ' ὅσον
ἐνδέχεται ἀθανατίζειν καὶ πάντα ποιεῖν
πρὸς τὸ ζῆν κατὰ τὸ κράτιστον τῶν ἐν
αὑτῷ. A sentence worthy of Plato himself.

4. εὐδαίμονα] i.e. εὐδαίμων signifies ὁ ἔχων τὸν δαίμονα εὖ κεκοσμημένον.

θεράπεια δὲ δὴ παντὶ] sc. τῆς ψυχῆς εἴδει.

6. ξυγγενεῖς εἰσὶ κινήσεις] cf. 47 B τὰς περιφορὰς τῆς παρ' ἡμῖν διανοήσεως ξυγγενεῖς ἐκείναις οὔσας, ἀκινήτοις τεταραγμένας. Plato frequently fuses in his language the symbol with what it symbolises, the περιφορά with the διανόησις.

7. τὰς περὶ τὴν γένεσιν] The περίοδοι are distorted by the inflowing and outflowing stream of nutrition; see 43 A foll.

10. κατὰ τὴν ἀρχαίαν φύσιν] i.e. according to its original and proper nature qua soul, before contamination by contact with matter: the priority being of course logical.

90 E—92 C, c. xliv. And now our tale is well-nigh told. For in the first generation the gods made men, and in the second women: and they caused love to arise between man and woman and a desire of continuing their race. And afterwards from such as followed not after wisdom and truth sprang the fowls of the air and the beasts of the field, whose heads are turned earthwards, because

the truth; and so far as it lies in human nature to possess immortality, he lacks nothing thereof; and seeing that he ever cherishes the divinest part and keeps in good estate the guardian spirit that dwells in him, he must be happy above all. And the care of this is always the same for every man, to wit that he assign to every part its proper exercise and nourishment. To the divine part of us are akin the thoughts and revolutions of the All: these every man should follow, restoring the revolutions in the head, that are marred through our earthly birth, by learning to know the harmonies and revolutions of the All, so as to render the thinking soul like the object of its thought according to her primal nature: and when he has made it like, so shall he have the fulfilment of that most excellent life that was set by the gods before mankind for time present and time to come.

XLIV. Thus then the task laid upon us at the beginning, to set forth the nature of the universe down to the generation of man, seems wellnigh to have reached its fulfilment. For the manner of the generation of other animals we may deal with in brief, so there be no need to speak at length: thus shall we in our own eyes preserve due measure in our account of them. Let us then state it in this way. Of those who were born as men, such as were cowardly and spent their life in unrighteousness, were, according to the probable account, transformed into women at the second incarnation. At that time the gods for these reasons invented the love of sexual intercourse, in that they created one kind of animate nature in men they have let their reason sleep. And below these were creatures of many legs, and worms that crawl on their belly; and, yet lower, the fish that for their foolishness may not even breathe pure air, and all living things whose habitation is in the water. Yet these are ever changing their rank, rising or falling as their understanding grows more or less. And so was the universe completed and all that is therein, one and only-begotten, the most fair and perfect image of its eternal maker.

19. ἀδίκως διῆλθον] Compare *Laws*

781 A ὃ καὶ ἄλλως γένος ἡμῶν τῶν ἀνθρώπων λαθραιότερον μᾶλλον καὶ ἐπικλοπώτερον, τὸ θῆλυ, διὰ τὸ ἀσθενές. Assuredly women treated on the Athenian system would have been either more or less than human, had they not developed some tendency in this direction. Plato however is apparently the only Greek thinker who saw the cause of the evil and proposed a remedy.

21. [ζῷον] This curious quasi-personification of sexual impulse as an animate being is manifestly to be understood as mythical.

340   ΠΛΑΤΩΝΟΣ   [91 A—

ἡμῖν, τὸ δ' ἐν ταῖς γυναιξὶ συστήσαντες ἔμψυχον, τοιῷδε τρόπῳ ποιήσαντες ἑκάτερον. τὴν τοῦ ποτοῦ διέξοδον, ᾗ διὰ τοῦ πλεύμονος τὸ πόμα ὑπὸ τοὺς νεφροὺς εἰς τὴν κύστιν ἐλθὸν καὶ τῷ πνεύματι θλιφθὲν ξυνεκπέμπει δεχομένη, ξυνέτρησαν εἰς τὸν ἐκ τῆς κεφαλῆς
5 κατὰ τὸν αὐχένα καὶ διὰ τῆς ῥάχεως μυελὸν ξυμπεπηγότα, ὃν δὴ B σπέρμα ἐν τοῖς πρόσθεν λόγοις εἴπομεν· ὁ δέ, ἅτ' ἔμψυχος ὢν καὶ λαβὼν ἀναπνοὴν τοῦθ' ἧπερ ἀνέπνευσε, τῆς ἐκροῆς ζωτικὴν ἐπιθυμίαν ἐμποιήσας αὐτῷ τοῦ γεννᾶν ἔρωτα ἀπετέλεσε. διὸ δὴ τῶν μὲν ἀνδρῶν τὸ περὶ τὴν τῶν αἰδοίων φύσιν ἀπειθές τε καὶ αὐτο-
10 κρατὲς γεγονός, οἷον ζῷον ἀνυπήκοον τοῦ λόγου, πάντων δι' ἐπιθυμίας οἰστρώδεις ἐπιχειρεῖ κρατεῖν· αἱ δ' ἐν ταῖς γυναιξὶν αὖ μῆτραί τε καὶ ὑστέραι λεγόμεναι διὰ τὰ αὐτὰ ταῦτα, ζῷον ἐπιθυμητικὸν C ἐνὸν τῆς παιδοποιίας, ὅταν ἄκαρπον παρὰ τὴν ὥραν χρόνον πολὺν γίγνηται, χαλεπῶς ἀγανακτοῦν φέρει, καὶ πλανώμενον πάντῃ κατὰ
15 τὸ σῶμα, τὰς τοῦ πνεύματος διεξόδους ἀποφράττον, ἀναπνεῖν οὐκ ἐῶν, εἰς ἀπορίας τὰς ἐσχάτας ἐμβάλλει καὶ νόσους παντοδαπὰς ἄλλας παρέχει· μέχριπερ ἂν ἑκατέρων ἡ ἐπιθυμία καὶ ὁ ἔρως ξυνδυάζοντες, οἷον ἀπὸ δένδρων καρπὸν καταδρέψαντες, ὡς εἰς D ἄρουραν τὴν μήτραν ἀόρατα ὑπὸ σμικρότητος καὶ ἀδιάπλαστα
20 ζῷα κατασπείραντες καὶ πάλιν διακρίναντες μεγάλα ἐντὸς ἐκθρέψωνται καὶ μετὰ τοῦτο εἰς φῶς ἀγαγόντες ζῴων ἀποτελέσωσι γένεσιν. γυναῖκες μὲν οὖν καὶ τὸ θῆλυ πᾶν οὕτω γέγονε· τὸ δὲ τῶν ὀρνέων φῦλον μετερρυθμίζετο, ἀντὶ τριχῶν πτερὰ φύον, ἐκ τῶν ἀκάκων ἀνδρῶν, κούφων δέ, καὶ μετεωρολογικῶν μέν, ἡγου-
25 μένων δὲ δι' ὄψεως τὰς περὶ τούτων ἀποδείξεις βεβαιοτάτας εἶναι E

3 πόμα : πῶμα SZ.   18 ξυνδυάζοντες scripsi ex Hermanni coniectura. ξυνδιαγαγόντες H, et, teste Bastio, A : Bekkerus autem ξυναγαγόντες in A legisse videtur. ἐξαγαγόντες SZ.   καταδρέψαντες : κᾆτα δρέψαντες ASZ.   21 μετὰ τοῦτο : μετὰ ταῦτα S.

2. διὰ τοῦ πλεύμονος] See 70 C.
6. ἐν τοῖς πρόσθεν λόγοις] 73 C, 74 A; cf. 86 C: and in the contrary sense Aristotle *de partibus animalium* II vi 651[b] 20.
7. λαβὼν ἀναπνοὴν τοῦθ'] It is possible that some error may lurk here: but if we alter τοῦθ' to ταύτῃ, as Stallbaum proposes, αὐτῷ is left without any reference.
13. παρὰ τὴν ὥραν] I think Stallbaum is certainly mistaken in paraphrasing this 'per tempus, quo vires maxime vigent'. Lindau more correctly gives 'praeter pubertatem': compare *Critias* 113 D ἤδη δ' ἐς ἀνδρὸς ὥραν ἡκούσης τῆς κόρης, i. e. when she was old enough to be married.
14. πλανώμενον] This refers to the metaphorical ζῷον above. Compare 88 E τά τε περὶ τὸ σῶμα πλανώμενα παθήματα.
18. ξυνδυάζοντες] This correction of Hermann's appears to me a happy one.

and another in women, which two they formed in the following way. To the channel of the drink, where it receives the fluid passing down through the lungs beneath the kidneys into the bladder and sends it forth by pressure of the air, they opened a passage into the column of marrow which runs from the head down the neck and along the spine, and which we have already termed the seed. This, being quick with soul and finding an outlet, gave to the part where it found the outlet a lively desire of egress and produced a longing to generate. Wherefore the nature of the generative part in man is disobedient and headstrong, like a creature that will not listen to reason, and endeavours to have all its will because of its frantic passions; and again for the same reason what is called the matrix and womb in women, which is in them a living nature appetent of childbearing, when it is a long time fruitless beyond the due season, is distressed and sorely disturbed, and straying about in the body and cutting off the passages of the breath it impedes respiration and brings the sufferer into the extremest anguish and provokes all manner of diseases besides; until the passion and love of both unite them, and, as it were plucking fruit from a tree, sow in the womb, as if in a field, living things invisible for smallness and unformed, and again separating them nourish them within till they grow large, and finally bringing them to light complete the birth of a living creature. Such is the nature of women and all that is female. The tribe of birds was transformed, by growing feathers instead of hair, from men that were harmless but light-minded; who were students of the heavenly bodies, but fancied in their simpleness that the demonstrations were most sure concerning them which they obtained through

---

The reading of A, ξυνδιαγαγόντες, is senseless, and equally so is ἐξαγαγόντες. As to συναγαγόντες, which would otherwise suit well enough, the aorist can hardly be tolerated, nor has this reading very good authority. The word in A is an easy corruption of ξυνδυάζοντες, and the other readings look like attempts at correcting it.

22. τὸ δὲ τῶν ὀρνέων] In birds are incarnate the souls of harmless silly people, astronomers who fancy that astronomy means nothing more than what they see with their eyes. The class of persons indicated is clearly enough shown by *Republic* 529 A foll. I can see no reason for supposing with Martin that the Ionian philosophers are meant. With the epithet κούφων compare Sophocles *Antigone* 343 κουφονόων τε φῦλον ὀρνίθων.

25. δι' ὄψεως] Cf. *Republic* 529 A κιν-

δι' εὐήθειαν. τὸ δ' αὖ πεζὸν καὶ θηριῶδες γέγονεν ἐκ τῶν μηδὲν προσχρωμένων φιλοσοφίᾳ μηδὲ ἀθρούντων τῆς περὶ τὸν οὐρανὸν φύσεως πέρι μηδέν, διὰ τὸ μηκέτι ταῖς ἐν τῇ κεφαλῇ χρῆσθαι περιόδοις, ἀλλὰ τοῖς περὶ τὰ στήθη τῆς ψυχῆς ἡγεμόσιν ἕπεσθαι μέρεσιν. ἐκ τούτων οὖν τῶν ἐπιτηδευμάτων τά τ' ἐμπρόσθια κῶλα καὶ τὰς κεφαλὰς εἰς γῆν ἑλκόμενα ὑπὸ ξυγγενείας ἤρεισαν, προμήκεις τε καὶ παντοίας ἔσχον τὰς κορυφάς, ὅπῃ συνεθλίφθησαν ὑπὸ ἀργίας ἑκάστων αἱ περιφοραί· τετράπουν τε τὸ γένος αὐτῶν 92 A ἐκ ταύτης ἐφύετο καὶ πολύπουν τῆς προφάσεως, θεοῦ βάσεις ὑποτιθέντος πλείους τοῖς μᾶλλον ἄφροσιν, ὡς μᾶλλον ἐπὶ γῆν ἕλκοιντο. τοῖς δ' ἀφρονεστάτοις αὐτῶν τούτων καὶ παντάπασι πρὸς γῆν πᾶν τὸ σῶμα κατατεινομένοις ὡς οὐδὲν ἔτι ποδῶν χρείας οὔσης, ἄποδα αὐτὰ καὶ ἰλυσπώμενα ἐπὶ γῆς ἐγέννησαν. τὸ δὲ τέταρτον γένος ἔνυδρον γέγονεν ἐκ τῶν μάλιστα ἀνοητοτάτων καὶ ἀμαθεστάτων, οὓς οὐδ' ἀναπνοῆς καθαρᾶς ἔτι ἠξίωσαν οἱ μεταπλάττοντες, ὡς τὴν ψυχὴν ὑπὸ πλημμελείας πάσης ἀκαθάρτως ἐχόντων, ἀλλ' ἀντὶ λεπτῆς καὶ καθαρᾶς ἀναπνοῆς ἀέρος εἰς ὕδατος θολερὰν καὶ βαθεῖαν ἔωσαν ἀνάπνευσιν· ὅθεν ἰχθύων ἔθνος καὶ τὸ B τῶν ὀστρέων ξυναπάντων τε ὅσα ἔνυδρα γέγονε, δίκην ἀμαθίας

δυνεύεις γὰρ καὶ εἴ τις ἐν ὀροφῇ ποικίλματα θεωμένος ἀνακύπτων καταμανθάνοι τι, ἡγεῖσθαι ἂν αὐτὸν νοήσει, ἀλλ' οὐκ ὄμμασι θεωρεῖν.

It is remarkable that the compiler of the *Timaeus Locrus* treats transmigration and retribution as a mere fable, though a fable which is useful as a deterrent from vice: cf. 104 D εἰ δέ κά τις σκλαρὸς καὶ ἀπειθής, τῷ δ' ἐπέσθω κόλασις ἅ τ' ἐκ τῶν νόμων καὶ ἁ ἐκ τῶν λόγων, σύντονα ἐπάγοισα δείματά τε ὑπουράνια καὶ τὰ καθ' Ἄιδεω, ὅθι κολάσιες ἀπαραίτητοι ἀπόκεινται δυσδαίμοσι νερτέροις, καὶ τἆλλα ὅσα ἐπαινέω τὸν Ἰωνικὸν ποιητὰν ἐκ παλαιᾶς ποιεῦντα τοὺς ἐναγέας· ὡς γὰρ τὰ σώματα νοσώδεσί ποκα ὑγιάζομες, αἴ κα μὴ εἴκῃ τοῖς ὑγιεινοτάτοις, οὕτω τὰς ψυχὰς ἀπείργομες ψευδέσι λόγοις, εἴ κα μὴ ἄγηται ἀλαθέσι. λέγοιντο δ' ἂν ἀναγκαίως τιμωρίαι ξέναι, ὡς μετενδυομένᾱν τᾶν ψυχᾶν τῶν μὲν δειλῶν ἐς γυναικέας σκάνεα ποθ' ὕβριν ἐκδιδόμενα, τῶν δὲ μιαιφόνων ἐς θηρίων σώματα ποτὶ κόλασιν, λάγνων δὲ ἐς συῶν ἢ κάπρων μορφάς, κούφων δὲ καὶ μετεώρων ἐς πτηνῶν ἀεροπόρων, ἀργῶν δὲ καὶ ἀπράκτων ἀμαθῶν τε καὶ ἀνοήτων ἐς τὰν τῶν ἐνύδρων ἰδέαν. Compare *Phaedo* 81 E foll.

5. **ἐκ τούτων οὖν τῶν ἐπιτηδευμάτων**] There is an interesting parallel in Aristotle *de partibus animalium* IV x 686[a] 25 ὁ μὲν οὖν ἄνθρωπος ἀντὶ σκελῶν καὶ ποδῶν τῶν προσθίων βραχίονας καὶ τὰς καλουμένας ἔχει χεῖρας, ὀρθὸν γάρ ἐστι μόνον τῶν ζῴων διὰ τὸ τὴν φύσιν αὐτοῦ καὶ τὴν οὐσίαν εἶναι θείαν· ἔργον δὲ τοῦ θειοτάτου τὸ νοεῖν καὶ φρονεῖν· τοῦτο δὲ οὐ ῥᾴδιον πολλοῦ τοῦ ἄνωθεν ἐπικειμένου σώματος· τὸ γὰρ βάρος δυσκίνητον ποιεῖ τὴν διάνοιαν καὶ τὴν κοινὴν αἴσθησιν. διὸ πλείους γενομένου τοῦ βάρους καὶ τοῦ σωματοειδοῦς ἀνάγκη ῥέπειν τὰ σώματα εἰς τὴν γῆν, ὥστε πρὸς τὴν ἀσφάλειαν ἀντὶ βραχιόνων καὶ χειρῶν τοὺς προσθίους πόδας ὑπέθηκεν ἡ φύσις τοῖς τετράποσιν. τοὺς μὲν γὰρ ὀπισθίους δύο πᾶσιν ἀναγκαῖον τοῖς πορευτικοῖς ἔχειν, τὰ δὲ τοιαῦτα τετράποδα ἐγένετο οὐ δυναμένης φέρειν τὸ βάρος τῆς ψυχῆς.

the sight. And the race of brutes that walk on dry land comes from those who sought not the aid of philosophy at all nor inquired into the nature of the universe, because they used no longer the revolutions in the head, but followed as their guides the parts of the soul that are in the breast. From these practices their front limbs and their heads were by their natural affinity drawn towards the ground and there supported; and their heads were lengthened out and took all sorts of forms, just as the orbits in each were crushed out of shape through disuse. For the same reason such races were made four-footed and many-footed; for God gave many props to the more senseless creatures, that they might the more be drawn earthward. As to the most senseless of all, whose whole bodies were altogether stretched at length on the earth, seeing they had no longer any need of feet, God made them footless to crawl upon the ground. And the fourth class that lives in the water was formed of the most utterly foolish and senseless of all, whom they that transfigured them thought not worthy even of pure respiration, because their soul was polluted with all manner of iniquity; but in place of inhaling the fine pure element of air they were thrust into the turbid and lowly respiration of water. Hence is the tribe of fishes and of all shell-fish that live in the water; which have the

6. προμήκεις τε καὶ παντοίας] Their heads were elongated, because the circles of the brain were distorted into an elliptical form: the proper and typical shape of the head is spherical, emulating the figure of the universe: see 44 D, 73 C, E; and for the effect of the κῦμα τῆς τροφῆς upon the shape of the head see note on 76 A.

12. πᾶν τὸ σῶμα κατατεινομένοις] Plato's theory pays small regard to the 'wisdom of the serpent': however, as the serpent has an exceptional gift of holding its head upright, perhaps we may allow it to be promoted a few grades on that account.

15. οὓς οὐδ' ἀναπνοῆς καθαρᾶς ἔτι ἠξίωσαν] It seems a little hard upon an animal so highly organised as the fish to be placed nearly at the bottom of the scale merely because it respires under water; and water-snails are probably as intelligent as land-snails. It is possible, as Martin suggests, that Plato may have taken the hint from Diogenes of Apollonia: see Theophrastos *de sensu* § 44 φρονεῖν δέ, ὥσπερ ἐλέχθη, τῷ ἀέρι καθαρῷ καὶ ξηρῷ· κωλύειν γὰρ τὴν ἰκμάδα τὸν νοῦν, διὸ καὶ ἐν τοῖς ὕπνοις καὶ ἐν ταῖς μέθαις καὶ ἐν ταῖς πλησμοναῖς ἧττον φρονεῖν. ὅτι δὲ ἡ ὑγρότης ἀφαιρεῖται τὸν νοῦν σημεῖον, ὅτι τὰ ἄλλα ζῷα χείρω τὴν διάνοιαν· ἀναπνεῖν τε γὰρ τὸν ἀπὸ τῆς γῆς ἀέρα καὶ τροφὴν ὑγροτέραν προσφέρεσθαι. τοὺς δὲ ὄρνιθας ἀναπνεῖν μὲν καθαρόν, φύσιν δὲ ὁμοίαν ἔχειν τοῖς ἰχθύσι· καὶ γὰρ τὴν σάρκα στιφρὰν καὶ τὸ πνεῦμα οὐ διιέναι διὰ παντὸς ἀλλὰ ἱστάναι περὶ τὴν κοιλίαν. Compare Herakleitos fr. 74 Bywater αὔη ψυχὴ σοφωτάτη καὶ ἀρίστη: and fr. 73.

ἐσχάτης ἐσχάτας οἰκήσεις εἰληχότων. καὶ κατὰ ταῦτα δὴ πάντα τότε καὶ νῦν διαμείβεται τὰ ζῷα εἰς ἄλληλα, νοῦ καὶ ἀνοίας ἀποβολῇ καὶ κτήσει μεταβαλλόμενα.

Καὶ δὴ καὶ τέλος περὶ τοῦ παντὸς νῦν ἤδη τὸν λόγον ἡμῖν C
5 φῶμεν ἔχειν· θνητὰ γὰρ καὶ ἀθάνατα ζῷα λαβὼν καὶ ξυμπληρωθεὶς ὅδε ὁ κόσμος οὕτω, ζῷον ὁρατὸν τὰ ὁρατὰ περιέχον, εἰκὼν τοῦ ποιητοῦ, θεὸς αἰσθητός, μέγιστος καὶ ἄριστος κάλλιστός τε καὶ τελεώτατος γέγονεν, εἷς οὐρανὸς ὅδε μονογενὴς ὤν.

7 ποιητοῦ dedi cum A. νοητοῦ HSZ.

1. **ἐσχάτας οἰκήσεις**] This means not the habitation of the ζῷα in the water, but the habitation of the soul in the bodies of fishes, molluscs and the like. It is plain from this passage also that Plato did not contemplate the entrance of a soul which had once been human into any vegetable form: not that there is any physical reason against this, but for the cause pointed out on 77 A.

2. **διαμείβεται τὰ ζῷα**] This passage is important, as clearly indicating that Plato does not admit any state of hopeless degradation. The animals are perpetually changing places as they advance or recede in intelligence: what is a bird in one incarnation may become a fish in another, and *vice versa*. Even the oyster may, in course of ages of evolution, become once more a human being. Hence it is evident that the everlasting vengeance wreaked upon desperate criminals in the *Republic*, *Phaedo* and *Gorgias* is merely part of the pictorial representation. How far the present scheme of transmigration is intended to be accepted literally is a matter exceedingly difficult of determination. It has no essential connexion with the Platonic ontology; nor again is it obviously inconsistent therewith. The continuance of individual personalities which it presumes is not material to Plato's theory, which requires that all soul shall be eternal and shall exist in a multitude of separate conscious beings, as well as in its universal unity; but it does not require that the same consciousness shall exist as such in successive embodiments. The question belongs to that mythical borderland of the Platonic philosophy where it is not always possible to draw the line with certainty between the literal and the allegorical.

6. **εἰκὼν τοῦ ποιητοῦ**] About the genuineness of this reading, which has the support, besides A, of Vat. 173, I can feel no doubt whatsoever. Had Plato written νοητοῦ, it is in the last degree improbable that a phrase so familiar and constantly recurring should have been altered into the far more difficult ποιητοῦ. On the other hand, assuming Plato to have written ποιητοῦ, the word was, I may venture to say, positively certain to be altered in some way: for, the scribe or annotator would argue, the κόσμος is not the image of its maker, but of the νοητὸν ζῷον from which the maker copied it: therefore νοητοῦ is the word. Add to this the probability that some readers would suppose it to be the genitive of ποιητός (a supposition which Lindau actually entertains), and we have so potent causes of corruption that it is surprising that a single manuscript has preserved the true reading. The word ποιητοῦ must necessarily be unintelligible to any student of the dialogue who had not arrived at some such conclusion about the nature of the δημιουργός as that which I have done my best to defend. Adopting then ποιητοῦ, we have of course but one possible inference to draw :

uttermost dwelling-place in penalty for the uttermost folly. In such manner then and now all creatures change places one with another, rising or falling with the loss or gain of understanding or of folly.

And now let us declare that our discourse concerning this All has reached its end. Having received all mortal and immortal creatures and being therewithal replenished, this universe hath thus come into being, living and visible, containing all things that are visible, the image of its maker, a god perceptible, most mighty and good, most fair and perfect, even this one and only-begotten world that is.

the δημιουργὸς and the αὐτὸ ζῷον are one and the same; the δημιουργὸς being simply a mythical duplicate of the αὐτὸ ζῷον, the introduction of which was necessitated by the poetical and narrative form of the exposition. Both the δημιουργὸς and the αὐτὸ ζῷον represent the primal unity, considered as though not yet pluralised, which must evolve and manifest itself under the form of plurality and so be a truly existent One. And surely nothing can be more thoroughly characteristic of Plato, than that, after talking parables throughout, he should at the very end of the dialogue drop one single word, φωνᾶεν συνετοῖσι, which was to open our eyes to the fact that he did speak in parables; that if we desire to understand the philosopher, we must be in sympathy with the poet.

8. **εἰς οὐρανὸς ὅδε μονογενὴς ὤν**] It is worth while to note how closely the phraseology of the concluding five lines corresponds with that of 30 C—31 D: compare especially the words in 31 B εἷς ὅδε μονογενὴς οὐρανὸς γεγονὼς ἔστι τε καὶ ἔτ' ἔσται. Plato doubtless designs by thus echoing his former language to assure us that the promise made in the beginning has been fulfilled, that the nature of the universe has been expounded precisely to the effect indicated in the sixth chapter, and that not a single point has been omitted. This very minute correspondence serves to render the one important deviation, εἰκὼν τοῦ ποιητοῦ, all the more strikingly significant. Mark too the emphatic stress which falls upon the two closing words of the dialogue, μονογενὴς ὤν. In them is virtually summed up Plato's whole system of idealistic monism: this one universe γίγνεταί τε καὶ ἔστι, it is create and uncreate, temporal and eternal, the sum total and unity of all modes of existence; in the words of the Platonic Parmenides πάντα πάντως ἐστί τε καὶ οὐκ ἔστι.

INDEX I.

A

ἀγαθῶν φυγάς, 256
ἀγγεῖα ἀέρος, 242
ἀδάμας, 214
ἀδιάπλαστα ζῷα, 340
ἀήθει λόγῳ, 188
ἀθάνατα φρονεῖν, 336
ἀιδίων θεῶν γεγονὸς ἄγαλμα, 118
αἰθήρ, 212
αἴσθησις, Plato's etymology of, 148
αἴσθησις and πάθημα distinguished, 235
αἰσθητὰ or αἰσθητικά, 226
αἰσθητικόν, not αἰσθητόν, 246
αἰτία ἑλομένου, 325
αἰώνιον εἰκόνα, 118
αἰωρήσεις, 332
ἀκάκων ἀνδρῶν, κούφων δέ, 340
ἀκοή, 246
ἀκολασία due to physical causes, 324
ἄκρατα καὶ πρῶτα σώματα, 206
ἀκρόπολις = the head, 258
ἀλείμματα, 178
ἀληθὴς δόξα, 180
ἅλμα μαλακόν, 260
ἄλογον μετέχον πῃ λόγου, 262
ἀλουργόν, 250
ἁλυκόν, 240
ἀλύτως ὕδατι, 218
ἄλφους, 318
ἁλῶν θεοφιλὲς σῶμα, 220
ἅμα or αἷμα, 314
ἀμαθία, 322
ἀμβλυτάτη τῶν ἐπιπέδων γωνιῶν, 194
ἀμέριστος οὐσία, 106
ἅμμα τῶν φλεβῶν, 258
ἀμφημερινοὶ πυρετοί, 322
ἄν omitted with the relative, 324

ἄν, position of, 82
ἀναγκαῖα καὶ ἄριστα, 280
ἀναγκαῖον and θεῖον, 252
ἀνάγκη, 166
ἀναίσθητον πάθημα, 234
ἀναλικμώμενα or ἀνικμώμενα, 186
ἀναλογία, 96
ἀνάμνησις, 142
ἀνάπαλιν ἡ γένεσις, 310
ἀναπνοή, 294
ἀναπνοή = free egress, 314
ἀναψυχῆς, 316
ἄνοια, 322
ἀντίας...πλαγίας...ὑπτίας, 150
ἀντίπους, 230
ἄνω, 228
ἄνω κοιλία, 270
ἀνωμαλότης, 208
ἀξύμμετρον ταῖς μεγίσταις ξυμμετρίαις, 328
ἀπάθεια of the ὑποδοχή, 176
'Απατούρια, 66
ἄπειρον, 24
ἄπειρον πνεῦμα, 101
ἀπείρους, play on, 198
ἀπήρξατο or ἀπειργάζετο, 130
ἀπορροαί, 157
ἀποτομή, 111
ἄρθρα a form of θάτερον, 274
ἀρτηρία, 260
ἀρχαί, three, maintained by Galen, 257
ἀρχὴν εἴτε ἀρχάς, 168
αὐλῶνος, 294
αὔξησις καὶ φθίσις, 206
ἄυπνος φύσις, 184
αὐστηρόν, 240
αὐτὸ ἐν μέσῳ, 230
αὐτὸ ζῷον, 41, 95

αὐτὸ τοῦτο ἐφ' ᾧ γέγονεν ἑαυτῆς, 184
ἀφομοιωθέντα ἐντός, 158
ἄφρονε ξυμβούλω, 256

B

βαρύ, 228
Βενδίδεια, 55
βιᾶται, 232
βίου ζωή, 152
βλεφάρων φύσις, 158
βραχέος, 80

Γ

γένη and εἴδη, 206
γλαυκόν, 252
γλισχρὸν καὶ λιπαρόν, 310
γλυκύ, 242
γόμφοι, 146

Δ

δαίμων = the intellect, 336
δάκρυον, 250
δεξαμένης or δεξαμενῆς, 188
δέρμα, 280
δεσμοὶ τῆς ψυχῆς, 270
δεχομένη, 206
δημιουργός, 37
διαιωνίας φύσεως, 122
διαζωγραφῶν, 196
διάμετρος, 194
διαρροίας καὶ δυσεντερίας, 320
διαφανῆ, 248
διάφραγμα, 256
διάφραγμά τ' ἴσχον, 316
διαχεῖν, 158
διαχυτικὸν μέχρι φύσεως, 216
διεγένοντο, 70
διέξοδος, 168
διιστάμενοι or διεσταμένοι, 306
δι' οὖν or δύ' οὖν, 244
διόψεως or τῶν δι' ὄψεως, 136
διυλασμένα, 254
διχῇ κατὰ τὰ ἐναντία προϊέναι, 126
δοκιμεῖα, δοκίμια, 240
δόξα ἀληθής, 180
δορυφορικὴ οἴκησις, 258
δοχάς, 264
δριμύ, 240
δρυόχων, 306
δύναμις = square root, 192

E

ἐγκαύματα, 82
ἐγκέφαλος, 272
ἐγκύρτια, 292
εἴδεσί τε καὶ ἀριθμοῖς, 188
εἰδῶν φίλοι, 22
εἰκὼν αἰῶνος, 118
εἰκὼν τοῦ ποιητοῦ, 344
εἴλλεσθαι, εἰλεῖσθαι, 132
εἰσιόντα καὶ ἐξιόντα, 176
ἐκγόνῳ, 176
ἐκμαγεῖον, 176
  „ in sense of napkin, 268
ἐκόλλησεν ὁμοιότητι, 278
ἐκ τρίτου, 192
ἔλαιον, 216
ἕλικα, 126
ἐμφαίνεσθαι, ἔμφασις, 159
ἐναντία καὶ πλάγια, 232
ἐναντίαν εἰληχότας αὐτῷ δύναμιν, 124
ἔνθεν καὶ ἔνθεν ὕψη λαβοῦσα, 160
ἐνιαυτὸς ὁ τέλεος, 128
ἐν μέρους εἴδει, 94
ἔντερα, 270
ἐξαρπασθέν, peculiar construction of, 220
ἐπανακυκλήσεις, 134
ἐπ' ἀνθρώπους, 74
ἐπίπνοια, 264
ἐπίρρυτον σῶμα καὶ ἀπόρρυτον, 148
ἐπιτόνους, 318
ἐπίτριτα, ἐπόγδοα, 108
ἐπιχειρητῇ παντὸς ἔρωτι, 256
Ἑρμοῦ ἱερόν, 122
ἐρυθρόν, 250
ἔστι, wrongly applied to γένεσις, 120
ἔσχατα, 230
εὐδαίμονα = εὖ τὸν δαίμονα ἔχοντα, 338
εὐπαραγωγόν, 256
εὑρεῖν τε ἔργον, said of the maker, 86
εὐφροσύνη, 300
ἔφυγον κακόν, εὗρον ἄμεινον, 152
ἑωσφόρος, 122

Z

ζέσιν τε καὶ ζύμωσιν, 242
ζυγόν, 232
ζύμωμα, 276
ζῷα, ideas of, 34
ζῷον ἐπιθυμητικὸν τῆς παιδοποιίας, 340

INDEX I.

H

ἡγεμονοῦν, ἡγεμονικόν, 140
ἡδοναὶ καὶ λῦπαι, 234
ἡδονὴ contrasted with εὐφροσύνη, 300
ᾗ φέρειν πέφυκε, 166
ἤλεκτρον, 300
ἥμερα δένδρα, 286
ἥμερον—ἡμέρα, play on, 154
ἡμιόλια, 108
ἡμιτριταῖοι πυρετοί, 322
ἦν, ἔσται, wrongly applied to ἀΐδιος οὐσία, 120
ἥπατος ἰδέα, 262
Ἡρακλεῖοι λίθοι, 302
Ἡρακλέους στῆλαι, 78

Θ

θάτερον, 43, 44, 106, 274
θεῖον and ἀναγκαῖον, 252
θεοὶ θεῶν, 136
θεοσεβέστατον ζῴων, 142
θερμόν, etymology of, 226
θνητὸν εἶδος ψυχῆς, 256
θρέμμα ἄγριον, said of the ἐπιθυμητικόν, 260
θρεπτικὴ ψυχή, 263
θρίξ, 282
θυματικὰ σώματα, 224
θώραξ, 256

I

ἰατρικαὶ σικύαι, 298
ἰδρὼς καὶ δάκρυον, 314
ἱερὸν νόσημα, 318
ἴλλεσθαι, 132
ἰλυσπώμενα, 342
ἶνες, 310
ἰός, 214
ἶσατις, 250
ἰσοπαλές, 230
ἰσόπλευρα τρίγωνα, 192
ἰχθύες, 342
ἴχνη αὐτῶν ἄττα, 188
ἰχώρ, 312

K

καθ' ὃν εἴληχε, 230
καλόν = ξύμμετρον, 328
καπνός, 244

κατὰ διάμετρον, 194
κατακορές, 250
κατὰ μῆκος στραφὲν τοῦ προσώπου, 160
κατὰ νοῦν, 114
κατὰ πλευράν, κατὰ διάμετρον, 112
κατεσκευασμένα γράμμασι, 72
κάτω, 228
κάτω κοιλία, 270
κενόν, Pythagorean, 101
κέραμος, 220
κερματίζειν, 226
κεφαλή, astronomical term, 129
κίκι, 216
κίνησις and στάσις, 208
κόσμος, play on, 130
κουρεῶτις, 66
κοῦφον, 228
κρύσταλλος, 216
κυανοῦν, 250
κῦμα ὃ τὴν τροφὴν παρέχει, 148
κύρτος, 290
κύτος τῆς ψυχῆς, 150

Λ

λαμπρόν, 250
λεῖμμα, 111
λεῖον, 234
λέμμα μεῖζον, 280
λεύκας, 318
λευκόν, 248
λίθων χυτὰ εἴδη, 224
λιπαραῖος λίθος, 220
λίτρον, 220, 240
λοβόν, 264
λογισμῷ τινὶ νόθῳ, 184
λογιστικόν = object of reason, 116
λόγος, 116, 180
λόγων νᾶμα, 280
λυόμενος, 70
λῦπαι καὶ ἡδοναί, 234

M

μάλαγμα, Hermann's conjecture, 260
μαλακόν, 228
μανία, 322
μαντεία, 264, 266
μάντεις and προφῆται distinguished, 266
μαρμαρυγαί, 250
μειζόνως διῃρημένη, 170

μείς, 128
μέλαν, 248
μέλαν χρῶμα ἔχων λίθος, 220
μέλι, 218
μέρη καὶ μέλη, 284
μετ᾽ ἀναισθησίας ἁπτόν, 184
μεταξὺ κορυφῆς τοῦ τε ὀμφαλοῦ, 246
μεταξὺ τιθεμένου, 176
μεταρρυθμισθέντος, 158
μετεωρολογικοί, 340
μέτριος παιδιά, said of science, 214
μητέρα καὶ ὑποδοχήν, 178
μίμησις instead of μέθεξις, 171
μιμητικὸν ἔθνος, 62
μονογενὴς ὤν, significance of, 345
μουσικῆς φωνῇ χρήσιμον, 164
μυελός, 270
μυλῖται λίθοι, 220

N

νεῦρα, 274
νεὼς πείσματα, 320
νοητοῦ or ποιητοῦ, 41, 344
νοητῶν, peculiar use of, 114
νόθος λογισμός, 184
νόσοι, 308
νόσων ξύστασις, 334
νοῦς βασιλεύς, 39
νοῦς καὶ δόξα ἀληθής, 180

Ξ

ξανθόν, 250

O

ὄγκων εἴτε δυνάμεων, 96
ὁδὸς ἄνω κάτω, 4
ὁδὸς = μέθοδος, 189
ὁδὸς πρὸς τὸ ξυγγενές, 232
ὀδυρόμενος ἂν θρηνοῖ μάτην, 164
ὄξος χρυσοῦ, 214
οἱ κρείττους = the gods, 286
οἶνος, 216
ὀλιγίστας or ὀλιγοστάς, 200
ὀλισθήμασιν ὕδατος, 148
ὁλκὴ denied, 302
ὁλόκληρος, 152
ὁμαλότης, 208
ὁμίχλη, 212, 244
ὁμοιότητι not = ὁμοίως, 279

ὁμώνυμον, 182
ὀνείρωξις, 184
ὄνυχες, 284
ὀξύ, 242
ὀπισθότονοι, 318
ὀπός, 218
ὄργανα χρόνου, 142, 146
ὅρος ὁρισθεὶς μέγας διὰ βραχέων, 180
ὀρφνινον, 250
ὀσμαί, 244
ὀστοῦν, 272
ὄστρεα, 342
οὐδεὶς ἑκὼν ἄδικος, 324
οὐρανῷ ξυγγένεια, 336
οὐσία, 106
οὐσίας ἀμῶς γέ πως ἀντεχομένην, 184
οὗ τι μὲν δή, 138
ὄχημα, 142, 254

Π

πάθημα, peculiar sense of, 225, 249
  „   distinguished from αἴσθησις, 235
παλίναιρετα, 312
πανσπερμία, 272
παραβολάς, 134
παραδείγματα, 31
παράλλαξις, 70
παρεμφαῖνον, 178
πάχνη, 216
πειθοῦς ἔμφρονος, 166
πέμπτη οὐσία, 199
πεπηγὸς γένος, 212
πέρας, πέρας ἔχον, 24, 177
περίβολον or περὶ ὅλον, 272
περιφερόμενον ὁμοίως or ὅμοιον, 174
περίωσις, 294
πηλοῦ κάρτα βραχέος, 80
πικρόν, 240
πλῆσις, 210
πιστόν, 184
πλανητὰ or πλανῆται, 122
πλανωμένη αἰτία, 166
πλάστιγγας, 232
πλάττειν σῶμα, 330
πλέγμα, 290
ποιητοῦ or νοητοῦ, 21, 344
ποικίλου πάσας ποικιλίας 176
πομφόλυγες, 242
πορείαν or πορεῖα, 154

INDEX I.

ποταμὸς = γένεσις, 148
πράσιον, 252
προλελεπτυσμένων ὑπὸ σηπεδόνος, 240
πρόμηκες, 192
πρόνοια τῆς ψυχῆς, 154
προοίμιον—νόμον, 90
προστυχόντος τε καὶ εἰκῇ, 104
προσχωρήσεις, 134
προφῆται, 266
πυθμένες σαρκός, 314
πύλας, 264
πυραμίς, 200
πυρετοί, 322
πυρρόν, 250

Ρ

ῥαφαί, 280
ῥῖγος, 228
ῥίζα τῶν τριγώνων, 306
ῥίζαν ἡμῶν = the brain, 336
ῥοή, 212
ῥυθμός, 164
ῥυπτικά, 240

Σ

σάρξ, 274
σικύαι, 298
σκεδαστὴ οὐσία, 116
σκληρόν, 228
σπληνὸς μανότης, 268
στὰς or πᾶς, 198.
στάσις and κίνησις, 208
στάσεις καὶ νόσους, 308
στέγειν = keep in, 290
στοιχεῖα, 168
στρυφνόν, 240
σύγκρισις καὶ διάκρισις, 240
συλλαβῆς εἴδεσι, 268
συμμεταίτια, 162
συμφυὲς with genitive, 236
συνάψεσιν, 234
συνδυάζοντες, 340
συνηρμόσθαι, middle, 202
συννόμου οἴκησιν ἄστρου, 144
σύνοδος = constitution, 334
συστάντα, 202
συστάσεις πρῶται, δεύτεραι, 308
συστάτῳ σώματι, 100
σφακελίζειν, 274, 316

σφίγγει, 208
σχῇ κεφαλήν, 130
σῶμα ἐπιμελῶς πλάττοντα, 330

Τ

τὰ δι' ἀνάγκης γιγνόμενα, 166
τὰ διὰ νοῦ δεδημιουργημένα, 166
ταλαντουμένην, 186
ταὐτόν, 106
τε, displacement of, 135
τέλεον ἐνιαυτόν, 128
τέτανοι, 318
τεταρταῖοι πυρετοί, 322
τετρακτύς, geometrical, 107
τὴν τοῦ κρατίστου φρόνησιν, 130
τιθασῶς πρὸς ἡμᾶς ἔσχε, 286
τιθήνη γενέσεως, 172, 186
τὸ τῆς χώρας ἀεί, 182
τὸ τυχὸν ἄτακτον, 162
τόδε καὶ τοῦτο καὶ τῷδε, 174
τοιοῦτον opposed to τοῦτο, 172
τραχύ, 234
τρία τριχῇ, 186, 336
τριπλῆν κατὰ δύναμιν, 192
τριταῖοι πυρετοί, 322
τρίτος ἄνθρωπος, 20
τρόμος, 228
τροφὴν τὴν ἑαυτοῦ φθίσιν, 102
τροχοῦ περιαγομένου, 296
τύχη, 164
τῷ ἑαυτοῦ κατὰ τρόπον ἤθει, 146
τῷ λόγῳ γεγονότας, 84
τῶν πάντων ἀεί τε ὄντων, 178

Υ

ὕαλος, 224
ὑγρόν, etymology of, 216
ὑγρὸν ὕδωρ, 212
ὑδραγωγίαν, 290
ὕλη, the Aristotelian, 183
„ = wood, 254
ὑπαρξάμενος, 140
ὑπερσκελές, 328
ὑπέρψυχον, 330
ὑποδοχή, 170
ὑποθετέον, ὑποτεθέντα, 226
ὕψη λαβοῦσα, 160

Φ

φαιόν, 250
φάτνη, 260
φθόγγοι, 300
φθόνος, 90
φλέβια perform the function of nerves, 240
φλέγμα, 314
φρένες, 256
φρόνησις τοῦ κρατίστου, 130
φρόνιμον, τό, 236
φυτά, 286
φωνή, 246
φωρᾶσαι, 232

X

χάλαζα, 216
χῖ, 111
χιών, 216
χλοῶδες or χολῶδες, 312
χολή, 312
χορείας, 134
χρόαι, 248
χρόνου εἴδη, μέρη, 120
χρυσοῦ ὄζος, 214
χυμοί, 216, 238
χυτὸν ὕδωρ, 212
χώρα, 44, 182

Ψ

ψυχὰς ἰσαρίθμους τοῖς ἄστροις, 140
ψυχὴ τοῦ κόσμου, 42
ψυχή, relation to νοῦς, 92
ψυχῆς ὅσον θνητόν, 224
    ,,   θνητὸν εἶδος, 256
ψυχρόν, 228

Ω

ὡς ἀποδοθησόμενα πάλιν, 146
ὠχρόν, 250

INDEX II.

A

Absolute knowledge impossible, 48
Accords of sound, 300
Acids, action of, 242
Acoustics, 108
Adamant, 214
Air, varieties of, 212
Alkali, 220, 240
Allegory, 36
Amasis, 68
Amber, phenomena of, 302
Anaxagoras, 9
    ,,   and causation, 10
    ,,   defects of, 11
    ,,   his deficiencies made good, 188
Animal and vegetable, 286
Aorta, 289
Apaturia, 67
Appuleius on the origin of the world, 92
Aristotle, his incorrect criticism of Plato, 105, 175, 179, 184, 190, 202, 229
    ,,   his classification of the soul's functions, 262
    ,,   his views on dreams, 264
    ,, ,, ,, the brain, 278
Arteries, supposed to be filled with air, 311
    ,,   not distinguished from veins, 258
Artificer, the, 37
    ,,   looks to the eternal type, 88
    ,,   his works immortal, 139
Astringent, 240
Astronomers become birds, 340
Athenion quoted concerning salt, 222
Athens, the ancient, 74
Atlantis, 78
Attraction, 228

B

Bastard reasoning, 184
Becoming, 4
Bile, several kinds of, 312
Black, 248
Black stone, 220
Blood, why it is red, 304
    ,,   will not coagulate if the fibrine be removed, 320
Blue, 250, 252
Bones, formation of, 252
Bones and hair insensible, 236
Brain, 272
Birds, origin of, 340
Bronze, 214

C

Cataclysms and conflagrations, 70
Catarrhs, 318
Categories, 116
Cathartics, worst means of κίνησις, 332
Causation, 10, 160, 167
Causes, divine and necessary, 232
Chaotic motion, 92, 254
Circle of the Same, of the Other, 112
Circular impulsion, 296
Classes without corresponding ideas, 25
Climate, its influence on character, 77
Cold, explanation of, 226
Colours, 248
Concave mirrors, 160
Consonance, 300
Constant sum of things, 147
Constriction of the universe, 208
Continuity first conceived by Herakleitos, 4

Cooling, 212
Cosmic soul, 42
Creation, motive of, 91
Cube, 196
Cupping instruments, 298

D

Daremberg, 291
Death, natural, is painless, 306
Demokritos, his notion of ἀνάγκη, 167
" his theory that like seeks like, 187
" his infinity of κόσμοι, 198
Destruction of records, 72
Determinism, 324
Diaphragm, 256
Digestion, 294
Diseases, origin of, 308 foll.
" have their natural course, 334
Dissolution, divers modes of, 222
Divination, 266
Dodecahedron, 197
Downward and upward, 228
Dreams, 158
Drink passes through the lungs, 258
Dropides, 64

E

Earth, question of her rotation, 132
" forms of, 218
East on the right hand, 212
Education, effects of bad, 326
Egypt considered part of Asia, 76
Egyptian institutions, similar to those of ancient Athens, 76
Eleaticism, 6
" defects of, 8
" complementary to Herakleiteanism, 8
Elements, ideas of the four, 34
" before generation of universe, 168
" interchangeability of, 172, 192, 204
" why not completely sorted, 208
Empedokles, 13
" his doctrine of vision, 157
" his theory of respiration, 294
Envy of the gods, 91

Epilepsy, 318
Equilateral triangles, 192
Essence, 106
Eternity contrasted with time, 120
Euripides *Phoenissae*, allusion to, 164
Evil, defect in presentation of the type, 33
" inevitable, 92
" responsibility for, 144
Exercise, three modes of, 332
Extension, 45
Eyes, 154

F

Fever relieves tetanus, 318
Fevers, quotidian, &c., 322
Fibrine, 310
Fire, varieties of, 210
" its properties explained, 226
Fishes, condemned to respire in water, 342
Five κόσμοι or one? 198
Flesh, formation of, 276
Freewill and necessity, 324
Front and back, 154
Frost, 216
Fusion, 212

G

Galen on Plato's theory of respiration, 291
Genealogy of Plato, 65
God irresponsible for evil, 144
'Gods of Gods', 137
Gold, 214
Glass, 224
Gravity, Plato's theory of, 228
Green, 252
Grey, 250

H

Hail, 216
Hair, 282
Hard, 228
Harmony allied to the proportions of the soul, 164
Head, spherical shape of, 154
" the seat of intelligence, 256
" why not covered with flesh, 278
Heads of beasts, why elongated, 342
Hearing, final cause of, 164
" theory of, 246

INDEX II.

Heat, explanation of, 226
Herakleitean language, 174
Herakleitos, 4
Hermokrates, 63
Honey, 218

I

Ice, 216
Icosahedron, 196
Idea with but one particular, 94
Ideal theory, earlier, 16
  ,,   ,,   ,, deficiencies thereof, 18
Ideas, six classes of, in the *Republic*, 16
  ,, of σκευαστά, 23
  ,, of things evil, 25
  ,, regarded as types, 31
  ,, their plurality inevitable, 34
  ,, restricted to ζῷα? 34
  ,, and the cosmic soul, 44
  ,, question of their existence raised, 180
  ,, pass not into aught else, 182
Image, must exist in another, 184
Immateriality conceived first by Plato, 17
Indirect interrogation, double, 135
Inexact language, 120
Infants, abeyance of reason with, 148
Interchange of elements, 172
Interpreters, 266
Intestines, functions of, 270
Inversion of right and left, 150
  ,,   ,,   ,, in mirrors, 160
Ion, the, quoted about the magnet, 302
Ionian school, 3
Isosceles, the primal, 292

J

Juices, 216

K

Knowledge, province of, 48

L

Lava, 220
Leprosies, 318
Light and heavy, 228
Like does not alter like, 204
Limit to duration of species and individual, 334

Liver, its functions, 262
  ,, its connexion with divination, 266
Loadstone, 302
Lokris, the Epizephyrian, 62
Lungs, function of, 258
Lymph, 312

M

Madness a condition of inspiration, 266
Magnet, 302
Maladies of the soul, 322
Man, the most god-fearing of animals, 142
Many and one, unknown to man, 252
Marrow, 270
  ,, disease of, fatal, 322
Martin, his theory of Plato's ἐπίπεδα, 203
Matter, subjectivity of, 30
  ,, requires an αἰτία, 86
  ,, no independent power, 166
  ,, resolved into space, 183
Means, 98
  ,, harmonical and arithmetical, 108
Medicines, use of, discouraged, 332
Melting, 212
Mercury, 124
Metals, forms of water, 212
Metempsychosis, 144, 342
Microcosm, the human body, 304
Midriff, 256
Mind, universal, 28, 29
Mirrors, 158
  ,, concave, 160
Mixed substances, 222
Molten gold, simile of, 174
Mortal kind of soul, 256
Motion, how continued, 208
  ,, of the ὑποδοχή, 106
Motions, the seven, 102, 148
Mouth, functions of, 280
  ,, compared to roots of plant, 337
Moving bodies, how propelled, 299
Music, use of, 164
Musical intervals, 108

N

Nails, 283
Natural science, place of, 214
Neith, 68
Nerves, unknown to Plato, 274

Nile, cause of its inundation, 71
Numbers, plane and solid, 96
　,,　derived from the heavenly bodies, 162

O

Octahedron, 196
Odours, sweet, are pure pleasures, 238
　,,　not to be classified, 244
Oil, 216
Old age explained, 306
One in the many, 252
Opinion, 180
Other, 43, 44, 106

P

Pain and pleasure, explanation of, 236
Panathenaia, time of celebrating, 66
Parmenides, 6
Parmenides, the, 20
Pentagon, 197
Petron of Himera, 198
Phaethon, myth of, 70
Philebus, the, 24
Philosophy, chief result of sight, 162
Phoroneus, 70
Physics, 46
Planes, 202
Plane numbers, 96
Planets, their relative distance from the earth, 113
　,,　their names, 123
Plants, nature of, 284
Plato, two stages of his thought, 14
　,,　his Herakleiteanism, 15, 36
　,,　his teleology, 17
　,,　always confines himself to subject in hand, 171
　,,　his corpuscular theory, 192
　,,　his determinism necessitated by his ontology, 325
Politicus, myth in the, 102
Portal vein, 264
Pottery, 220
Predication, 16
Preplatonic contribution to Platonism, 2, 12
Projection explained, 299
Prophecy, 266

Proportion, 96
Protagoras, 13
Punishment, theory of, 325
Purple, 250
Pyramid, 194
Pythagoreans, 12

Q

Quadrupeds, origin of, 342
Quickness of apprehension incompatible with dense flesh, 278
Quotidian, tertian and quartan fevers, 322

R

Reason and true opinion shown to be different, 180
Red, 250
Reflections, 158
Relativity, 21
Republic, its metaphysical side ignored, 56
Respiration, 292
Rhythm, 164
Ritual terms, 152
Rotation of the earth, 132
Rough, 234

S

Sais, 68
Salt, 220
Same, 106
　,,　applied to ὑποδοχή, 176
Saps, 216
Scale, the Platonic, 109
Scalene, the rectangular, 192
Seers cannot interpret their own sayings, 266
Self-consciousness, 256
Seneca quoted on mirrors, 159
Sensation, physical theory of, 234
Sex, origin of, 340
Sifting of the elements, 186
Sinews, 274
Singular, instead of plural, 154
Skin, the, 280
Sleep, 158
Smell, theory of, 242
Smooth, 234
Snow, 216

Soft, 228
Sokrates, 14
Solid numbers, 96
Solidification, 212
Solon, his poetry, 68
  ,, his Egyptian travels, 68
Sophist, the, 22
Sophists, 13
Soul, prior to body, 104
  ,, mortal kind of, 256
  ,, diseases of, 322
  ,, and body, symmetry of, 328
  ,, ,, to be equally exercised, 330
Souls assigned to the stars, 141
  ,, all have an equal chance, 142
  ,, sowed among the planets, 146
Sound, 246, 300
Space, 44, 182
Speech, final cause of, 164
Spirals, 126
Spleen, 268
Stone, how formed, 218
  ,, soluble, 224
Subsidiary causes, 160
Substrate, 172
  ,, its permanence, 176
  ,, its formlessness, 176
  ,, compared to a mother, 176
Sutures, Hippokrates on the, 281
Sweet, 242
Symbolical apprehension, 30, 32
Symmetry, 328
Synovial fluid, 311

T

Tastes, various, 240
Tetanus, 318
Tetrahedron, 195
Theaetetus, the, 21
Theogony, 136
Thought, the sole existence, 28
  ,, pluralised, 29
  ,, identified with its object, 115
Threefold division of existence, 176
Thunderbolts, 301
Timaeus, 63
Timaeus, the, importance attached to it by ancient authorities, 1

Timaeus, the, key to the Platonic system, 2
  ,, questions left to be answered by, 27
  ,, metaphysic of, 28 foll.
  ,, allegorical method of, 36
  ,, physical theories of, 46
Time, in a sense eternal, 119
  ,, contrasted with eternity, 120
  ,, coeval with the universe, 120, 122
Transmigration, 144, 342
Transparent, 248
Triangles, the primal, 190
  ,, variation in their size, 206
Type cannot exist in the image, 184

U

Unguents, simile of, 178
Unity and plurality, 29
Universe has no beginning in time, 86
  ,, the copy of a type, 88
  ,, its unity, 94, 198
Up and down, 228
Upright posture of man, 336

V

Variation in size of triangles, 206
Vegetable food, 285, 304
Veins, the channel of communication, 259
  ,, of the head cross, 289
Vena cava, 264
Venus and Mercury, motions of, 124
Vengeance, not admitted by Plato, 325
Verjuice, 218
Vertebral column, 272
Vibration of the ὑποδοχή, 186
Vice involuntary, 324
Violet, 250
Vision, 154
Visual current, 156
  ,, ,, not subject to pleasure or pain, 236
Void, absence of, 210
Volcanic stones, 220

W

Water, liquid and fusible, 212
Wax, 224
Weels, 291
Weighing, 232

White, 248
Wine, 216
Winnowing fans, 186
Woman, false position of, in Plato's theory, 144
Words, their relation to their subject, 88
World-soul, 42

X

Xenophanes, 6

Y

Year, the great, 129
Yellow, 250

Z

Zeller's theory of ideas and particulars, 182
,, platonische Studien quoted, 184
Zeno, 6
Zodiac, 197

www.ingramcontent.com/pod-product-compliance
Lightning Source LLC
Chambersburg PA
CBHW020229240426
43672CB00006B/463